Springer INdAM Series

Volume 14

More information about this volume at http://www.springer.com/series/10283

Filippo Callegaro • Frederick Cohen •
Corrado De Concini • Eva Maria Feichtner •
Giovanni Gaiffi • Mario Salvetti
Editors

Configuration Spaces

Geometry, Topology and Representation Theory

Springer

Editors
Filippo Callegaro
Dipartimento di Matematica
Università di Pisa
Pisa, Italy

Frederick Cohen
Department of Mathematics
University of Rochester
Rochester, USA

Corrado De Concini
Dipartimento di Matematica
Università di Roma La Sapienza
Roma, Italy

Eva Maria Feichtner
Department of Mathematics
University of Bremen
Bremen, Germany

Giovanni Gaiffi
Dipartimento di Matematica
Università di Pisa
Pisa, Italy

Mario Salvetti
Dipartimento di Matematica
Università di Pisa
Pisa, Italy

ISSN 2281-518X ISSN 2281-5198 (electronic)
Springer INdAM Series
ISBN 978-3-319-31579-9 ISBN 978-3-319-31580-5 (eBook)
DOI 10.1007/978-3-319-31580-5

Library of Congress Control Number: 2016943640

Printed on acid-free paper

This Springer imprint is published by Springer Nature
The registered company is Springer International Publishing AG Switzerland

Preface

Configuration spaces: In by now classical notation following the seminal paper of Fadell and Neuwirth in the early 1960s, they read as

$$F_n(X) = \{(x_1, \ldots, x_n) \in X^n \mid x_i \neq x_j \text{ for } i \neq j\},$$

where X is a topological space and n a natural number. Beautifully simple at the outset and studied for decades, configuration spaces continue to thrive at the crossroads of algebra, geometry, and topology, and even attract the interest of applied fields such as phylogenetics, robotics, and distributed computing.

The present volume is a result of the INdAM Conference "Configuration Spaces: Geometry, Topology, and Representation Theory", which took place at the Palazzone in Cortona, Italy, from 31 August to 6 September 2014. This conference was an initiative of the Istituto Nazionale di Alta Matematica (INdAM) and of the Futuro in Ricerca project "Perspectives in Lie Theory", a satellite event of the INdAM research program "Perspectives in Lie Theory" held in Pisa from December 2014 to February 2015.

This volume reflects much of the spirit of the INdAM Conference. It provides a snapshot of the current developments in configuration spaces and a glimpse on what the near future holds in store.

Conference participants and readers alike have the opportunity to explore the landscape of configuration spaces, knowledgeably guided by experts in the field. Let us draw a road map of this volume:

Arrangement complements have for long been understood as special, though most instructive cases of configuration spaces. Rooted in the world of classical configuration spaces with a prototype example, the braid arrangement, a prominent part of this volume is devoted to arrangement theory. The volume begins with a series of chapters "Around the Tangent Cone Theorem", "Higher Resonance Varieties of Matroids", "Local Asymptotic Euler-Maclaurin Expansion for Riemann Sums Over a Semi-Rational Polyhedron", "Leray Spectral Sequence for Complements of Certain Arrangements of Smooth Submanifolds", and "Higher Topological Complexity of Artin Type Groups" that cover arrangement theory from its purest parts (resonance and characteristic varieties) over the integral geometry of polyhedra to the application-inspired realm of (higher) topological complexity. From there the volume ventures into actual applications, both within other parts of mathematics, in fact, in probabilistic topology (chapter "Random Simplicial Complexes"), and beyond, notably in theoretical distributed computing (chapter "Combinatorial Topology of the Standard Chromatic Subdivision and Weak Symmetry Breaking for 6 Processes").

There have been many instances over time where combinatorial data have been crucial for a better understanding of configuration spaces and their invariants. Matroid theory has for long been the right degree of abstraction to deal with arrangements, arrangement compactifications and the like. More recently, the classical concept of nets appeared, which is believed, if not proven, to be crucial for the understanding of Milnor fiber cohomology. Chapters "Cryptomorphisms for Abstract Rigidity Matroids" and "Induced and Complete Multinets" are of mostly combinatorial nature and pay tribute to matroids and nets as key concepts in arrangement theory. In chapters "Cohomology of the Milnor Fibre of a Hyperplane Arrangement with Symmetry" and "On the Twisted Cohomology of Affine Line Arrangements", which address the topology of Milnor fibers of arrangement complements, we turn back to the heart of arrangement theory. Geometric group theory is considered in chapter "Braid Groups and Euclidean Simplices", and the volume closes with three chapters providing the link to representation theory.

We are very grateful to the participants of the Cortona conference for creating a unique atmosphere of interaction, and we are indebted to the authors of this volume for capturing this spirit. May this volume inspire further thought to the benefit of a multifaceted field.

Finally, we wish to thank the Palazzone and INdAM staff for their very efficient job in keeping the organizational matters of the conference afloat.

Pisa, Italy — Filippo Callegaro
Rochester, NY, USA — Fred Cohen
Roma, Italy — Corrado De Concini
Bremen, Germany — Eva Maria Feichtner
Pisa, Italy — Giovanni Gaiffi
Pisa, Italy — Mario Salvetti
September 2015

Contents

Contributors

Jeremiah Bartz Francis Marion University, Florence, SC, USA

Nicole Berline École Polytechnique, Centre de Mathématiques Laurent Schwartz, Palaiseau Cedex, France

Frederick R. Cohen University of Rochester, Rochester, NY, USA

Armindo Costa School of Mathematical Sciences, Queen Mary University of London, London, UK

Emanuele Delucchi Department of Mathematics, University of Fribourg, Fribourg, Switzerland

Graham Denham Department of Mathematics, University of Western Ontario, London, Canada

Alexandru Dimca University of Nice Sophia Antipolis, CNRS, LJAD, UMR 7351, Nice, France

Michael Farber School of Mathematical Sciences, Queen Mary University of London, London, UK

Matthias Franz Department of Mathematics, University of Western Ontario, London, Canada

Rita Jiménez Rolland Centro de Ciencias Matemáticas, Universidad Nacional Autónoma de México, Morelia, México

Dmitry N. Kozlov Department of Mathematics, University of Bremen, Bremen, Germany

Gus Lehrer School of Mathematics and Statistics F07, University of Sydney, Sydney, NSW, Australia

Elizabeth Leyton Chisholm Department of Mathematics, University of California, Santa Barbara, CA, USA

Tim Lindemann Department of Mathematics, University of Bremen, Bremen, Germany

Jon McCammond Department of Mathematics, University of California, Santa Barbara, CA, USA

Mario Salvetti Department of Mathematics, University of Pisa, Pisa, Italy

Matteo Serventi Department of Mathematics, University of Pisa, Pisa, Italy

Mentor Stafa Tulane University, New Orleans, LA, USA

Alexander I. Suciu Department of Mathematics, Northeastern University, Boston, MA, USA

Michèle Vergne Université Paris 7 Diderot, IMJ-PRG, Sophie Germain, Paris, France

Andrzej Weber Department of Mathematics, University of Warsaw, Warszawa, Poland

Sergey Yuzvinsky University of Oregon, Eugene, OR, USA

About the Editors

Filippo Callegaro was born in 1980. He received a Ph.D. in Mathematics from Scuola Normale Superiore of Pisa under the direction of Mario Salvetti in 2007. In 2008, he was offered a research position at the Scuola Normale. In 2013, he switched to the University of Pisa, where his work focuses on the topology of configuration spaces, cohomology of braid groups and generalizations, hyperplane and toric arrangements.

Frederick R. Cohen was born in 1945. He received a Ph.D. in Mathematics from the University of Chicago under the direction of Peter May. He has been a Professor of Mathematics at the University of Rochester since 1988. He has been a visiting scholar at the Institute for Advanced Studies, the Institute for Mathematics and its Applications and the Centro di Ricerca Matematica Ennio De Giorgi, as well as the University of Pennsylvania, MIT, and Stanford. He has served on the editorial boards of Algebraic and Geometric Topology, Forum Mathematicum, and the Journal of Homotopy Theory and Related Structures. He has worked in homotopy theory, the topology of configuration spaces, the cohomology and related properties of braid groups on surfaces, and toric topology, as well as applied problem concerning sensor counting and robotic motion.

Corrado De Concini was born in 1949. He received a Ph.D. in Mathematics from the University of Warwick under the direction of George Lusztig in 1975. He has been a Professor of Mathematics since 1983. Since 1996, he has been a professor at the University of Roma, La Sapienza. He has been a visiting scholar at Brandeis University, Mittag Leffler Institute, Tata Institute of Fundamental Research, Harvard University, MIT, Université de Paris VI, IHES, MSRI, and the École Normale Supérieure of Paris. He serves on the editorial boards of Rendiconti Accademia Nazionale dei Lincei, Journal of Transformation Groups, IMRN, Rendiconti di Matematica, Rendiconti del Seminario Matematico-Padova, and Communications in Contemporary Mathematics. He was a plenary speaker for the European Congress of Mathematics (Paris) in 1992. He has made important contributions to several areas of algebra and algebraic geometry, e.g., invariant theory, commutative algebra, algebraic and quantic group theory, the Schottky problem, and hyperplane arrangements.

Eva Maria Feichtner was born in 1972. She received a Ph.D. in Mathematics from TU Berlin in 1997. After postdoctoral appointments at MIT and at the Institute for Advanced Studies, she was an assistant professor at ETH Zurich and was awarded a research professorship from the Swiss National Science Foundation. In 2006, she was appointed professor of topology at the University of Stuttgart; since 2007, she has served as a professor of algebra at the University of Bremen, where she co-directs the Institute for Algebra, Geometry, Topology and their Applications (ALTA). She has been a visiting scholar at many research institutions, notably at MSRI, Berkeley, as research professor and as program organizer. Her research interests lie in the interplay of algebra, geometry, topology, and combinatorics, notably in arrangements of hyperplanes, toric varieties, and tropical geometry. She serves on the editorial board of Advances in Applied Mathematics.

Giovanni Gaiffi was born in 1970. He received his degree in Mathematics from the University of Pisa in 1993 and attended the Scuola Normale, where he also completed his Ph.D. in Mathematics under the direction of Corrado De Concini in 1999. In 1997, he was offered a research position at the Scuola Normale, before switching to the University of Pisa as an associate professor in 2005. His field of interest mainly concerns algebra, in particular Lie theory, representation theory, and their interplay with combinatorics and algebraic topology. He has organized international conferences and is co-author of the educational book "Giochi e Percorsi Matematici" released by Springer-Italia.

Mario Salvetti was born in 1957. He received a degree in Mathematics from the University of Pisa and was a "perfezionando" at the Scuola Normale in Pisa. He has been a Professor of Mathematics since 1992, and in Pisa since 2000. He has been a visiting scholar at Columbia University, Moscow State University (at the "Gelfand Seminar"), Graduate School of Mathematical Sciences in Tokyo, and MSRI. He has served on the organizing and scientific committees of international conferences and meetings in Italy, and edited a number of books. He organized a mini-symposium on "Braids" at the European Congress of Mathematics in Cracow in 2012. He has worked in the topology and combinatorics of hyperplane arrangements, braid groups and Artin groups, Coxeter groups, configuration spaces, and to some extent in the applied topology related to image understanding.

Around the Tangent Cone Theorem

Alexander I. Suciu

Abstract A cornerstone of the theory of cohomology jump loci is the Tangent Cone theorem, which relates the behavior around the origin of the characteristic and resonance varieties of a space. We revisit this theorem, in both the algebraic setting provided by CDGA models, and in the topological setting provided by fundamental groups and cohomology rings. The general theory is illustrated with several classes of examples from geometry and topology: smooth quasi-projective varieties, complex hyperplane arrangements and their Milnor fibers, configuration spaces, and elliptic arrangements.

Keywords Algebraic model • Characteristic variety • Cohomology ring • Configuration space • Elliptic arrangement • Formality • Hyperplane arrangement • Milnor fiber • Quasi-projective variety • Resonance variety • Tangent cone

1 Introduction

The Tangent Cone theorem relates two seemingly disparate sets of cohomology jump loci associated to a space X: the resonance varieties, which are constructed from information encoded in either the cohomology ring of X, or an algebraic model for this space, and the characteristic varieties, which depend on the *a priori* much more subtle information carried by the cohomology of X with coefficients in rank 1 local systems. We focus here on the interplay between these two sets of jump loci, which are even more tightly related under certain algebraic (positivity of weights), topological (formality), or geometric (quasi-projectivity) assumptions.

A.I. Suciu (✉)
Department of Mathematics, Northeastern University, Boston, MA 02115, USA
e-mail: a.suciu@neu.edu

F. Callegaro et al. (eds.), *Configuration Spaces*, Springer INdAM Series 14,
DOI 10.1007/978-3-319-31580-5_1

1.1 Resonance Varieties

We start in Sect. 2 with a description of the various resonance varieties associated to a commutative, differential graded algebra (for short, a CDGA). We continue in Sect. 3 with the resonance varieties associated to a space X, using as input either its cohomology algebra or a suitable algebraic model, and discuss the algebraic version of the Tangent Cone theorem.

We will assume throughout that X is a reasonably nice space, to wit, a connected CW-complex with finitely many cells in each dimension. To such a space, we associate two types of resonance varieties. The classical ones are obtained from the cohomology algebra $A = H^*(X, \mathbb{C})$, by setting

$$\mathcal{R}^i(X) = \{a \in A^1 \mid H^i(A, \delta_a) \neq 0\}, \tag{1}$$

where, for each $a \in A^1$, we denote by (A, δ_a) the cochain complex with differentials $\delta_a: A^i \to A^{i+1}$ given by left-multiplication by a. Each set $\mathcal{R}^i(X)$ is a homogeneous subvariety of the complex affine space $A^1 = H^1(X, \mathbb{C})$.

Lately, an alternate definition of resonance has emerged (in works such as [23–25, 45, 59]), whereby one replaces the cohomology algebra by an algebraic model for X, that is, a commutative differential graded algebra (A, d) weakly equivalent to the Sullivan model of polynomial forms on X, as defined in [70]. We may then set up a cochain complex (A, δ_a) as above, but now with differentials given by $\delta_a(u) = au + \mathrm{d}u$, and define the resonance varieties $\mathcal{R}^i(A) \subset H^1(A)$ just as before.

Assuming now that each graded piece A^i is finite-dimensional, the sets $\mathcal{R}^i(A)$ are subvarieties of the affine space $H^1(A)$, which depend only on the isomorphism type of A. These varieties are not necessarily homogeneous; nevertheless, as shown in [45], the following inclusion holds,

$$\mathrm{TC}_0(\mathcal{R}^i(A)) \subseteq \mathcal{R}^i(X), \tag{2}$$

where TC_0 denotes the tangent cone at $0 \in H^1(A)$.

Under some additional hypothesis, one can say more. Suppose our finite-type model (A, d) admits a $\mathbb{Q}$-structure compatible with that of the Sullivan model, and also has positive weights, in the sense of [51, 70]. Then, $\mathcal{R}^i(A)$ is a finite union of rationally defined linear subspaces of $H^1(A)$, and

$$\mathcal{R}^i(A) \subseteq \mathcal{R}^i(X). \tag{3}$$

1.2 Characteristic Varieties

We turn in Sect. 4 to the characteristic varieties of a space X, and to the two types of tangent cones associated to them. This sets the stage for the topological version of the Tangent Cone theorem, which is treated in Sect. 5.

Unlike the resonance varieties, which arise from an algebraic model, the characteristic varieties arise from the chain complex of the universal abelian cover of the space. Let $\pi = \pi_1(X)$ be the fundamental group of X, let $\pi_{\mathrm{ab}} = H_1(X, \mathbb{Z})$ be its abelianization, and let $\mathrm{Char}(X) = \mathrm{Hom}(\pi_{\mathrm{ab}}, \mathbb{C}^*)$ be its group of complex-valued characters. Then

$$\mathcal{V}^i(X) = \{\rho \in \mathrm{Char}(X) \mid H_i(X, \mathbb{C}_\rho) \neq 0\}, \tag{4}$$

where $\mathbb{C}_\rho$ denotes the complex vector space $\mathbb{C}$, viewed as a module over the group algebra $\mathbb{C}[\pi_{\mathrm{ab}}]$ via $g \cdot z = \rho(g)z$, for $g \in \pi$ and $z \in \mathbb{C}$.

The relationship between the characteristic and resonance varieties of a space goes through the tangent cone construction. Let us start by identifying the tangent space at the identity to the complex algebraic group $\mathrm{Char}(X)$ with the complex affine space $H^1(X, \mathbb{C})$. Then, as shown in [24, 40], we have the following chain of inclusions:

$$\tau_1(\mathcal{V}^i(X)) \subseteq \mathrm{TC}_1(\mathcal{V}^i(X)) \subseteq \mathcal{R}^i(X), \tag{5}$$

where τ_1 denotes the 'exponential tangent cone' at the identity $1 \in \mathrm{Char}(X)$ (a finite union of rationally defined linear subspaces), and TC_1 denotes the usual tangent cone at 1 (a homogeneous subvariety).

The crucial property that bridges the gap between the two types of tangent cones to a characteristic variety and the corresponding resonance variety is that of formality, in the sense of Sullivan [70]. Given a 1-formal space, one of the main results from [24] establishes an isomorphism between the analytic germ of $\mathcal{V}^1(X)$ at 1 and the analytic germ of $\mathcal{R}^1(X)$ at 0.

More generally, if X has an algebraic model A with good finiteness properties, then, as shown in [23], the characteristic varieties $\mathcal{V}^i(X)$ may be identified around the identity with the resonance varieties $\mathcal{R}^i(A)$. Consequently, if X is formal (that is, the cohomology algebra of X, endowed with the zero differential, is weakly equivalent to the Sullivan model), then the following 'Tangent Cone formula' holds:

$$\tau_1(\mathcal{V}^i(X)) = \mathrm{TC}_1(\mathcal{V}^i(X)) = \mathcal{R}^i(X). \tag{6}$$

Consequently, if either one of the two inclusions in (5) fails to be an equality, the space X is not formal. Viewed this way, the Tangent Cone theorem can be thought of as a (quite powerful) formality obstruction.

1.3 Quasi-Projective Varieties

We conclude our overview of cohomology jump loci with an exploration of the Tangent Cone theorem in the framework of complex algebraic geometry. We start in Sect. 6 with the general theory of jump loci of smooth, quasi-projective varieties. We then specialize in Sect. 7 to complements of hyperplane arrangements and their Milnor fibers, and in Sect. 8 to complements of elliptic arrangements.

Let X be a smooth, complex quasi-projective variety. Work of Arapura [2], as recently sharpened by Budur and Wang [9], reveals a profound fact about the characteristic varieties $\mathcal{V}^i(X)$: they are all finite unions of torsion-translated subtori of the character group $\mathrm{Char}(X)$.

Every quasi-projective variety as above can be realized as the complement, $X = \overline{X} \setminus D$, of a normal-crossings divisor D in a smooth, complex projective variety $\overline{X}$. Given such a 'good' compactification, Morgan associates in [51] an algebraic model for our variety, $A(X) = A(\overline{X}, D)$. This 'Gysin' model is a finite-dimensional, rationally defined CDGA with positive weights, which is weakly equivalent to Sullivan's model for X.

Using the aforementioned work of Arapura and Budur–Wang, as well as work of Dimca and Papadima [23], we obtain the following formulation of the Tangent Cone theorem for smooth, quasi-projective varieties X:

$$\tau_1(\mathcal{V}^i(X)) = \mathrm{TC}_1(\mathcal{V}^i(X)) = \mathcal{R}^i(A(X)) \subseteq \mathcal{R}^i(X). \tag{7}$$

In degree $i = 1$, the irreducible components of $\mathcal{V}^1(X)$ which pass through the identity are in one-to-one correspondence with the set $\mathcal{E}_X$ of 'admissible' maps $f{:}X \to \Sigma$, where Σ is a smooth complex curve with $\chi(\Sigma) < 0$. This leads to a concrete description of the variety $\mathcal{R}^1(A(X))$, and of the variety $\mathcal{R}^1(X)$ when X is 1-formal.

Especially interesting is the case when $X = M(\mathcal{A})$ is the complement of an arrangement $\mathcal{A}$ of hyperplanes in some complex vector space. The cohomology algebra $A = H^*(X, \mathbb{C})$ admits a combinatorial description, in terms of the intersection lattice of $\mathcal{A}$. Moreover, the CDGA $(A, 0)$ is a model for X; thus, formula (7) holds with equalities throughout.

For an arrangement complement as above, work of Falk and Yuzvinsky [29] identifies the set $\mathcal{E}_X$ with the set of multinets on sub-arrangements of $\mathcal{A}$, up to relabeling (see also [60]). This yields a completely combinatorial description of the resonance variety $\mathcal{R}^1(X)$, and of the components of the characteristic variety $\mathcal{V}^1(X)$ passing through the identity.

Another smooth variety associated to an arrangement $\mathcal{A}$ is the Milnor fiber $F = F(\mathcal{A})$, defined as the level set $Q = 1$, where Q is a defining polynomial for $\mathcal{A}$. The topology of this variety (even its first Betti number!) is much less understood. As shown by Zuber [74], though, the inclusion $\mathrm{TC}_1(\mathcal{V}^1(F)) \subset \mathcal{R}^1(F)$ can be strict; hence, F can be non-formal. Further understanding of how the Tangent

Cone formula works in this context hinges on finding a good compactification for F, and then computing the corresponding Gysin model and its resonance varieties.

The machinery of cohomology jump loci can also be brought to bear in the study of elliptic arrangements. Let $E^{\times n}$ be the n-fold product of an elliptic curve E. An elliptic arrangement in $E^{\times n}$ is a finite collection of fibers of group homomorphisms $E^{\times n} \to E$. Assuming that all subspaces in the intersection poset of $\mathcal{A}$ are connected, Bibby constructs in [7] a finite-dimensional, algebraic model for the complement, which can be thought of as a concrete version of the Gysin model.

A special case of this construction is the configuration space $\mathrm{Conf}(E, n)$ of n distinct, ordered points on E, itself a classifying space for the n-stranded pure braid group on the torus. We illustrate the general theory in a simple, yet instructive example. Direct computation shows that, for $X = \mathrm{Conf}(E, 3)$, the resonance variety $\mathcal{R}^1(A(X))$ is properly contained in $\mathcal{R}^1(X)$, thereby establishing the non-formality of X.

2 The Resonance Varieties of a CDGA

We start with the resonance varieties associated to a commutative differential graded algebra, some of their properties, and various ways to compute them.

2.1 Commutative Differential Graded Algebras

Let $A = (A^{\boldsymbol{\cdot}}, \mathrm{d})$ be a commutative, differential graded algebra over the field $\mathbb{C}$. That is, $A = \bigoplus_{i\geq 0} A^i$ is a graded $\mathbb{C}$-vector space, endowed with a multiplication map $\cdot\colon A^i \otimes A^j \to A^{i+j}$ satisfying $u \cdot v = (-1)^{ij} v \cdot u$, and a differential $\mathrm{d}\colon A^i \to A^{i+1}$ satisfying $\mathrm{d}(u \cdot v) = \mathrm{d}u \cdot v + (-1)^i u \cdot \mathrm{d}v$, for all $u \in A^i$ and $v \in A^j$.

Unless otherwise stated, we will assume throughout that A is connected, i.e., $A^0 = \mathbb{C}$, and of finite-type, i.e., A^i is finite-dimensional, for all $i \geq 0$.

Using only the underlying cochain complex structure of the CDGA, we let $Z^i(A) = \ker(\mathrm{d}\colon A^i \to A^{i+1})$ and $B^i(A) = \mathrm{im}(\mathrm{d}\colon A^{i-1} \to A^i)$, and set $H^i(A) = Z^i(A)/B^i(A)$. The direct sum of the cohomology groups, $H^{\boldsymbol{\cdot}}(A) = \bigoplus_{i\geq 0} H^i(A)$, inherits an algebra structure from A.

A morphism between two CDGAs, $\varphi\colon A \to B$, is both an algebra map and a cochain map. Consequently, φ induces a morphism $\varphi^*\colon H^{\boldsymbol{\cdot}}(A) \to H^{\boldsymbol{\cdot}}(B)$ between the respective cohomology algebras. We say that φ is a quasi-isomorphism if φ^* is an isomorphism. Likewise, we say φ is a q-isomorphism (for some $q \geq 1$) if φ^* is an isomorphism in degrees up to q and a monomorphism in degree $q + 1$.

Two CDGAs A and B are *weakly equivalent* (or just *q-equivalent*) if there is a zig-zag of quasi-isomorphisms (or q-isomorphisms) connecting A to B, in which case we write $A \simeq B$ (or $A \simeq_q B$).

A CDGA (A, d) is said to be *formal* (or just *q-formal*) if it is weakly equivalent (or just q-equivalent) to its cohomology algebra, $H^{\bullet}(A)$, endowed with the zero differential.

Finally, we say that (A, d) is rationally defined if A is the complexification of a graded $\mathbb{Q}$-algebra $A_{\mathbb{Q}}$, and the differential d preserves $A_{\mathbb{Q}}$.

We will also consider the dual vector spaces $A_i = (A^i)^{\vee} := \mathrm{Hom}(A^i, \mathbb{C})$, and the chain complex $(A_{\bullet}, \partial)$, where $\partial: A_{i+1} \to A_i$ is the dual to $\mathrm{d}: A^i \to A^{i+1}$. If $H_i(A)$ are the homology groups of this chain complex, then, by the Universal Coefficients theorem, $H_i(A) \cong (H^i(A))^{\vee}$.

2.2 Resonance Varieties

Our connectivity assumption on the CDGA (A, d) allows us to identify the vector space $H^1(A)$ with the cocycle space $Z^1(A)$. For each element $a \in Z^1(A) \cong H^1(A)$, we turn A into a cochain complex,

$$(A^{\bullet}, \delta_a)\colon A^0 \xrightarrow{\delta_a^0} A^1 \xrightarrow{\delta_a^1} A^2 \xrightarrow{\delta_a^2} \cdots, \tag{8}$$

with differentials given by $\delta_a^i(u) = a \cdot u + \mathrm{d}u$, for all $u \in A^i$. The cochain condition is verified as follows: $\delta_a^{i+1}\delta_a^i(u) = a^2u + a \cdot \mathrm{d}u + \mathrm{d}a \cdot u - a \cdot \mathrm{d}u + \mathrm{dd}u = 0$.

Computing the homology of these chain complexes for various values of the parameter a, and keeping track of the resulting Betti numbers singles out certain *resonance varieties* inside the affine space $H^1(A)$. More precisely, for each non-negative integer i, define

$$\mathcal{R}^i(A) = \{a \in H^1(A) \mid H^i(A^{\bullet}, \delta_a) \neq 0\}. \tag{9}$$

These sets can be defined for any connected CDGA. If A is of finite-type (as we always assume), the sets $\mathcal{R}^i(A)$ are, in fact, algebraic subsets of the ambient affine space $H^1(A)$. Clearly, $H^i(A^{\bullet}, \delta_0) = H^i(A)$; thus, the point $0 \in H^1(A)$ belongs to the variety $\mathcal{R}^i(A)$ if and only if $H^i(A) \neq 0$. Moreover, $\mathcal{R}^0(A) = \{0\}$.

When the differential of A is zero, the resonance varieties $\mathcal{R}^i(A)$ are homogeneous subsets of $H^1(A) = A^1$. In general, though, the resonance varieties of a CDGA are not homogeneous: see [45, Example 2.7] and Example 2 below.

The following lemma follows quickly from the definitions (see [45, Lemma 2.6] for details).

Lemma 1 (Măcinic et al. [45]) *Let $\varphi: A \to A'$ be a* CDGA *morphism, and assume φ is an isomorphism up to degree q, and a monomorphism in degree $q+1$, for some $q \geq 0$. Then the induced isomorphism in cohomology, $\varphi^*: H^1(A') \to H^1(A)$, identifies $\mathcal{R}^i(A)$ with $\mathcal{R}^i(A')$ for each $i \leq q$, and sends $\mathcal{R}^{q+1}(A)$ into $\mathcal{R}^{q+1}(A')$.*

Corollary 1 *If A and A' are isomorphic* CDGAs, *then their resonance varieties are ambiently isomorphic.*

The conclusions of Lemma 1 do not follow if we only assume that $\varphi: A \to A'$ is a q-isomorphism. This phenomenon is illustrated in [45, Example 2.7] and also in Example 2 below.

As shown in [56, 59], the resonance varieties behave reasonably well under tensor products:

$$\mathcal{R}^i(A \otimes A') \subseteq \bigcup_{p+q=i} \mathcal{R}^p(A) \times \mathcal{R}^q(A'). \tag{10}$$

Moreover, if the differentials of both A and A' are zero, then equality is achieved in the above product formula.

In a similar manner, we can define a homological version of resonance varieties, by considering the chain complexes $(A_{\bullet}, \partial^{\alpha})$ with differentials $\partial_i^{\alpha} = (\delta_a^i)^{\vee}$ for $\alpha \in H_1(A)$ dual to $a \in H^1(A)$, and setting

$$\mathcal{R}_i(A) = \{\alpha \in H_1(A) \mid H_i(A_{\bullet}, \partial^{\alpha}) \neq 0\}. \tag{11}$$

Lemma 2 *For each $i \geq 0$, the duality isomorphism $H^1(A) \cong H_1(A)$ identifies the resonance varieties $\mathcal{R}^i(A)$ and $\mathcal{R}_i(A)$.*

Proof By the Universal Coefficients theorem (over the field $\mathbb{C}$), we have that $H_i(A^{\bullet}, \delta_a) \cong H_i(A_{\bullet}, \partial^{\alpha})$. The claim follows. □

2.3 A Generalized Koszul Complex

Let us fix now a basis $\{e_1, \dots, e_n\}$ for the complex vector space $H^1(A)$, and let $\{x_1, \dots, x_n\}$ be the Kronecker dual basis for the vector space $H_1(A) = (H^1(A))^{\vee}$. In the sequel, we shall identify the symmetric algebra $\operatorname{Sym}(H_1(A))$ with the polynomial ring $S = \mathbb{C}[x_1, \dots, x_n]$, and we shall view S as the coordinate ring of the affine space $H^1(A)$.

Consider now the cochain complex of free S-modules,

$$(A^{\bullet} \otimes S, \delta)\colon \quad \cdots \longrightarrow A^i \otimes S \xrightarrow{\delta^i} A^{i+1} \otimes S \xrightarrow{\delta^{i+1}} A^{i+2} \otimes S \longrightarrow \cdots, \tag{12}$$

where the differentials are the S-linear maps defined by

$$\delta^i(u \otimes s) = \sum_{j=1}^{n} e_j u \otimes s x_j + du \otimes s \tag{13}$$

for all $u \in A^i$ and $s \in S$. As before, the fact that this is a cochain complex is easily verified. Indeed, $\delta^{i+1}\delta^i(u \otimes s)$ equals

$$\sum_k e_k\Big(\sum_j e_j u \otimes sx_j + du \otimes s\Big) \otimes x_k + d\Big(\sum_j e_j u \otimes sx_j + du \otimes s\Big)$$
$$= \sum_{j,k} e_k e_j u \otimes sx_j x_k + \sum_k e_k du \otimes sx_k - \sum_j e_j du \otimes sx_j$$
$$= 0,$$

where we used the fact that $e_k e_j = -e_j e_k$.

Remark 1 The cochain complex (12) is independent of the choice of basis $\{e_1, \dots, e_n\}$ for $H^1(A)$. Indeed, under the canonical identification $H^1(A) \otimes H_1(A) \cong \operatorname{Hom}(H^1(A), H^1(A))$, the element $\sum_{j=1}^n e_j \otimes x_j$ used in defining the differentials δ^i corresponds to the identity map of $H^1(A)$.

Example 1 Let $E = \bigwedge(e_1, \dots, e_n)$ be the exterior algebra (with zero differential), and let $S = \mathbb{C}[x_1, \dots, x_n]$ be its Koszul dual. Then the cochain complex $(E^{\boldsymbol{\cdot}} \otimes S, \delta)$ is simply the Koszul complex $K_{\boldsymbol{\cdot}}(x_1, \dots, x_n)$.

More generally, if the CDGA A has zero differential, each boundary map $\delta^i \colon A^i \otimes S \to A^{i+1} \otimes S$ is given by a matrix whose entries are linear forms in the variables $x_1, \dots, x_n$. In general, though, the entries of δ^i may also have non-zero constant terms, as can be seen in Examples 2, 3, and 21 below.

The relationship between the cochain complexes (8) and (12) is given by the following lemma (for a more general statement, we refer to the proof of Lemma 8.8(1) from [23]).

Lemma 3 *The specialization of the cochain complex $A \otimes S$ at an element $a \in H^1(A)$ coincides with the cochain complex (A, δ_a).*

Proof Write $a = \sum_{j=1}^n a_j e_j \in H^1(A)$, and let $\mathfrak{m}_a = (x_1 - a_1, \dots, x_n - a_n)$ be the maximal ideal at a. The evaluation map $\operatorname{ev}_a \colon S \to S/\mathfrak{m}_a = \mathbb{C}$ is the ring morphism given by $g \mapsto g(a_1, \dots, a_n)$. The resulting cochain complex, $A(a) = A \otimes_S S/\mathfrak{m}_a$, has differentials $\delta^i(a)$ given by

$$\delta^i(a)(u) = \sum_{j=1}^n e_j u \otimes \operatorname{ev}_a(x_j) + du = \sum_{j=1}^n e_j u \cdot a_j + du = a \cdot u + du. \tag{14}$$

Thus, $A(a) = (A, \delta_a)$, as claimed. □

In a completely analogous fashion, we may define a chain complex

$$(A_{\bullet}\otimes S, \partial)\colon \quad \cdots \longrightarrow A_{i+1}\otimes S \xrightarrow{\partial_{i+1}} A_i \otimes S \xrightarrow{\partial_i} A_{i-1}\otimes S \longrightarrow \cdots \tag{15}$$

by essentially transposing the differentials of $(A^{\bullet}\otimes S, \delta)$. The previous lemma shows that the specialization of $(A_{\bullet} \otimes S, \partial)$ at an element $\alpha \in H_1(A)$ coincides with the chain complex $(A_{\bullet}, \partial^{\alpha})$.

2.4 Alternate Views of Resonance

As is well-known, the classical Koszul complex is exact. For an arbitrary CDGA, though, the cochain complex $(A^{\bullet} \otimes S, \delta)$ is not: its non-exactness is measured by the cohomology groups $H^i(A \otimes S)$, which are finitely generated modules over the polynomial ring S. This leads us to consider the support loci of these cohomology modules,

$$\widetilde{\mathcal{R}}^i(A) = \operatorname{supp}(H^i(A^{\bullet} \otimes S, \delta)), \tag{16}$$

viewed again as algebraic subsets of the affine space $H^1(A)$.

For instance, if $(E^{\bullet} \otimes S, \delta) = K_{\bullet}(x_1, \dots, x_n)$ is the Koszul complex from Example 1, the support loci $\widetilde{\mathcal{R}}^i(E)$ vanish, for all $0 \leq i \leq n$.

We may also identify the polynomial ring S with the symmetric algebra on $H^1(A)$, viewed as the coordinate ring of $H_1(A)$. In this case, we have the support loci of the corresponding homology modules,

$$\widetilde{\mathcal{R}}_i(A) = \operatorname{supp}(H_i(A_{\bullet} \otimes S, \partial)), \tag{17}$$

which are algebraic subsets of the affine space $H_1(A)$. For a detailed discussion of support loci of chain complexes over an affine algebra, we refer to [58].

Since the ring S is no longer a field (or even a PID, unless $H_1(A) = 0$ or $\mathbb{C}$), the relation between these two types of support loci is not as straightforward as the one between the corresponding jump loci (see Example 2 below). Nevertheless, the cohomology jump loci and the homology support loci may be related in a filtered way, as follows.

Theorem 1 *For any finite-type* CDGA (A, d), *and for any* $q \geq 0$, *the duality isomorphism* $H^1(A) \cong H_1(A)$ *restricts to an isomorphism*

$$\bigcup_{i\leq q} \mathcal{R}^i(A) \cong \bigcup_{i\leq q} \widetilde{\mathcal{R}}_i(A).$$

Proof As noted previously, the duality isomorphism $H^1(A) \cong H_1(A)$ identifies $\mathcal{R}^i(A)$ with $\mathcal{R}_i(A)$, for each $i \geq 0$.

On the other hand, we know that $(A_{\bullet} \otimes S, \partial)$ is a chain complex of free, finitely generated modules over the affine $\mathbb{C}$-algebra S. Therefore, by Theorem 2.5 from [58], we have that $\bigcup_{i\leq q} \mathcal{R}_i(A) = \bigcup_{i\leq q} \widetilde{\mathcal{R}}_i(A)$, and the conclusion follows. □

As noted previously, when the CDGA A has differential $\mathrm{d} = 0$, the boundary maps δ and ∂ from the chain complexes (12) and (15) have entries which are linear forms in the variables of S. Consequently, the sets $\mathcal{R}^i(A)$ and $\widetilde{\mathcal{R}}_i(A)$ are homogeneous subvarieties of the affine space $\mathbb{A}^1 = H^1(A)$.

Corollary 2 *If A has zero differential, the resonance variety $\mathcal{R}^1(A) \subset \mathbb{A}^1$ is the vanishing locus of the codimension 1 minors of the matrix of S-linear forms $\partial_2: A_2 \otimes S \to A_1 \otimes S$, or of its transpose, $\delta^1: A^1 \otimes S \to A^2 \otimes S$.*

Proof Using Theorem 1 and the above discussion, we obtain the equality $\mathcal{R}^1(A) = \widetilde{\mathcal{R}}_1(A)$. By definition, $\widetilde{\mathcal{R}}_1(A)$ is the support locus of the S-module $H_1(A_{\bullet} \otimes S) = \ker \partial_1 / \operatorname{im} \partial_2$. Writing $A_1 = \mathbb{C}^n$ and $S = \mathbb{C}[x_1, \dots, x_n]$, we have that $\partial_1 = \begin{pmatrix} x_1 & \cdots & x_n \end{pmatrix}$. The conclusion follows. □

We illustrate the theory with a simple, yet meaningful example, variants of which can also be found in [25, 45].

Example 2 Let A be the exterior algebra on generators a, b in degree 1, endowed with the differential given by $\mathrm{d}a = 0$ and $\mathrm{d}b = b \cdot a$. Then $H^1(A) = \mathbb{C}$, generated by a. Writing $S = \mathbb{C}[x]$, the chain complex (15) takes the form

$$A_{\bullet} \otimes S\colon\ S \xrightarrow{\partial_2 = \binom{0}{x-1}} S^2 \xrightarrow{\partial_1 = (x\ 0)} S\,. \tag{18}$$

Hence, $H_1(A_{\bullet} \otimes S) = S/(x-1)$, and so $\widetilde{\mathcal{R}}_1(A) = \{1\}$. Using the above theorem, we conclude that $\mathcal{R}^1(A) = \{0, 1\}$.

Note that $\mathcal{R}^1(A)$ is a non-homogeneous subvariety of $\mathbb{C}$. Note also that $H^1(A_{\bullet} \otimes S) = S/(x)$, and so $\widetilde{\mathcal{R}}^1(A) = \{0\}$, which differs from $\widetilde{\mathcal{R}}_1(A)$.

Finally, let A' be the sub-CDGA generated by a. Clearly, the inclusion map, $\iota: A' \hookrightarrow A$, induces an isomorphism in cohomology. Nevertheless, $\mathcal{R}^1(A') = \{0\}$, and so the resonance varieties of A and A' differ, although A and A' are quasi-isomorphic.

Problem 1 Can the resonance varieties of a CDGA have positive-dimensional irreducible components not passing through 0?

3 The Resonance Varieties of a Space

There are two basic types of resonance varieties that one can associate to a space, depending on which CDGA is used to approximate it. In this section, we discuss both types of resonance varieties, and several ways in which these varieties can be related.

3.1 *The Cohomology Algebra*

Throughout this section, X will be a connected, finite-type CW-complex. The first approach (which has been in use since the 1990s) is to take the cohomology algebra $H^{\bullet}(X,\mathbb{C})$, endowed with the zero differential, and let $\mathcal{R}^i(X)$ be the resonance varieties of this CDGA. As indicated previously, these sets are homogeneous algebraic subvarieties of the affine space $H^1(X,\mathbb{C})$.

These varieties have been much studied in recent years, and have many practical applications, see e.g. [24, 56–58, 64, 65, 67] and the references therein. Let us just mention here two of their naturality properties.

First, the resonance varieties are homotopy-type invariants. More precisely, if $f\colon X \to Y$ is a homotopy equivalence, then the induced homomorphism $f^*\colon H^1(Y,\mathbb{C}) \to H^1(X,\mathbb{C})$ restricts to an isomorphism $f^*\colon \mathcal{R}^i(Y) \xrightarrow{\simeq} \mathcal{R}^i(X)$, for all $i \geq 0$, see e.g. [64].

Next, if $p\colon Y \to X$ is a finite, regular cover, then the induced homomorphism, $p^*\colon H^1(X,\mathbb{C}) \hookrightarrow H^1(Y,\mathbb{C})$, maps each resonance variety $\mathcal{R}^i(X)$ into $\mathcal{R}^i(Y)$, with equality if the group of deck transformations acts trivially on $H^*(Y,\mathbb{C})$, see e.g. [22, 67].

Yet the resonance varieties $\mathcal{R}^i(X)$ do not always provide accurate enough information about the space X, since the cohomology algebra may not be a (rational homotopy) model for X. It is thus important to look for alternate definitions of resonance in the non-formal setting.

3.2 *The Sullivan Model*

The second approach is to use Sullivan's model of polynomial forms, $A_{\mathrm{PL}}(X)$. This is a rationally defined CDGA, whose construction is inspired by the de Rham algebra of differential forms on a smooth manifold (see [32, 70]). In particular, the cohomology algebra $H^{\bullet}(A_{\mathrm{PL}}(X))$ is isomorphic as a graded algebra to $H^{\bullet}(X,\mathbb{C})$, via an isomorphism preserving $\mathbb{Q}$-structures. For a finite simplicial complex K, the model $A_{\mathrm{PL}}(K)$ admits a nice combinatorial description, closely related to the Stanley–Reisner ring of K (see [33]).

A connected, finite-type CW-complex X is said to be *formal* if its Sullivan model is formal, i.e., there is a weak equivalence $A_{\mathrm{PL}}(X) \simeq (H^{\bullet}(X, \mathbb{C}), 0)$ preserving $\mathbb{Q}$-structures. The notion of q-formality of a space is defined analogously. Of course, if X is formal, then it is q-formal, for all q. As a partial converse, if X is q-formal and $\dim X \leq q + 1$, then X is formal (see [44]).

Particularly interesting is the notion of 1-formality. It turns out that a space X as above is 1-formal if and only if its fundamental group, $\pi = \pi_1(X, x_0)$, is 1-formal, that is, if the Malcev–Lie algebra of π is the degree completion of a quadratic Lie algebra.

For instance, if $H^*(X, \mathbb{Q})$ is the quotient of a free CDGA by an ideal generated by a regular sequence, then X is a formal space (see [70]). In particular, if X has the rational cohomology of a torus, then X is formal. For more on these formality notions, we refer to [24, 44, 54, 55, 68].

When X is non-formal, the Sullivan model may have infinite-dimensional graded pieces. In particular, the sets $\mathcal{R}^i(A_{\mathrm{PL}}(X))$ are not *a priori* algebraic sets. Thus, we will restrict our attention to spaces X for which $A_{\mathrm{PL}}(X)$ can be replaced (up to weak equivalence) by a finite-type model (A, d).

For this class of spaces, which includes many interesting examples of non-formal spaces, the resonance varieties $\mathcal{R}^i(A)$ may be viewed as algebraic subsets of the affine space $H^1(X, \mathbb{C}) \cong H^1(A)$.

3.3 An Algebraic Tangent Cone Theorem

Before proceeding, let us briefly recall a standard notion in algebraic geometry. Let $W \subset \mathbb{C}^n$ be a Zariski closed subset, defined by an ideal I in the polynomial ring $S = \mathbb{C}[z_1, \ldots, z_n]$. The *tangent cone* of W at 0 is the algebraic subset $\mathrm{TC}_0(W) \subset \mathbb{C}^n$ defined by the ideal $\mathrm{in}(I) \subset S$ generated by the initial forms of all non-zero elements from I. This set is a homogeneous subvariety of $\mathbb{C}^n$, which depends only on the analytic germ of W at zero. In particular, $\mathrm{TC}_0(W) \neq \emptyset$ if and only if $0 \in W$.

In the previous two sections, we associated two types of resonance varieties to a space X having a finite-type model A. The next theorem, which may be viewed as an algebraic analogue of the Tangent Cone theorem, establishes a tight relationship between these two kinds of varieties.

Theorem 2 (Măcinic et al. [45]) *Let X be a finite-type CW-complex, and suppose there is a finite-type* CDGA (A, d) *such that* $A_{\mathrm{PL}}(X) \simeq A$*. Then, for each $i \geq 0$, the tangent cone at* 0 *to the resonance variety $\mathcal{R}^i(A)$ is contained in $\mathcal{R}^i(X)$.*

As we shall see in Example 3 below, the inclusion $\mathrm{TC}_0(\mathcal{R}^i(A)) \subseteq \mathcal{R}^i(X)$ may well be strict.

It seems natural to ask whether one can dispense in the above theorem with the hypothesis that the CDGA (A, d) be realized by a space X, and distill a purely algebraic statement from it.

Problem 2 Let (A, d) be a finite-type CDGA. For each $i \geq 0$, determine whether the tangent cone at 0 to $\mathcal{R}^i(A)$ is contained in $\mathcal{R}^i(H^{\bullet}(A))$.

3.4 Positive Weights

Under some additional hypothesis on the CDGA under consideration, one can say more about the nature of its resonance varieties.

Following Sullivan [70] and Morgan [51], we say that a rationally defined CDGA (A, d) has *positive weights* if each graded piece can be decomposed into weighted pieces, with positive weights in degree 1, and in a manner compatible with the CDGA structure. That is,

1. For each $i \geq 0$, there is a vector space decomposition, $A^i = \bigoplus_{\alpha \in \mathbb{Z}} A^i_{\alpha}$.
2. $A^1_{\alpha} = 0$, for all $\alpha \leq 0$.
3. If $a \in A^i_{\alpha}$ and $b \in A^j_{\beta}$, then $ab \in A^{i+j}_{\alpha+\beta}$ and $\mathrm{d}a \in A^{i+1}_{\alpha}$.

A space X is said to have positive weights if its Sullivan model does. If X is formal, then X does have positive weights: simply set the weight of a cohomology class in $A = H^{\bullet}(X, \mathbb{C})$ equal to its degree. On the other hand, as we shall see in Sects. 6–8, the converse is far from true, even when X is a smooth, complex algebraic variety.

The existence of positive weights on a CDGA model A for X imposes stringent conditions on the resonance varieties of A, and leads to an even tighter relationship between the resonance varieties of the space and its model.

Theorem 3 (Dimca and Papadima [23] and Măcinic et al. [45]) *Let X be finite-type CW-complex, and suppose there is a rationally defined, finite-type* CDGA (A, d) *with positive weights, and a q-equivalence between $A_{\mathrm{PL}}(X)$ and A preserving $\mathbb{Q}$-structures. Then, for each $i \leq q$,*

1. *$\mathcal{R}^i(A)$ is a finite union of rationally defined linear subspaces of $H^1(A)$.*
2. *$\mathcal{R}^i(A) \subseteq \mathcal{R}^i(X)$.*

Once again, it seems natural to ask whether one can dispense with the hypothesis that (A, d) be a model for a finite-type CW-complex X.

Problem 3 Let (A, d) be a finite-type CDGA with positive weights. For each $i \geq 0$, determine whether $\mathcal{R}^i(A)$ is contained in $\mathcal{R}^i(H^{\bullet}(A))$, and whether $\mathcal{R}^i(A)$ is a finite union of rationally defined linear subspaces.

Example 3 Let X be the 3-dimensional Heisenberg nilmanifold, i.e., the circle bundle over the torus, with Euler number 1. Then $H^1(X, \mathbb{C}) = \mathbb{C}^2$, and all cup products of degree 1 classes vanish; thus, $\mathcal{R}^1(X) = H^1(X, \mathbb{C})$.

On the other hand, X admits as a model (A, d) the exterior algebra on generators a, b, c in degree 1, with differential $\mathrm{d}a = \mathrm{d}b = 0$ and $\mathrm{d}c = a \wedge b$. Clearly, this is a finite-dimensional model, with positive weights: simply assign weight 1 to a and b, and weight 2 to c.

Writing $S = \mathbb{C}[x, y]$, the chain complex (15) takes the form

$$A.\otimes S\colon \quad \cdots \longrightarrow S^3 \xrightarrow{\begin{pmatrix} y & 0 & 0 \\ -x & 0 & 0 \\ 1 & -x & -y \end{pmatrix}} S^3 \xrightarrow{(x\ y\ 0)} S\ . \tag{19}$$

It follows that $H_1(A.\otimes S) = S/(x, y)$, and so $\mathcal{R}^1(A) = \{0\}$, a proper subset of $\mathcal{R}^1(X) = \mathbb{C}^2$.

4 Characteristic Varieties

We now turn to another type of homological jump loci associated to a space: the characteristic varieties, which keep track of jumps in the homology with coefficients in rank 1 local systems. Closely related objects are the support loci for the Alexander modules.

4.1 Homology Jump Loci for Rank 1 *Local Systems*

As before, let X be a finite-type, connected CW-complex. Fix a basepoint x_0, and let $\pi = \pi_1(X, x_0)$ be its fundamental group. Finally, let $\mathrm{Char}(X) = \mathrm{Hom}(\pi, \mathbb{C}^*)$ be the algebraic group of complex-valued, multiplicative characters on π, with identity 1 corresponding to the trivial representation. The identity component of this group, $\mathrm{Char}(X)^0$, is an algebraic torus of dimension $n = b_1(X)$; the other components are translates of this torus by characters corresponding to the torsion subgroup of $\pi_{\mathrm{ab}} = H_1(X, \mathbb{Z})$.

For each character $\rho\colon \pi \to \mathbb{C}^*$, let $\mathbb{C}_\rho$ be the corresponding rank 1 local system on X. The *characteristic varieties* of X are the jump loci for homology with coefficients in such local systems,

$$\mathcal{V}_i(X) = \{\rho \in \mathrm{Char}(X) \mid H_i(X, \mathbb{C}_\rho) \neq 0\}. \tag{20}$$

In more detail, let $X^{\mathrm{ab}} \to X$ be the maximal abelian cover, with group of deck transformations π_{ab}. Upon lifting the cell structure of X to this cover, we obtain a chain complex of $\mathbb{C}[\pi_{\mathrm{ab}}]$-modules,

$$\cdots \longrightarrow C_{i+1}(X^{\mathrm{ab}}, \mathbb{C}) \xrightarrow{\partial^{\mathrm{ab}}_{i+1}} C_i(X^{\mathrm{ab}}, \mathbb{C}) \xrightarrow{\partial^{\mathrm{ab}}_i} C_{i-1}(X^{\mathrm{ab}}, \mathbb{C}) \longrightarrow \cdots\ . \tag{21}$$

Tensoring this chain complex with the $\mathbb{C}[\pi_{\mathrm{ab}}]$-module $\mathbb{C}_\rho$, we obtain a chain complex of $\mathbb{C}$-vector spaces,

$$\cdots \longrightarrow C_{i+1}(X,\mathbb{C}_\rho) \xrightarrow{\partial^{\mathrm{ab}}_{i+1}(\rho)} C_i(X,\mathbb{C}_\rho) \xrightarrow{\partial^{\mathrm{ab}}_{i}(\rho)} C_{i-1}(X,\mathbb{C}_\rho) \longrightarrow \cdots, \tag{22}$$

where the evaluation of ∂^{ab}_i at ρ is obtained by applying the ring morphism $\mathbb{C}[\pi] \to \mathbb{C}$, $g \mapsto \rho(g)$ to each entry. Taking homology in degree i of this chain complex, we obtain the twisted homology groups $H_i(X,\mathbb{C}_\rho)$, whose jumps in dimension the variety $\mathcal{V}_i(X)$ keeps track of.

In a similar fashion, we may define the cohomology jump loci $\mathcal{V}^i(X)$ by the condition $H^i(X,\mathbb{C}_\rho) \neq 0$. Note that $H^i(X,\mathbb{C}_\rho) \cong H_i(X,\mathbb{C}_{\rho^{-1}})$; thus, the inversion automorphism $\rho \mapsto \rho^{-1}$ of the character group of X identifies $\mathcal{V}^i(X)$ with $\mathcal{V}_i(X)$, for each $i \geq 0$.

4.2 Some Properties of the Characteristic Varieties

The sets $\mathcal{V}_i(X)$ are algebraic subsets of the character group $\mathrm{Char}(X)$. Clearly, $1 \in \mathcal{V}_i(X)$ if and only if the i-th Betti number $b_i(X)$ is non-zero. In degree 0, we have $\mathcal{V}_0(X) = \{1\}$. In degree 1, the variety $\mathcal{V}_1(X)$ depends only on the fundamental group $\pi = \pi_1(X,x_0)$—in fact, only on its maximal metabelian quotient, π/π''—so we shall sometimes denote it as $\mathcal{V}_1(\pi)$.

The characteristic varieties are homotopy-type invariants of our space. More precisely, if $f: X \to Y$ is a homotopy equivalence, then the induced morphism between character group, $f^*: \mathrm{Char}(Y) \to \mathrm{Char}(X)$, restrict to an isomorphism $f^*: \mathcal{V}_i(Y) \xrightarrow{\simeq} \mathcal{V}_i(X)$, for all $i \geq 0$; see [66] for more details.

If $p: Y \to X$ is a finite, regular cover, then the induced morphism between character groups, $p^*: \mathrm{Char}(X) \hookrightarrow \mathrm{Char}(Y)$, maps each characteristic variety $\mathcal{V}_i(X)$ into $\mathcal{V}_i(Y)$; see [22, 67] for details.

As noted in [56], the characteristic varieties behave well under finite direct products. More precisely, let X_1 and X_2 be two connected, finite-type CW-complexes. Identifying the character group of the product $X = X_1 \times X_2$ with $\mathrm{Char}(X_1) \times \mathrm{Char}(X_2)$, we have

$$\mathcal{V}_i(X_1 \times X_2) = \bigcup_{p+q=i} \mathcal{V}_p(X_1) \times \mathcal{V}_q(X_2). \tag{23}$$

The proof of this formula is straightforward: For each character $\rho = (\rho_1,\rho_2) \in \mathrm{Char}(X)$, the chain complex $C_\bullet(X,\mathbb{C}_\rho)$ decomposes as the tensor product of the chain complexes $C_\bullet(X_1,\mathbb{C}_{\rho_1})$ and $C_\bullet(X_2,\mathbb{C}_{\rho_2})$. Taking homology, we see that $H_i(X,\mathbb{C}_\rho) = \bigoplus_{p+q=i} H_p(X_1,\mathbb{C}_{\rho_1}) \otimes_{\mathbb{C}} H_q(X_2,\mathbb{C}_{\rho_2})$, and the claim follows.

4.3 Alexander Varieties

An alternative approach, going back to the definition of the Alexander polynomials of knots and links [1], uses the homology modules of the universal abelian cover of our space X. As before, let $\pi = \pi_1(X)$, and let

$$H_i(X^{\mathrm{ab}}, \mathbb{C}) = H_1(X, \mathbb{C}[\pi_{\mathrm{ab}}]) \tag{24}$$

be the homology groups of the chain complex (21). These *Alexander invariants* are in a natural way modules over the group ring $\mathbb{C}[\pi_{\mathrm{ab}}]$. Identifying this commutative, Noetherian ring with the coordinate ring of the character group of π, we let

$$\widetilde{\mathcal{V}}_i(X) = \mathrm{supp}(H_i(X^{\mathrm{ab}}, \mathbb{C})) \tag{25}$$

be the subvariety of $\mathrm{Char}(X)$ defined by the annihilator ideal of the respective homology module. One may also consider the cohomology modules $H^i(X^{\mathrm{ab}}, \mathbb{C})$ and their support varieties, $\widetilde{\mathcal{V}}^i(X)$; we will not pursue this approach here, but refer instead to [43] for details.

As shown in [56, 58], the characteristic varieties and their homology support loci counterparts are related in the following way:

$$\bigcup_{i \le q} \mathcal{V}_i(X) = \bigcup_{i \le q} \widetilde{\mathcal{V}}_i(X). \tag{26}$$

Of special interest is the first characteristic variety, $\mathcal{V}_1(\pi) = \mathcal{V}_1(X)$. Suppose π admits a finite presentation, say, $F/R = \langle x_1, \dots, x_n \mid r_1, \dots, r_m \rangle$, and let $\phi: F \twoheadrightarrow \pi$ be the presenting homomorphism. Let $\partial_j r_i \in \mathbb{Z}[F]$ be the Fox derivatives of the relators, and let $\partial_2^{\mathrm{ab}} = (\partial_j r_i)^{\mathrm{ab} \circ \phi}$ be the corresponding Alexander matrix, with entries in $\mathbb{Z}[\pi_{\mathrm{ab}}]$. It follows from (26) that $\mathcal{V}_1(\pi)$ coincides (at least away from 1) with the zero locus of the ideal $E_1(\pi)$ of codimension 1 minors of ∂_2^{ab}, a result due to Hironaka [36].

As shown in [47], the first resonance variety $\mathcal{R}_1(\pi)$ admits a similar description, at least when π is a commutator-relators group, i.e., when all the relators r_i belong to the commutator subgroup $[F, F]$. In this case, $\mathcal{R}_1(\pi)$ is the zero locus of the codimension 1 minors of the 'linearized' Alexander matrix, $(\partial_2^{\mathrm{ab}})^{\mathrm{lin}}$, which is the m by n matrix over the polynomial ring $S = \mathbb{Z}[y_1, \dots, y_n]$ with ij-entries equal to $\sum_{k=1}^{n} \epsilon(\partial_k \partial_j r_i) y_k$, where $\epsilon: \mathbb{Z}[F] \to \mathbb{Z}$ is the augmentation map.

Remark 2 The characteristic varieties can be arbitrarily complicated. For instance, let $f \in \mathbb{Z}[t_1^{\pm 1}, \dots, t_n^{\pm 1}]$ be an integral Laurent polynomial. Then, as shown in [69], there is a finitely presented group π with $\pi_{\mathrm{ab}} = \mathbb{Z}^n$ and $\mathcal{V}_1(\pi) = V(f) \cup \{1\}$.

More generally, let Z be an algebraic subset of $(\mathbb{C}^*)^n$, defined over $\mathbb{Z}$, and let k be a positive integer. Then, as shown in [72], there is a finite, connected CW-complex X with $\mathrm{Char}(X) = (\mathbb{C}^*)^n$ such that $\mathcal{V}_i(X) = \{1\}$ for $i < q$ and $\mathcal{V}_q(X) = Z \cup \{1\}$.

5 The Tangent Cone Theorem

We are now ready to state a key result in the theory of cohomology jump loci: given a space X, and an algebraic model A with good finiteness properties, the characteristic varieties $\mathcal{V}^i(X) = \mathcal{V}_i(X)$ may be identified around the identity with the resonance varieties $\mathcal{R}^i(A)$. The resulting Tangent Cone theorem imposes strong restrictions on the nature of the resonance varieties $\mathcal{R}^i(X)$ of a formal space X.

5.1 Two Types of Tangent Cones

We start by reviewing two constructions which yield approximations to a subvariety W of a complex algebraic torus $(\mathbb{C}^*)^n$. The first one is the classical tangent cone, a variant of the construction described in Sect. 3.3, while the second one is the exponential tangent cone, a construction first introduced in [24] and further studied in [66, 69].

Let I be an ideal in the Laurent polynomial ring $\mathbb{C}[t_1^{\pm 1}, \dots, t_n^{\pm 1}]$ such that $W = V(I)$. Picking a finite generating set for I, and multiplying these generators with suitable monomials if necessary, we see that W may also be defined by the ideal $I \cap R$ in the polynomial ring $R = \mathbb{C}[t_1, \dots, t_n]$. Let J be the ideal in the polynomial ring $S = \mathbb{C}[z_1, \dots, z_n]$, generated by the polynomials $g(z_1, \dots, z_n) = f(z_1 + 1, \dots, z_n + 1)$, for all $f \in I \cap R$.

The *tangent cone* of W at 1 is the algebraic subset $\mathrm{TC}_1(W) \subset \mathbb{C}^n$ defined by the ideal $\mathrm{in}(J) \subset S$ generated by the initial forms of all non-zero elements from J. As before, the set $\mathrm{TC}_1(W)$ is a homogeneous subvariety of $\mathbb{C}^n$, which depends only on the analytic germ of W at the identity. In particular, $\mathrm{TC}_1(W) \neq \emptyset$ if and only if $1 \in W$. Moreover, TC_1 commutes with finite unions.

On the other hand, the *exponential tangent cone* to W at the origin is the set

$$\tau_1(W) = \{z \in \mathbb{C}^n \mid \exp(\lambda z) \in W, \text{ for all } \lambda \in \mathbb{C}\}. \tag{27}$$

As shown in [24, 66], this set is a finite union of rationally defined linear subspaces of the affine space $\mathbb{C}^n$. An alternative interpretation of this construction is given in [69, Sect. 6.3].

It is readily seen that τ_1 commutes with finite unions and arbitrary intersections. Clearly, the exponential tangent cone of W only depends on the analytic germ of W at the identity $1 \in (\mathbb{C}^*)^n$. In particular, $\tau_1(W) \neq \emptyset$ if and only if $1 \in W$.

Example 4 Suppose W is an algebraic subtorus of $(\mathbb{C}^*)^n$. Then $\tau_1(W)$ equals $\mathrm{TC}_1(W)$, and both coincide with $T_1(W)$, the tangent space at the identity to the Lie group W.

More generally, there is always an inclusion between the two types of tangent cones associated to an algebraic subset $W \subset (\mathbb{C}^*)^n$, namely,

$$\tau_1(W) \subseteq \mathrm{TC}_1(W). \tag{28}$$

But, as we shall see in several examples spread through this paper, this inclusion is far from being an equality for an arbitrary W. For instance, the tangent cone $\mathrm{TC}_1(W)$ may be a non-linear, irreducible subvariety of $\mathbb{C}^n$, or $\mathrm{TC}_1(W)$ may be a linear space containing the exponential tangent cone $\tau_1(W)$ as a union of proper linear subspaces.

5.2 Germs of Jump Loci

As before, let X be a connected, finite-type CW-complex. Recall that, for each $i \geq 0$, we have a characteristic variety $\mathcal{V}^i(X)$ inside the abelian, complex algebraic group $\mathrm{Char}(X)$. Furthermore, the identity component of this algebraic group, $\mathrm{Char}(X)^0$, is isomorphic to $(\mathbb{C}^*)^n$, where $n = b_1(X)$.

Now suppose we have a finite-type CDGA model (A, d) for our space X. Then, for each $i \geq 0$, we have a resonance variety $\mathcal{R}^i(A)$ inside the affine space $H^1(A) = H^1(X, \mathbb{C})$. Furthermore, this affine space may be identified with $\mathbb{C}^n$, the tangent space at 1 to $(\mathbb{C}^*)^n$.

The next result, due to Dimca and Papadima [23], relates the two types of cohomology jump loci around the origins of the respective ambient spaces.

Theorem 4 (Dimca and Papadima [23]) *Suppose Sullivan's model $A_{\mathrm{PL}}(X)$ is q-equivalent to a finite-type* CDGA (A, d). *Then, for all $i \leq q$, the germ at* 1 *of $\mathcal{V}^i(X)$ is isomorphic to the germ at* 0 *of $\mathcal{R}^i(A)$.*

It is important to note that all the above isomorphisms are induced by an analytic isomorphism $\mathrm{Char}(X)^0_{(1)} \cong H^1(A)_{(0)}$, the inverse of which is obtained by suitably restricting the exponential map $\exp\colon \mathbb{C}^n \to (\mathbb{C}^*)^n$.

Theorem 4 shows that, at least around the origin, the resonance varieties of a finite-type CDGA model for X depend only on the characteristic varieties of X, and thus, only on the homotopy type of X. This observation leads to the following corollary.

Corollary 3 *Let X be finite-type CW-complex. Suppose (A, d) and (A', d') are two finite-type* CDGAs, *both q-equivalent to the Sullivan model $A_{\mathrm{PL}}(X)$, for some $q \geq 1$. There is then an isomorphism $H^1(A) \cong H^1(A')$ restricting to isomorphisms $\mathcal{R}^i(A)_{(0)} \cong \mathcal{R}^i(A')_{(0)}$, for all $i \leq q$.*

A particular case of Theorem 4 is worth singling out.

Corollary 4 *If X is a q-formal space, then, for all $i \leq q$, the germ at* 1 *of $\mathcal{V}^i(X)$ is isomorphic to the germ at* 0 *of $\mathcal{R}^i(X)$.*

A precursor to this corollary can be found in the pioneering work of Green and Lazarsfeld [34, 35] on the cohomology jump loci of compact Kähler manifolds. The case when $q = 1$ was first established in [24, Theorem A]. For further developments in this direction, we refer to [10, 45].

5.3 Tangent Cones and Jump Loci

Returning now to the general situation, consider an arbitrary connected, finite-type CW-complex X. We then have the following relationship (due to Libgober) between the characteristic and resonance varieties of such a space.

Theorem 5 (Libgober [40]) *For all $i \geq 0$,*

$$\mathrm{TC}_1(\mathcal{V}^i(X)) \subseteq \mathcal{R}^i(X). \tag{29}$$

Putting together these inclusions with those from (28), we obtain the following immediate corollary.

Corollary 5 *For all $i \geq 0$,*

$$\tau_1(\mathcal{V}^i(X)) \subseteq \mathrm{TC}_1(\mathcal{V}^i(X)) \subseteq \mathcal{R}^i(X). \tag{30}$$

In particular, if $\mathcal{R}^i(X) = \{0\}$, then $\tau_1(\mathcal{V}^i(X)) = \mathrm{TC}_1(\mathcal{V}^i(X)) = \{0\}$.

In general, though, each of the inclusions from (30), or both inclusions can be strict, as examples to follow will show.

We now turn to spaces which admit finite-type algebraic models, and to the relations that hold between cohomology jump loci in this framework.

Theorem 6 *Let X be a connected, finite-type CW-complex, and suppose the Sullivan model $A_{\mathrm{PL}}(X)$ is q-equivalent to a finite-type* CDGA *A. Then, for all $i \leq q$,*

1. $\mathrm{TC}_1(\mathcal{V}^i(X)) = \mathrm{TC}_0(\mathcal{R}^i(A))$.
2. *If, moreover, A has positive weights, and the q-equivalence between $A_{\mathrm{PL}}(X)$ and A preserves $\mathbb{Q}$-structures, then* $\mathrm{TC}_1(\mathcal{V}^i(X)) = \mathcal{R}^i(A)$.

Proof The first assertion follows at once from Theorem 4. The second assertion follows from the first one, when coupled with Theorem 3. □

The following examples show that the positive-weights assumption in Theorem 6, part 2 is really necessary. That is, we cannot always replace $\mathrm{TC}_0(\mathcal{R}^i(A))$ with $\mathcal{R}^i(A)$ in part 1.

Example 5 Let $X = S^1$. We can take as a finite-dimensional model for the circle the CDGA (A, d) from Example 2, with $A = \bigwedge(a, b)$ and $\mathrm{d}a = 0$, $\mathrm{d}b = b \cdot a$. Since $\mathcal{V}^1(S^1) = \{1\}$, the resonance variety $\mathcal{R}^1(A) = \{0, 1\}$ properly contains $\mathrm{TC}_1(\mathcal{V}^1(S^1)) = \{0\}$. Of course, we can also take as a model for S^1 its cohomology algebra, $A' = \bigwedge(a)$, endowed with the zero differential, in which case the conclusion of part 2 is satisfied.

Example 6 Let Γ be a discrete, co-compact subgroup of a simply-connected, solvable, real Lie group G, and let $M = G/\Gamma$ be the corresponding solvmanifold. As shown in [37, 53], all the characteristic varieties of M are finite subsets of $\mathrm{Char}(M)$. Moreover, as shown by Papadima and Păunescu in [53], if (A, d) is any finite-dimensional model for M (such as the one constructed by Kasuya [37]), then all the resonance varieties $\mathcal{R}^i(A)$ contain 0 as an isolated point; in particular, $\mathrm{TC}_1(\mathcal{V}^i(M)) = \mathrm{TC}_0(\mathcal{R}^i(A)) = \{0\}$.

Now suppose G is completely solvable, and (A, d) is the classical Hattori model for the solvmanifold $M = G/\Gamma$. Work of Millionschikov [49], as reprised in [53], shows that $\mathcal{R}^i(A)$ is also a finite set. Furthermore, there are examples of solvmanifolds of this type where $\mathcal{R}^i(A)$ is different from $\{0\}$.

5.4 *The Influence of Formality*

The main connection between the formality property of a space and its cohomology jump loci is provided by the following theorem. (Again, the case $q = 1$ was established in [24], and the general case in [23].)

Theorem 7 (Dimca et al. [24] and Dimca and Papadima [23]) *If X is a q-formal space, the following "tangent cone formula" holds, for all $i \leq q$,*

$$\tau_1(\mathcal{V}^i(X)) = \mathrm{TC}_1(\mathcal{V}^i(X)) = \mathcal{R}^i(X). \tag{31}$$

As an application of this theorem, we have the following characterization of the irreducible components of the cohomology jump loci in the formal setting.

Corollary 6 *If X is a q-formal space, then, for all $i \leq q$,*

1. *All irreducible components of the resonance variety $\mathcal{R}^i(X)$ are rationally defined subspaces of $H^1(X, \mathbb{C})$.*
2. *All irreducible components of the characteristic variety $\mathcal{V}^i(X)$ which pass through the origin are algebraic subtori of $\mathrm{Char}(X)^0$, of the form $\exp(L)$, where L runs through the linear subspaces comprising $\mathcal{R}^i(X)$.*

Even when the space X is formal, the characteristic varieties $\mathcal{V}^i(X)$ may have irreducible components which do not pass through the identity of $\mathrm{Char}(X)^0$, and thus are not detected by the resonance varieties $\mathcal{R}^i(X)$.

Example 7 Let K be a non-trivial knot in the 3-sphere, with complement $X = S^3 \setminus K$. Then $H^*(X, \mathbb{Z}) \cong H^*(S^1, \mathbb{Z})$; therefore, X is formal and $\mathcal{R}^1(X) = \{0\}$. The characteristic variety $\mathcal{V}^1(X) \subset \mathbb{C}^*$ consists of 1, together with all the roots of the Alexander polynomial, $\Delta_K \in \mathbb{Z}[t^{\pm 1}]$. Thus, if $\Delta_K \not\equiv 1$, then $\mathcal{V}^1(X)$ has components which do not contain 1.

Example 8 Let X be the 2-complex obtained by gluing a Möbius band to a 2-torus along a meridian circle. Then X has the same rational cohomology ring as the 2-torus; thus, X is formal and $\mathcal{R}^1(X) = \{0\}$. On the other hand, $\pi_1(X) = \langle x_1, x_2 \mid x_1x_2^2 = x_2^2x_1\rangle$; hence, the variety $\mathcal{V}^1(X) \subset (\mathbb{C}^*)^2$ consists of the identity together with the translated subtorus $t_1t_2^{-1} = -1$.

5.5 Formality Tests

The next several examples illustrate the various ways in which the inclusions from Corollary 5 may fail to hold as equalities, thereby showing how Theorem 7 and Corollary 6 can be used to detect non-formality. More examples will be given in Sects. 6–8.

Example 9 Let X be the presentation 2-complex for the group $\pi = \langle x_1, x_2 \mid [x_1, [x_1, x_2]]\rangle$. In this case, $\mathcal{V}^1(X) = \{t_1 = 1\}$, and so $\tau_1(\mathcal{V}^1(X)) = \mathrm{TC}_1(\mathcal{V}^1(X)) = \{x_1 = 0\}$. On the other hand, $\mathcal{R}^1(X) = \mathbb{C}^2$, and so X is not 1-formal.

The next example (adapted from [24]) shows how the rationality property from Corollary 6 can be used as a formality test.

Example 10 Let X be the presentation 2-complex for the group π with generators $x_1, \dots, x_4$ and relators $r_1 = [x_1, x_2]$, $r_2 = [x_1, x_4][x_2^{-2}, x_3]$, and $r_3 = [x_1^{-1}, x_3][x_2, x_4]$. Computing the linearized Alexander matrix of this presentation by the method described in Sect. 4.3, we see that $\mathcal{R}^1(X)$ is the quadric hypersurface in $\mathbb{C}^4$ given by the equation $z_1^2 - 2z_2^2 = 0$. This quadric splits into two linear subspaces defined over $\mathbb{R}$, but it is irreducible over $\mathbb{Q}$. Thus, X is not 1-formal.

Example 11 In view of Remark 2, there is a finitely presented group π with abelianization $\mathbb{Z}^3$ and characteristic variety

$$\mathcal{V}^1(\pi) = \{(t_1, t_2, t_3) \in (\mathbb{C}^*)^3 \mid (t_2 - 1) = (t_1 + 1)(t_3 - 1)\}.$$

As noted in [69], this variety is irreducible, and its exponential tangent cone at the origin splits as a union of two (rationally defined) lines in $\mathbb{C}^3$,

$$\tau_1(\mathcal{V}^1(\pi)) = \{x_2 = x_3 = 0\} \cup \{x_1 - x_3 = x_2 - 2x_3 = 0\}.$$

The variety $\mathcal{V}^1(\pi)$ is a complex, 2-dimensional torus passing through the origin. Nevertheless, this torus does not embed as an algebraic subgroup in $(\mathbb{C}^*)^3$; indeed, if it did, $\tau_1(\mathcal{V}^1(\pi))$ would be a single plane. Consequently, the group π is not 1-formal.

6 Smooth Quasi-Projective Varieties

We now switch our focus, from the general theory of cohomology jump loci to some of the applications of this theory within the class of smooth, complex quasi-projective varieties. For such spaces, the cohomology jump loci are severely restricted by the extra structure imposed on their CDGA models and cohomology rings by the underlying algebraic geometry.

6.1 Compactifications and Formality

A complex projective variety is a subset of a complex projective space $\mathbb{CP}^n$, defined as the zero-locus of a homogeneous prime ideal in $\mathbb{C}[z_0, \dots, z_n]$. A Zariski open subvariety of a projective variety is called a quasi-projective variety. We will only consider here projective and quasi-projective varieties which are connected and smooth.

If M is a (compact, smooth) projective variety, then the Hodge decomposition on $H^*(M, \mathbb{C})$ puts strong constraints on the topological properties of M. For instance, as shown in [16], every such a manifold is formal.

Each smooth, quasi-projective variety X admits a good compactification. That is to say, there is a smooth, complex projective variety $\overline{X}$ and a normal-crossings divisor D such that $X = \overline{X} \setminus D$. By a well-known theorem of Deligne, each cohomology group of X admits a mixed Hodge structure. This additional structure puts definite constraints on the algebraic topology of such manifolds.

For instance, if X admits a smooth compactification $\overline{X}$ with $b_1(\overline{X}) = 0$, the weight 1 filtration on $H^1(X, \mathbb{C})$ vanishes; in turn, by work of Morgan [51], this implies the 1-formality of X. Thus, as noted by Kohno in [38], if X is the complement of a hypersurface in $\mathbb{CP}^n$, then $\pi_1(X)$ is 1-formal. Moreover, if $n = 2$, then X itself is formal, see [11, 44].

In general, though, smooth quasi-projective varieties need not be 1-formal. Furthermore, even when they are 1-formal, they still can be non-formal.

Example 12 Let $E^{\times n}$ be the n-fold product of an elliptic curve. The closed form $\frac{1}{2}\sqrt{-1}\sum_{i=1}^{n} dz_i \wedge d\bar{z}_i$ defines an integral cohomology class $\omega \in H^{1,1}(E^{\times n})$. By the Lefschetz theorem on $(1,1)$-classes, ω can be realized as the first Chern class of an algebraic line bundle over $E^{\times n}$. Let X_n be the complement of the zero-section of this bundle. Then X_n is a smooth, quasi-projective variety which is not formal. In fact, as we shall see in Example 14, X_1 is not 1-formal. On the other hand, X_n is 1-formal, for all $n > 1$.

6.2 Algebraic Models

As before, let X be a connected, smooth quasi-projective variety, and choose a smooth compactification $\overline{X}$ such that the complement is a finite union, $D = \bigcup_{j\in J} D_j$, of smooth divisors with normal crossings. There is then a rationally defined CDGA, $A = A(\overline{X}, D)$, called the *Gysin model* of the compactification, constructed as follows. As a $\mathbb{C}$-vector space, A^i is the direct sum of all subspaces

$$A^{p,q} = \bigoplus_{|S|=q} H^p\Big(\bigcap_{k\in S} D_k, \mathbb{C}\Big)(-q) \tag{32}$$

with $p + q = i$, where $(-q)$ denotes the Tate twist. Furthermore, the multiplication in A is induced by the cup-product in $\overline{X}$, and has the property that $A^{p,q} \cdot A^{p',q'} \subseteq A^{p+p',q+q'}$, while the differential, $\mathrm{d}\colon A^{p,q} \to A^{p+2,q-1}$, is constructed from the Gysin maps arising from intersections of divisors. The CDGA just constructed depends on the compactification $\overline{X}$; for simplicity, though, we will denote it by $A(X)$ when the compactification is understood.

An important particular case is when our variety X has dimension 1. That is to say, let Σ be a connected, possibly non-compact, smooth algebraic curve. Then Σ admits a canonical compactification, $\overline{\Sigma}$, and thus, a canonical Gysin model, $A(\Sigma)$. We illustrate the construction of this model in a simple situation, which we shall encounter again in Sect. 8.

Example 13 Let $\Sigma = E^*$ be a once-punctured elliptic curve. Then $\overline{\Sigma} = E$, and the Gysin model $A(\Sigma)$ is the algebra $A = \bigwedge(a, b, e)/(ae, be)$ on generators a, b in bidegree $(1, 0)$ and generator e in bidegree $(0, 1)$, with differential $\mathrm{d}\colon A \to A$ given by $\mathrm{d}a = \mathrm{d}b = 0$ and $\mathrm{d}e = ab$.

The above construction is functorial, in the following sense: If $f\colon X \to Y$ is a morphism of quasi-projective manifolds which extends to a regular map $\bar{f}\colon \overline{X} \to \overline{Y}$ between the respective good compactifications, then there is an induced CDGA morphism $f^!\colon A(Y) \to A(X)$ which respects the bigradings.

Morgan showed in [51] that the Sullivan model $A_{\mathrm{PL}}(X)$ is connected to the Gysin model $A(X)$ by a chain of quasi-isomorphisms preserving $\mathbb{Q}$-structures. Moreover, setting the weight of $A^{p,q}$ equal to $p + 2q$ defines a positive-weight decomposition on $(A^{\bullet}, d)$.

In [26], Dupont constructs a Gysin-type model for certain types of quasi-projective varieties, where the normal-crossing divisors assumption on the compactification can be relaxed. More precisely, let $\mathcal{A}$ be an arrangement of smooth hypersurfaces in a smooth, n-dimensional complex projective variety $\overline{X}$, and suppose $\mathcal{A}$ locally looks like an arrangement of hyperplanes in $\mathbb{C}^n$. There is then a CDGA model for the complement, $X = \overline{X} \setminus \bigcup_{L\in\mathcal{A}} L$, which builds on the combinatorial definition of the Orlik–Solomon algebra of a hyperplane arrangement (an algebra we will return to in Sect. 7.1).

6.3 Configuration Spaces

In a special situation, an alternate model is available. A construction due to Fadell and Neuwirth associates to a space X and a positive integer n the space of ordered configurations of n points in X,

$$\mathrm{Conf}(X,n) = \{(x_1,\dots,x_n) \in X^n \mid x_i \neq x_j \text{ for } i \neq j\}. \tag{33}$$

The E_2-term of the Leray spectral sequence for the inclusion $\mathrm{Conf}(X,n) \hookrightarrow X^n$ was described concretely by Cohen and Taylor in the late 1970s. If X is a smooth, complex projective variety of dimension m, then $\mathrm{Conf}(X,n)$ is a smooth, quasi-projective variety. Moreover, as shown by Totaro in [71], the Cohen–Taylor spectral sequence collapses at the E_{m+1}-term, and the E_m-term is a CDGA model for the configuration space $\mathrm{Conf}(X,n)$.

The most basic example is the configuration space of n ordered points in $\mathbb{C}$, which is a classifying space for P_n, the pure braid group on n strings, whose cohomology ring was computed by Arnol'd in the late 1960s. We shall come back to this example in Sect. 7 in the setting of arrangements of hyperplanes, and we shall look at configuration space of n points on an elliptic curve E in Sect. 8, in the setting of elliptic arrangements.

More generally, following Eastwood and Huggett [27], one may consider the "graphic configuration spaces"

$$\mathrm{Conf}(X,\Gamma) = \{(x_1,\dots,x_n) \in X^{\times n} \mid x_i \neq x_j \text{ for } \{i,j\} \in E(\Gamma)\} \tag{34}$$

associated to a space X and a simple graph Γ with vertex set $[n]$ and edge set $E(\Gamma)$. Especially interesting is the case when X is a Riemann surface Σ_g. For such a space, the naive compactification, $\overline{\mathrm{Conf}(X,n)} = X^{\times n}$, satisfies the hypothesis which permit the construction of the Dupont model, [26]. For recent work exploiting this model, we refer to [5].

6.4 Characteristic Varieties

The structure of the jump loci for cohomology in rank 1 local systems on smooth, complex projective and quasi-projective varieties (and, more generally, on Kähler and quasi-Kähler manifolds) was determined through the work of Beauville [4], Green and Lazarsfeld [34, 35], Simpson [61], and Arapura [2].

In the quasi-projective setting, further improvements and refinements were given in [3, 39, 66]. The definitive structural result was obtained by Budur and Wang in [9], building on work of Dimca and Papadima [23].

Theorem 8 (Budur and Wang [9]) *Let X be a smooth quasi-projective variety. Then each characteristic variety $\mathcal{V}^i(X)$ is a finite union of torsion-translated subtori of* $\operatorname{Char}(X)$.

Work of Arapura [2] explains how the non-translated subtori occurring in the above decomposition of $\mathcal{V}^1(X)$ arise. Let us say that a holomorphic map $f: X \to \Sigma$ is *admissible* if f is surjective, has connected generic fiber, and the target Σ is a connected, smooth complex curve with negative Euler characteristic. Up to reparametrization at the target, the variety X admits only finitely many admissible maps; let $\mathcal{E}_X$ be the set of equivalence classes of such maps.

If $f: X \to \Sigma$ is an admissible map, it is readily verified that $\mathcal{V}^1(\Sigma) = \operatorname{Char}(\Sigma)$. Thus, the image of the induced morphism between character groups, $f^*: \operatorname{Char}(\Sigma) \to \operatorname{Char}(X)$, is an algebraic subtorus of $\operatorname{Char}(X)$.

Theorem 9 (Arapura [2]) *The correspondence $f \mapsto f^*(\operatorname{Char}(\Sigma))$ establishes a bijection between the set $\mathcal{E}_X$ of equivalence classes of admissible maps from X to curves and the set of positive-dimensional, irreducible components of $\mathcal{V}^1(X)$ containing* 1.

The positive-dimensional, irreducible components of $\mathcal{V}^1(X)$ which do not pass through 1 can be similarly described, by replacing the admissible maps with certain "orbifold fibrations," whereby multiple fibers are allowed. For more details and further explanations, we refer to [3, 66].

6.5 Resonance Varieties

We now turn to the resonance varieties associated to a quasi-projective manifold, and how they relate to the characteristic varieties. The Tangent Cone theorem takes a very special form in this algebro-geometric setting.

Theorem 10 *Let X be a smooth, quasi-projective variety, and let $A(X)$ be a Gysin model for X. Then, for each $i \geq 0$,*

$$\tau_1(\mathcal{V}^i(X)) = \operatorname{TC}_1(\mathcal{V}^i(X)) = \mathcal{R}^i(A(X)) \subseteq \mathcal{R}^i(X). \tag{35}$$

Moreover, if X is q-formal, the last inclusion is an equality, for all $i \leq q$.

Proof By Theorem 8, each irreducible component of $\mathcal{V}^i(X)$ passing through 1 is a complex algebraic subtorus $W \subset \operatorname{Char}(X)$. As noted in Example 4, $\tau_1(W) = \operatorname{TC}_1(W)$. Since both τ_1 and TC_1 commute with finite unions, the first equality in (35) follows.

Next, recall that $A(X)$ is a finite-type, rationally defined CDGA which admits positive weights. Moreover, there is a weak equivalence between $A(X)$ and $A_{\mathrm{PL}}(X)$ preserving the respective $\mathbb{Q}$-structures. The second equality now follows from Theorem 6, part 2, while the last inclusion follows from Theorem 3. □

In particular, the resonance varieties $\mathcal{R}^i(A(X))$ are finite unions of rationally defined linear subspaces of $H^1(X,\mathbb{C})$. On the other hand, the varieties $\mathcal{R}^i(X)$ can be much more complicated: for instance, they may have non-linear irreducible components. If X is q-formal, though, Theorem 35 guarantees this cannot happen, as long as $i \leq q$.

6.6 Resonance in Degree 1

Once again, let X be a smooth, quasi-projective variety, and let $A(X)$ be the Gysin model associated to a good compactification $\overline{X}$. The degree 1 resonance varieties $\mathcal{R}^1(A(X))$, and, to some extent, $\mathcal{R}^1(X)$, admit a much more precise description than those in higher degrees.

As in the setup from Theorem 9, let $\mathcal{E}_X$ be the set of equivalence classes of admissible maps from X to curves, and let $f: X \to \Sigma$ be such map. Recall from Sect. 6.2 that the curve Σ admits a canonical Gysin model, $A(\Sigma)$. As noted in [23], the induced CDGA morphism, $f^!: A(\Sigma) \to A(X)$, is injective. Let $f^*: H^1(A(\Sigma)) \to H^1(A(X))$ be the induced homomorphism in cohomology.

Theorem 11 (Dimca and Papadima [23] and Măcinic et al. [45]) *For a smooth, quasi-projective variety X, the decomposition of $\mathcal{R}^1(A(X))$ into (linear) irreducible components is given by*

$$\mathcal{R}^1(A(X)) = \bigcup_{f \in \mathcal{E}_X} f^*(H^1(A(\Sigma))). \tag{36}$$

If X admits no admissible maps, i.e., $\mathcal{E}_X = \emptyset$, formula (36) should be understood to mean $\mathcal{R}^1(A(X)) = \{0\}$ if $b_1(X) > 0$ and $\mathcal{R}^1(A(X)) = \emptyset$ if $b_1(X) = 0$.

Example 14 Let $X = X_1$ be the complex, smooth quasi-projective surface constructed in Example 12. Clearly, this manifold is a $\mathbb{C}^*$-bundle over $E = S^1 \times S^1$ which deform-retracts onto the Heisenberg manifold from Example 3. Hence, $\mathcal{V}^1(X) = \{1\}$, and so $\tau_1(\mathcal{V}^1(X)) = \mathrm{TC}_1(\mathcal{V}^1(X)) = \{0\}$. On the other hand, $\mathcal{R}^1(X) = \mathbb{C}^2$, and so X is not 1-formal.

Under a 1-formality assumption, the usual resonance varieties $\mathcal{R}^1(X)$ admit a similar description.

Theorem 12 (Dimca et al. [24]) *Let X be a smooth, quasi-projective variety, and suppose X is 1-formal. The decomposition into irreducible components of the first resonance variety is then given by*

$$\mathcal{R}^1(X) = \bigcup_{f \in \mathcal{E}_X} f^*(H^1(\Sigma,\mathbb{C})), \tag{37}$$

with the same convention as before when $\mathcal{E}_X = \emptyset$. Moreover, all the (rationally defined) linear subspaces in this decomposition have dimension at least 2*, and any two distinct ones intersect only at* 0*.*

If X is compact, then the formality assumption in the above theorem is automatically satisfied, due to [16]. Furthermore, the conclusion of the theorem can also be sharpened in this case: each (non-trivial) irreducible component of $\mathcal{R}^1(X)$ is even-dimensional, of dimension at least 4.

In general, though, the resonance varieties $\mathcal{R}^1(X)$ can have non-linear components. For instance, if $X = \mathrm{Conf}(E, n)$ is the configuration space of $n \geq 3$ points on an elliptic curve E, then $\mathcal{R}^1(X)$ is an irreducible, non-linear variety (in fact, a rational normal scroll), see [24] and also Example 21 below.

7 Hyperplane Arrangements and the Milnor Fibration

Next, we turn our focus to a class of quasi-projective varieties which are obtained by deleting finitely many hyperplanes from a complex affine space. These hyperplane arrangement complements are formal spaces, yet the associated Milnor fibers may fail the Tangent Cone test for formality.

7.1 Complement and Intersection Lattice

A hyperplane arrangement $\mathcal{A}$ is a finite collection of codimension 1 linear subspaces in a complex affine space $\mathbb{C}^n$. Its complement, $M(\mathcal{A}) = \mathbb{C}^n \setminus \bigcup_{H\in\mathcal{A}} H$, is a connected, smooth, quasi-projective variety. This manifold is a Stein domain, and thus has the homotopy-type of a finite CW-complex of dimension n. Moreover, $M(\mathcal{A}) \cong U(\mathcal{A}) \times \mathbb{C}^*$, where $U(\mathcal{A})$ is the complement in $\mathbb{CP}^{n-1}$ of the projectivized arrangement.

The topological invariants of the complement are intimately tied to the combinatorics of the arrangement. The latter is encoded in the *intersection lattice*, $L(\mathcal{A})$, which is the poset of all intersections of $\mathcal{A}$, ordered by reverse inclusion. The rank of the arrangement, denoted $\mathrm{rk}(\mathcal{A})$, is the codimension of the intersection $\Sigma(\mathcal{A}) = \bigcap_{H\in\mathcal{A}} H$.

Example 15 The braid arrangement of rank $n-1$ consists of the diagonal hyperplanes $H_{ij} = \{z_i - z_j = 0\}$ in $\mathbb{C}^n$. The complement of this arrangement is the configuration space $\mathrm{Conf}(\mathbb{C}, n)$, while the intersection lattice is the lattice of partitions of $[n] = \{1, \dots, n\}$, ordered by refinement.

For each hyperplane $H \in \mathcal{A}$, pick a linear form $f_H \in \mathbb{C}[z_0, \dots, z_n]$ such that $\ker(f_H) = H$. The homogeneous polynomial $Q(\mathcal{A}) = \prod_{H\in\mathcal{A}} f_H$, then, is a defining polynomial for the arrangement.

Building on work of Arnol'd on the cohomology ring of $\mathrm{Conf}(\mathbb{C}, n)$, Brieskorn showed in [8] that the closed 1-forms $\mathrm{d}f_H/f_H$ generate the de Rham cohomology of $M(\mathcal{A})$. Moreover, the inclusion of the subalgebra generated by those forms into the de Rham algebra $\Omega^*_{\mathrm{dR}}(M(\mathcal{A}))$ induces an isomorphism in cohomology; consequently, the complement $M(\mathcal{A})$ is a formal space.

In [52], Orlik and Solomon gave a simple combinatorial description of the cohomology ring of the complement. Let $E = \bigwedge(\mathcal{A})$ be the exterior algebra (over $\mathbb{Z}$) on degree-one classes e_H dual to the meridians around the hyperplanes $H \in \mathcal{A}$, and set $e_{\mathcal{B}} = \prod_{H\in\mathcal{B}} e_H$ for each sub-arrangement $\mathcal{B} \subset \mathcal{A}$. Next, define a differential $\partial\colon E \to E$ of degree -1, starting from $\partial(e_H) = 1$, and extending it to a linear map on E, using the graded Leibniz rule. Then

$$H^*(M(\mathcal{A}), \mathbb{Z}) = \bigwedge(\mathcal{A})/I(\mathcal{A}), \tag{38}$$

where $I(\mathcal{A})$ is the (homogeneous) ideal generated by all elements of the form

$$\begin{aligned} &e_{\mathcal{B}}, && \text{if } \Sigma(\mathcal{B}) = \emptyset,\\ &\partial e_{\mathcal{B}}, && \text{if } \operatorname{codim} \Sigma(\mathcal{B}) < |\mathcal{B}|. \end{aligned} \tag{39}$$

More generally, suppose $\mathcal{A}$ is an arrangement of complex linear subspaces in $\mathbb{C}^n$. Using a blow-up construction, De Concini and Procesi gave in [15] a 'wonderful' CDGA model for the complement of such an arrangement. Based on a simplication of this model due to Yuzvinsky [73], Feichtner and Yuzvinsky showed in [30] the following: If $L(\mathcal{A})$ is a geometric lattice, then the complement of $\mathcal{A}$ is a formal space. In general, however, the complement of a complex subspace arrangement need not be formal. This phenomenon is illustrated in [17], within the class of coordinate subspace arrangements, and in [48], within the class of diagonal subspace arrangements.

7.2 Cohomology Jump Loci of the Complement

Once again, let $\mathcal{A}$ be a complex hyperplane arrangement. The resonance varieties of the arrangement, $\mathcal{R}^i(\mathcal{A}) := \mathcal{R}^i(M(\mathcal{A}))$, live inside the affine space $H^1(M(\mathcal{A}), \mathbb{C}) = \mathbb{C}^{|\mathcal{A}|}$. These varieties depend only on the Orlik–Solomon algebra of $\mathcal{A}$, and thus, only on the intersection lattice $L(\mathcal{A})$.

In [28], Falk asked whether the resonance varieties $\mathcal{R}^i(\mathcal{A})$ are finite unions of linear subspaces. A special case of the Tangent Cone theorem, proved in [13] specifically for arrangement complements and in degree $i = 1$, led to a positive answer to this question, at least for $\mathcal{R}^1(\mathcal{A})$. With the technology provided by the general version of the Tangent Cone theorem, it is now easy to answer Falk's question in full generality. Indeed, since the complement $M(\mathcal{A})$ is a formal space,

Corollary 6 shows that $\mathcal{R}^i(\mathcal{A})$ is, in fact, a finite union of rationally defined linear subspaces, for each $i \geq 0$.

In degree $i = 1$, these linear spaces can be described much more precisely. Indeed, as shown by Falk and Yuzvinsky in [29] (see also [42, 46]), each component of $\mathcal{R}^1(\mathcal{A})$ corresponds to a multinet on a sub-arrangement of $\mathcal{A}$.

Briefly, a *k-multinet* on $\mathcal{A}$ is a partition into $k \geq 3$ subsets $\mathcal{A}_\alpha$, together with an assignment of multiplicities m_H to each $H \in \mathcal{A}$, and a choice of rank 2 flats, called the base locus. All these data must satisfy certain compatibility conditions. For instance, any two hyperplanes from different parts of the partition intersect in the base locus, while the sum of the multiplicities over each part is constant. Furthermore, if X is a flat in the base locus, then the sum $n_X = \sum_{H\in\mathcal{A}_\alpha\cap\mathcal{A}_X} m_H$ is independent of α. The multinet is *reduced* if all the m_H's are equal to 1. If, moreover, all the n_X's are equal to 1, the multinet is, in fact, a *net*, a classical notion from combinatorial geometry.

Every k-multinet on $\mathcal{A}$ gives rise to an admissible map $M(\mathcal{A}) \to \Sigma$, where $\Sigma = \mathbb{CP}^1 \setminus \{k \text{ points}\}$, and the converse also holds. Moreover, the set $\mathcal{E}_{M(\mathcal{A})}$ of admissible maps (up to reparametrization at the target) from $M(\mathcal{A})$ to curves coincides with the set of multinets (up to the natural S_k-permutation action on k-multinets) on subarrangements of $\mathcal{A}$, see [29, 60]. The *essential* components of $\mathcal{R}^1(A)$ are those corresponding to multinets fully supported on $\mathcal{A}$.

Example 16 Let $\mathcal{A}$ be a generic 3-slice of the braid arrangement of rank 3, with defining polynomial $Q(\mathcal{A}) = z_0z_1z_2(z_0-z_1)(z_0-z_2)(z_1-z_2)$. Take a generic plane section, and label the corresponding lines as 1 to 6. Then, the variety $\mathcal{R}^1(\mathcal{A}) \subset \mathbb{C}^6$ has 4 'local' components, corresponding to the triple points 124, 135, 236, 456, and one essential component, corresponding to the 3-net (16|25|34).

From Theorem 9, we know that the characteristic varieties $\mathcal{V}^i(\mathcal{A}) := \mathcal{V}^i(M(\mathcal{A}))$ consists of subtori in $(\mathbb{C}^*)^n$, possibly translated by roots of unity, together with a finite number of torsion points. By Theorem 7, we have that $\mathrm{TC}_1(\mathcal{V}^i(\mathcal{A})) = \mathcal{R}^i(A)$. Thus, the components of $\mathcal{V}^i(\mathcal{A})$ passing through the origin are completely determined by $L(\mathcal{A})$.

As pointed out in [62], though, the characteristic variety $\mathcal{V}^1(\mathcal{A})$ may contain translated subtori, that is, components not passing through 1. It is still not known whether such components are combinatorially determined.

7.3 The Milnor Fibration

Once again, let $\mathcal{A}$ be a hyperplane arrangement in $\mathbb{C}^n$, with complement $M = M(\mathcal{A})$ and defining polynomial $Q = Q(\mathcal{A})$. As shown by Milnor [50] in a more general context, the restriction of the polynomial map $Q: \mathbb{C}^n \to \mathbb{C}$ to the complement is a smooth fibration, $Q: M \to \mathbb{C}^*$.

The typical fiber of this fibration, $Q^{-1}(1)$, is called the *Milnor fiber* of the arrangement, and is denoted by $F = F(\mathcal{A})$. The Milnor fiber is a Stein domain of complex dimension n, and thus has the homotopy type of a finite CW-complex of dimension n. Furthermore, the monodromy homeomorphism $h\colon F \to F$ is given by $h(z) = e^{2\pi i/m}z$, where $m = |\mathcal{A}|$, and thus has order m.

Example 17 The Boolean arrangement consists of the coordinate hyperplanes in $\mathbb{C}^n$; its complement is the complex algebraic torus $(\mathbb{C}^*)^n$. The map $Q\colon (\mathbb{C}^*)^n \to \mathbb{C}^*$, $z \mapsto z_1 \cdots z_n$ is a morphism of algebraic groups. Hence, the Milnor fiber $F = \ker Q$ is an algebraic subgroup, isomorphic to $(\mathbb{C}^*)^{n-1}$.

Example 18 Consider a pencil of m lines in $\mathbb{C}^2$, with defining polynomial $Q = z_1^m - z_2^m$ and complement $M = \mathbb{C}^* \times \mathbb{C} \setminus \{m \text{ points}\}$. The Milnor fiber, then, is a smooth complex curve of genus $\binom{m-1}{2}$ with m punctures.

In general, though, the polynomial map $Q\colon \mathbb{C}^n \to \mathbb{C}$ will have a non-isolated singularity at 0, and the topology of the Minor fiber $F = F(\mathcal{A})$ will be much more difficult to ascertain. In particular, it is a long-standing open problem to decide whether the first Betti number $b_1(F)$ is determined by the intersection lattice of $\mathcal{A}$, and, if so, to find an explicit combinatorial formula for it, see for instance [12, 18, 21, 22, 41, 67]. Marked progress towards a positive solution to this problem was made recently in [60], using in an essential way the relationship between the varieties $\mathcal{V}^1(\mathcal{A})$ and $\mathcal{R}^1(\mathcal{A})$ provided by the Tangent Cone theorem, as well as the multinet interpretation of the components of $\mathcal{R}^1(\mathcal{A})$.

To make this machinery work, one starts by viewing the Minor fiber F as the regular, cyclic m-fold cover of the projectivized complement, $U = \mathbb{P}(M)$, defined by the homomorphism $\delta\colon \pi_1(U) \to \mathbb{Z}_m$ which takes each meridian generator to 1, see [12] and also [63, 67]. Embedding $\mathbb{Z}_m$ into $\mathbb{C}^*$ by sending $1 \mapsto e^{2\pi i/m}$, we may view δ as a character on $\pi_1(U)$. The relative position of this character with respect to the variety $\mathcal{V}^1(U) \cong \mathcal{V}^1(\mathcal{A})$ determines the first Betti number of F, as well as the characteristic polynomial of the algebraic monodromy, $h_*\colon H_1(F, \mathbb{C}) \to H_1(F, \mathbb{C})$.

7.4 Cohomology Jump Loci of the Milnor Fiber

Very little is known about the homology with coefficients in rank 1 local systems of the Milnor fiber of an arrangement $\mathcal{A}$. Since $M(\mathcal{A})$ is a smooth, quasi-projective variety, Theorem 8 guarantees that the characteristic varieties $\mathcal{V}^i(F(\mathcal{A}))$ are finite unions of torsion-translated subtori.

Let $\pi\colon F(\mathcal{A}) \to U(\mathcal{A})$ be the restriction of the Hopf fibration to the Milnor fiber. Since π is a finite, regular cover, we have that $\pi^*(\mathcal{V}^i(U(\mathcal{A}))) \subseteq \mathcal{V}^i(F(\mathcal{A}))$. In general, though, this inclusion may well be strict. For instance, suppose $\mathcal{A}$ admits a non-trivial, reduced multinet, and let T be the corresponding component of $\mathcal{V}^1(\mathcal{A})$. Then, as shown in [22], the variety $\mathcal{V}^1(F(\mathcal{A}))$ has an irreducible component passing through the identity and containing $\pi^*(T)$ as a proper subset.

Example 19 Let $\mathcal{A}$ be the braid arrangement from Example 16. Recall that $\mathcal{V}^1(\mathcal{A})$ has four 2-dimensional components, $T_1, \ldots, T_4$, corresponding to the triple points, and also an essential, 2-dimensional component T, corresponding to a 3-net. The characteristic variety $\mathcal{V}^1(F(\mathcal{A})) \subset (\mathbb{C}^*)^7$ has four 2-dimensional components, $\pi^*(T_1), \dots, \pi^*(T_4)$, as well as 4-dimensional component W which properly includes the 2-torus $\pi^*(T)$.

Returning to the general situation, let again $\mathcal{A}$ be a complex hyperplane arrangement, and let $F = F(\mathcal{A})$ be its Milnor fiber. By Theorem 10, $\mathrm{TC}_1(\mathcal{V}^i(F)) = \mathcal{R}^i(A(F))$, where $A(F)$ is a Gysin model for F. Thus, to better understand the topology of the Milnor fiber, it would help a lot to address the following two problems.

Problem 4 Find a smooth compactification $\overline{F}$ such that $\overline{F} \setminus F$ is a normal-crossings divisor. Does the monodromy $h\colon F \to F$ extend to a diffeomorphism $\bar{h}\colon \overline{F} \to \overline{F}$?

Problem 5 Given a compactification $\overline{F}$ as above, write down an explicit presentation for the resulting Gysin model, $A(F)$. Furthermore, compute the resonance varieties $\mathcal{R}^i(A(F))$, and decide whether these varieties depend only on the intersection lattice $L(\mathcal{A})$.

7.5 *Formality of the Milnor Fiber*

The following question was raised in [55], in a more general context: Is the Milnor fiber of a hyperplane arrangement $\mathcal{A}$ always formal? Of course, if $\mathrm{rank}(\mathcal{A}) = 2$, then $F(\mathcal{A})$ has the homotopy type of a wedge of circles, and so it is formal.

If $\mathrm{rank}(\mathcal{A}) = 3$, formality and 1-formality are equivalent for the Milnor fiber, since in this case $F(\mathcal{A})$ has the homotopy type of a 2-complex. As noted by Dimca and Papadima [22], if the monodromy map acts as the identity on $H_1(F(\mathcal{A}), \mathbb{C})$, then $F(\mathcal{A})$ is formal. In general, though, the Milnor fiber of an arrangement is not formal, as the following example of Zuber [74] shows.

Example 20 Let $\mathcal{A}$ be the arrangement associated to the complex reflection group $G(3,3,3)$, and defined by the polynomial $Q = (z_1^3 - z_2^3)(z_1^3 - z_3^3)(z_2^3 - z_3^3)$. The resonance variety $\mathcal{R}_1(\mathcal{A}) \subset \mathbb{C}^9$ has 12 local components, corresponding to the triple points, and 4 essential components corresponding to 3-nets.

Consider the 3-net whose associated rational map $\mathbb{CP}^2 \dashrightarrow \mathbb{CP}^1$ is given by $(z_1, z_2, z_3) \mapsto (z_1^3 - z_2^3, z_2^3 - z_3^3)$. This map restricts to an admissible map $U(\mathcal{A}) \to \mathbb{CP}^1 \setminus \{(1,0),(0,1),(1,-1)\}$. Let T be the essential, 2-dimensional component of $\mathcal{V}_1(U(\mathcal{A}))$ obtained by pullback along this pencil. Further pulling back T via the covering projection $\pi\colon F(\mathcal{A}) \to U(\mathcal{A})$ produces a 4-dimensional subtorus inside $\mathrm{Char}(F(\mathcal{A})) = (\mathbb{C}^*)^{12}$.

The subtorus $\pi^*(T)$ is of the form $\exp(L)$, for some linear subspace $L \subset H^1(F(\mathcal{A}), \mathbb{C})$. Using the mixed Hodge structure on the cohomology of the Milnor fiber, Zuber showed in [74] that L cannot possibly be a component of the resonance variety $\mathcal{R}^1(F(\mathcal{A}))$. Thus, the tangent cone formula from Theorem 7 is violated, and so the Milnor fiber $F(\mathcal{A})$ is not 1-formal.

In related work, Fernández de Bobadilla gave in [14] an example of a quasi-homogeneous polynomial whose Milnor fibration has trivial geometric monodromy and whose Milnor fiber is simply-connected, yet non-formal.

Zuber's example naturally leads to the following problem.

Problem 6 Given a rank 3 arrangement $\mathcal{A}$, decide whether the tangent cone formula holds for the Milnor fiber $F(\mathcal{A})$. Is this enough to guarantee that $F(\mathcal{A})$ is formal?

8 Elliptic Arrangements

We conclude with another class of arrangements, this time lying in a product of elliptic curves. An especially convenient algebraic model is available for complements of 'unimodular' elliptic arrangements. Comparing the resonance varieties of this model to those of its cohomology algebra shows that complements of elliptic arrangements may be non-formal.

8.1 Complements of Elliptic Arrangements

Let $E = \mathbb{C}/\mathbb{Z}^2$ be an elliptic curve. We denote by $E^{\times n}$ be the n-fold product of such a curve. This is an abelian variety, with group law inherited from addition in $\mathbb{C}^n$.

An *elliptic arrangement* in $E^{\times n}$ is a finite collection of fibers of group homomorphisms from $E^{\times n}$ to E. Each "elliptic hyperplane" $H \subset E^{\times n}$ may be written as $H = f^{-1}(\zeta)$, for some point $\zeta \in E$ and some homomorphism $f\colon E^{\times n} \to E$ given by

$$f(z_1, \dots, z_n) = \sum_{j=1}^{n} c_j z_j, \tag{40}$$

where $c_j \in \mathbb{Z}$. Thus, an arrangement $\mathcal{A} = \{H_1, \dots, H_m\}$ in $E^{\times n}$ is determined by an integral $m \times n$ matrix $C = (c_{ij})$ and a point $\zeta = (\zeta_1, \dots, \zeta_m) \in E^{\times m}$. We will write $\operatorname{corank}(\mathcal{A}) := n - \operatorname{rank}(C)$ and say that $\mathcal{A}$ is *essential* if its corank is zero.

Let $L(\mathcal{A})$ denote the collection of all connected components of intersections of zero or more elliptic hyperplanes from $\mathcal{A}$. Then $L(\mathcal{A})$ forms a finite poset under inclusion. We say that $\mathcal{A}$ is *unimodular* if all subspaces in $L(\mathcal{A})$ are connected.

Now let $M(\mathcal{A}) = E^{\times n} \setminus \bigcup_{H \in \mathcal{A}} H$ be the complement of our elliptic arrangement. This space is a smooth, quasi-projective variety. Moreover, as shown in [19], the complement $M(\mathcal{A})$ has the homotopy type of a CW-complex of dimension $n + r$, where $r = \operatorname{corank}(\mathcal{A})$. Furthermore, if $r = 0$, then $M(\mathcal{A})$ is a Stein manifold.

8.2 An Algebraic Model

Using the spectral sequence analyzed by Totaro in [71], Bibby constructs in [7] an algebraic model for the complement of a unimodular elliptic arrangement. (An alternative approach is given by Dupont in [26].) Let us briefly review this construction, which generalizes the Gysin model of $E^* = E \setminus \{0\}$ described in Example 13.

Let a, b be the standard generators of $H^1(E, \mathbb{Z}) = \mathbb{Z}^2$. Applying the Künneth formula, we may identify the cohomology ring $H^*(E^{\times n}, \mathbb{Z})$ with the exterior algebra $\bigwedge(a_1, b_1, \ldots, a_n, b_n)$. For a homomorphism $f \colon E^{\times n} \to E$ as in (40), the induced homomorphism in cohomology, $f^* \colon H^*(E, \mathbb{Z}) \to H^*(E^{\times n}, \mathbb{Z})$, is given by

$$f^*(a) = \sum_{j=1}^{n} c_j a_j, \quad f^*(b) = \sum_{j=1}^{n} c_j b_j. \tag{41}$$

Given an arrangement $\mathcal{A} = \{H_1, \ldots, H_m\}$ in $E^{\times n}$, realize each elliptic hyperplane H_i as a coset of the kernel of a homomorphism $f_i \colon E^{\times n} \to E$. Next, consider the graded algebra

$$A_{\mathbb{Z}}(\mathcal{A}) = \bigwedge(a_1, b_1, \ldots, a_n, b_n, e_1, \ldots, e_m)/I(\mathcal{A}), \tag{42}$$

where $I(\mathcal{A})$ is the (homogeneous) ideal generated by the Orlik–Solomon relations (39) among the generators e_i, together with the elements

$$f_i^*(a)e_i,\ f_i^*(b)e_i, \qquad 1 \le i \le m. \tag{43}$$

Define a degree 1 differential d on $A_{\mathbb{Z}}(\mathcal{A})$ by setting $\mathrm{d}a_i = \mathrm{d}b_i = 0$ and

$$\mathrm{d}e_i = f_i^*(a) \wedge f_i^*(b), \tag{44}$$

and extending d to the whole algebra by the graded Leibniz rule. Finally, let $A(\mathcal{A}) = A_{\mathbb{Z}}(\mathcal{A}) \otimes \mathbb{C}$, and extend d to $A(\mathcal{A})$ in the obvious way.

Theorem 13 (Bibby [7]) *Let $\mathcal{A}$ be a unimodular elliptic arrangement, and let $(A(\mathcal{A}), d)$ be the (rationally defined)* CDGA *constructed above. There is then a weak equivalence $A_{\mathrm{PL}}(M(\mathcal{A})) \simeq A(\mathcal{A})$ preserving $\mathbb{Q}$-structures.*

In particular, we have an isomorphism $H^{\boldsymbol{\cdot}}(M(\mathcal{A}), \mathbb{C}) \cong H^{\boldsymbol{\cdot}}(A(\mathcal{A}), \mathrm{d})$. Using this result, we obtain the following form of the tangent cone theorem for elliptic arrangements (the analogue of Theorem 10 in this context).

Theorem 14 *Let $\mathcal{A}$ be a unimodular elliptic arrangement. Then, for each $i \geq 0$,*

$$\tau_1(\mathcal{V}^i(M(\mathcal{A}))) = \mathrm{TC}_1(\mathcal{V}^i(M(\mathcal{A}))) = \mathcal{R}^i(A(\mathcal{A})) \subseteq \mathcal{R}^i(M(\mathcal{A})), \tag{45}$$

with equality for $i \leq q$ if $M(\mathcal{A})$ is q-formal.

Proof The CDGA model $(A(\mathcal{A}), \mathrm{d})$ is finite-dimensional, since the underlying graded algebra $A(\mathcal{A})$ is a quotient of a finitely-generated exterior algebra. Furthermore, this model has positive weights: simply assign weight 1 to the generators a_i, b_i and weight 2 to the generators e_i.

Using now Theorem 13, the rest of the argument from Theorem 10 goes through, once we replace the Gysin model $A(M(\mathcal{A}))$ with the Bibby model $A(\mathcal{A})$. □

As a consequence, each resonance variety $\mathcal{R}^i(A(\mathcal{A}))$ is a union of rationally defined linear subspaces. As we shall see in Sect. 8.3, that's not always true for the resonance variety $\mathcal{R}^i(M(\mathcal{A}))$, in which case the last inclusion from (45) fails to be an equality, and $M(\mathcal{A})$ fails to be i-formal.

It is worth noting that the Orlik–Solomon-type relations for the model $A(\mathcal{A})$ are combinatorially determined, yet the relations (43) depend on the actual defining equations for the arrangement. This observation leads to the following natural question.

Problem 7 Let $\mathcal{A}$ be a unimodular elliptic arrangement. Are the resonance varieties $\mathcal{R}^i(A(\mathcal{A}))$ and $\mathcal{R}^i(M(\mathcal{A}))$ determined by the intersection lattice of $\mathcal{A}$? Furthermore, is there a combinatorial criterion to decide whether the two varieties coincide, and, if so, whether the complement $M(\mathcal{A})$ is formal?

8.3 *Ordered Configurations on an Elliptic Curve*

The configuration space of n points on an elliptic curve, $\mathrm{Conf}(E, n)$, is the complement of the elliptic braid arrangement, which is the arrangement in $E^{\times n}$ defined by the equations $z_i = z_j$ for $1 \leq i < j \leq n$. This space is a $K(\pi, 1)$, with $\pi = PE_n$, the elliptic pure braid group on n strings.

The resonance varieties $\mathcal{R}^1(\mathrm{Conf}(E, n))$ were computed in [24], while the positive-dimensional components of $\mathcal{V}^1(\mathrm{Conf}(E, n))$ were computed by Dimca in [20]. An alternate way to perform this computation is to use work of Feler, namely, [31, Theorem 3.1].

Since $E = S^1 \times S^1$ is a topological group, the space $\mathrm{Conf}(E, n)$ splits up to homeomorphism as a direct product, $\mathrm{Conf}(E^*, n-1) \times E$, where E^* denotes the elliptic curve E with the identity removed. Thus, for our purposes here it is enough to consider the configuration spaces $\mathrm{Conf}(E^*, n-1)$. In the next example, we work out in detail the case when $n = 3$.

Example 21 Let $X = \mathrm{Conf}(E^*, 2)$ be the configuration space of 2 labeled points on a punctured elliptic curve. This is the complement of the arrangement $\mathcal{A}$ in $E^{\times 2}$ defined by the polynomial $f = z_1 z_2 (z_1 - z_2)$.

By Theorem 13, the space X admits as a model the CDGA (A, d), where A is the exterior algebra on generators $a_1, b_1, a_2, b_2, e_1, e_2, e_3$ in degree 1, modulo the ideal generated by the quadrics

$$a_1e_1, b_1e_1, a_2e_2, b_2e_2, (a_1 - a_2)e_3, (b_1 - b_2)e_3, (e_1 - e_2)(e_1 - e_3),$$

while the differential $\mathrm{d}: A \to A$ is given by $\mathrm{d}a_1 = \mathrm{d}b_1 = \mathrm{d}a_2 = \mathrm{d}b_2 = 0$ and

$$\mathrm{d}e_1 = a_1b_1,\ \mathrm{d}e_2 = a_2b_2,\ \mathrm{d}e_3 = (a_1 - a_2)(b_1 - b_2).$$

Identify $H^1(A) = \mathbb{C}^4$, with basis the classes represented by a_1, b_1, a_2, b_2, and let $S = \mathbb{C}[x_1, y_1, x_2, y_2]$ be the corresponding polynomial ring. Fixing bases as above for $A^1 = \mathbb{C}^7$ and $\{a_1b_1, a_1a_2, a_1b_2, b_1a_2, b_1b_2, a_2b_2, a_2e_1, b_2e_1, a_1e_2, b_1e_2, a_1e_3, b_1e_3, e_1e_2, e_1e_3\}$ for $A^2 = \mathbb{C}^{14}$, we find that the boundary maps for the chain complex $A_\bullet \otimes S$ are given by

$$\partial_2 = \begin{pmatrix} -y_1 & -x_2 & -y_2 & 0 & 0 & 0 & 0 & 0 & 0 & 0 & 0 & 0 & 0 & 0 \\ x_1 & 0 & 0 & -x_2 & -y_2 & 0 & 0 & 0 & 0 & 0 & 0 & 0 & 0 & 0 \\ 0 & x_1 & 0 & y_1 & 0 & -y_2 & 0 & 0 & 0 & 0 & 0 & 0 & 0 & 0 \\ 0 & 0 & x_1 & 0 & y_1 & x_2 & 0 & 0 & 0 & 0 & 0 & 0 & 0 & 0 \\ 1 & 0 & 0 & 0 & 0 & 0 & x_2 & y_2 & 0 & 0 & 0 & 0 & 0 & 0 \\ 0 & 0 & 0 & 0 & 0 & 1 & 0 & 0 & x_1 & y_1 & 0 & 0 & 0 & 0 \\ 1 & 0 & -1 & 1 & 0 & 1 & 0 & 0 & 0 & 0 & x_1 + x_2 & y_1 + y_2 & 0 & 0 \end{pmatrix}$$

and $\partial_1 = \begin{pmatrix} x_1 & y_1 & x_2 & y_2 & 0 & 0 & 0 \end{pmatrix}$. Computing homology, we find that $H_1(A_\bullet \otimes S)$ is presented by the S-linear map $\varphi: S^7 \to S^3$ with matrix

$$\varphi = \begin{pmatrix} y_2 & x_2 & y_2 & x_2 & -y_2 & -x_2 \\ y_1 & x_1 & 0 & 0 & 0 & 0 \\ 0 & 0 & y_2 & x_2 & y_1 & x_1 \end{pmatrix}.$$

By Theorem 1, the resonance variety $\mathcal{R}^1(A)$ is the zero locus of the ideal of 3×3 minors of φ. An easy computation shows that this variety is the union of three planes in $\mathbb{C}^4$,

$$\mathcal{R}^1(A) = \{x_1 = y_1 = 0\} \cup \{x_2 = y_2 = 0\} \cup \{x_1 + x_2 = y_1 + y_2 = 0\}.$$

On the other hand, the ring $H^{\cdot}(A) \cong H^{\cdot}(X,\mathbb{C})$ is the exterior algebra on generators a_1, a_2, b_1, b_2 in degree 1, modulo the ideal spanned by $a_1b_2 + a_2b_1$, a_1b_1, and a_2b_2. Proceeding as above, we see that

$$H_1(H_{\bullet}(A) \otimes S) = \operatorname{coker} \begin{pmatrix} y_2 & x_2 & -y_1 & -x_1 \\ y_1 & x_1 & 0 & 0 \\ 0 & 0 & y_2 & x_2 \end{pmatrix}.$$

Hence, the first resonance variety of X is an irreducible quadric hypersurface in $\mathbb{C}^4$, given by

$$\mathcal{R}^1(X) = \{x_1y_2 - x_2y_1 = 0\}.$$

It follows from Corollary 6 that the configuration space $X = \mathrm{Conf}(E^*, 2)$ is not 1-formal, a result already known from [6, 24].

Turning now to homology with coefficients in rank 1 local systems, direct computation (recorded in [65, Example 8.2]) shows that the first characteristic variety of X consists of three 2-dimensional algebraic tori inside $(\mathbb{C}^*)^4$,

$$\mathcal{V}^1(X) = \{t_1 = s_1 = 1\} \cup \{t_2 = s_2 = 1\} \cup \{t_1t_2 = s_1s_2 = 1\}.$$

As noted in [20, Proposition 5.1], these three subtori arise by pullback along the fibrations $\mathrm{Conf}(E^*, 2) \to E^*$ obtained by sending a point (z_1, z_2) to z_2, z_1, and $z_1z_2^{-1}$, respectively. Likewise, according to Theorem 11, the three planes comprising $\mathcal{R}^1(A)$ are obtained by pulling back the linear space $H^1(A(E^*)) = \mathbb{C}^2$ along the same fibrations. In particular,

$$\tau_1(\mathcal{V}^1(X)) = \mathrm{TC}_1(\mathcal{V}^1(X)) = \mathcal{R}^1(A),$$

as predicted by Theorem 14.

All three varieties are 2-dimensional; thus, they are all properly contained in the 3-dimensional variety $\mathcal{R}^1(X)$. Therefore, the Tangent Cone theorem shows, once again, that X is not 1-formal.

Acknowledgements I wish to thank Stefan Papadima for several useful conversations regarding this work, and also the referee, for a careful reading of the manuscript. This work was supported in part by National Security Agency grant H98230-13-1-0225.

References

1. J.W. Alexander, Topological invariants of knots and links. Trans. Am. Math. Soc. **30**(2), 275–306 (1928)
2. D. Arapura, Geometry of cohomology support loci for local systems I. J. Algebraic Geom. **6**(3), 563–597 (1997)
3. E. Artal Bartolo, J.I. Cogolludo-Agustín, D. Matei, Characteristic varieties of quasi-projective manifolds and orbifolds. Geom. Topol. **17**(1), 273–309 (2013)
4. A. Beauville, Annulation du H^1 pour les fibrés en droites plats, in *Complex Algebraic Varieties (Bayreuth, 1990)*. Lecture Notes in Mathematics, vol. 1507 (Springer, Berlin, 1992), pp. 1–15
5. B. Berceanu, A. Măcinic, S. Papadima, R. Popescu, On the geometry and topology of partial configuration spaces of Riemann surfaces (2015). arXiv:1504.04733v1
6. R. Bezrukavnikov, Koszul DG-algebras arising from configuration spaces. Geom. Funct. Anal. **4**(2), 119–135 (1994)
7. C. Bibby, Cohomology of abelian arrangements. Proc. Am. Math. Soc. **144**(7), 3093–3104 (2016)
8. E. Brieskorn, Sur les groupes de tresses (d'après V. I. Arnol'd). *Séminaire Bourbaki, 24ème année (1971/1972)*. Lecture Notes in Mathematics, vol. 317 (Springer, Berlin, 1973), Exp. No. 401, pp. 21–44
9. N. Budur, B. Wang, Cohomology jump loci of quasi-projective varieties. Ann. Sci. École Norm. Sup. **48**(1), 227–236 (2015)
10. N. Budur, B. Wang, Cohomology jump loci of differential graded Lie algebras. Compos. Math. **151**(8), 1499–1528 (2015)
11. J.I. Cogolludo-Agustín, D. Matei, Cohomology algebra of plane curves, weak combinatorial type, and formality. Trans. Am. Math. Soc. **364**(11), 5765–5790 (2012)
12. D.C. Cohen, A.I. Suciu, On Milnor fibrations of arrangements. J. London Math. Soc. (2) **51**(1), 105–119 (1995)
13. D.C. Cohen, A.I. Suciu, Characteristic varieties of arrangements. Math. Proc. Camb. Philos. Soc. **127**(1), 33–53 (1999)
14. J.F. de Bobadilla, On homotopy types of complements of analytic sets and Milnor fibres, in *Topology of Algebraic Varieties and Singularities*. Contemporary Mathematics, vol. 538 (American Mathematical Society, Providence, 2011), pp. 363–367
15. C. De Concini, C. Procesi, Wonderful models of subspace arrangements. Selecta Math. (N.S.) **1**(3), 459–494 (1995)
16. P. Deligne, P. Griffiths, J. Morgan, D. Sullivan, Real homotopy theory of Kähler manifolds. Invent. Math. **29**(3), 245–274 (1975)
17. G. Denham, A.I. Suciu, Moment-angle complexes, monomial ideals, and Massey products. Pure Appl. Math. Q. **3**(1), 25–60 (2007)
18. G. Denham, A.I. Suciu, Multinets, parallel connections, and Milnor fibrations of arrangements. Proc. Lond. Math. Soc. **108**(6), 1435–1470 (2014)
19. G. Denham, A.I. Suciu, S. Yuzvinsky, Combinatorial covers and vanishing cohomology (2014). arXiv:1411.7981v1
20. A. Dimca, Characteristic varieties and logarithmic differential 1-forms. Compos. Math. **146**(1), 129–144 (2010)
21. A. Dimca, Monodromy of triple point line arrangements, in *Singularities in Geometry and Topology 2011*. Advanced Studies in Pure Mathematics. vol. 66 (Mathematical Society of Japan, Tokyo, 2015), pp. 71–80
22. A. Dimca, S. Papadima, Finite Galois covers, cohomology jump loci, formality properties, and multinets. Ann. Sc. Norm. Super. Pisa Cl. Sci. **10**(2), 253–268 (2011)
23. A. Dimca, S. Papadima, Non-abelian cohomology jump loci from an analytic viewpoint. Commun. Contemp. Math. **16**(4), 1350025 (47 p) (2014)
24. A. Dimca, S. Papadima, A.I. Suciu, Topology and geometry of cohomology jump loci. Duke Math. J. **148**(3), 405–457 (2009)

25. A. Dimca, S. Papadima, A. Suciu, Algebraic models, Alexander-type invariants, and Green-Lazarsfeld sets. Bull. Math. Soc. Sci. Math. Roum. **58**(3), 257–269 (2015)
26. C. Dupont, Hypersurface arrangements and a global Brieskorn-Orlik-Solomon theorem. Ann. Inst. Fourier. **65**(6), 2507–2545 (2015)
27. M. Eastwood, S. Huggett, Euler characteristics and chromatic polynomials. Eur. J. Comb. **28**(6), 1553–1560 (2007)
28. M. Falk, Arrangements and cohomology. Ann. Comb. **1**(2), 135–157 (1997)
29. M. Falk, S. Yuzvinsky, Multinets, resonance varieties, and pencils of plane curves. Compos. Math. **143**(4), 1069–1088 (2007)
30. E.M. Feichtner, S. Yuzvinsky, Formality of the complements of subspace arrangements with geometric lattices. Zap. Nauchn. Sem. S.-Peterburg. Otdel. Mat. Inst. Steklov **326**, 235–247 (2005)
31. Y. Feler, Configuration spaces of tori. Atti Accad. Naz. Lincei Rend. Lincei Mat. Appl. **18**(2), 139–151 (2007)
32. Y. Félix, S. Halperin, J.-C. Thomas, *Rational Homotopy Theory*. Graduate Texts in Mathematics, vol. 205 (Springer, New York, 2001)
33. Y. Félix, B. Jessup, P.-E. Parent, The combinatorial model for the Sullivan functor on simplicial sets. J. Pure Appl. Algebra **213**(2), 231–240 (2009)
34. M. Green, R. Lazarsfeld, Deformation theory, generic vanishing theorems and some conjectures of Enriques, Catanese and Beauville. Invent. Math. **90**(2), 389–407 (1987)
35. M. Green, R. Lazarsfeld, Higher obstructions to deforming cohomology groups of line bundles. J. Am. Math. Soc. **4**(1), 87–103 (1991)
36. E. Hironaka, Alexander stratifications of character varieties. Ann. l'Inst. Fourier (Grenoble) **47**(2), 555–583 (1997)
37. H. Kasuya, Minimal models, formality, and hard Lefschetz properties of solvmanifolds with local systems. J. Differ. Geom. **93**(2), 269–297 (2013)
38. T. Kohno, On the holonomy Lie algebra and the nilpotent completion of the fundamental group of the complement of hypersurfaces. Nagoya Math. J. **92**, 21–37 (1983)
39. A. Libgober, Characteristic varieties of algebraic curves, in *Applications of Algebraic Geometry to Coding Theory, Physics and Computation (Eilat, 2001)*. Nato Science Series II: Mathematics, Physics and Chemistry, vol. 36 (Kluwer Academic, Dordrecht, 2001), pp. 215–254
40. A. Libgober, First order deformations for rank one local systems with a non-vanishing cohomology. Topol. Appl. **118**(1–2), 159–168 (2002)
41. A. Libgober, On combinatorial invariance of the cohomology of Milnor fiber of arrangements and Catalan equation over function field, in *Arrangements of hyperplanes (Sapporo 2009)*. Advanced Studies in Pure Mathematics, vol. 62 (The Mathematical Society of Japan, Tokyo, 2012), pp. 175–187
42. A. Libgober, S. Yuzvinsky, Cohomology of the Orlik-Solomon algebras and local systems. Compos. Math. **21**(3), 337–361 (2000)
43. Y. Liu, L. Maxim, Characteristic varieties of hypersurface complements (2014). arXiv:1411.7360v1
44. A. Măcinic, Cohomology rings and formality properties of nilpotent groups. J. Pure Appl. Algebra **214**(10), 1818–1826 (2010)
45. A. Măcinic, S. Papadima, R. Popescu, A.I. Suciu, Flat connections and resonance varieties: from rank one to higher ranks. Trans. Am. Math. Soc. Available at http://dx.doi.org/10.1090/tran/6799 (to appear)
46. M. Marco-Buzunáriz, A description of the resonance variety of a line combinatorics via combinatorial pencils. Graphs Combinatorics **25**(4), 469–488 (2009)
47. D. Matei, A.I. Suciu, Cohomology rings and nilpotent quotients of real and complex arrangements, in *Arrangements–Tokyo 1998*. Advanced Studies in Pure Mathematics, vol. 27 (The Mathematical Society of Japan, Tokyo, 2000), pp. 185–215
48. M. Miller, M. Wakefield, Edge colored hypergraphic arrangements. Pure Appl. Math. Q. **8**(3), 757–779 (2012)

49. D.V. Millionschikov, Cohomology of solvmanifolds with local coefficients and problems of the Morse–Novikov theory. Russ. Math. Surv. **57**(4), 813–814 (2002)
50. J. Milnor, *Singular Points of Complex Hypersurfaces*. Annals of Mathematics Studies, vol. 61 (Princeton University Press, Princeton, 1968)
51. J.W. Morgan, The algebraic topology of smooth algebraic varieties. Inst. Hautes Études Sci. Publ. Math. **48**, 137–204 (1978)
52. P. Orlik, L. Solomon, Combinatorics and topology of complements of hyperplanes. Invent. Math. **56**(2), 167–189 (1980)
53. S. Papadima, L. Păunescu, Rank two jump loci for solvmanifolds and Lie algebras (2014). arXiv:1411.0447v1
54. S. Papadima, A.I. Suciu, Chen Lie algebras. Int. Math. Res. Not. **2004**(21), 1057–1086 (2004)
55. S. Papadima, A. Suciu, Geometric and algebraic aspects of 1-formality. Bull. Math. Soc. Sci. Math. Roum. **52**(3), 355–375 (2009)
56. S. Papadima, A.I. Suciu, Bieri–Neumann–Strebel–Renz invariants and homology jumping loci. Proc. Lond. Math. Soc. **100**(3), 795–834 (2010)
57. S. Papadima, A.I. Suciu, Vanishing resonance and representations of Lie algebras. J. Reine Angew. Math. (Crelle's Journal) **706**, 83–101 (2015)
58. S. Papadima, A.I. Suciu, Jump loci in the equivariant spectral sequence. Math. Res. Lett. **21**(4), 863–883 (2014)
59. S. Papadima, A.I. Suciu, Non-abelian resonance: product and coproduct formulas, in *Bridging Algebra, Geometry, and Topology*. Springer Proceedings in Mathematics & Statistics, vol. 96 (Springer, Cham, 2014), pp. 269–280
60. S. Papadima, A.I. Suciu, The Milnor fibration of a hyperplane arrangement: from modular resonance to algebraic monodromy (2014). arXiv:1401.0868v2
61. C. Simpson, Higgs bundles and local systems. Inst. Hautes Études Sci. Publ. Math. **75**, 5–95 (1992)
62. A.I. Suciu, Translated tori in the characteristic varieties of complex hyperplane arrangements. Topol. Appl. **118**(1–2), 209–223 (2002)
63. A.I. Suciu, Fundamental groups, Alexander invariants, and cohomology jumping loci, in *Topology of Algebraic Varieties and Singularities*. Contemporary Mathematics, vol. 538 (American Mathematical Society, Providence, 2011), pp. 179–223
64. A.I. Suciu, Resonance varieties and Dwyer–Fried invariants, in *Arrangements of Hyperplanes (Sapporo 2009)*. Advanced Studies Pure Mathematics, vol. 62 (Kinokuniya, Tokyo, 2012), pp. 359–398
65. A.I. Suciu, Geometric and homological finiteness in free abelian covers, in *Configuration Spaces: Geometry, Combinatorics and Topology*. CRM Series, vol. 14 (Ed. Norm., Pisa, 2012), pp. 461–501
66. A.I. Suciu, Characteristic varieties and Betti numbers of free abelian covers. Int. Math. Res. Not. **2014**(4), 1063–1124 (2014)
67. A.I. Suciu, Hyperplane arrangements and Milnor fibrations. Ann. Faculté Sci. Toulouse **23**(2), 417–481 (2014)
68. A.I. Suciu, H. Wang, Formality properties of finitely generated groups and Lie algebras (2015). arXiv:1504.08294v2
69. A.I. Suciu, Y. Yang, G. Zhao, Homological finiteness of abelian covers. Ann. Sc. Norm. Super. Pisa Cl. Sci. **14**(1), 101–153 (2015)
70. D. Sullivan, Infinitesimal computations in topology. Inst. Hautes Études Sci. Publ. Math. **47**, 269–331 (1977)
71. B. Totaro, Configuration spaces of algebraic varieties. Topology **35**(6), 1057–1067 (1996)
72. B. Wang, Examples of topological spaces with arbitrary cohomology jump loci (2013). arXiv: 1304.0239v2
73. S. Yuzvinsky, Small rational model of subspace complement. Trans. Am. Math. Soc. **354**(5), 1921–1945 (2002)
74. H. Zuber, Non-formality of Milnor fibers of line arrangements. Bull. Lond. Math. Soc. **42**(5), 905–911 (2010)

Higher Resonance Varieties of Matroids

Graham Denham

Abstract We present some new results about the resonance varieties of matroids and hyperplane arrangements. Though these have been the objects of ongoing study, most work so far has focused on cohomological degree 1. We show that certain phenomena become apparent only by considering all degrees at once.

Keywords Hyperplane arrangement • Matroid • Resonance variety

1 Introduction

Resonance varieties are a cohomological invariant that first appeared in the study of the cohomology of one-dimensional fundamental group representations. Though they were first considered for topological spaces, they are algebraic in nature, and they may be defined for any (differential) graded-commutative algebra [31, 32].

The resonance varieties associated with the complement of a complex hyperplane arrangement are an interesting special case, and we mention in particular the surveys [19, 36] for their description of the history and motivation. Here, the underlying topological space is a complex, quasi-projective variety; however, it follows from the Brieskorn-Orlik-Solomon Theorem [23] that the resonance varieties depend only on combinatorial data coming from a matroid.

With that in mind, it is tempting to ask for a formula or an efficient algorithm that expresses the resonance varieties in terms of the matroid. However, this seems to be a difficult problem. In cohomological degree 1, Falk and Yuzvinsky [20] have given a characterization, building on work of Libgober and Yuzvinsky [22] as well as Falk [17]. In degree greater than 1, there has been some progress (in particular [4, 8]), but comparatively little is known. By way of contrast, Papadima and Suciu gave a closed formula for the resonance varieties of exterior Stanley-Reisner rings (and right-angled Artin groups) in [26], building on work of Aramova et al. [1]. Although the situation for matroids has some similarities, an analogous definitive result is so far out of reach.

G. Denham (✉)
Department of Mathematics, University of Western Ontario, London, ON, Canada N6A 5B7
e-mail: gdenham@uwo.ca

F. Callegaro et al. (eds.), *Configuration Spaces*, Springer INdAM Series 14,
DOI 10.1007/978-3-319-31580-5_2

Our main goal here is to develop some basic tools systematically. We consider the behaviour of resonance varieties of matroids and arrangements under such constructions as weak maps, Gale duality, and the deletion/contraction constructions. Combining these basic ingredients allows us to compute the resonance varieties of some graphic arrangements explicitly. A broader range of phenomena appear in moving from degree 1 to 2: as an example, we find a straightforward way to make hyperplane arrangements for which the Milnor fibre F has nontrivial monodromy eigenspaces in $H^2(F, \mathbb{C})$.

A hyperplane arrangement is a matroid realization over $\mathbb{C}$. Some results about resonance from the literature are known for all matroids, while others depend on realizability. Our approach is combinatorial, so matroids (rather than arrangements) seem to be the appropriate objects for this paper. It was recently shown in [28] that realizability imposes a non-trivial qualitative restriction on resonance varieties, at least in positive characteristic. This encourages us to keep track of the role of realizability.

1.1 Outline

The paper is organized as follows. We begin by recalling the definition of the Orlik-Solomon algebra, viewed as a matroid invariant. We would like to make use of the naturality of the construction; however, not all weak maps of matroids induce homomorphisms of Orlik-Solomon algebras. In Sect. 2.3 we impose a condition on weak maps to define a category $\mathcal{M}$ of matroids on which the Orlik-Solomon construction is functorial.

In Sect. 3, we define resonance varieties and review known results about some qualitative properties that distinguish the resonance varieties of Orlik-Solomon algebras from the general case. For example, for a matroid M of rank ℓ, the resonance varieties are known to satisfy

$$\mathcal{R}^p(\mathsf{M}) \subseteq \mathcal{R}^{p+1}(\mathsf{M})$$

for $0 \leq p < \ell$ [15]. At least in the realizable, characteristic-zero case, resonance varieties are unions of linear subspaces. In Sect. 3.4, we construct subspace arrangements $\mathcal{S}^p(\mathsf{M})$ that contain them, based on a result of Cohen et al. [7]. We find that in some interesting cases this upper envelope is tight: i.e., $\mathcal{R}^p(\mathsf{M}) = \mathcal{S}^p(\mathsf{M})$.

In Sect. 4, we examine the effect of standard matroid operations on resonance. Some results are known, some folklore, and others new. We use these to compute some examples and find that certain special properties of components of $\mathcal{R}^p(\mathsf{M})$ for $p = 1$ no longer hold for $p \geq 2$.

In the last section, we revisit the combinatorics of multinets and singular subspaces in terms of maps of Orlik-Solomon algebras as another attempt to characterize components of resonance varieties. The results are inconclusive, although the last example strongly suggests that some interesting combinatorics remains to be uncovered.

2 Background

2.1 Arrangements and Matroids

We refer to the books of Orlik and Terao [24] and Oxley [25] for basic facts about hyperplane arrangements and matroids, respectively. If M is a matroid on the set $[n] := \{1, 2, \ldots, n\}$ and $\Bbbk$ is a field, let $V = \Bbbk^n$, a vector space with a distinguished basis we will call $e_1, \ldots, e_n$. The Orlik-Solomon algebra $A_\Bbbk(\mathsf{M})$ is the quotient of an exterior algebra $E := \Lambda(V)$ by an ideal $I = I(\mathsf{M})$ generated by homogeneous relations indexed by circuits in M. More explicitly, let ∂ be the derivation on E defined by $\partial(e_i) = 1$ for all $1 \leq i \leq n$. Then I is generated by

$$\{\partial(e_C) : \text{circuits } C \subseteq [n] \text{ of } \mathsf{M}\}, \tag{1}$$

where $e_C := \prod_{i \in C} e_i$, with indices taken in increasing order. We will omit the M or $\Bbbk$ from the notation $A_\Bbbk(\mathsf{M})$ where no confusion arises. If $i \in \mathsf{M}$ is a loop, then $C = \{i\}$ is a circuit and $A(\mathsf{M}) = 0$.

We regard hyperplane arrangements as linear representations of loop-free matroids. For us, an arrangement $\mathcal{A}$ over a field F is an ordered n-tuple of (nonzero) linear forms $(f_1, \ldots, f_n)$, where $f_i \in W^*$ for $1 \leq i \leq n$, and W is a vector space over F. We let $\mathsf{M}(\mathcal{A})$ denote the matroid on $[n]$ whose dependent sets index the linear dependencies of the f_i's. In particular, our arrangements are all central, and we explicitly allow repeated hyperplanes.

If $\mathcal{A}$ is an arrangement, let $H_i = \ker(f_i)$, a hyperplane in W, for $1 \leq i \leq n$. Let $M(\mathcal{A}) = W - \bigcup_{i=1}^n H_i$, and $U(\mathcal{A}) = \mathbb{P}W - \bigcup_{i=1}^n \mathbb{P}H_i$. If $\mathcal{A}$ is an arrangement, we abbreviate $A_\Bbbk(\mathcal{A}) := A_\Bbbk(\mathsf{M}(\mathcal{A}))$. If $F = \mathbb{C}$, the complement $M(\mathcal{A})$ is a complex manifold, and the Brieskorn-Orlik-Solomon Theorem states that $A_\Bbbk(\mathcal{A}) \cong H^*(M(\mathcal{A}), \Bbbk)$ as graded algebras.

2.2 Projectivization

Suppose $A(\mathsf{M}) = E/I$ is the Orlik-Solomon algebra of a matroid M. Since $\partial^2 = 0$, it follows $\partial(I) = 0$, so ∂ induces a derivation on A as well, which we denote by ∂_A. Let

$$\overline{V} = \ker(\partial|_V) = \Big\{v \in \Bbbk^n : \sum_{i=1}^n v_i = 0\Big\},$$

and let $\overline{A}(\mathsf{M})$ denote the subalgebra of A generated by $\overline{V}$.

Lemma 1 *We have* $\overline{A}(\mathsf{M}) = \ker \partial_A$.

Proof Clearly $\overline{A}(\mathsf{M}) \subseteq \ker \partial_A$. By Orlik and Terao [24, Lemma 3.13], the chain complex (A, ∂) is exact, so if $\partial_A(x) = 0$, then $x = \partial_A(y)$ for some $y \in A$. If $e_I = e_{i_1} \cdots e_{i_k} \in E$, then $\partial(e_I) = (e_{i_1} - e_{i_k})(e_{i_2} - e_{i_k}) \cdots (e_{i_{k-1}} - e_{i_k})$, which implies $\partial(e_I)$ is in the subalgebra generated by $\overline{V}$. By applying ∂_A to a representative in E for y, we see $x \in \overline{A}(\mathsf{M})$. □

Together with the exactness of (A, ∂_A), this gives a short exact sequence of graded $\Bbbk$-modules,

$$0 \longrightarrow \overline{A}(\mathsf{M}) \longrightarrow A(\mathsf{M}) \xrightarrow{\partial} \overline{A}(\mathsf{M})[-1] \longrightarrow 0. \tag{2}$$

Of course, if $\mathsf{M} = \mathsf{M}(\mathcal{A})$ where $\mathcal{A}$ is a complex arrangement, this sequence has a well-known origin: the quotient map $M(\mathcal{A}) \to U(\mathcal{A})$ makes $M(\mathcal{A})$ a split $\mathbb{C}^*$-bundle over $U(\mathcal{A})$, so the induced algebra homomorphism $H^*(U(\mathcal{A}), \Bbbk) \to H^*(M(\mathcal{A}), \Bbbk)$ is injective. In fact, under the isomorphism $H^*(M(\mathcal{A}), \Bbbk) \cong A(\mathcal{A})$, the Gysin map is identified with ∂_A, and $\overline{A}(\mathcal{A}) \cong H^*(U(\mathcal{A}), \Bbbk)$: see [9] or [13, §6.1] for details. With this in mind, we will call $\overline{A}(\mathsf{M})$ the *projective Orlik-Solomon algebra* even if M does not have a complex realization.

2.3 A Category of Matroids

We would like to make use of maps of Orlik-Solomon algebras, so it will be useful to have a functorial construction. For this, we recall the definition of a weak map of matroids from [34, Chap. 9]. If M_1 and M_2 are matroids on sets S_1 and S_2, respectively, we add a disjoint loop "0" to M_i to make a matroid M_i^+, for $i = 1, 2$.

Definition 1 A *weak map* $f: \mathsf{M}_1 \to \mathsf{M}_2$ is a map of sets $f: S_1 \cup \{0\} \to S_2 \cup \{0\}$ with the following properties: $f(0) = 0$, and for all $I \subset S_1$, if $f|_I$ is injective and $f(I)$ is independent in M_2^+, then I is independent in M_1.

We will say a weak map $f: \mathsf{M}_1 \to \mathsf{M}_2$ is *non-degenerate* if $f^{-1}(0) = \{0\}$ and *complete* if, for every circuit C of M_1, we have $|C \cap f^{-1}(0)| \neq 1$. Clearly, non-degenerate weak maps are complete.

Proposition 1 *If* $f: \mathsf{M}_1 \to \mathsf{M}_2$ *and* $g: \mathsf{M}_2 \to \mathsf{M}_3$ *are complete weak maps, so is* $g \circ f$. *If* f *and* g *are non-degenerate weak maps, so is* $g \circ f$.

Proof Weak maps are closed under composition, so the second assertion is trivial, and it is enough to check that the composite of complete maps is complete. Suppose instead that there is a circuit C in M_1 and a unique element $i \in C$ for which $g \circ f(i) = 0$. Since f is complete, $f(i) \neq 0$. Then $f(i)$ is contained in a circuit $C' \subseteq f(C)$ of M_2. By assumption, $j = f(i)$ is the only element of C' which $g(j) = 0$. But g is complete, a contradiction. □

In view of the previous result, matroids on finite sets form a category with morphisms taken to be the complete weak maps, which we will denote by $\mathcal{M}$. Let $\overline{\mathcal{M}}$ denote the (wide) subcategory whose morphisms are non-degenerate weak maps. Our hypotheses (complete and non-degenerate) are designed to make the Orlik-Solomon algebra and its projective version functorial, respectively.

First, we note a categorical formality. If $f\colon \mathsf{M}_1 \to \mathsf{M}_2$ is a complete weak map and $f(S_1) \supseteq S_2$, then it is easy to see f is an epimorphism. While we avoid trying to characterize epimorphisms, here is a convenient necessary condition.

Lemma 2 *If $f\colon \mathsf{M}_1 \to \mathsf{M}_2$ is an epimorphism in $\mathcal{M}$, then any element $i \in S_2 - f(S_1)$ is either a loop or it is parallel to an element of $f(S_1)$.*

Proof Suppose instead that some $i \in S_2 - f(S_1)$ is neither a loop nor parallel to an element of $f(S_1)$. Define two maps $g, h\colon \mathsf{M}_2 \to \mathsf{U}_{1,1}$: let $g(j) = 0$ for all $j \in S_2$ and $h(j) = 0$ for all $j \neq i$, but $h(i) = 1$. Then g and h are complete weak maps, but $g \circ f = h \circ f$, so f is not an epimorphism. □

By the obvious action on the distinguished basis, a weak map $f\colon \mathsf{M}_1 \to \mathsf{M}_2$ induces a linear map $V(\mathsf{M}_1) \to V(\mathsf{M}_2)$ which we will also denote by f.

Lemma 3 *Suppose $f\colon \mathsf{M}_1 \to \mathsf{M}_2$ is a complete weak map of matroids. Then $\Lambda(f)$ induces a well-defined map of $\Bbbk$-algebras, $A(\mathsf{M}_1) \to A(\mathsf{M}_2)$. If f is also non-degenerate, then it restricts to a map of projective Orlik-Solomon algebras.*

Proof Write $A(\mathsf{M}_i) = E_i/I_i$ for $i = 1, 2$, and consider the homomorphism $\Lambda(f)\colon E_1 \to E_2$. For any circuit C in M_1, we consider two cases. If $0 \notin f(C)$, then $f(C) \cap S_2$ is dependent, so the image of $\Lambda(f)(\partial(e_C)) \in I_2$. If $0 \in f(C)$, by completeness, $f(e_i) = f(e_j)$ for some distinct $i, j \in C$, so $\Lambda(f)(\partial(e_C)) = 0$. So $\Lambda(f)(I_1) \subseteq I_2$, as required.

If, moreover, f is non-degenerate, we see $\partial|_{V(\mathsf{M}_2)} \circ f = f \circ \partial|_{V(\mathsf{M}_1)}$, by evaluating on the distinguished basis. Then $\operatorname{im}(f|_{\overline{V}(\mathsf{M}_1)}) \subseteq \overline{V}(\mathsf{M}_2)$. Since $\overline{A}(\mathsf{M}_1)$ is generated in degree 1, the image of $\Lambda(f)|_{\overline{A}(\mathsf{M}_1)}$ lies in $\overline{A}(\mathsf{M}_2)$. □

If f is a complete weak map, put $A(f) = \Lambda(f)\colon A(\mathsf{M}_1) \to A(\mathsf{M}_2)$. If f is also non-degenerate, let $\overline{A}(f)\colon \overline{A}(\mathsf{M}_1) \to \overline{A}(\mathsf{M}_2)$ denote the restriction.

Theorem 1 *A and $\overline{A}$ are functors from $\mathcal{M}$ and $\overline{\mathcal{M}}$, respectively, to the category of graded-commutative $\Bbbk$-algebras. Moreover, A preserves epimorphisms.*

Proof If f is a morphism, $A(f)$ is determined by its action in degree 1, where A obviously preserves composition, so A is functorial. Now suppose f is an epimorphism. The algebra $A(\mathsf{M}_2)$ is spanned by monomials e_I, where I is independent.

If $I \subseteq \operatorname{im}(f)$, there exists a subset $J \subseteq f^{-1}(I)$ with $|J| = |I|$. If not, by Lemma 2, we may replace some elements of I with parallel elements to form a set $I' \subseteq \operatorname{im}(f)$ and find J as above with $f(J) = I'$.

Since $e_i = e_j$ for parallel elements in $A(\mathsf{M}_2)$, in both cases we have $A(f)(e_J) = e_I$. So $A(f)$ is a surjective ring homomorphism. □

Example 1 The map $f\colon \mathsf{U}_{2,3} \to \mathsf{U}_{2,2}$ given by $f(i) = i$ for $i = 1, 2$ and $f(3) = 0$ is a weak map. Taking $C = [3]$, we see f fails to be complete, and $\Lambda(f)$ fails to give a map of Orlik-Solomon algebras.

On the other hand, if M is the matroid of the graph

$G =$ 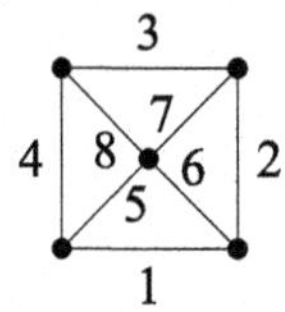

, $\mathsf{M}(G) \to \mathsf{U}_{3,4}$: 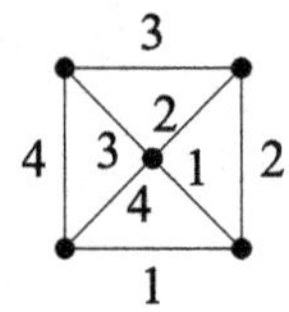

(3)

then the map from the edges of G to the set $[4]$ given by the edge labels on the right defines a non-degenerate weak map $\mathsf{M}(G) \to \mathsf{U}_{3,4}$, because G contains no three-cycles with distinct edge labels.

We will end this section with two easy but useful observations. Suppose M is a matroid on $[n]$ and π is a partition of $[n]$ into k parts. Let $p_\pi\colon [n] \to [k]$ be the map that sends i to s whenever $i \in \pi_s$.

Proposition 2 *$p_\pi\colon \mathsf{M} \to \mathsf{U}_{2,k}$ is a morphism of $\overline{\mathcal{M}}$ if and only if, whenever i and j are parallel in M, i and j are in the same block of π.*

Definition 2 If M is a matroid on $[n]$, we define an equivalence relation on $[n]$ by letting $i \sim j$ if and only if $\{i, j\}$ is dependent. The *simplification* of M, denoted M_s is, by definition, the induced matroid on the equivalence classes. The natural map $s\colon \mathsf{M} \to \mathsf{M}_s$ is a morphism of $\overline{\mathcal{M}}$.

The map $A(s)\colon A(\mathsf{M}) \to A(\mathsf{M}_s)$ is easily seen to be an isomorphism, where $I(\mathsf{M})$ contains relations $e_i - e_j$ if $\{i, j\}$ is dependent in M.

Remark 1 Clearly the complement of a hyperplane arrangement is unaffected by the presence of repeated hyperplanes, so for topological purposes there is no loss in assuming that the underlying matroid is simple. However, we will make some use of the fact that the isomorphism $A(s)$ is not an equality in Theorem 7.

3 Resonance Varieties

Now suppose $E = \Lambda(V)$ is the exterior algebra over a (finite-dimensional) $\Bbbk$-vector space V. Suppose A and B are graded E-modules.

3.1 Definitions

For any $v \in V$, we have $v \cdot v = 0$, so there is a cochain complex of $\Bbbk$-modules, $(A, \cdot v)$, in which the differential is by right-multiplication by v. This construction is natural, in the sense that if $f: A \to B$ is a graded E-module homomorphism, then for any $v \in V$,

$$f: (A, \cdot v) \to (B, \cdot f(v)) \tag{4}$$

is clearly a homomorphism of cochain complexes.

The resonance varieties of A are defined for all integers $p, d \geq 0$ to be

$$\mathcal{R}^p_d(A) = \{v \in V: \dim_{\Bbbk} H^p(A, \cdot v) \geq d\},$$

and we abbreviate $\mathcal{R}^p(A) := \mathcal{R}^p_1(A)$. We note that our definition varies slightly from the usual one (see, e.g., [27]) in that we do not assume either that A itself is a $\Bbbk$-algebra or that $V = A^1$. The modules of greatest interest are, in fact, algebras $A = A(\mathsf{M}) = E/I$; however, we do allow I to contain relations of degree 1, accommodating parallel elements in M. We suggest distinguishing the two parameters by referring to $\mathcal{R}^p_d(A)$ with $p > 1$ as "higher" resonance, versus "deeper" for $d > 1$. Our focus here is on the former.

For any nonzero $v \in V$, clearly $v \in \mathcal{R}^p_d(A)$ if and only if $\lambda v \in \mathcal{R}^p_d(A)$ for any $\lambda \in \Bbbk^\times$, so they determine projective subvarieties of $\mathbb{P}V$.

3.2 Resonance of Orlik-Solomon Algebras

From now on, we restrict our attention to Orlik-Solomon algebras, and abbreviate: $\mathcal{R}^p(\mathsf{M}) := \mathcal{R}^p(A(\mathsf{M}))$ and $\mathcal{R}^p(\mathcal{A}) := \mathcal{R}^p(\mathsf{M}(\mathcal{A}))$ for matroids M and arrangements $\mathcal{A}$, respectively. If G is a graph, let $\mathsf{M}(G)$ denote its matroid, and $\mathcal{R}^p(G) := \mathcal{R}^p(\mathsf{M}(G))$.

First we mention some properties of resonance varieties that specific to Orlik-Solomon algebras. One such feature is a nestedness property discovered by Eisenbud et al. [15] and studied further in [4, 11].

Theorem 2 *Let* M *be a matroid of rank* ℓ. *Then*

$$\{0\} \subseteq \mathcal{R}^0(\mathsf{M}) \subseteq \mathcal{R}^1(\mathsf{M}) \subseteq \cdots \subseteq \mathcal{R}^\ell(\mathsf{M}) \subseteq \overline{V}. \tag{5}$$

Proof The inclusions $\mathcal{R}^p(\mathsf{M}) \subseteq \mathcal{R}^{p+1}(\mathsf{M})$ for $0 \leq p \leq \ell - 1$ were proven in [15, Theorem 4.1(b)]. The authors work with arrangements, but their arguments apply to all matroids. The inclusion $\mathcal{R}^p(\mathsf{M}) \subseteq \overline{V}$ for all $p \geq 0$ is due to Yuzvinsky, [35, Proposition 2.1]. □

Another is the following. By contrast, this result depends on complex geometry and a result due to Arapura [2]: for a full explanation, we refer to [14].

Theorem 3 *Let $\mathcal{A}$ be a complex hyperplane arrangement, and $\Bbbk$ a field of characteristic zero. Then $\mathcal{R}^p_{\Bbbk}(\mathcal{A})$ is a union of linear components, for $0 \le p \le \operatorname{rank}(\mathcal{A})$.*

Remark 2 In characteristic zero, then, resonance varieties of realizable matroids are subspace arrangements. Falk [18] has shown that this is not in general the case for $\operatorname{char} \Bbbk \neq 0$: see also [19, Example 4.24]. So even for hyperplane arrangements, the resonance varieties depend on the characteristic of the field (unlike the Orlik-Solomon algebra itself). For a striking application of resonance in characteristic 3, we refer to Papadima and Suciu [28].

A component W of a resonance variety is called *essential* if $W \cap (\Bbbk^{\times})^n \neq \emptyset$.

Question 1 Assume $\operatorname{char} \Bbbk = 0$. Then the components of $\mathcal{R}^1(\mathsf{M})$ are linear for any matroid M [22, Corollary 3.6]. Is $\mathcal{R}^p_{\Bbbk}(\mathsf{M})$ a union of linear components for all matroids M, for $p > 1$?

The next result, due to Libgober and Yuzvinsky [22], is a qualitative property of $\mathcal{R}^1(\mathsf{M})$ which is both special to matroids and, we will see, to degree $p = 1$.

Theorem 4 *Assume* $\operatorname{char} \Bbbk = 0$. *If $\mathcal{R}^1(\mathsf{M})$ contains a component W of dimension $k > 0$, there is an injective homomorphism $\overline{A}(\mathsf{U}_{2,k+1}) \to \overline{A}(\mathsf{M})$. Conversely, the image of such a homomorphism in degree 1 lies in $\mathcal{R}^1(\mathsf{M})$.*

Proof Multiplication in $\overline{A}^1(\mathsf{U}_{2,k+1})$ is zero, so this is just a reformulation of the following result from [22]: if W is a component of $\mathcal{R}^1(\mathsf{M})$, then for any $v, w \in W$, we have $vw = 0$. □

Next, we see that questions of resonance can be reduced to the projective Orlik-Solomon algebra, via the short exact sequence (2).

Lemma 4 *For any matroid M on $[n]$ and $v \in \overline{V}$, there is a short exact sequence of cochain complexes*

$$0 \longrightarrow (\overline{A}(\mathsf{M}), v) \longrightarrow (A(\mathsf{M}), v) \xrightarrow{\partial} (\overline{A}(\mathsf{M}), v)[-1] \longrightarrow 0. \tag{6}$$

If $\operatorname{char} \Bbbk \nmid n$, *the sequence* (6) *is split.*

Proof The inclusion $\overline{A}(\mathsf{M}) \to A(\mathsf{M})$ makes $A(\mathsf{M})$ an $\overline{A}(\mathsf{M})$-module. Using Lemma 1, it is easily checked that ∂ is a $\overline{A}(\mathsf{M})$-module homomorphism. With this, we see that the maps in the sequence (2) commute with multiplication by v.

If n is nonzero in $\Bbbk$, (left) multiplication by $\frac{1}{n}\sum_{i=1}^n e_i$ gives a right inverse to ∂, proving the second assertion. □

With this, we see that the resonance of A and $\overline{A}$ differ only by a trivial factor.

Proposition 3 *Let M be a matroid on $[n]$ of rank ℓ. If $\operatorname{char} \Bbbk \nmid n$, then $\mathcal{R}^p(\overline{A}(\mathsf{M})) = \mathcal{R}^p(A(\mathsf{M}))$ for all $0 \le p \le \ell - 1$, and $\mathcal{R}^\ell(A(\mathsf{M})) = \mathcal{R}^{\ell-1}(A(\mathsf{M}))$. For all $d \ge 0$, we* *also have*

$$\mathcal{R}^p_d(A(\mathsf{M})) = \bigcup_{j \le d} \mathcal{R}^p_j(\overline{A}(\mathsf{M}) \cap \mathcal{R}^{p-1}_{d-j}(\overline{A}(\mathsf{M})). \tag{7}$$

Proof The equality (7) is a direct consequence of Lemma 4. In particular, for $d = 1$, we have

$$\mathcal{R}^p(A(\mathsf{M})) = \mathcal{R}^p(\overline{A}(\mathsf{M})) \cup R^{p-1}(\overline{A}(\mathsf{M})) \tag{8}$$

for $0 \le p \le \ell$. We prove $\mathcal{R}^p(\overline{A}(\mathsf{M})) = \mathcal{R}^p(A(\mathsf{M}))$ by induction. The case $p = 0$ follows from (8). The induction step is obtained by combining (8) with Theorem 2. □

3.3 Top and Bottom

The two ends of the resonance filtration (5) have straightforward descriptions. First, it will be convenient to have some notation.

Definition 3 If π is a partition of $[n]$ with k parts, let P_π denote the codimension-k subspace of $\mathbb{k}^n$ given by equations

$$\sum_{j \in \pi_i} x_j = 0$$

for $1 \le i \le k$. If $k = 1$, we recover $\overline{V} = P_{\{[n]\}}$. At the other extreme, if each block of π is a singleton, then $P_\pi = \{0\}$.

Dually, let Q_π denote the k-dimensional subspace of $\mathbb{k}^n$ given as the span of vectors

$$\sum_{j \in \pi_i} e_j$$

for $1 \le i \le k$. Clearly, P_π and Q_π are complementary subspaces (with respect to the distinguished basis in V.)

Proposition 4 *For any matroid* M *on* $[n]$*, let* π *denote the partition of* $[n]$ *given by simplification (Definition 2.) Then* $\mathcal{R}^0_1(\mathsf{M}) = P_\pi$.

Proof The simplification map $s\colon \mathsf{M} \to \mathsf{M}_s$ (Definition 2) gives an isomorphism of complexes

$$A(s)\colon (A(\mathsf{M}), v) \to (A(\mathsf{M}_s), s(v))$$

for all $v \in V$. For a simple matroid, clearly $\mathcal{R}^0(\mathsf{M}_s) = \{0\}$, so $\mathcal{R}^0(\mathsf{M}) = s^{-1}(0)$, which is the subspace P_π. □

At the other extreme, by the invariance of Euler characteristic, we have

$$\sum_{p=0}^{\ell-1}(-1)^p \dim_{\Bbbk} H^p(\overline{A}(\mathsf{M}), v) = (-1)^{\ell-1}\beta(\mathsf{M}),$$

for any $v \in \overline{V}$, where $\beta(\mathsf{M})$ denotes Crapo's beta invariant. We recall that $\beta(\mathsf{M}) \neq 0$ if and only if M is connected. If $\mathsf{M} = \mathsf{M}(\mathcal{A})$ for an arrangement $\mathcal{A}$, it is usual to say $\mathcal{A}$ is irreducible to mean $\mathsf{M}(\mathcal{A})$ is connected.

If $\mathsf{M}(\mathcal{A})$ is connected, then for any $v \in \overline{V}$, the nonzero Euler characteristic implies that $H^p(\overline{A}(\mathsf{M}), v) \neq 0$ for some p. By Theorem 2, $H^{\ell-1}(\overline{A}(\mathsf{M}), v) \neq 0$, which proves the following.

Proposition 5 ([35]) *If* M *is connected of rank* ℓ, *then* $\mathcal{R}^{\ell-1}(\overline{A}(\mathsf{M})) = \overline{V} = P_{\{[n]\}}$.

Example 2 Consider the matroid $\mathsf{M}(G)$, where

$G =$ 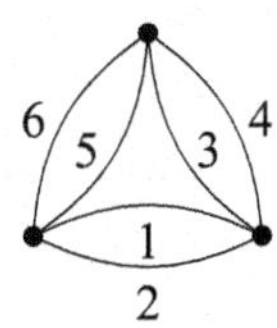 .

Using Proposition 4, we see $\mathcal{R}^0(G) = P_{12|34|56} = V(x_1 + x_2, x_3 + x_4, x_5 + x_6)$, a 3-dimensional subspace. Since $\mathsf{M}(G)$ is connected of rank 2, $\mathcal{R}^1(G) = R^2(G) = \overline{V} = V(x_1 + \cdots + x_6)$.

3.4 Upper Bounds

Most of the results we present in this paper give lower bounds for the resonance varieties: that is, conditions which imply nonvanishing cohomology. Here, we give two upper bounds. The first uses a result of Schechtman and Varchenko [29] to give a weak but easily stated upper bound.

Theorem 5 *If* M *is a matroid of rank* ℓ, *then*

$$\mathcal{R}^p(\mathsf{M}) \subseteq \bigcup_{X \in L^{\mathrm{irr}}_{\leq p+1}(\mathsf{M})} P_{\{X,[n]-X\}}$$

for all $0 \leq p \leq \ell$.

In other words, if $H^p(A(\mathsf{M}), v) \neq 0$, then there exists an irreducible flat X of rank at most $p + 1$ for which $v \in P_{\{X,[n]-X\}}$; equivalently, for which $v \in \overline{V}(\mathsf{M})$ and $\sum_{i \in X} v_i = 0$.

Proof In Schechtman and Varchenko [29], the authors construct a map of cochain complexes

$$(F(\mathsf{M}), d) \xrightarrow{S(v)} (A(\mathsf{M}), v) \tag{9}$$

for each $v \in V(\mathsf{M})$. The complex $(F(\mathsf{M}), d)$ is isomorphic to the $\Bbbk$-dual of $(A(\mathsf{M}), \partial_A)$; in particular it is exact [29, Corollary 2.8] and does not depend on v.

Let $v(X) := \sum_{i \in X} v_i$. The determinant formula [29, Theorem 3.7] expresses the determinant of $S(v)$ in terms of powers of products of $v(X)$'s. In particular, it follows that $S^p(v)$ is an isomorphism if $v(X) \neq 0$ for all $X \in L^{\mathrm{irr}}_{\leq p}(\mathsf{M})$. Suppose, then, that $v(X) \neq 0$ for all irreducible X of rank at most $p + 1$. This implies (9) is an isomorphism up to degree $p + 1$, so $H^q(A(\mathsf{M}), v) = 0$ for all $q \leq p$, from which the claim follows. □

Our second result gives a more refined bound for the largest (nontrivial) resonance variety, based on the main result of Cohen et al. [7]. Since their result applies to the cohomology of local systems on a hyperplane complement, the proof of this bound requires that the matroid have a complex realization.

Definition 4 Let us say that a subset of flats C *covers* M if there is a surjective function $f \colon [n] \to C$ for which $i \in f(i)$ for all $1 \leq i \leq n$. For a given cover C, let $P_C = \bigcap_{X \in C} P_{\{X,[n]-X\}}$, a linear subspace of $\overline{V}$.

For each $p \geq 0$, we define a subspace arrangement in $\overline{V}$ using M: let

$$\mathcal{S}^p(\mathsf{M}) = \bigcup_C P_C,$$

where the union is over all subsets $C \subseteq L^{\mathrm{irr}}_{\leq p+1}(\mathsf{M})$ that cover M.

Finally, say a cover C of M is *essential* if $|X| > 1$ for all $X \in C$, and let

$$\mathcal{S}^p_{\mathrm{ess}}(\mathsf{M}) = \bigcup_C P_C,$$

where the union is over essential covers $C \subseteq L^{\mathrm{irr}}_{\leq p+1}(\mathsf{M})$.

We note that if M is irreducible of rank ℓ, then $\mathcal{S}^0(\mathsf{M}) \subseteq \cdots \subseteq \mathcal{S}^\ell(\mathsf{M}) = \overline{V}$.

Theorem 6 *If* M *is a complex-realizable matroid, then* $\mathcal{R}^p(\mathsf{M}) \subseteq \mathcal{S}^p(\mathsf{M})$, *for all* $p \geq 0$.

Proof Suppose $\mathcal{A}$ is a complex arrangement of rank ℓ and $\mathsf{M} = \mathsf{M}(\mathcal{A})$. The main result of [30] allows us to translate [7, Theorem 1] into the following statement about resonance. That is, for $v \in \overline{V}(\mathsf{M})$, suppose that for some $i \in [n]$, whenever $i \in X \in L^{\mathrm{irr}}_{<\ell}(\mathsf{M})$, we have $v(X) \neq 0$. Then $H^p(\overline{A}(\mathsf{M}), v) = 0$ for all $p \neq \ell - 1$.

If $v \in \mathcal{R}^{\ell-2}(\mathsf{M})$, then, for all $i \in [n]$, there exists some X for which $i \in X$ and $v \in P_{\{X,[n]-X\}}$, since $P_{\{X,[n]-X\}} = \{v \in \overline{V}: v(X) = 0\}$. So

$$\begin{aligned} \mathcal{R}^{\ell-2}(\mathsf{M}) &\subseteq \bigcap_{i=1}^{n} \bigcup_{\substack{X\in L^{\mathrm{irr}}_{<\ell}:\\ i\in X}} P_{\{X,[n]-X\}}, \\ &= \bigcup_{C} \bigcap_{X\in C} P_{\{X,[n]-X\}}, \qquad (10) \\ &= \mathcal{S}^{\ell-2}(M) \end{aligned}$$

since C runs over all covers in $L^{\mathrm{irr}}_{\leq \ell-1}$. To obtain the result for $p < \ell - 1$, we consider the truncation $T_{p+1}\mathsf{M}$ of M to rank $p + 1$. This matroid is also realizable (by intersecting $\mathcal{A}$ with a generic linear space of codimension $p + 1$) and $L_{\leq p}(\mathsf{M}) = L_{\leq p}(T_{p+1}\mathsf{M})$, so it is enough to apply (10) to $T_{p+1}\mathsf{M}$. □

Corollary 1 *If W is a component of $\mathcal{R}^p(\mathsf{M})$ for a complex-realizable matroid M and $p \geq 0$, then there exists a cover $C \subseteq L^{\mathrm{irr}}_{\leq p+1}(\mathsf{M})$ for which $W \subseteq P_C$.*

Corollary 2 *If M is a complex-realizable matroid of rank ℓ, then for all $p \geq 0$,*

$$\mathcal{R}^p(\mathsf{M}) \cap (\Bbbk^{\times})^n \subseteq \mathcal{S}^p_{\mathrm{ess}}(\mathsf{M}). \qquad (11)$$

Proof If $X = \{i\}$, then $P_{\{X,[n]-X\}}$ is contained in the coordinate hyperplane $x_i = 0$, so its intersection with $(\Bbbk^{\times})^n$ is empty. □

We will see examples in the next section in which the upper bound above is sharp: that is, the containment (11) is an equality.

Question 2 The bound given in Theorem 6 is, of course, completely combinatorial, although the result we use from [7] is topological in nature. Can the hypothesis that M is complex-realizable be dropped?

4 Matroid Operations and Resonance

In this section, we systematically examine the behaviour of resonance under several familiar matroid operations. In some cases, apparently nontrivial components of resonance varieties can be obtained from tautological components belonging to other matroids.

4.1 Naturality

In some cases, resonance behaves well under the morphisms of $\mathcal{M}$ from Sect. 2.3. Combining Theorem 1 with (4), we see that if $f\colon \mathsf{M}_1 \to \mathsf{M}_2$ is morphism, then for all $v \in V(\mathsf{M}_1)$, there is a map of complexes

$$A(f)\colon (A(\mathsf{M}_1), v) \to (A(\mathsf{M}_2), f(v)). \tag{12}$$

Here is an important special case.

Proposition 6 *If X is a flat of a matroid M, for any $v \in V(\mathsf{M}_X)$, the complex $(A(\mathsf{M}_X), v)$ is a split subcomplex of $(A(\mathsf{M}), v)$.*

Proof The inclusion $j\colon \mathsf{M}_X \to \mathsf{M}$ and projection $\phi\colon \mathsf{M} \to \mathsf{M}_X$ given by

$$\phi(i) = \begin{cases} i & \text{if } i \in X; \\ 0 & \text{otherwise} \end{cases}$$

are both complete weak maps. Since $\phi \circ j = \mathrm{id}$, the result follows by naturality. □

Lemma 5 *Suppose that $f\colon \mathsf{M}_1 \to \mathsf{M}_2$ is a morphism induced by the identity map on underlying sets. If the posets $L_{\leq k}(\mathsf{M}_1) \cong L_{\leq k}(\mathsf{M}_2)$ for some integer $k \geq 1$, then $\mathcal{R}^p(\mathsf{M}_1) \cong \mathcal{R}^p(\mathsf{M}_2)$ for $0 \leq p \leq k-1$.*

Proof Consider the map $A(f)\colon (A(\mathsf{M}_1), v) \to (A(\mathsf{M}_2), v)$, for some $v \in V(\mathsf{M}_1)$. The hypotheses imply this is an isomorphism in degrees $\leq k$, so $H^p(A(\mathsf{M}_1), v) \cong H^p(A(\mathsf{M}_2), v)$ for all $0 \leq p \leq k-1$. □

We say that a matroid M is k-generic for some $k \geq 0$ if M has no circuits of size k or smaller.

Proposition 7 *Suppose M is k-generic. Then $\mathcal{R}^p(\mathsf{M}) = \{0\}$ for all i, $0 \leq p \leq k-1$.*

Proof Let F_n denote the free matroid on $[n]$ (the underlying matroid of the Boolean arrangement.) Clearly $A(F_n) = E$, the exterior algebra, and it is straightforward to check that $\mathcal{R}^p(E) = \{0\}$ for $0 \leq p \leq n$. The claim then follows from Lemma 5. □

Example 3 (Uniform Matroids) Consider the uniform matroid $\mathsf{U}_{\ell,n}$ of rank ℓ on $[n]$. Assume $n > \ell$. Then $\mathsf{U}_{\ell,n}$ is k-generic for all $k < \ell$, so $\mathcal{R}^p(\mathsf{U}_{\ell,n}) = \{0\}$ for $0 \leq p < \ell - 1$, and $\mathcal{R}^{\ell-1}(\mathsf{U}_{\ell,n}) = \overline{V}$ because $\mathsf{U}_{\ell,n}$ is connected.

4.2 Sums, Submatroids and Duals

Let us single out three particularly well-behaved constructions. For convenience, if $X \in L(\mathsf{M})$, we abuse notation and identify $V(\mathsf{M}_X)$ with its image as a coordinate subspace in $V(\mathsf{M})$, supported in the coordinates indexed by X.

Theorem 7 *Suppose* M *is a matroid of rank* ℓ *on* $[n]$. *Let* X *be any flat of* M, *and* $\mathsf{M}^{\perp}$ *the dual matroid. Then:*

	Construction	Resonance
(1)	$\mathsf{M} = \mathsf{M}_1 \oplus \mathsf{M}_2$	$\mathcal{R}^k(\mathsf{M}) = \bigcup_{p+q=k} \mathcal{R}^p(\mathsf{M}_1) \times \mathcal{R}^q(\mathsf{M}_2)$
(2)	submatroids	$\mathcal{R}^p(\mathsf{M}_X) \subseteq \mathcal{R}^p(\mathsf{M}) \cap V(\mathsf{M}_X)$ for all $p \geq 0$
(3)	duality	$\mathcal{R}^{\ell-p}(\mathsf{M}) \cap (\Bbbk^{\times})^n = \mathcal{R}^{n-\ell-p}(\mathsf{M}^{\perp}) \cap (\Bbbk^{\times})^n$, for $p \geq 0$

Proof (1) is due to Papadima and Suciu [27, Proposition 13.1], since if $\mathsf{M} = \mathsf{M}_1 \oplus \mathsf{M}_2$, then $A(\mathsf{M}) = A(\mathsf{M}_1) \otimes_{\Bbbk} A(\mathsf{M}_2)$. Claim (2) follows directly from Proposition 6. Claim (3) appears as Theorem 27 in [10]. □

4.3 Local Components

If X is a flat of an arrangement M on $[n]$, we define a partition $\pi(X)$: let $X = X_1 \vee X_2 \cdots \vee X_r$ be the decomposition of X into connected components. Let $\pi(X)$ be the partition given by $i \sim j \Leftrightarrow \{i,j\} \subseteq X_k$, for all i, j, k, and $i \sim i$ if $i \notin X$.

Proposition 8 *For any matroid* M, *then for any flat* $X \in L_q(\mathsf{M})$, *there is an equality* $P_{\pi(X)} = V_X \cap R^p(\mathsf{M})$ *for all* p *with* $q - r \leq p \leq \ell$, *where* $r = |\pi(X)|$.

Proof Since $\mathsf{M}_X = \bigoplus_{i=1}^r \mathsf{M}_{X_i}$, in view of the product formula of Theorem 7(1), it is enough to prove the claim when X is connected. In this case, $P_{\pi(X)} = \overline{V}_X$. By Proposition 5, $\mathcal{R}^{q-1}(\mathsf{M}_X) = \overline{V}_X$, which is contained in $\mathcal{R}^{q-1}(\mathsf{M}) \cap V_X$ by Theorem 7(2). By propagation (Theorem 2),

$$\overline{V}_X \subseteq \mathcal{R}^p(\mathsf{M}) \cap V_X \text{ for } q - 1 \leq p \leq \ell. \tag{13}$$

On the other hand, $\mathcal{R}^p(\mathsf{M}) \subseteq \overline{V}$ and $\overline{V} \cap V_X = \overline{V}_X$, so (13) is an equality for all p with $q - 1 \leq p \leq \ell$. □

Example 4 Let us consider the dual to Example 2. That is, $\mathsf{M}^{\perp} = \mathsf{M}(G^{\perp})$, where

$G^{\perp} =$ 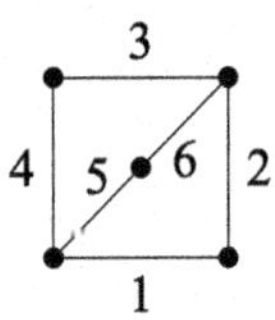 $\cdot$

By Proposition 7, we see $\mathcal{R}^0(\mathsf{M}^{\perp}) = \mathcal{R}^1(\mathsf{M}^{\perp}) = \{0\}$. By Proposition 8, each of the circuits of size 4 contribute local components, so each of the 3-dimensional linear spaces $P_{1|2|3456}$, $P_{3|4|1256}$, $P_{5|6|1234}$ are contained in $\mathcal{R}^2(\mathsf{M}^{\perp})$.

By Theorem 7(3), we see $\mathcal{R}^2(\mathsf{M}^\perp)$ also contains the component $P_{12|34|56}$. With the help of Macaulay2 [21], we find that the upper bound from Theorem 6 is sharp, so

$$\mathcal{R}^2(\mathsf{M}^\perp) = \mathcal{S}^2(\mathsf{M}^\perp) = P_{1|2|3456} \cup P_{3|4|1256} \cup P_{5|6|1234} \cup P_{12|34|56}.$$

Since $\mathsf{M}^\perp$ is irreducible of rank 4, we have $\mathcal{R}^3(\mathsf{M}^\perp) = \mathcal{R}^4(\mathsf{M}^\perp) = P_{123456}$, by Proposition 4.

Remark 3 As an aside, let's recall an application to the cohomology of the Milnor fibration of a hyperplane arrangement, a full treatment of which can be found in the paper of Papadima and Suciu [28]. The matroid $\mathsf{M}^\perp$ is realized by the arrangement defined by

$$Q = (x_1 - x_2)(x_2 - x_3)(x_3 - x_4)(x_1 - x_4)(x_1 - x_5)(x_3 - x_5).$$

Each part of $\pi = 12|34|56$ has even length, so the vector $d = (1,1,1,1,1,1) \in \mathcal{R}^2_{\Bbbk}(\mathsf{M}^\perp)$ if $\operatorname{char}\Bbbk = 2$. From an application of the tangent cone formula (see, for example, [12]), it follows -1 is a monodromy eigenvalue of $H^2(F,\mathbb{C})$, where $F = Q^{-1}(1)$, the Milnor fibre of the arrangement. Our point here is that an apparently nontrivial cohomological phenomenon can result from matroid operations applied to a rather trivial example (Example 2).

4.4 Deletion-Contraction

In the case of complex hyperplane arrangements, Cohen [6] showed that there is a long exact sequence relating cohomology of a local system on the complement of an arrangement, its deletion, and its restriction. We will use the combinatorial analogue here in order to examine the behaviour of resonance varieties under deletion and contraction.

If M is a matroid and i_0 is an element which is not a loop, let $\mathsf{M}' := \mathsf{M} - \{i_0\}$ and $\mathsf{M}'' := \mathsf{M}/i_0$ denote the deletion and contraction matroids, respectively. From Orlik and Terao [24, Theorem 3.65], there is a short exact sequence of $\Bbbk$-modules

$$0 \longrightarrow A(\mathsf{M}') \xrightarrow{\ j\ } A(\mathsf{M}) \xrightarrow{\ \phi\ } A(\mathsf{M}'')[-1] \longrightarrow 0 \tag{14}$$

where the map ϕ is defined on monomials by

$$\phi(e_{i_0} e_S) = e_S,$$

and $\phi(e_S) = 0$ if $i_0 \notin S$. The inclusion j is a ring homomorphism, induced by the morphism $\mathsf{M}' \hookrightarrow \mathsf{M}$, so (14) is a sequence of $A(\mathsf{M}')$-modules.

For any $v \in \overline{V}(\mathsf{M}')$, then, we obtain a short exact sequence of complexes,

$$0 \longrightarrow (A(\mathsf{M}'), v) \xrightarrow{\;j\;} (A(\mathsf{M}), v) \xrightarrow{\;\phi\;} (A(\mathsf{M}''), v)[-1] \longrightarrow 0 \tag{15}$$

where we identify $\overline{V}(\mathsf{M}')$ and $\overline{V}(\mathsf{M}'')$ with the coordinate hyperplane $H_{i_0} := P_{i_0|1,\ldots,\hat{i_0},\ldots,n} \subseteq \overline{V}(\mathsf{M})$.

Proposition 9 *Let* $(\mathsf{M}, \mathsf{M}', \mathsf{M}'')$ *be a deletion-contraction triple. Then, for all integers* $p \geq 0$ *and* $k \geq j \geq 0$, *we have the following inclusions:*

$$H_{i_0} \cap \mathcal{R}^p_k(\mathsf{M}) - \mathcal{R}^p_{j+1}(\mathsf{M}') \subseteq \mathcal{R}^{p-1}_{k-j}(\mathsf{M}''); \tag{16}$$

$$\mathcal{R}^{p-1}_k(\mathsf{M}'') - \mathcal{R}^p_{j+1}(\mathsf{M}) \subseteq \mathcal{R}^{p+1}_{k-j}(\mathsf{M}'); \tag{17}$$

$$\mathcal{R}^p_k(\mathsf{M}') - \mathcal{R}^{p-2}_{j+1}(\mathsf{M}'') \subseteq \mathcal{R}^p_{k-j}(\mathsf{M}). \tag{18}$$

Proof Given $v \in H_{i_0}$, consider the long exact sequence in cohomology of (15). By exactness, we have

$$\dim_{\Bbbk} H^p(A(\mathsf{M}), v) \leq \dim_{\Bbbk} H^p(A(\mathsf{M}'), v) + \dim_{\Bbbk} H^{p-1}(A(\mathsf{M}''), v),$$

from which the first containment easily follows. The remaining two are obtained in the same way. □

We note that, by putting $j = 0$ and $k = 1$, we can erase the subscripts that keep track of depth.

Corollary 3 *Suppose* M *is k-generic. Then for any element* i_0,

$$\mathcal{R}^p(\mathsf{M} - \{i_0\}) \subseteq \mathcal{R}^p(\mathsf{M}),$$

for all $0 \leq p \leq k$.

Proof If M is k-generic, the contraction M/i_0 is $(k-1)$-generic. The result then follows by Propositions 9(18) and 7. □

Since all simple matroids are 1-generic, the situation for $p = 1$ is particularly nice. The next example shows that, in general, $\mathcal{R}^p(\mathsf{M}') \not\subseteq \mathcal{R}^p(\mathsf{M})$ for $p > 1$.

Example 5 Let $\mathsf{M} = \mathsf{M}(G_1)$ be the matroid of the graph shown in Fig. 1. Deleting the horizontal edge gives the graphic matroid of Example 4, which we denote for the moment by M'. We saw $W := P_{12|34|56|7} \subseteq \mathcal{R}^2(\mathsf{M}')$: let us parameterize

$$W = \{(-a, a, b, -b, -c, c, 0) : a, b, c \in \Bbbk\}.$$

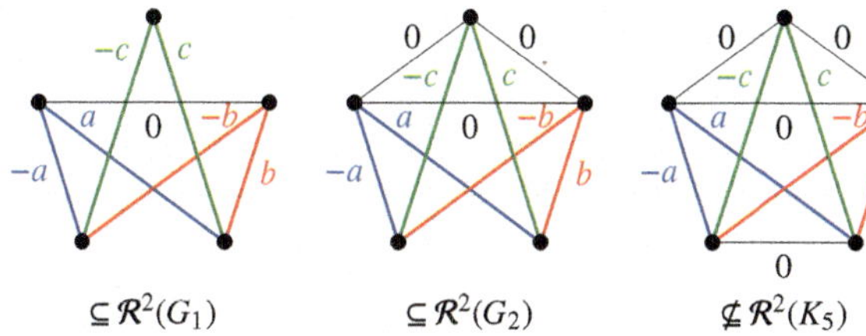

Fig. 1 Adding edges may not preserve resonance (Example 5)

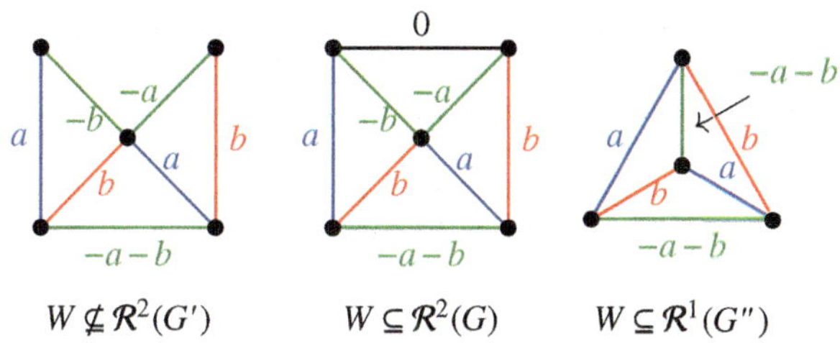

Fig. 2 Deletion may not preserve resonance (Example 6)

If $v \in W$, then $v \notin \mathcal{R}^0(\mathsf{M}'')$ provided either $a + b \neq 0$ or $c \neq 0$. From Proposition 9(18), we see $W \subseteq \mathcal{R}^2(\mathsf{M})$. We can repeat this twice to find $W \subseteq \mathcal{R}^2(G_2)$ as well, where G_2 is the graph obtained from K_5 by deleting an edge.

However, the same argument does not allow us to conclude $W \subseteq \mathcal{R}^2(K_5)$: if we contract the bottom edge, we see W is now contained in $\mathcal{R}^0$ of the contraction (Proposition 4), so Proposition 9 does not apply. Indeed, it turns out that $W \not\subseteq \mathcal{R}^2(K_5)$. Since $W \subseteq \mathcal{S}^2(\mathsf{M}(K_5))$, though, our upper bound is of no use here, and we are forced to verify this with a direct calculation.

Along the same lines, we see also that a component of $\mathcal{R}^2(\mathsf{M})$ contained in a coordinate hyperplane need not be a component of the deletion $\mathcal{R}^2(\mathsf{M}')$.

Example 6 Consider the matroid of the graph G from Example 1, ordered as shown in (3). With the help of Macaulay 2 and Theorem 6, we see

$$W := \{(-a-b, b, 0, a, b, a, -a, -b) : a, b \in \Bbbk\}$$

is a component of $\mathcal{R}^2(G)$. Let G' be the graph obtained by deleting the edge with label 0 (see Fig. 2). One argues that $W \not\subseteq \mathcal{R}^2(G')$ using Theorem 6 as follows. If $v \in W$ and $a, b \neq 0$, we choose an edge incident to a degree vertex, and check that the only irreducible flats X that contain it have $v(X) \neq 0$. It follows that $v \notin \mathcal{S}^2(G')$.

4.5 Parallel Connections

Here, we consider another matroid construction through which resonance varieties can be traced. The underlying data is the following. Suppose M_1 and M_2 are matroids on ground sets S_1 and S_2, respectively, and $X = S_1 \cap S_2$ is a modular flat of M_1 and $(\mathsf{M}_1)_X = (\mathsf{M}_2)_X$. The (generalized) parallel connection $\mathsf{M}_1 \parallel_X \mathsf{M}_2$ is the matroid on $S_1 \cup S_2$ obtained from $\mathsf{M}_1 \oplus \mathsf{M}_2$ by identifying the common copy of X—see [25] for details, and Fig. 3 for an example. Let $\mathsf{M}_{12} = (\mathsf{M}_1)_X = (\mathsf{M}_2)_X$.

The Orlik-Solomon complex of the parallel connection can be described as follows. We begin with the degree-1 part. Let $i_j\colon \mathsf{M}_{12} \to \mathsf{M}_j$ denote the inclusions, for $j = 1, 2$. The identification map $\phi\colon S_1 \sqcup S_2 \to S_1 \cup S_2$. These are all morphisms of $\mathcal{M}$, and they fit in a short exact sequence:

$$0 \longrightarrow V(\mathsf{M}_{12}) \xrightarrow{i_1 \oplus -i_2} V(\mathsf{M}_1) \oplus V(\mathsf{M}_2) \xrightarrow{\phi} V(\mathsf{M}_1 \parallel_X \mathsf{M}_2) \longrightarrow 0.$$

By restricting, we also obtain:

$$0 \longrightarrow \overline{V}(\mathsf{M}_{12}) \longrightarrow \overline{V}(\mathsf{M}_1) \oplus V(\mathsf{M}_2) \xrightarrow{\phi} \overline{V}(\mathsf{M}_1 \parallel_X \mathsf{M}_2) \longrightarrow 0. \tag{19}$$

Now let P denote the pushout of graded $\Bbbk$-algebras:

$$\begin{array}{ccc} A(\mathsf{M}_{12}) & \xrightarrow{A(i_1)} & A(\mathsf{M}_1) \\ {\scriptstyle -A(i_2)}\downarrow & \ulcorner & \downarrow \\ A(\mathsf{M}_2) & \longrightarrow & P \end{array} \quad . \tag{20}$$

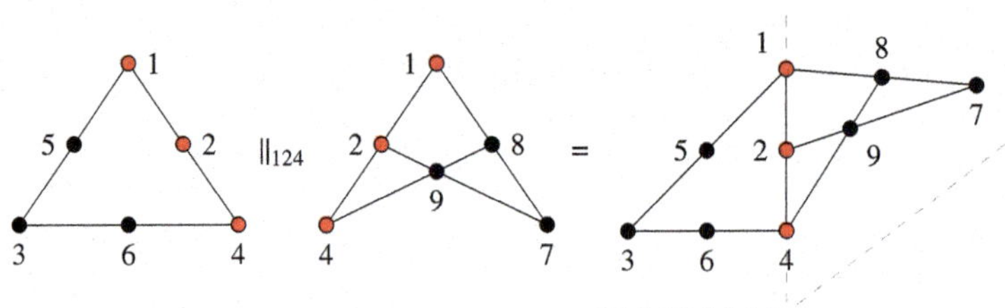

Fig. 3 Parallel connection

We can express the algebra P variously as

$$\begin{aligned} P &\cong \Bbbk \otimes_{A(\mathsf{M}_{12})} \big(A(\mathsf{M}_1) \otimes_{\Bbbk} A(\mathsf{M}_2)\big), && \\ &\cong \Bbbk \otimes_{A(\mathsf{M}_{12})} A(\mathsf{M}_1 \oplus \mathsf{M}_2) && \text{by [27, Proposition 13.1];} \\ &\cong A(\mathsf{M}_1 \oplus \mathsf{M}_2)/\!\!/A(\mathsf{M}_{12}) && \text{see, e.g., [5, p. 349];} \\ &\cong A(\mathsf{M}_1 \parallel_X \mathsf{M}_2). && \end{aligned}$$

Remark 4 We note that, since X is modular in M_1, $A(\mathsf{M}_1)$ is a free $A(\mathsf{M}_{12})$-module (see [33]). This implies $A(\mathsf{M}_1 \oplus \mathsf{M}_2)$ is also a free $A(\mathsf{M}_{12})$-module. Taking Hilbert series, we obtain

$$h(A(\mathsf{M}_1 \parallel_X \mathsf{M}_2, t) = h(A(\mathsf{M}_1), t)h(A(\mathsf{M}_2), t)/h(A(\mathsf{M}_{12}), t). \tag{21}$$

This amounts to the classical formula relating the characteristic polynomials of the four matroids, so the diagram (20) can be taken as an algebraic refinement.

We now restrict to the classical parallel connection, where X consists of a single element. The description above is particularly straightforward in this case: in particular, the linear map $\overline{\phi}$ in (19) is an isomorphism, and we show next that it induces an algebra isomorphism. In the case of complex hyperplane arrangements, the maps below come from maps of spaces: see [16] and [12, §7]. If $X = \{i\}$ for some $i \in S_1 \cup S_2$, let i' denote its image in a copy of S_2 disjoint from S_1.

Theorem 8 *If $X = \{i\}$, there is a short exact sequence*

$$0 \longrightarrow A(\mathsf{M}_1 \parallel_{\{i\}} \mathsf{M}_2)[-1] \longrightarrow A(\mathsf{M}_1 \oplus \mathsf{M}_2) \xrightarrow{A(\phi)} A(\mathsf{M}_1 \parallel_{\{i\}} \mathsf{M}_2) \longrightarrow 0.$$

and an isomorphism $\overline{A}(\mathsf{M}_1) \otimes_{\Bbbk} \overline{A}(\mathsf{M}_2) \cong \overline{A}(\mathsf{M}_1 \parallel_{\{i\}} \mathsf{M}_2)$.

Proof Since ϕ is surjective, so is $A(\phi)$ (Theorem 1), and $A(\mathsf{M}_1 \parallel_{\{i\}} \mathsf{M}_2) \cong A(\mathsf{M}_1 \oplus \mathsf{M}_2)/(r)$, where $r = e_i - e_{i'}$. On the other hand, multiplication by r gives a degree-1 map $A(\mathsf{M}_1 \oplus \mathsf{M}_2)/(r) \to A(\mathsf{M}_1 \oplus \mathsf{M}_2)$. Since r is easily seen to be nonresonant, this map is injective: that is, an isomorphism onto its image, $(r) = \ker A(\phi)$. It follows that the sequence is exact.

To prove the second claim, note the image of the restriction of $A(\phi)$ to $\overline{A}(\mathsf{M}_1) \otimes_{\Bbbk} \overline{A}(\mathsf{M}_2)$ is contained in $\overline{A}(\mathsf{M}_1 \parallel_{\{i\}} \mathsf{M}_2)$, since the source is generated in degree 1, where the situation is that of (19). Since the target is also generated in degree 1, it follows the map is surjective. To conclude it is an isomorphism, we compare Hilbert series using (21) and (2). □

The effect on resonance varieties is immediate.

Corollary 4 *If* $X = \{i\}$*, the map* $\overline{\phi}\colon \overline{V}(\mathsf{M}_1) \oplus \overline{V}(\mathsf{M}_2) \to \overline{V}(\mathsf{M}_1 \|_X \mathsf{M}_2)$ *restricts to an isomorphism for each* $p \geq 0$*:*

$$\mathcal{R}^p(\mathsf{M}_1 \oplus \mathsf{M}_2) \cong \mathcal{R}^p(\mathsf{M}_1 \|_X \mathsf{M}_2).$$

5 Singular Subspaces and Multinets

So far, we have seen that the resonance varieties of a matroid in top and bottom degrees are easy to account for, and that various resonance components can be obtained from these by comparing with submatroids, duals, deletion and contraction. The lower bounds obtained in this way sometimes match the upper bound given by Theorem 6.

Some quite special matroids are known to have additional components in $\mathcal{R}^1(\mathsf{M})$, however. We will assume from now on that $\operatorname{char} \Bbbk = 0$. Building on work of Libgober and Yuzvinsky [22] as well as Falk [17], Falk and Yuzvinsky [20] have characterized these in terms of auxiliary combinatorics. This is the notion of a multinet, and we briefly recall the construction from [20] in Sect. 5.1 with a view to higher-degree generalizations. We refer to Yuzvinsky's survey [36] in particular for a complete introduction.

Some first steps generalizing this theory to $\mathcal{R}^p(\mathsf{M})$ for $p > 1$ appear in [8], in the case of complex hyperplane arrangements, as well as in forthcoming work of Falk [3]. We interpret these constructions in terms of maps of Orlik-Solomon algebras, as in Sect. 4.

5.1 $\mathcal{R}^1(\mathsf{M})$*: Multinets*

Definition 5 If M is a matroid on $[n]$ and k is an integer with $k \geq 3$, a (k,d)-multinet is a partition $\mathcal{L} := \{\mathcal{L}_1, \ldots, \mathcal{L}_k\}$ of $[n]$ together with a set $\mathcal{X} \subseteq L_2^{\mathrm{irr}}(\mathsf{M})$ with the following four properties:

1. $|\mathcal{L}_s| = d$ for all $1 \leq s \leq k$.
2. If $i,j \in [n]$ belong to different parts of the partition $\mathcal{L}$, then they span a flat in $\mathcal{X}$.
3. For any $X \in \mathcal{X}$, the number $|\mathcal{L}_s \cap X|$ is independent of s.
4. For any $i,j \in \mathcal{L}_s$ for some s, there is a sequence $i = i_0, i_1, \ldots, i_r = j$ for which $\operatorname{span}\{i_{q-1}, i_q\} \notin \mathcal{X}$ for all $1 \leq q \leq r$.

The original (equivalent) formulation in [20] replaces M with its simplification M_s and records the number of parallel elements with a multiplicity function. We will say that a simple matroid supports a multinet if it is the simplification of a matroid with a partition as in Definition 5.

Theorem 9 (Theorems 2.3 and 2.4, [20]) $\mathcal{R}^1_{\Bbbk}(\mathsf{M})$ *contains an essential component if and only if* M *supports a multinet.*

Explicitly, this component is the linear space $Q_{\mathcal{L}} \cap \overline{V}$ (Definition 3). Their construction can be interpreted in terms of maps of Orlik-Solomon algebras as follows. First, we note that if $\mathcal{L}$ is a multinet, the partition satisfies the hypothesis of Proposition 2, so there is a morphism $p_{\mathcal{L}}: \mathsf{M} \to \mathsf{U}_{2,k}$ and a surjection

$$A(p_{\mathcal{L}}): \overline{A}(\mathsf{M}) \to \overline{A}(\mathsf{U}_{2,k}).$$

The more interesting aspect is the existence of a right inverse to $A(p_{\mathcal{L}})$. Multinets give the following construction [20].

Proposition 10 *If* $(\mathcal{L}, \mathcal{X})$ *is a multinet on* M, *there is a ring homomorphism* $i_{\mathcal{L}}: A(\mathsf{U}_{2,k}) \to A(\mathsf{M})$ *defined by*

$$i_{\mathcal{L}}(e_s) = \frac{1}{|\mathcal{L}_s|} \sum_{i \in \mathcal{L}_s} e_i \text{ for all } 1 \le s \le k,$$

which restricts to a map $i_{\mathcal{L}}: \overline{A}(\mathsf{U}_{2,k}) \to \overline{A}(\mathsf{M})$.

We note that $A(p_{\mathcal{L}}) \circ i_{\mathcal{L}} = \mathrm{id}$. From this it follows that $i_{\mathcal{L}}$ is injective in cohomology. From Example 3, $\mathcal{R}^1(\mathsf{U}_{2,k}) = \overline{V}(\mathsf{U}_{2,k})$, and its image in $\mathcal{R}^1(\mathsf{M})$ is just $Q_{\mathcal{L}} \cap \overline{V}$.

Example 7 ([20]) Let M be the matroid of the B_3 root system, giving each of the short roots multiplicity two. The corresponding hyperplane arrangement is defined by the polynomial $x^2y^2z^2(x-y)(x+y)(x-z)(x+z)(y-z)(y+z)$. Numbering the points of the matroid $\{1, 1', 2, 2', 3, 3', 4, 5, 6, 7, 8, 9\}$, the dependencies are shown in Fig. 4. The multinet $\mathcal{L} = \{11'89|22'67|33'45\}$ has $\mathcal{X} = L_2^{\mathrm{irr}}(\mathsf{M})$, so $Q_{\mathcal{L}} \cap \overline{V}$ is a 2-dimensional, essential component of $\mathcal{R}^1(\mathsf{M})$.

Returning to the upper bound of Corollary 2, let $C = L_2^{\mathrm{irr}}(B_3)$, an essential cover. By direct computation, P_C has dimension 2, so it equals the essential component computed above.

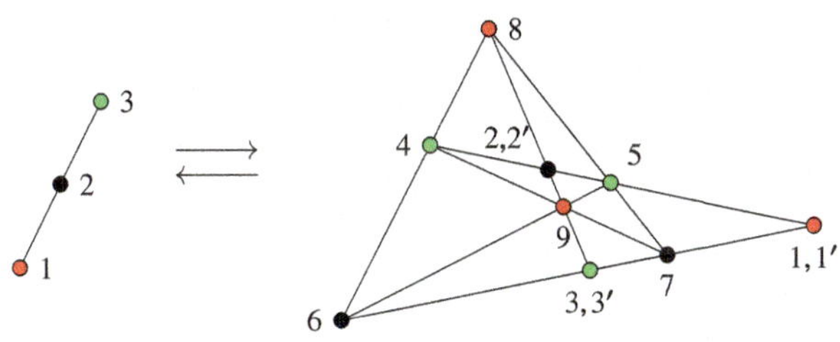

Fig. 4 The (3, 4)-multinet for the B_3 root system

5.2 $\mathcal{R}^{\geq 1}(\mathsf{M})$: *Singular Subspaces*

The Multinet Theorem 9 does not yet have a complete higher analogue. Here, we indicate some first steps in that direction, beginning with a definition from [8, §3].

Definition 6 A subspace $W \subseteq \overline{V}(\mathsf{M})$ is called *singular* if the multiplication map $\Lambda^k(W) \to A^k(\mathsf{M})$ is zero, where $k = \dim W$. The *rank* of W is the largest q for which $\Lambda^q(W) \to A^q(\mathsf{M})$ is not the zero map.

Proposition 11 *If* $\phi: \overline{A}(\mathsf{U}_{q+1,k+1}) \to \overline{A}(\mathsf{M})$ *is a graded homomorphism which is injective in degree* 1, *then the degree-*1 *part of* $\operatorname{im}\phi$ *is a singular subspace of dimension* k *and rank at most* q.

Conversely, if W *is a singular subspace of rank* q *in* $\overline{V}(\mathsf{M})$, *there exists a map* ϕ *as above for which* $W = (\operatorname{im}\phi)^1$.

Proof Let $W \subseteq \overline{V}$ be a subspace of dimension k. By inspecting the Orlik-Solomon relations (1), we can identify $\overline{A}(\mathsf{U}_{q+1,k+1})$ with a truncated exterior algebra $\Lambda(W)/(\Lambda^{q+1}(W))$.

If a map $\phi: \overline{A}(\mathsf{U}_{q+1,k+1}) \to \overline{A}(\mathsf{M})$ is given, let $W = \phi(\overline{V}(\mathsf{U}_{q+1,k+1}))$. Since $\overline{A}(\mathsf{U}_{q+1,k})^p = 0$ for $p > q$, its image W is singular of rank at most q.

Conversely, the hypothesis implies that the natural map $\Lambda(W) \to A(\mathsf{M})$ factors through $\overline{A}(\mathsf{U}_{q+1,k+1}) \cong \Lambda(W)/(\Lambda^{q+1}(W))$. □

If $\operatorname{char}\Bbbk = 0$, then components of $\mathcal{R}^1(\mathsf{M})$ are just the same as rank-1 singular subspaces, by Theorem 4. For higher rank, the situation is more subtle. If W is a singular subspace of rank q and dimension k, the condition implies that the natural homomorphism $\Lambda(W) \to \overline{A}(\mathsf{M})$ factors through $\overline{A}(U_{q+1,k+1})$, a truncated exterior algebra. If the resulting map $\overline{A}(U_{q+1,k+1}) \to \overline{A}(\mathsf{M})$ is injective in cohomology, then $W \subseteq \mathcal{R}^q(\mathsf{M})$.

Example 8 The graph G from Example 6 provides an interesting example of a singular subspace. $\mathcal{R}^1(G)$ consists of the four local components from the three-element flats. $\mathcal{R}^2(G)$ is more complicated. We find two essential components by first constructing a singular subspace. We label the vertices of G with $\{1, 2, \ldots, 5\}$ so that edge i has vertices $(i, i+1)$ for $i = 1, 2, 3$. If edge $i = \{s, t\}$ and $s < t$, let $f_i = x_t - x_s \in \Bbbk[x_1, \ldots, x_5]$. The linear forms $\{f_i : 1 \leq i \leq 8\}$ define the graphic arrangement $\mathcal{A}(G)$ with matroid $\mathsf{M}(G)$.

Following the approach of Cohen et al. [8], we observe that there is a linear relation of cubic polynomials

$$f_1f_7f_8 + f_2f_5f_8 + f_3f_5f_6 = f_4f_6f_7.$$

This implies that the polynomial mapping $\Phi: U(\mathcal{A}(G)) \subseteq \mathbb{P}^4 \to \mathbb{P}^2$ given by

$$\Phi(x) = [f_1f_7f_8 : f_2f_5f_8 : f_3f_5f_6]$$

has its image inside the projective complement of the arrangement consisting of the three coordinate hyperplanes together with the projective hyperplane orthogonal to $[1 : 1 : 1]$. The existence of the induced map in cohomology $\Phi^*: \overline{A}(\mathsf{U}_{3,4}) \to \overline{A}(G)$ shows that W is a rank-2 singular subspace, by Proposition 11, where

$$W = \{v \in \Bbbk^8 : v = (a, b, c, d, b + c, c + d, d + a, a + b) \text{ and } a + b + c + d = 0\}.$$

Below, we will show Φ^* is split, which implies that $H^2(\Phi^*)$ is injective, and $W \subseteq \mathcal{R}^2(G)$.

We proceed indirectly to show that W is maximal (i.e., a component.) First, using the cover $C = L_2^{\mathrm{irr}}(\mathsf{M}) = \{156, 267, 378, 458\}$, we obtain a 3-dimensional linear space

$$P_C = \{v \in \Bbbk^8 : v = (c + d, d + a, a + b, b + c, a, b, c, d) \text{ and } a + b + c + d = 0\}.$$

Note P_C is not maximal in $\mathcal{S}^2(G)$: for example, if $C' = \{1234, 156, 378\}$, $P_C \subsetneq P_{C'}$. Up to symmetry, though, this is the only subspace in $\mathcal{S}^2(G)$ that properly contains P_C, and it has dimension 4. By checking a single $v \in P_{C'} - P_C$, we see $P_{C'} \not\subseteq \mathcal{R}^2(G)$.

Now we note that the matroid $\mathsf{M}(G)$ is self-dual, and we may identify $\mathsf{M}(G)$ with its dual via the permutation $\sigma = [56, 784, 123]$. Since W is essential, $\sigma(W) \subset \mathcal{R}^2(G)$ as well, by Theorem 7(3). However, $\sigma(W) = P_C$, so $P_C \subseteq \mathcal{R}^2(G)$ as well. It follows that W and $\sigma(W)$ are both (maximal) linear components of $\mathcal{R}^2(G)$.

In order to try to imitate the multinet construction (Theorem 9), we give the inner four edges of G multiplicity 2, and denote the (non-simple) matroid by $\widetilde{\mathsf{M}}$. Let $\mathcal{L}$ be the partition of $\{1, 2, 3, 4, 5, 5', 6, 6', 7, 7', 8, 8'\}$ given by

$\mathcal{L} =$ 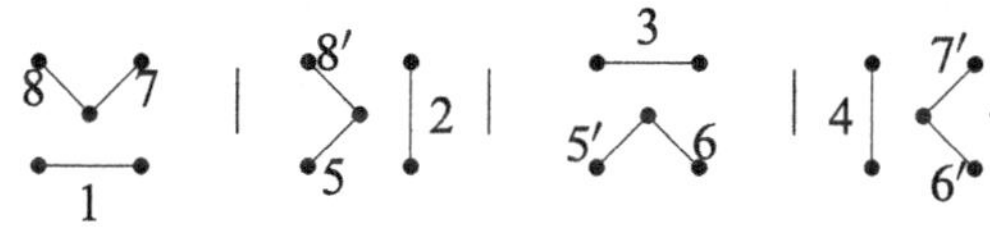

Then $W \cong Q_{\mathcal{L}} \cap \overline{V}(\widetilde{\mathsf{M}})$.

The map $A(\mathsf{U}_{3,4}) \to A(\widetilde{\mathsf{M}})$ given by sending the ith generator to the sum of elements in the ith block of $\mathcal{L}$, for $1 \leq i \leq 4$, restricts to Φ^* above. To construct a left-inverse, recall that in Example 1 we found a morphism $f: \mathsf{M} \to \mathsf{U}_{3,4}$ in $\overline{\mathcal{M}}$. A simple check shows that $-A(f) \circ \Phi^*$ is the identity on $\overline{A}(\mathsf{U}_{34})$, so Φ^* is indeed split.

We continue this example by observing that not every essential component of $\mathcal{R}^2(G)$ is a singular subspace.

Example 9 (Example 8, Continued) The second component $P_C = \sigma(W)$ can also be expressed in terms of a partition: $\sigma(W) \cong Q_{\mathcal{L}^*} \cap \overline{V}(\widetilde{\mathsf{M}})$ for another non-simple matroid $\widetilde{M}$, where

$\mathcal{L}^* =$ 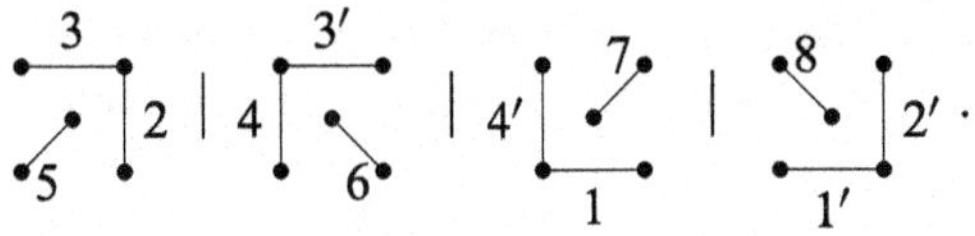

The image of $\Lambda^3(\sigma(W))$ in $\overline{A}(G)$ is nonzero, so this subalgebra generated by $\sigma(W)$ does not factor through $\overline{A}(\mathsf{U}_{3,4})$.

However, the partitions $\mathcal{L}$ and $\mathcal{L}^*$ above look qualitatively rather similar. One might hope, then, that Theorem 9 admits a combinatorial generalization that treats both components above equally.

Question 3 By the theory of multinets, every component of $\mathcal{R}^1(\mathsf{M})$ comes from the tautological resonance of a rank-2 matroid, via a split surjection of Orlik-Solomon algebras. Does every component of $\mathcal{R}^p(\mathsf{M})$ come from a matroid of rank $p+1$, for all $p \geq 1$?

Our last example shows that we cannot always find a uniform matroid with this property for $p = 2$, unlike for $p = 1$; however, since the only simple matroids of rank 2 are uniform, this should not necessarily be seen evidence that the answer is negative.

Acknowledgements The author would like to thank Hal Schenck for the ongoing conversations from which the main ideas for this paper emerged. This work was partially supported by a grant from NSERC of Canada.

References

1. A. Aramova, L.L. Avramov, J. Herzog, Resolutions of monomial ideals and cohomology over exterior algebras. Trans. Am. Math. Soc. **352**(2), 579–594 (2000)
2. D. Arapura, Geometry of cohomology support loci for local systems. I. J. Algebr. Geom. **6**(3), 563–597 (1997)
3. C. Bibby, M. Falk, I. Williams, Decomposable cocycles for p-generic arrangements, in preparation
4. N. Budur, Complements and higher resonance varieties of hyperplane arrangements. Math. Res. Lett. **18**(5), 859–873 (2011)
5. H. Cartan, S. Eilenberg, *Homological Algebra*. Princeton Landmarks in Mathematics (Princeton University Press, Princeton, 1999); With an appendix by David A. Buchsbaum, Reprint of the 1956 original
6. D.C. Cohen, Triples of arrangements and local systems. Proc. Am. Math. Soc. **130**(10), 3025–3031 (2002) (electronic)
7. D.C. Cohen, A. Dimca, P. Orlik, Nonresonance conditions for arrangements. Ann. Inst. Fourier (Grenoble) **53**(6), 1883–1896 (2003)
8. D.C. Cohen, G. Denham, M. Falk, A. Varchenko, Vanishing products of one-forms and critical points of master functions. *Arrangements of hyperplanes—Sapporo 2009*. Advanced Studies in Pure Mathematics, vol. 62 (Mathematical Society of Japan, Tokyo, 2012), pp. 75–107

9. D. Cohen, G. Denham, M. Falk, H. Schenck, A. Suciu, S. Yuzvinsky, *Complex Arrangements: Algebra, Geometry, Topology*, in preparation
10. G. Denham, The combinatorial Laplacian of the Tutte complex. J. Algebr. **242**(1), 160–175 (2001)
11. G. Denham, A. Suciu, S. Yuzvinsky, Abelian duality and propagation of resonance. arXiv:1512.07702
12. G. Denham, A.I. Suciu, Multinets, parallel connections, and Milnor fibrations of arrangements. Proc. Lond. Math. Soc. (3) **108**(6), 1435–1470 (2014)
13. A. Dimca, *Singularities and Topology of Hypersurfaces*. Universitext (Springer, New York, 1992)
14. A. Dimca, Ş. Papadima, A.I. Suciu, Topology and geometry of cohomology jump loci. Duke Math. J. **148**(3), 405–457 (2009)
15. D. Eisenbud, S. Popescu, S. Yuzvinsky, Hyperplane arrangement cohomology and monomials in the exterior algebra. Trans. Am. Math. Soc. **355**(11), 4365–4383 (2003)
16. C.J. Eschenbrenner, M.J. Falk, Orlik-Solomon algebras and Tutte polynomials. J. Algebr. Combin. **10**(2), 189–199 (1999)
17. M. Falk, Arrangements and cohomology. Ann. Comb. **1**(2), 135–157 (1997)
18. M.J. Falk, Resonance varieties over fields of positive characteristic. Int. Math. Res. Not. **2007**(3), Art. ID rnm009, 25 (2007)
19. M. Falk, Geometry and combinatorics of resonant weights. *Arrangements, Local Systems and Singularities*. Progress in Mathematics, vol. 283 (Birkhäuser, Basel, 2010), pp. 155–176
20. M. Falk, S. Yuzvinsky, Multinets, resonance varieties, and pencils of plane curves. Compos. Math. **143**(4), 1069–1088 (2007)
21. D. Grayson, M. Stillman, Macaulay2—a software system for algebraic geometry and commutative algebra. Available at http://www.math.uiuc.edu/Macaulay2/Citing/
22. A. Libgober, S. Yuzvinsky, Cohomology of the Orlik-Solomon algebras and local systems. Compos. Math. **121**(3), 337–361 (2000)
23. P. Orlik, L. Solomon, Combinatorics and topology of complements of hyperplanes. Invent. Math. **56**(2), 167–189 (1980)
24. P. Orlik, H. Terao, *Arrangements of Hyperplanes*. Grundlehren der Mathematischen Wissenschaften (Fundamental Principles of Mathematical Sciences), vol. 300 (Springer, Berlin, 1992)
25. J. Oxley, *Matroid Theory*. Oxford Graduate Texts in Mathematics, vol. 21, 2nd edn. (Oxford University Press, Oxford, 2011)
26. S. Papadima, A.I. Suciu, Toric complexes and Artin kernels. Adv. Math. **220**(2), 441–477 (2009)
27. S. Papadima, A.I. Suciu, Bieri-Neumann-Strebel-Renz invariants and homology jumping loci. Proc. Lond. Math. Soc. (3) **100**(3), 795–834 (2010)
28. S. Papadima, A.I. Suciu, The Milnor fibration of a hyperplane arrangement: from modular resonance to algebraic monodromy (2014, preprint). arXiv:1401.0868
29. V.V. Schechtman, A.N. Varchenko, Arrangements of hyperplanes and Lie algebra homology. Invent. Math. **106**(1), 139–194 (1991)
30. V. Schechtman, H. Terao, A. Varchenko, Local systems over complements of hyperplanes and the Kac-Kazhdan conditions for singular vectors. J. Pure Appl. Algebra **100**(1–3), 93–102 (1995)
31. A.I. Suciu, Around the tangent cone theorem, in *Configuration Spaces: Geometry, Topology and Representation Theory*. INdAM Series, vol. 14 (Springer, Berlin, 2016, to appear)
32. A.I. Suciu, Fundamental groups, Alexander invariants, and cohomology jumping loci. *Topology of Algebraic Varieties and Singularities*. Contemporary Mathematics, vol. 538 (American Mathematical Society, Providence, 2011), pp. 179–223
33. H. Terao, Modular elements of lattices and topological fibration. Adv. Math. **62**(2), 135–154 (1986)
34. N. White (ed.), *Theory of Matroids*. Encyclopedia of Mathematics and its Applications, vol. 26 (Cambridge University Press, Cambridge, 1986)

35. S. Yuzvinsky, Cohomology of the Brieskorn-Orlik-Solomon algebras. Commun. Algebra **23**(14), 5339–5354 (1995)
36. S. Yuzvinsky, Resonance varieties of arrangement complements. *Arrangements of Hyperplanes—Sapporo 2009*. Advanced Studies in Pure Mathematics, vol. 62 (Mathematical Society of Japan, Tokyo, 2012), pp. 553–570

Local Asymptotic Euler-Maclaurin Expansion for Riemann Sums over a Semi-Rational Polyhedron

Nicole Berline and Michèle Vergne

Abstract Consider the Riemann sum of a smooth compactly supported function $h(x)$ on a polyhedron $\mathfrak{p} \subseteq \mathbb{R}^d$, sampled at the points of the lattice $\mathbb{Z}^d/t$. We give an asymptotic expansion when $t \to +\infty$, writing each coefficient of this expansion as a sum indexed by the faces $\mathfrak{f}$ of the polyhedron, where the $\mathfrak{f}$ term is the integral over $\mathfrak{f}$ of a differential operator applied to the function $h(x)$. In particular, if a Euclidean scalar product is chosen, we prove that the differential operator for the face $\mathfrak{f}$ can be chosen (in a unique way) to involve only normal derivatives to $\mathfrak{f}$.
Our formulas are valid for a semi-rational polyhedron and a real sampling parameter t, if we allow for *step-polynomial* coefficients, instead of just constant ones.

Keywords Asymptotic Riemann sum • Euler-Maclaurin • Polyhedron • Polytope

1 Introduction

Let $\mathfrak{p} \subset \mathbb{R}^d$ be a convex polyhedron of dimension ℓ. We assume that $\mathfrak{p}$ is *semi-rational* (its facets are affine hyperplanes parallel to rational ones). Let $h(x)$ be a smooth compactly supported function on $\mathbb{R}^d$. In this article, we give an asymptotic expansion of the Riemann sum

$$\langle R_t(\mathfrak{p}), h\rangle = \frac{1}{t^\ell} \sum_{x \in t\mathfrak{p} \cap \mathbb{Z}^d} h\left(\frac{x}{t}\right),$$

when $t \to +\infty$, $t \in \mathbb{R}$.

N. Berline
École Polytechnique, Centre de Mathématiques Laurent Schwartz, 91128 Palaiseau Cedex, France
e-mail: nicole.berline@math.cnrs.fr

M. Vergne (✉)
Université Paris 7 Diderot, IMJ-PRG, Sophie Germain, 75013 Paris, France
e-mail: michele.vergne@imj-prg.fr

F. Callegaro et al. (eds.), *Configuration Spaces*, Springer INdAM Series 14,
DOI 10.1007/978-3-319-31580-5_3

The basic example is given by the classical Euler-Maclaurin expansion on a half-line $[s, +\infty[$ with $s \in \mathbb{R}$. For $x \in \mathbb{R}$, denote by $\{x\} \in [0, 1[$ the fractional part of x. For $n > 0$, and any real number $t > 0$, we have

$$\frac{1}{t}\sum_{x\in\mathbb{Z}, x\geq ts} h\left(\frac{x}{t}\right) =$$

$$\int_s^\infty h(x)dx - \sum_{k=1}^{n-1} \frac{1}{t^k}\frac{B_k(\{-ts\})}{k!} h^{(k-1)}(s) + O\left(\frac{1}{t^n}\right). \quad (1)$$

This formula can be considered as an asymptotic expansion when $t \to +\infty$, $t \in \mathbb{R}$, [with a closed expression for the remainder, see Formula (31)], where we allow the coefficient of $\frac{1}{t^k}$ to be a so-called *step-polynomial function* of t, (here, the Bernoulli polynomial computed at the fractional part $\{-ts\}$). If the end-point s is an integer and if the parameter t is also restricted to integers, (1) becomes the more familiar asymptotic expansion with constant coefficients

$$\frac{1}{t}\sum_{x\in\mathbb{Z}, x\geq ts} h\left(\frac{x}{t}\right) = \int_s^\infty h(x)dx - \sum_{k=1}^{n-1} \frac{1}{t^k}\frac{b_k}{k!} h^{(k-1)}(s) + O\left(\frac{1}{t^n}\right) \quad (2)$$

where $b_k = B_k(0)$ are the Bernoulli numbers.

The asymptotic expansion which we obtain in this paper is a generalization of (1) to any semi-rational polyhedron $\mathfrak{p}$. We prove that $R_t(\mathfrak{p})$ has an asymptotic expansion

$$R_t(\mathfrak{p}) \sim \int_{\mathfrak{p}} h(x) + \sum_{k\geq 1} a_k(t)\frac{1}{t^k}$$

where $a_k(t)$ is a step-polynomial function of t (see Definition 3), and, given a Euclidean scalar product, we write explicitly $a_k(t)$ as a sum of integrals over all the (proper) faces of $\mathfrak{p}$,

$$a_k(t) = \sum_{\mathfrak{f}} \int_{\mathfrak{f}} D_{k,\mathfrak{f},t} \cdot h, \quad (3)$$

where $D_{k,\mathfrak{f},t}$ is a differential operator of degree $k + \dim \mathfrak{f} - \dim \mathfrak{p}$ involving only derivatives normal to the face $\mathfrak{f}$, with coefficients which are step-polynomials functions of t. With these conditions, the operators $D_{k,\mathfrak{f},t}$ are unique.

If $\mathfrak{p}$ is rational, then the coefficients of the operators $D_{k,\mathfrak{f},t}$ are periodic functions of t of period q, where q is an integer, and the coefficients $a_k(t)$ of the asymptotic expansion are periodic functions of t. If moreover $\mathfrak{p}$ is a lattice polyhedron (every face contains an integral point), the period q is 1, so if, in addition, the parameter t is restricted to integral values, then the coefficients $a_k(t)$ are just constants. If $\mathfrak{p}$ is a polytope (compact polyhedron) and if $h(x)$ is a polynomial, the asymptotic formula is an exact finite expansion and coincides with our previous *local Euler-Maclaurin expansion* for polynomial functions [3]. However, we do not use this previous result, and we give an elementary proof directly in the C^∞-context.

Let us explain our approach. If $\mathfrak{p}$ is compact, the Fourier transform of the distribution $R_t(\mathfrak{p})$ is given by the holomorphic function of ξ

$$\mathcal{F}(R_t(\mathfrak{p}))(\xi) = \frac{1}{t^\ell} \sum_{x \in t\mathfrak{p} \cap \mathbb{Z}^d} \mathrm{e}^{-i\langle \frac{\xi}{t}, x\rangle}. \tag{4}$$

If $\mathfrak{p}$ is not compact, but pointed (that is $\mathfrak{p}$ does not contain any affine space), the function

$$S(\mathfrak{p})(\xi) = \sum_{x \in \mathfrak{p} \cap \mathbb{Z}^d} \mathrm{e}^{\langle \xi, x\rangle}$$

can still be defined as a meromorphic function with simple real hyperplane singularities around $\xi = 0$. The Fourier transform $\mathcal{F}(R_t(\mathfrak{p}))(\xi)$ is a generalized function equal to a boundary value of the meromorphic function $\frac{1}{t^\ell} S(t\mathfrak{p})(-i\frac{\xi}{t})$. The archetype is $\mathfrak{p} = [0, \infty[$ where

$$\mathcal{F}(R_t(\mathfrak{p}))(\xi) = \frac{1}{t} \lim_{\epsilon \to 0, \epsilon > 0} \frac{1}{1 - \mathrm{e}^{-i\frac{\xi}{t} - \epsilon}}. \tag{5}$$

When $\mathfrak{p}$ is a pointed affine cone, we obtain a canonical asymptotic expansion of the Fourier transform $\mathcal{F}(R_t(\mathfrak{p}))(\xi)$ in terms of the Laurent series of $S(t\mathfrak{p})(\xi)$ at $\xi = 0$, (Theorem 1 in the case of cone with vertex 0, Theorem 6 for an affine cone with any real vertex). This result is easily reduced to the one-dimensional Euler-Maclaurin formula, by subdividing a cone into unimodular ones. Nevertheless, it is the most important observation of this paper.

If the support of the test function is small enough, the Riemann sum for a polyhedron is actually a Riemann sum on one of its supporting cones. Such a cone is a product of a linear space and of a pointed affine cone. So, by a partition of unity argument, the basic result we need is for a pointed affine cone $\mathfrak{c}$. In this case, by Fourier transform, Eq. (3) is equivalent to a decomposition (depending of a choice of a scalar product) of $S(\mathfrak{c})$ into a sum of terms indexed by the faces $\mathfrak{f}$ of $\mathfrak{c}$, where each term is a product of a term involving $\mathfrak{f}$ and a term involving the transverse cone to $\mathfrak{c}$ along $\mathfrak{f}$. This expression for $S(\mathfrak{c})$ (Theorem 2) was already obtained in [3], but we give here a simpler proof in the Appendix. The easier case of simplicial cones is explained in Sect. 4.2.

Our motivation was to clarify and simplify some previous results on such asymptotic expansions, [10–12]. In [12], Tate obtained an asymptotic expansion with normal derivatives as in (3) for a lattice polytope and an integral parameter t. In [11], Le Floch and Pelayo showed in small dimensions how to derive Tate's result from the Todd operator asymptotic formula of V. Guillemin and S. Sternberg [10].

Along the way, we realized that our method permits nice generalizations.

First, our formula is valid for any polyhedron, we do not need to assume that it is compact, nor even pointed. Another generalization is to consider a real parameter t instead of an integer for the Riemann sum. Even if we start with a rational polyhedron $\mathfrak{p}$, the dilated polyhedron $t\mathfrak{p}$ is only semi-rational, so we assume only that $\mathfrak{p}$ is semi-rational.

The proofs in the case of a semi-rational polyhedron and a real parameter t are very similar to those for a lattice polyhedron and an integral parameter, so we give the details only in that case.

The differential operators in our formula are those which we constructed in [3]. There, for a face $\mathfrak{f}$ of the polyhedron $\mathfrak{p}$, the symbol of operator corresponding to the face $\mathfrak{f}$ is the μ-function of the transverse cone to $\mathfrak{p}$ along the face $\mathfrak{f}$. S. Paycha observed that this μ-function can be defined by a more intuitive method of algebraic renormalization [8]. Although this renormalization construction is easy for the generating function of a simplicial cone, we explain it, in the Appendix, in the more general context of rational functions with hyperplane singularities, on any base field. For computing the μ-function of a cone, we wrote a Maple program [2, with V. Baldoni].

We had to leave open the problem of writing a closed formula for the remainder at order n of the asymptotic expansion, similar to the one dimensional remainder in (31), in the spirit of the present article, (a remainder for a simple polytope, in the spirit of the Todd operator formula, is obtained in [10]). We only give a small computation in Example 10.

2 Notations and Basic Facts

2.1 Various Notations

2.1.1 Vector Spaces

V is a finite dimensional vector space over $\mathbb{R}$ with a lattice Λ. The dimension of V is denoted by d. The Lebesgue measure dx on V is determined by Λ.

Elements of V are denoted by Latin letters $x, y, v \ldots$ and elements of the dual space V^* are denoted by Greek letters $\xi, \gamma \ldots$. The pairing between V and V^* is denoted by $\langle \xi, x \rangle$.

2.1.2 Bernoulli Numbers and Polynomials

$$\frac{z e^{sz}}{e^z - 1} = \sum_{n=0}^{\infty} B_n(s) \frac{z^n}{n!}, \quad \frac{z}{e^z - 1} = \sum_{n=0}^{\infty} b_n \frac{z^n}{n!}.$$

$$B_0(s) = 1, B_1(s) = s - \frac{1}{2}, B_2(s) = s^2 - s + \frac{1}{6}, B_3(s) = s^3 - \frac{3}{2}s^2 + \frac{1}{2}s.$$

$$b_0 = 1, \quad b_1 = -\frac{1}{2}, \quad b_2 = \frac{1}{6}, \quad b_{2n+1} = 0 \text{ for } n \geq 1.$$

2.1.3 Fractional Part of a Real Number

For $s \in \mathbb{R}$, the fractional part $\{s\}$ is defined by $\{s\} \in [0, 1[, s - \{s\} \in \mathbb{Z}$.

2.1.4 Fourier Transform

The Fourier transform of a distribution on V is a generalized function on V^*. Our convention for the Fourier transform of a test density $\phi(x)dx$ is

$$\mathcal{F}(\phi dx)(\xi) = \int_V \mathrm{e}^{-i\langle \xi, x\rangle} \phi(x) dx.$$

2.2 *Polyhedra, Cones*

2.2.1 Convex Polyhedron

A *convex polyhedron* $\mathfrak{p}$ in V (we will simply say *polyhedron*) is, by definition, the intersection of a finite number of closed half spaces bounded by affine hyperplanes. If the hyperplanes are rational, we say that the polyhedron is *rational*. If the hyperplanes have rational directions, we say that the polyhedron is *semi-rational*. For instance, if $\mathfrak{p} \subset V$ is a rational polyhedron, t is a real number and s is any point in V, then the dilated polyhedron $t\mathfrak{p}$ and the translated polyhedron $s + \mathfrak{p}$ are semi-rational.

The *lineality space* of a polyhedron $\mathfrak{p}$ is the subspace of $y \in V$ such that $x + \mathbb{R}y \subseteq \mathfrak{p}$ for all $x \in \mathfrak{p}$. A polyhedron is called pointed when its lineality space is $\{0\}$.

In this article, a cone is a rational polyhedral cone (with vertex 0) and an affine cone is a translated set $s + \mathfrak{c}$ of a cone $\mathfrak{c}$ by any element $s \in V$. A cone is called *simplicial* if it is generated by independent elements of Λ. A cone is called *unimodular* if it is generated by independent elements $v_1, \dots, v_k$ of Λ such that $\{v_1, \dots, v_k\}$ can be completed to an integral basis of Λ. An affine cone $\mathfrak{a}$ is called *simplicial* (resp. *simplicial unimodular*) if the associated cone is.

For a polyhedron $\mathfrak{f}$, the affine span of $\mathfrak{f}$ is denoted by $\langle \mathfrak{f} \rangle$ and the corresponding linear space is denoted by $\mathrm{lin}(\mathfrak{f})$.

If $\mathfrak{f}$ is a face of a polyhedron $\mathfrak{p}$, the lineality space of the supporting cone of $\mathfrak{p}$ along $\mathfrak{f}$ is just $\mathrm{lin}(\mathfrak{f})$. The projected cone in the quotient space $V/\mathrm{lin}(\mathfrak{f})$ is a pointed cone called the *transverse cone* of $\mathfrak{p}$ along $\mathfrak{f}$.

Definition 1 The supporting cone of $\mathfrak{p}$ along $\mathfrak{f}$ is denoted by $C(\mathfrak{p}, \mathfrak{f})$. The transverse cone of $\mathfrak{p}$ along $\mathfrak{f}$ is denoted by $\mathfrak{N}(\mathfrak{p}, \mathfrak{f})$.

The Lebesgue measure on $\langle \mathfrak{f} \rangle$ is determined by the intersection lattice $\mathrm{lin}(\mathfrak{f}) \cap \Lambda$. It is denoted by $dm_{\mathfrak{f}}$.

2.2.2 Subdivision of a Cone into Unimodular Ones

A subdivision of a cone $\mathfrak{c}$ is a family of cones $\mathfrak{c}_\alpha$ contained in $\mathfrak{c}$, such that the intersection of any two of them is a face of both and belongs to the family and such that $\mathfrak{c}$ is the union of the family. Then the indicator $[\mathfrak{c}]$ is equal to a linear combination $[\mathfrak{c}] = \sum_\alpha n_\alpha [\mathfrak{c}_\alpha]$ with $n_\alpha \in \mathbb{Z}$. It follows from Minkowski's theorem that any cone has a subdivision into unimodular cones.

2.3 Discrete and Continuous Generating Functions of a Pointed Polyhedron $S(\mathfrak{p}), I(\mathfrak{p})$

If $\mathfrak{p} \subset V$ is a pointed polyhedron, there is an $a > 0$ and a non empty open set $U \subseteq V^*$ such that for $\xi \in U$ and $x \in \mathfrak{p}$ large enough, $\langle \xi, x \rangle \leq -a\|x\|$. Then the functions

$$S(\mathfrak{p})(\xi) = \sum_{x \in \mathfrak{p} \cap \Lambda} \mathrm{e}^{\langle \xi, x \rangle}, \quad I(\mathfrak{p})(\xi) = \int_{\mathfrak{p}} \mathrm{e}^{\langle \xi, x \rangle} dm_{\mathfrak{p}}(x)$$

are defined and holomorphic on $U + iV^* \subseteq V_{\mathbb{C}}^*$, and have meromorphic continuation to the whole of $V_{\mathbb{C}}^*$. If $\mathfrak{p}$ is bounded, these functions are holomorphic.

If $\mathfrak{c}$ is a simplicial cone with edge generators $v_1, \ldots, v_\ell \in \Lambda$, and $s \in V$, one has

$$I(s + \mathfrak{c})(\xi) = (-1)^\ell \mathrm{e}^{\langle \xi, s \rangle} \frac{|\det_\Lambda(v_j)|}{\prod_j \langle \xi, v_j \rangle},$$

where the determinant is relative to the intersection lattice $\mathrm{lin}(\mathfrak{c}) \cap \Lambda$, and

$$S(s + \mathfrak{c})(\xi) = S(s + \mathfrak{b})(\xi) \prod_j \frac{1}{1 - \mathrm{e}^{\langle \xi, v_j \rangle}}$$

where $S(s + \mathfrak{b})(\xi)$ is the holomorphic function

$$S(s + \mathfrak{b})(\xi) = \sum_{x \in (s + \sum_j [0,1[v_j) \cap \Lambda} \mathrm{e}^{\langle \xi, x \rangle}.$$

If moreover $\mathfrak{c}$ is unimodular and the v_j's are primitive vectors, then, for the vertex $s = 0$,

$$S(\mathfrak{c})(\xi) = \frac{1}{\prod_j (1 - e^{\langle \xi, v_j \rangle})}, \quad I(\mathfrak{c})(\xi) = (-1)^\ell \frac{1}{\prod_j \langle \xi, v_j \rangle}.$$

If $\mathfrak{c}$ is simplicial, the function $g(\xi) = (\prod_{j=1}^{\ell} \langle \xi, v_j \rangle) S(s + \mathfrak{c})(\xi)$ is holomorphic near $\xi = 0$, in other words, $S(s + \mathfrak{c})(\xi)$ has *simple hyperplane singularities* defined by its edges, near $\xi = 0$. If ξ is not in any of these singular hyperplanes, the function $z \mapsto S(s + \mathfrak{c})(z\xi)$ is meromorphic, with Laurent series around $z = 0$

$$S(s + \mathfrak{c})(z\xi) = \sum_{m=-\ell}^{\infty} S(s + \mathfrak{c})_{[m]}(\xi) z^m,$$

where $S(s + \mathfrak{c})_{[m]}(\xi) = \frac{g_{[m+\ell]}(\xi)}{\prod_{j=1}^{\ell} \langle \xi, v_j \rangle}$ is a homogeneous rational fraction of degree m which we call the homogeneous component of degree m of $S(s + \mathfrak{c})(\xi)$. We write formally

$$S(s + \mathfrak{c})(\xi) = \sum_{m=-\ell}^{\infty} S(s + \mathfrak{c})_{[m]}(\xi). \tag{6}$$

If $\mathfrak{c}$ is no longer assumed simplicial, then $\mathfrak{c}$ can be subdivided into simplicial cones without adding edges. Therefore, again, $S(s + \mathfrak{c})(\xi)$ has simple hyperplane singularities near $\xi = 0$ (defined by the edges) and the decomposition into homogeneous components (6) still holds.

3 Asymptotic Expansions for Riemann Sums over a Cone

3.1 Dimension One

We recall the dimension one Euler-Maclaurin summation formula for a test function $h(x)$ on a half line, [5, Theorem 9.2.2]. For the moment, we consider only the half-line $[0, +\infty[$.

$$\sum_{x \in \mathbb{Z}, x \geq 0} h(x) =$$

$$\int_0^\infty h(x) dx - \sum_{k=1}^{n} \frac{b_k}{k!} h^{(k-1)}(0) - \int_0^\infty \frac{B_n(\{-x\})}{n!} h^{(n)}(x)\, dx. \tag{7}$$

We write the whole right hand side as an integral over the half line.

$$\sum_{x\in\mathbb{Z},x\geq 0} h(x) = \int_0^\infty \left(h(x)+\sum_{k=1}^{n}\frac{b_k}{k!}h^{(k)}(x)-\frac{B_n(\{-x\})}{n!}h^{(n)}(x)\right)dx. \quad (8)$$

For $t>0$, consider the scaled function

$$h_t(x)=\frac{1}{t}h\left(\frac{x}{t}\right).$$

Substituting h_t for h in this formula, and changing variables in the integral on the right-hand-side, we obtain, for any $t>0$,

$$\frac{1}{t}\sum_{x\in\mathbb{Z},x\geq 0} h\left(\frac{x}{t}\right) =$$

$$\int_0^\infty \left(h(x)+\sum_{k=1}^{n-1}\frac{1}{t^k}\frac{b_k}{k!}h^{(k)}(x)+\frac{1}{t^n}\frac{b_n-B_n(\{-tx\})}{n!}h^{(n)}(x)\right)dx. \quad (9)$$

(9) gives the asymptotic expansion when $t\to+\infty$,

$$\frac{1}{t}\sum_{x\in\mathbb{Z},x\geq 0} h\left(\frac{x}{t}\right)\sim\sum_{k=0}^{\infty}\frac{1}{t^k}\int_0^\infty\frac{b_k}{k!}h^{(k)}(x),$$

with a closed formula for the remainder.

3.2 Fourier Transforms and Boundary Values

For a polyhedron $\mathfrak{p}\subset V$, consider the distributions given on a test function h on V by $\int_{\mathfrak{p}} h(x)dx$ and $\sum_{x\in\mathfrak{p}\cap\Lambda} h(x)$. Clearly, they are tempered distributions, therefore we can consider their Fourier transforms.

We recall a well known result on Fourier transforms.

Definition 2 If $F(\xi)$ is a rational function on V^* whose denominator $g(\xi)$ is a product of linear forms and if $\lambda\in V^*$ is such that $g(\lambda)\neq 0$, the following formula defines a tempered generalized function on V^*:

$$\lim_{\lambda}(F(\xi)) = \lim_{\epsilon\to 0,\epsilon>0} F(\xi+i\epsilon\lambda).$$

Example 1 Let v be a non zero vector in V and let λ be such that $\langle \lambda, v \rangle < 0$. Then, $\lim_\lambda(\frac{1}{i\langle \xi, v \rangle})$ is the Fourier transform of the Heaviside distribution of the half-line $\mathbb{R}_{\geq 0} v$, given by $\int_0^\infty h(tv)dt$. When $V = \mathbb{R}$, the Fourier transform of the Heaviside distribution of the half-line $\mathbb{R}_{\geq 0}$ is more often denoted by $\frac{1}{i(\xi - i0)}$.

Of course, such boundary values are defined for more general types of meromorphic functions on $V_{\mathbb{C}}^*$. In this paper we will need only the rational functions of Definition 2 and the generating functions of cones. The following result is immediate.

Proposition 1 *Let $\mathfrak{c} \subset V$ be a pointed cone (with vertex 0). Let $\lambda \in V^*$ be any element such that $-\lambda$ lies in the dual cone of $\mathfrak{c}$. Then,*

(i) The boundary value

$$\lim_\lambda (I(\mathfrak{c})(-i\xi)) = \lim_{\epsilon \to 0, \epsilon > 0} I(\mathfrak{c})(-i\xi + \epsilon\lambda)$$

exists and it is the Fourier transform of the distribution $\int_{\mathfrak{c}} h(x)dx$.

(ii) The boundary value

$$\lim_\lambda (S(\mathfrak{c})(-i\xi)) = \lim_{\epsilon \to 0, \epsilon > 0} S(\mathfrak{c})(-i\xi + \epsilon\lambda)$$

exists and it is the Fourier transform of the distribution $\sum_{x \in \mathfrak{c} \cap \Lambda} h(x)$.

In particular, these boundary values do not depend on the choice of $-\lambda$ in the dual cone.

Example 2 For $\mathfrak{c} = \mathbb{R}_{\geq 0}$, the Fourier transform of the distribution $h \mapsto \sum_{n=0}^\infty h(n)$ is the boundary value $\lim_{\epsilon \to 0, \epsilon > 0} \frac{1}{1 - e^{-i\xi - \epsilon}}$.

3.3 Asymptotic Expansion of the Fourier Transform of a Riemann Sum over a Cone

For a polyhedron $\mathfrak{p} \subseteq V$, with $\dim \mathfrak{p} = \ell$, we consider the Riemann sum

$$\langle R_t(\mathfrak{p}), h \rangle = \frac{1}{t^\ell} \sum_{x \in t\mathfrak{p} \cap \Lambda} h\left(\frac{x}{t}\right). \tag{10}$$

If $\mathfrak{c}$ is a pointed cone of dimension ℓ, it follows immediately from Proposition 1 that the Fourier transform of $R_t(\mathfrak{c})$ is given by

$$\mathcal{F}(R_t(\mathfrak{c}))(\xi) = \frac{1}{t^\ell} \lim_\lambda \left(S(\mathfrak{c}) \left(-i\frac{\xi}{t} \right) \right). \tag{11}$$

When $t \to \infty$, $\frac{\xi}{t} \to 0$. Replacing ξ by $-i\frac{\xi}{t}$ in the formal expansion into homogeneous components $S(\mathfrak{c})(\xi) = \sum_{k=-\ell}^{\infty} S(\mathfrak{c})_{[k]}(\xi)$, we obtain formally

$$\frac{1}{t^\ell} S(\mathfrak{c})\left(-i\frac{\xi}{t}\right) = \sum_{k=0}^{\infty} \frac{1}{t^k} S(\mathfrak{c})_{[k-\ell]}(-i\xi). \tag{12}$$

It is a remarkable fact that this formal expansion leads to an asymptotic expansion of boundary values. We are going to derive it from the dimension one Euler-Maclaurin formula.

Theorem 1 *Let $\mathfrak{c} \subset V$ be a pointed cone of dimension ℓ. Let $\lambda \in V^*$ be such that $-\lambda$ lies in the dual cone of $\mathfrak{c}$. Consider the distribution on V given by the Riemann sum*

$$\langle R_t(\mathfrak{c}), h\rangle = \frac{1}{t^\ell} \sum_{x \in \mathfrak{c} \cap \Lambda} h\left(\frac{x}{t}\right).$$

It has an asymptotic expansion when $t \to \infty$, the Fourier transform of which is given by

$$\mathcal{F}(R_t(\mathfrak{c})) = t^{-\ell} \lim_{\lambda} \left(S(\mathfrak{c}) \left(-i\frac{\xi}{t}\right)\right) \sim \sum_{k=0}^{\infty} \frac{1}{t^k} \lim_{\lambda} S(\mathfrak{c})_{[k-\ell]}(-i\xi). \tag{13}$$

Proof Let us first look at the case where $\mathfrak{c} = \mathbb{R}_{\geq 0}$. By writing the Fourier transform of the asymptotic expansion (9), we obtain immediately (14) below. (This is the reason why we rewrote Euler-Maclaurin formula (8) as an integral over the half-line).

$$\mathcal{F}(R_t(\mathbb{R}_{\geq 0})) \sim \sum_{k=0}^{\infty} \frac{1}{t^k} \frac{b_k}{k!} \frac{(-i\xi)^k}{i(\xi - i0)}. \tag{14}$$

On the other hand, we write the well known Laurent series

$$S(\mathbb{R}_{\geq 0})(\xi) = \frac{1}{1-e^\xi} = \left(\frac{-1}{\xi}\right) \sum_{k=0}^{\infty} \frac{b_k}{k!} \xi^k,$$

hence

$$S(\mathbb{R}_{\geq 0})_{[k-1]}(-i\xi - \epsilon) = \frac{b_k}{k!} \frac{(-i\xi - \epsilon)^k}{i\xi + \epsilon}.$$

This proves the theorem in the case where $\mathfrak{c} = \mathbb{R}_{\geq 0}$, and so when $\mathfrak{c}$ has dimension one. For the general case, we consider a subdivision of $\mathfrak{c}$ into unimodular cones $\mathfrak{c}_\alpha$, so that $[\mathfrak{c}] = \sum_\alpha n_\alpha [\mathfrak{c}_\alpha]$. The cones $\mathfrak{c}_\alpha$ are contained in $\mathfrak{c}$, so that $-\lambda$ belongs to the dual cone of $\mathfrak{c}_\alpha$ as well. Let ℓ_α be the dimension of $\mathfrak{c}_\alpha$. For a test function h and $t > 0$, we have

$$\langle R_t(\mathfrak{c}), h\rangle = \frac{1}{t^\ell} \sum_{x\in\mathfrak{c}\cap\Lambda} h\left(\frac{x}{t}\right) = \sum_\alpha n_\alpha \frac{1}{t^\ell} \sum_{x\in\mathfrak{c}_\alpha\cap\Lambda} h\left(\frac{x}{t}\right)$$

$$= \sum_\alpha n_\alpha \frac{1}{t^{\ell-\ell_\alpha}} \langle R_t(\mathfrak{c}_\alpha), h\rangle.$$

Thus we may assume that $\mathfrak{c}$ is unimodular. In this case the theorem follows immediately from the dimension one case. □

3.4 Asymptotic Expansions of Riemann Sums over Cones in Terms of Differential Operators

In this section, we explain how one obtains formulas of Euler-Maclaurin type for the asymptotic expansion of the Riemann sum itself, by taking inverse Fourier transforms in Theorem 1.

By Theorem 1, the coefficients of the asymptotic expansion of the Riemann sum are given by

$$\langle R_t(\mathfrak{c}), h\rangle \sim \sum_{k=0}^{\infty} \frac{1}{t^k} \langle F_k, h\rangle, \text{ with } F_k = \mathcal{F}^{-1}(\lim_\lambda (S(\mathfrak{c})_{[k-\ell]}(-i\xi))). \tag{15}$$

For $k = 0$, we have $S(\mathfrak{c})_{[-\ell]} = I(\mathfrak{c})$, and, (Proposition 1,(i)),

$$\langle \mathcal{F}^{-1}(\lim_\lambda I(\mathfrak{c})(-i\xi)), h\rangle = \int_\mathfrak{c} h(x) dm_\mathfrak{c}(x).$$

So we recover the fact that $\langle F_0, h\rangle$ is the integral of h over $\mathfrak{c}$ (with respect to the Lebesgue measure $dm_\mathfrak{c}(x)$ defined by the intersection lattice $\mathrm{lin}(\mathfrak{c}) \cap \Lambda$). For $k = 1$, it is well known that

$$\langle F_1, h\rangle = \frac{1}{2} \int_{\partial\mathfrak{c}} h$$

where $\partial\mathfrak{c}$ denotes the boundary of $\mathfrak{c}$, union of the facets of $\mathfrak{c}$, and the Lebesgue measure on each facet $\mathfrak{f}$ is again defined by the intersection lattice $\mathrm{lin}(\mathfrak{f}) \cap \Lambda$. This formula is true for any lattice polyhedron. For a unimodular cone, it follows easily from (26). If the cone is not unimodular, we do a subdivision, as in the proof of Theorem 1.

Let $(v_1, \dots, v_n)$ be the generators of the edges of $\mathfrak{c}$. The homogeneous component $S(\mathfrak{c})_{[k-\ell]}(\xi)$ can be written (in many ways) as a sum

$$S(\mathfrak{c})_{[k-\ell]}(\xi) = \sum_J \frac{P_J(\xi)}{\prod_{j\in J}\langle \xi, v_j\rangle} \tag{16}$$

where $J \subseteq (1, \dots, n)$ is such that $(v_j, j \in J)$ are linearly independent, and P_J is a homogeneous polynomial of degree $k - \ell + |J|$. Taking inverse Fourier transform of boundary values, using again Proposition 1(i), we obtain

$$\langle F_k, h\rangle = \sum_J a_J \int_{\mathfrak{c}_J} P_J\left(\frac{\partial}{\partial x}\right) \cdot h(x) dm_{\mathfrak{c}_J}(x),$$

where $\mathfrak{c}_J$ is the cone generated by $(v_j, j \in J)$ and $a_J = \frac{1}{|\det(v_j, j\in J)|}$ (determinant with respect to the intersection lattice on the subspace spanned by $(v_j, j \in J)$).

Example 3 Let $V = \mathbb{R}^2$ with lattice $\mathbb{Z}^2$, standard basis (e_1, e_2) and coordinates x_1, x_2. Let $\mathfrak{c}$ be the cone generated by $v_1 = e_1$, $v_2 = e_1 + e_2$, with edges $\mathfrak{f}_1 = \mathbb{R}_{\geq 0} v_1$, $\mathfrak{f}_2 = \mathbb{R}_{\geq 0} v_2$. The homogeneous components of $S(\mathfrak{c})$ can be computed by taking the product of the Todd series

$$\frac{1}{1 - \mathrm{e}^{v_j}} = -\frac{1}{v_j}\left(1 + \sum_{k\geq 1} \frac{b_k}{k!} v_j^k\right).$$

$$\begin{aligned} S(\mathfrak{c})(\xi_1, \xi_2) &= \frac{1}{(1 - \mathrm{e}^{\xi_1})(1 - \mathrm{e}^{\xi_1+\xi_2})} = \frac{1}{\xi_1(\xi_1 + \xi_2)} - \frac{1}{2\xi_1} - \frac{1}{2(\xi_1 + \xi_2)} \\ &\quad + \frac{1}{4} + \frac{1}{12}\frac{\xi_1 + \xi_2}{\xi_1} + \frac{1}{12}\frac{\xi_1}{\xi_1 + \xi_2} + \cdots . \end{aligned}$$

We write the homogeneous component of degree 0 in two different ways.

$$S(\mathfrak{c})_{[0]}(\xi_1, \xi_2) = \frac{1}{4} + \frac{1}{12}\frac{\xi_1 + \xi_2}{\xi_1} + \frac{1}{12}\frac{\xi_1}{\xi_1 + \xi_2} \tag{17}$$

$$= \frac{3}{8} + \frac{1}{12}\frac{\xi_2}{\xi_1} + \frac{1}{24}\frac{\xi_1 - \xi_2}{\xi_1 + \xi_2}. \tag{18}$$

The corresponding expressions for the term F_2 in the asymptotic expansion (15) of $R_t(\mathfrak{c})$ are

$$\langle F_2, h\rangle = \frac{1}{4}h(0,0) - \frac{1}{12}\int_{\mathfrak{f}_1} \partial_{v_2} h - \frac{1}{12}\int_{\mathfrak{f}_2} \partial_{v_1} h \tag{19}$$

$$= \frac{3}{8}h(0,0) - \frac{1}{12}\int_{\mathfrak{f}_1} \partial_{x_2} h - \frac{1}{24}\int_{\mathfrak{f}_2} (\partial_{x_1} - \partial_{x_2})h. \tag{20}$$

4 Simplicial Cones and Normal Derivatives Formula

4.1 Asymptotic Expansions of Riemann Sums and Todd Operator

In this section, we compare Theorem 1 with Guillemin-Sternberg formula for a unimodular cone. This comparison will not be used in this article.

Let $\mathfrak{c}$ be a unimodular cone with primitive edge generators v_j. One can compute the distributions F_k by expanding $S(\mathfrak{c}) = \prod_j \frac{1}{1-e^{v_j}}$ as a product of Todd series. It is easy to see that the resulting expansion can be written in a unique way in the form

$$\sum_{m\geq 0, J\subseteq(1,\ldots,\ell)} \frac{G_{m,J}}{\prod_{j\in J} v_j},$$

where $G_{m,J}(\xi)$ is a homogeneous polynomial of degree m belonging to the algebra generated by $v_k, k \notin J$. Our Theorem 1 gives thus the following explicit formula:

$$\frac{1}{t^\ell}\sum_{x\in\mathfrak{c}\cap\Lambda} h\left(\frac{x}{t}\right) \sim \sum_{k=0}^{\infty}\frac{1}{t^k}\sum_{\substack{m\geq 0, J\\ m+\ell-|J|=k}} \int_{\mathfrak{f}_J} G_{m,J}\left(\frac{\partial}{\partial_x}\right)\cdot h(x)\, dm_{\mathfrak{f}_J}(x). \tag{21}$$

Here $\mathfrak{f}_J$ is the face generated by the edges $(v_j, j \in J)$, the corresponding constant coefficient differential operator $G_{m,J}(\partial)$ is homogeneous of degree m and belongs to the algebra generated by $(\partial_{v_k}, k \notin J)$.

Le Floch and Pelayo [11] observed that (21) follows from Guillemin-Sternberg formula for the asymptotic expansion on a Riemann sum over the unimodular cone $\mathfrak{c}$, in terms of a Todd operator [7, Theorem 3.2]. Let us recall this formula. The unimodular cone $\mathfrak{c}$ with primitive edge generators v_j is defined by the inequalities $\langle\lambda_j, x\rangle \leq 0$, where λ_j is the basis of the lattice Λ^* dual to v_j. Let $\mathfrak{c}(a)$ be the affine cone defined by the inequalities $\langle\lambda_j, x\rangle \leq a_j$. Let

$$Todd(z) = \frac{z}{1-e^{-z}} = \sum_{k=0}^{\infty}(-1)^k\frac{b_k}{k!}z^k.$$

Consider the formal power series (in powers of $\frac{1}{t}$) of constant coefficients differential operators:

$$Todd\left(\frac{1}{t}\frac{\partial}{\partial_a}\right)=\prod_{j=1}^{d} Todd\left(\frac{1}{t}\frac{\partial}{\partial_{a_j}}\right).$$

Let $h(x)$ be a test function on V. Then, when $t\to\infty$, Guillemin-Sternberg state that

$$\frac{1}{t^d}\sum_{x\in\mathfrak{c}\cap\Lambda} h\left(\frac{x}{t}\right)\sim\left(Todd\left(\frac{1}{t}\frac{\partial}{\partial_a}\right)\cdot\int_{\mathfrak{c}(a)} h(x)dx\right)|_{a=0}. \tag{22}$$

As $\mathfrak{c}(a)=\sum_j a_j v_j+\mathfrak{c}$, one sees easily that (22) is precisely the Fourier transform of (13).

4.2 Local Euler-Maclaurin Formula

Our method in the present article is to give expressions of type (20), in terms of derivatives normal to the faces with respect to a given Euclidean scalar product. The method is much simpler for simplicial cones, so we sketch it now in this case. We will discuss directly the non simplicial case in the next section, and the result of this section are not needed. However, it might be useful to understand this case.

Let $(v_1,v_2,\ldots,v_d)$ be a basis of V. For any subset J of $\{1,2,\ldots,d\}$, let L_J be the subspace of V spanned by $(v_j, j\in J)$ and let $C_J\subseteq V$ be its orthogonal component (for the given scalar product). Let K be a subset of $\{1,2,\ldots,d\}$. It is easy to see that any rational function of the form $R(\xi)=\frac{P(\xi)}{\prod_{j\in K}\langle\xi,v_j\rangle}$ can be written in a unique way as $R(\xi)=\sum_{J\subseteq K}\frac{P_J(\xi)}{\prod_{j\in J}\langle\xi,v_j\rangle}$ with $P_J\in \mathrm{Sym}(C_J)$, the symmetric algebra of C_J. Indeed, write $P=Q+\sum_{j\in K}Q_jv_j$, with $Q\in\mathrm{Sym}(C_K)$. Then $R=\frac{Q}{\prod_{j\in K}v_j}+\sum_{j\in K}\frac{Q_j}{\prod_{k\in K,k\neq j}v_k}$. Then iterate on each term $\frac{Q_j}{\prod_{k\in J,k\neq j}v_k}$. The uniqueness of the P_J's is also easily proved by induction on $|K|$, by taking partial residues.

Now if $\mathfrak{c}$ is a simplicial cone, with generators $v_1,\ldots,v_\ell$, any homogeneous component $S(\mathfrak{c})_{[k-\ell]}(\xi)$ is of the form $\frac{P(\xi)}{\prod_{j=1}^{\ell}\langle\xi,v_j\rangle}$. By the preceding discussion, $S(\mathfrak{c})_{[k-\ell]}(\xi)=\sum_{J\subseteq\{1,\ldots,\ell\}}\frac{P_{k,J}}{\prod_{j\in J}v_j}$ with $P_{k,J}\in\mathrm{Sym}(C_J)$. For any $J\subseteq\{1,2,\ldots,\ell\}$, the cone generated by $(v_j, j\in J)$ is a face $\mathfrak{f}$ of $\mathfrak{c}$ and $I(\mathfrak{f})$ is equal to $(-1)^{|J|}\frac{|\det(v_j,j\in J)|}{\prod_{j\in J}v_j}$. Gathering the homogeneous terms face by face, we obtain the following decomposition of $S(\mathfrak{c})(\xi)=\sum_{k\geq-\ell}S(\mathfrak{c})_{[k]}(\xi)$, indexed by all the faces $\mathfrak{f}$ of $\mathfrak{c}$.

Lemma 1

$$S(\mathfrak{c})(\xi) = \sum_{\mathfrak{f}} M_{\mathfrak{f}}(\xi) I(\mathfrak{f})(\xi)$$

where $M_{\mathfrak{f}}$ is a holomorphic function of ξ depending only of the orthogonal projection of ξ on the subspace $\mathfrak{f}^{\perp}$ of V^*.

Remark 1 Given our Euclidean scalar product, the decomposition of $S(\mathfrak{c})(\xi)$ in Lemma 1 is unique. Thus the holomorphic term $M_{\mathfrak{f}}$ is the μ-function of the transverse cone $\mathfrak{N}(\mathfrak{c}, \mathfrak{f})$ to the face $\mathfrak{f}$ defined in [3]. In particular, the term corresponding to $\mathfrak{f} = \{0\}$ is the μ-function of the cone $\mathfrak{c}$. In the Appendix (Theorem 9), we establish this formula directly for any cone, while in [3], we showed a valuation property for the μ function allowing us to deduce the case of a general cone from the case of a simplicial one.

Write $M_{\mathfrak{f}} = \sum M_{m,\mathfrak{f}}$ as a sum of its homogeneous terms. We obtain the following corollary.

Lemma 2 *Consider an Euclidean scalar product on V. If $\mathfrak{c}$ is a simplicial cone, there exists unique constant coefficients differential operators $M_{m,\mathfrak{f}}(\frac{\partial}{\partial x})$, homogeneous of degree m, involving only derivatives normal to the face $\mathfrak{f}$ such that*

$$\frac{1}{t^{\ell}} \sum_{x \in \mathfrak{c} \cap \Lambda} h\left(\frac{x}{t}\right) \sim \sum_{k=0}^{\infty} \frac{1}{t^k} \sum_{\substack{m \geq 0, \mathfrak{f} \in \mathcal{F}(\mathfrak{c}) \\ m+\ell-\dim \mathfrak{f}=k}} \int_{\mathfrak{f}} M_{m,\mathfrak{f}}\left(\frac{\partial}{\partial x}\right) \cdot h(x)\, dm_{\mathfrak{f}}(x).$$

In the next section, we will show that such an asymptotic formula is valid for any affine cone, with differential operators $M_{m,\mathfrak{f}}(\frac{\partial}{\partial x})$ expressed in terms of the μ-function [3] of the transverse cone to the face $\mathfrak{f}$.

5 Local Euler-Maclaurin Asymptotic Expansion for Riemann Sums over a Lattice Polyhedron

5.1 μ Function of a Pointed Cone

We fix a Euclidean scalar product Q on V. Thus V^* inherits also a scalar product. In [3], given the scalar product Q, we defined an analytic function $\mu_Q(\mathfrak{a})(\xi)$ on V^* for any semi-rational affine cone $\mathfrak{a}$ in a rational quotient space V/L of V. The function $\mu_Q(\mathfrak{a})(\xi)$ can also be defined by a renormalization procedure which is more natural in some respects than our inductive definition in [3]. Postponing this new definition (Definition 10) to the Appendix, we note some of its properties. In this section, we need only the case of a cone with a lattice vertex, the results for a semi-rational affine cone will be recalled in Sect. 6. We will often drop the subscript Q.

Proposition 2

1. *If s is a lattice point, then $\mu(s + \mathfrak{c}) = \mu(\mathfrak{c})$.*
2. *If $\mathfrak{a}$ is an affine cone in a quotient space V/L, then $\mu(\mathfrak{a})(\xi)$ depends only on the orthogonal projection of ξ on the subspace $L^\perp \subset V^*$.*

Example 4 In dimension one, $V = \mathbb{R}$ and $\Lambda = \mathbb{Z}$, for the cone $\mathbb{R}_{\geq 0}$, we have

$$\mu(\mathbb{R}_{\geq 0})(\xi) = \frac{1}{1 - e^\xi} + \frac{1}{\xi} = -\sum_{m \geq 0} \frac{b_{m+1}}{(m+1)!} \xi^m. \tag{23}$$

Recall the definition of the *transverse cone* $\mathfrak{N}(\mathfrak{c}, \mathfrak{f})$ of $\mathfrak{c}$ along a face $\mathfrak{f}$ (Definition 1). The following theorem is proven in [3] (a new proof is given in the Appendix). It generalizes Lemma 1 to any cone, not necessarily simplicial.

Theorem 2 *Let $\mathfrak{c} \subset V$ be a pointed cone. Fix a scalar product Q on V. Then*

$$S(\mathfrak{c}) = \sum_{\mathfrak{f}} \mu_Q(\mathfrak{N}(\mathfrak{c}, \mathfrak{f})) I(\mathfrak{f}), \tag{24}$$

where the sum runs over the set of faces of $\mathfrak{c}$.

More concretely, for each face $\mathfrak{f}$, write $V = L \oplus L^\perp$ where L is the linear span of $\mathfrak{f}$ and $L^\perp \subset V$ the orthogonal complement of L in V with respect to our scalar product. Write $\xi \in V^*$ as $\xi = \xi_0 + \xi_1$ with $\xi_0 \in L^*$ and $\xi_1 \in (L^\perp)^*$, then the term $\mu_Q(\mathfrak{N}(\mathfrak{c}, \mathfrak{f}))$ is a holomorphic function of ξ_1, while $I(\mathfrak{f})$ is a rational function of ξ_0. Equation (24) determines uniquely the functions $\mu_Q(\mathfrak{N}(\mathfrak{c}, \mathfrak{f}))(\xi_1)$.

For the case of a simplicial cone, we have seen that is was quite immediate to obtain such a decomposition.

5.2 Reduction to Pointed Polyhedra

As we continue, we will need to work with the supporting cones of a polyhedron, which are non pointed affine cones. Actually, we may consider more generally a non pointed polyhedron $\mathfrak{p}$ as well. To begin with, the case $\mathfrak{p} = \mathbb{R}$. We write the Euler-Maclaurin formula for a Riemann sum over the whole line. For any $n > 0$,

$$\frac{1}{t} \sum_{x \in \mathbb{Z}} h\left(\frac{x}{t}\right) = \int_{-\infty}^{\infty} h(x) - \frac{1}{t^n} \int_{-\infty}^{\infty} \frac{B_n(\{-tx\})}{n!} h^{(n)}(x). \tag{25}$$

Since the polyhedron $\mathfrak{p}$ is rational, its lineality space L is a rational subspace, meaning that $L \cap \Lambda$ is a lattice Λ_L. Let $\pi_{V/L}$ be the projection map $V \mapsto V/L$. Then the projected polyhedron $\pi_{V/L}(\mathfrak{p})$ is pointed in the quotient space V/L; it is rational with respect to the projected lattice $\Lambda_{V/L} = \pi_{V/L}(\Lambda)$.

Lemma 3 *Let $\mathfrak{p} \subset V$ be a polyhedron of dimension ℓ with lineality space L. If $h(x)$ is a test function on V, let $(\pi_{V/L})_* h(y)$ be the function on V/L obtained by averaging h (with respect to the Lebesgue measure corresponding to the lattice Λ_L),*

$$(\pi_{V/L})_* h(y) = \int_L h(y+x)dx.$$

Then

$$\frac{1}{t^\ell} \sum_{x \in t\mathfrak{p} \cap \Lambda} h\left(\frac{x}{t}\right) = \frac{1}{t^{\ell - \dim L}} \sum_{y \in t\pi_{V/L}(\mathfrak{p}) \cap \Lambda_{V/L}} (\pi_{V/L})_* h\left(\frac{y}{t}\right) + O(t^{-\infty}).$$

Proof If $\mathfrak{p} = L$, then $\frac{1}{t^\ell} \sum_{x \in \Lambda} h(\frac{x}{t}) = \int_L h(x)dx + O(t^{-\infty})$. This follows immediately from (25).

In the general case, we obtain a product situation by choosing a complementary rational subspace L' to L, $V = L \oplus L'$. We identify V/L and L'. Let $\Lambda' = (\Lambda + L) \cap L'$ be the projected lattice on L'. Let $\mathfrak{p}' \subset L'$ be the projected polyhedron. Then $\mathfrak{p}'$ is a pointed polyhedron. We have $\mathfrak{p} = L \oplus \mathfrak{p}'$ and $\mathfrak{p} \cap \Lambda = L \cap \Lambda \oplus \mathfrak{p}' \cap \Lambda'$. So the lemma follows from the case when $\mathfrak{p}$ is a vector space. □

5.3 Local Euler-Maclaurin Asymptotic Expansion for a Cone

Let us state the following asymptotic expansion which we call the *local Euler-Maclaurin asymptotic expansion* of the weighted sum on a cone. It depends on the choice of a scalar product on the ambient space.

Theorem 3 *Let $\mathfrak{c} \subseteq V$ be a rational cone of dimension ℓ. Let $\mathcal{F}(\mathfrak{c})$ be the set of faces of $\mathfrak{c}$. Let $h(x)$ be a test function on V. Then the following asymptotic expansion holds for $t \to \infty$, t real.*

$$\frac{1}{t^\ell} \sum_{x \in \mathfrak{c} \cap \Lambda} h\left(\frac{x}{t}\right) \sim \sum_{k=0}^{\infty} \frac{1}{t^k} \sum_{\substack{m \geq 0, \mathfrak{f} \in \mathcal{F}(\mathfrak{c}) \\ m + \ell - \dim \mathfrak{f} = k}} \int_{\mathfrak{f}} \mu(\mathfrak{N}(\mathfrak{c}, \mathfrak{f}))_{[m]} \left(\frac{\partial}{\partial x}\right) \cdot h(x)\, dm_{\mathfrak{f}}(x). \quad (26)$$

Here the constant coefficient differential operator $\mu(\mathfrak{N}(\mathfrak{c}, \mathfrak{f}))_{[m]}(\frac{\partial}{\partial x})$ involves only derivatives normal to the face $\mathfrak{f}$.

Remark 2 There is just one term for $k = 0$, which is of course $\int_{\mathfrak{c}} h(x)dx$. Indeed, for $k = 0$, since $m \geq 0$, if $m + \ell = \dim \mathfrak{f}$, then $m = 0$ and $\mathfrak{f} = \mathfrak{c}$. Moreover, for the face $\mathfrak{f} = \mathfrak{c}$ itself, we have $\mathfrak{N}(\mathfrak{c}, \mathfrak{c}) = \{0\}$ and $\mu(\{0\}) = 1$, so this face occurs only for $k = 0$.

Proof First, assume that $\mathfrak{c}$ is pointed. In that case, we take the inverse Fourier transform of (13), collect the homogeneous components in Formula (24), noting that $I(\mathfrak{f})$ is homogeneous of degree $(-\dim \mathfrak{f})$, and apply Proposition 1 (i). Thus we obtain (26) when $\mathfrak{c}$ is pointed.

If $\mathfrak{c}$ is not pointed, let L be its lineality subspace and let $\pi = \pi_{V/L}$ be the projection map. Then (26) holds for the pointed cone $\pi(\mathfrak{c})$. We write it for the averaged test function $\pi_* h$. The faces of $\pi(\mathfrak{c})$ are the projections of the faces of $\mathfrak{c}$. If $\mathfrak{f}$ is a face of $\mathfrak{c}$, the transverse cone $\mathfrak{N}(\pi(\mathfrak{c}), \pi(\mathfrak{f}))$ coincides with $\mathfrak{N}(\mathfrak{c}, \mathfrak{f})$ under the identification $(V/L)/\operatorname{lin}(\pi(\mathfrak{f})) = V/\operatorname{lin}(\mathfrak{f})$. Thus,

$$\frac{1}{t^{\ell-\dim L}} \sum_{y \in \pi(\mathfrak{c}) \cap \pi(\Lambda)} (\pi_* h)\left(\frac{y}{t}\right) \sim \sum_{k=0}^{\infty} \frac{1}{t^k}$$

$$\sum_{\substack{\{(\mathfrak{f},m)\},\, \mathfrak{f} \in \mathcal{F}(\mathfrak{c}) \\ m \geq 0,\, m + (\ell - \dim L) - (\dim \mathfrak{f} - \dim L) = k}} \int_{\pi(\mathfrak{f})} \mu(\mathfrak{N}(\mathfrak{c}, \mathfrak{f}))_{[m]} \left(\frac{\partial}{\partial y}\right) \cdot (\pi_* h)(y)\, dm_{\pi(\mathfrak{f})}(y). \tag{27}$$

Since $\mu(\mathfrak{N}(\mathfrak{c}, \mathfrak{f}))_{[m]}(\frac{\partial}{\partial y})$ is a differential operator with constant coefficients, we have

$$\mu(\mathfrak{N}(\mathfrak{c}, \mathfrak{f}))_{[m]} \left(\frac{\partial}{\partial y}\right) \cdot (\pi_* h) = \pi_* \left(\mu(\mathfrak{N}(\mathfrak{c}, \mathfrak{f}))_{[m]} \left(\frac{\partial}{\partial x}\right) \cdot h\right).$$

So, the $(k, \mathfrak{f}, m)$ term in the RHS of (27) is equal to

$$\int_{\mathfrak{f}} \mu(\mathfrak{N}(\mathfrak{c}, \mathfrak{f}))_{[m]} \left(\frac{\partial}{\partial x}\right) \cdot h(x)\, dm_{\mathfrak{f}}(x).$$

Using Lemma 3, we obtain (26) for the cone $\mathfrak{c}$. □

Example 5 If $\mathfrak{c} = V$, there is just one face $\mathfrak{c}$ itself, with transverse cone $\{0\}$, for which $\mu(\{0\}) = 1$, so all terms of the asymptotic expansion are 0, except for $k = 0$ which gives the integral over V.

Example 6 Let us describe (26) in dimension one, $\mathfrak{c} = \mathbb{R}_{\geq 0}$. The face $\mathfrak{f} = \{0\}$ occurs in the k term for $m = k-1$. We have $\mathfrak{N}(\mathfrak{c}, \{0\}) = \mathfrak{c} = \mathbb{R}_{\geq 0}$ and $\mu(\mathbb{R}_{\geq 0})_{[k-1]} = -\frac{b_k}{k!}\xi^{k-1}$. So (26) is

$$\frac{1}{t} \sum_{x \in \mathbb{Z}_{\geq 0}} h\left(\frac{x}{t}\right) \sim \int_0^{\infty} h(x) dx - \sum_{k=1}^{\infty} \frac{1}{t^k} \frac{b_k}{k!} h^{k-1}(0).$$

Of course, it is Formula (7) applied to $\frac{1}{t} h(\frac{x}{t})$.

Example 7 Let us describe (26) for the non pointed cone $\mathfrak{c} = \mathbb{R}_{\geq 0} e_1 \oplus \mathbb{R} e_2 \subset \mathbb{R}^2$, with the standard scalar product on $\mathbb{R}^2$. Besides $\mathfrak{c}$ itself, there is only one face, $\mathfrak{f} = \mathbb{R} e_2$, with transverse cone $\mathfrak{N}(\mathfrak{c}, \mathfrak{f}) = \mathbb{R}_{\geq 0} e_1$ and $\mu_Q(\mathfrak{N}(\mathfrak{c}, \mathfrak{f}))_{[k-1]}(\xi) = -\frac{b_k}{k!}\xi_1^{k-1}$.

So (26) is

$$\frac{1}{t^2}\sum_{x_1\in\mathbb{Z}_{\geq 0},x_2\in\mathbb{Z}} h\left(\frac{x_1}{t},\frac{x_2}{t}\right)\sim$$

$$\int_0^\infty\int_{\mathbb{R}} h(x_1,x_2)dx_2dx_1-\sum_{k=1}^{\infty}\frac{1}{t^k}\frac{b_k}{k!}\int_{\mathbb{R}}\frac{\partial^{k-1}h}{\partial x_1}(0,x_2)dx_2.$$

It is Example 6 applied to the averaged function $g(x_1)=\int_{\mathbb{R}} h(x_1,x_2)dx_2$.

Example 8 Let $V=\mathbb{R}^3$ with standard lattice. Let $\mathfrak{c}\subset\mathbb{R}^3$ be the non simplicial cone based on a square, with generators $v_1=e_3+e_1, v_2=e_3+e_2, v_3=e_3-e_1, v_4=e_3-e_2$. By subdivising $\mathfrak{c}$ into unimodular cones, we obtain

$$S(\mathfrak{c})(\xi)=\frac{1+e^{\xi_3}-e^{2\xi_3}-e^{3\xi_3}}{\left(1-e^{\xi_3+\xi_1}\right)\left(1-e^{\xi_3-\xi_1}\right)\left(1-e^{\xi_3+\xi_2}\right)\left(1-e^{\xi_3-\xi_2}\right)}.$$

We compute the first four homogeneous components of $S(\mathfrak{c})(\xi)$ from the formula $S(\mathfrak{c})=\sum_{\mathfrak{f}}\mu_Q(\mathfrak{N}(\mathfrak{c},\mathfrak{f}))I(\mathfrak{f})$ (where Q is the standard Euclidean product). We obtain

$$S(\mathfrak{c})_{[-3]}=I(\mathfrak{c})=-\frac{4\xi_3}{(\xi_3+\xi_1)(\xi_3-\xi_1)(\xi_3+\xi_2)(\xi_3+\xi_2)}.$$

$$S(\mathfrak{c})_{[-2]}=\frac{1}{(\xi_3+\xi_1)(\xi_3+\xi_2)}+\frac{1}{(\xi_3+\xi_1)(\xi_3-\xi_2)}+\frac{1}{(\xi_3-\xi_1)(\xi_3+\xi_2)}+\frac{1}{(\xi_3-\xi_1)(\xi_3-\xi_2)}.$$

$$S(\mathfrak{c})_{[-1]}=-\frac{2}{9}\sum_{j=1,2}\frac{1}{\xi_3+\xi_j}+\frac{1}{\xi_3-\xi_j}-\frac{1}{36}\frac{\xi_3-\xi_1-\xi_2}{(\xi_3+\xi_1)(\xi_3+\xi_2)}$$
$$-\frac{1}{36}\frac{\xi_3-\xi_1+\xi_2}{(\xi_3+\xi_1)(\xi_3-\xi_2)}-\frac{1}{36}\frac{\xi_3+\xi_1-\xi_2}{(\xi_3-\xi_1)(\xi_3+\xi_2)}-\frac{1}{36}\frac{\xi_3+\xi_1+\xi_2}{(\xi_3-\xi_1)(\xi_3-\xi_2)}.$$

$$S(\mathfrak{c})_{[0]}=\frac{1}{24}\frac{\xi_3-\xi_1}{\xi_3+\xi_1}+\frac{1}{24}\frac{\xi_3-\xi_2}{\xi_3+\xi_2}+\frac{1}{24}\frac{\xi_3+\xi_1}{\xi_3-\xi_1}+\frac{1}{24}\frac{\xi_3+\xi_2}{\xi_3-\xi_2}+\frac{1}{6}.$$

Thus the coefficients of the asymptotic expansion

$$\langle R_t(\mathfrak{c}), h\rangle \sim \sum_{k=0}^{\infty} \frac{1}{t^k}\langle F_k, h\rangle$$

are

$$\langle F_0, h\rangle = \int_{\mathfrak{c}} h, \ \ \langle F_1, h\rangle = \frac{1}{2}\int_{\partial\mathfrak{c}} h,$$

$$\langle F_2, h\rangle = \frac{2}{9}\int_{\text{edges}} h - \frac{1}{12}\int_{\partial\mathfrak{c}} \partial_u h, \langle F_3, h\rangle = \frac{1}{6}h(0,0,0) - \frac{1}{24}\int_{\text{edges}} \partial_v h,$$

where for each facet, u is the primitive vector (for the projected lattice) normal to the facet pointing inwards, and $v = e_3 - e_i$ for the edge $e_3 + e_i$, $v = e_3 + e_i$ for the edge $e_3 - e_i, i = 1, 2$.

5.4 *Local Euler-Maclaurin Asymptotic Expansion for a Lattice Polyhedron*

In this section, we will obtain an expansion similar to Theorem 3 first for an affine cone with lattice vertex, then for a polyhedron. Thus, in the remainder of this section, we assume that $\mathfrak{p}$ is a *lattice polyhedron*, meaning that each of its faces contains a lattice point. We also want $t\mathfrak{p}$ to be a lattice polyhedron so we will restrict to $t \in \mathbb{N}$.

If $\mathfrak{p}$ is a lattice polyhedron, the transverse cone along a face is a pointed affine cone with lattice vertex. Theorem 2 extends readily to such cones. If $s \in \Lambda$ and $\mathfrak{c}$ is a pointed cone, for the shifted cone $\mathfrak{a} = s + \mathfrak{c}$, we have

$$S(s+\mathfrak{c})(\xi) = e^{\langle \xi, s\rangle} S(\mathfrak{c})(\xi).$$

The μ function of an affine cone $s + \mathfrak{c}$, with $s \in \Lambda$, is equal to $\mu(\mathfrak{c})$.

For every face $\mathfrak{f}$ of $\mathfrak{c}$, we have $\mathfrak{N}(s+\mathfrak{c}, s+\mathfrak{f}) = \text{proj}_{V/\text{lin}(f)}(s) + \mathfrak{N}(\mathfrak{c}, \mathfrak{f})$, therefore

$$\mu(\mathfrak{N}(s+\mathfrak{c}, s+\mathfrak{f})) = \mu(\mathfrak{N}(\mathfrak{c}, \mathfrak{f})).$$

Moreover $I(s+\mathfrak{f})(\xi) = e^{\langle \xi, s\rangle} I(\mathfrak{f})(\xi)$, so Formula (24) still holds if we replace $\mathfrak{c}$ by an affine cone $\mathfrak{a}$ with lattice vertex.

If we dilate $\mathfrak{a}$ by a positive integer t, the vertex of $t\mathfrak{a}$ is still a lattice point. Therefore, we have the following extension of Theorem 3 to the case of an affine *lattice* cone (such that each of its faces contains a lattice point; we do not assume that it is pointed), provided the parameter t is an integer.

This theorem was proved by Tate [12].

Theorem 4 *Let $\mathfrak{a} \subseteq V$ be an affine lattice cone of dimension ℓ. Let $\mathcal{F}(\mathfrak{a})$ be the set of faces of $\mathfrak{a}$. Let $h(x)$ be a test function on V. Then the following asymptotic expansion holds when $t \to \infty$ with $t \in \mathbb{N}$.*

$$\frac{1}{t^\ell} \sum_{x \in t\mathfrak{a} \cap \Lambda} h\left(\frac{x}{t}\right) \sim \sum_{k=0}^{\infty} \frac{1}{t^k} \sum_{\substack{m \geq 0, \mathfrak{f} \in \mathcal{F}(\mathfrak{a}) \\ m+\ell-\dim \mathfrak{f}=k}} \int_{\mathfrak{f}} \mu(\mathfrak{N}(\mathfrak{a}, \mathfrak{f}))_{[m]} \left(\frac{\partial}{\partial x}\right) \cdot h(x)\, dm_{\mathfrak{f}}(x).$$

Proof We can write $\mathfrak{a} = s + \mathfrak{c}$ where $\mathfrak{c}$ is a cone and $s \in \Lambda$. We apply Theorem 3 to the cone $\mathfrak{c}$ and the shifted function $h(s + x)$. □

We will now show that this theorem leads to an asymptotic expansion for any lattice polyhedron. This theorem was obtained by Tate [12] (for a lattice polyhedron).

Theorem 5 *Let $\mathfrak{p} \subset V$ be a lattice polyhedron of dimension ℓ. Let $h(x)$ be a test function on V. Fix a Euclidean scalar product Q on V. For a face $\mathfrak{f}$ of $\mathfrak{p}$, let $\mathfrak{N}(\mathfrak{p}, \mathfrak{f})$ be the transverse cone of $\mathfrak{p}$ along $\mathfrak{f}$. Then the following asymptotic expansion holds when $t \to +\infty$, $t \in \mathbb{N}$,*

$$\frac{1}{t^\ell} \sum_{x \in t\mathfrak{p} \cap \Lambda} h\left(\frac{x}{t}\right) \sim$$

$$\int_{\mathfrak{p}} h(x)dx + \sum_{k \geq 1} \frac{1}{t^k} \sum_{\substack{m \geq 0, \mathfrak{f} \in \mathcal{F}(\mathfrak{p}) \\ m+\ell-\dim \mathfrak{f}=k}} \int_{\mathfrak{f}} \left(\mu\left(\mathfrak{N}(\mathfrak{p}, \mathfrak{f})_{[m]} \left(\frac{\partial}{\partial x}\right)\right) \cdot h\right)(x) dm_{\mathfrak{f}}(x). \tag{28}$$

Moreover, the differential operators which appear in (28) are unique in the sense that, if $\rho_{[m]}(\mathfrak{p}, \mathfrak{f})(\xi)$ is a family of polynomials on V^*, such that (28) holds for any test function $h(x)$ (with $\rho_{[m]}(\mathfrak{p}, \mathfrak{f})$ in place of $\mu(\mathfrak{N}(\mathfrak{p}, \mathfrak{f})_{[m]})$ and such that $\rho_{[m]}(\mathfrak{p}, \mathfrak{f})(\xi)$ is homogeneous of degree m, and depends only on the Q-projection of ξ on $(\mathrm{lin}(\mathfrak{f}))^\perp$, *then*

$$\rho_{[m]}(\mathfrak{p}, \mathfrak{f})(\xi) = \mu_{[m]}(\mathfrak{N}(\mathfrak{p}, \mathfrak{f}))(\xi).$$

Proof First, we prove (28). Let $h(x)$ be a test function. Using a partition of unity, we may assume that the support of h is contained in an open set U such that $U \cap \mathfrak{p} = U \cap C(\mathfrak{p}, \mathfrak{q})$ for some face $\mathfrak{q}$ of $\mathfrak{p}$ (recall that $C(\mathfrak{p}, \mathfrak{q})$ is the supporting cone of $\mathfrak{p}$ along $\mathfrak{q}$). If U is small enough and convex, we observe that the set of faces $\mathcal{F}(C(\mathfrak{p}, \mathfrak{q}))$ of $C(\mathfrak{p}, \mathfrak{q})$ is in 1–1 correspondence with the set of faces of $\mathfrak{p}$ which meet U. If $\mathfrak{d}$ is such a face, the corresponding face $\mathfrak{f}$ of $C(\mathfrak{p}, \mathfrak{q})$ is $\mathfrak{f} = \langle \mathfrak{d} \rangle \cap C(\mathfrak{p}, \mathfrak{q})$, where $\langle \mathfrak{d} \rangle$ denotes the affine span of $\mathfrak{d}$. Moreover, $\mathfrak{N}(C(\mathfrak{p},\mathfrak{q}),\mathfrak{f}) = \mathfrak{N}(\mathfrak{p},\mathfrak{d})$. Thus Theorem 3 gives

$$\frac{1}{t^\ell}\sum_{x\in t\mathfrak{p}\cap\Lambda} h\left(\frac{x}{t}\right) = \frac{1}{t^\ell}\sum_{x\in tC(\mathfrak{p},\mathfrak{q})} h\left(\frac{x}{t}\right) \sim$$

$$\sum_{k=0}^{\infty}\frac{1}{t^k}\sum_{\substack{\{(\mathfrak{d},m)\},\,\mathfrak{d}\in\mathcal{F}(\mathfrak{p}),\,\mathfrak{q}\subseteq\mathfrak{d}\\ m\geq 0,\, m+\ell-\dim\mathfrak{d}=k}} \int_{\langle\mathfrak{d}\rangle\cap C(\mathfrak{p},\mathfrak{q})} \mu(\mathfrak{N}(\mathfrak{p},\mathfrak{d}))_{[m]}\left(\frac{\partial}{\partial x}\right)\cdot h(x)\,dm_{\mathfrak{d}}(x). \tag{29}$$

So the right hand side of (29) is

$$\sum_{k=0}^{\infty}\frac{1}{t^k}\sum_{\substack{\{(\mathfrak{d},m)\},\,\mathfrak{d}\in\mathcal{F}(\mathfrak{p})\\ m\geq 0,\, m+\ell-\dim\mathfrak{d}=k}} \int_{\mathfrak{d}} \mu(\mathfrak{N}(\mathfrak{p},\mathfrak{d}))_{[m]}\left(\frac{\partial}{\partial x}\right)\cdot h(x)\,dm_{\mathfrak{d}}(x).$$

This is (28), except that the faces of $\mathfrak{p}$ are labeled $\mathfrak{d}$ instead of $\mathfrak{f}$.

Let us prove the uniqueness part of the theorem. By uniqueness of asymptotic expansions, each k term is a uniquely determined distribution

$$\sum_{\substack{m\geq 0,\mathfrak{f}\in\mathcal{F}(\mathfrak{p})\\ m+\ell-\dim\mathfrak{f}=k}} \int_{\mathfrak{f}}\left(\mu\left(\mathfrak{N}(\mathfrak{p},\mathfrak{f})_{[m]}\left(\frac{\partial}{\partial x}\right)\right)\cdot h\right)(x)dm_{\mathfrak{f}}(x).$$

By decreasing induction on $\dim\mathfrak{f}$, it follows that each $(\mathfrak{f},m)$ term is a uniquely determined distribution

$$\int_{\mathfrak{f}}\left(\mu\left(\mathfrak{N}(\mathfrak{p},\mathfrak{f})_{[m]}\left(\frac{\partial}{\partial x}\right)\right)\cdot h\right)(x)dm_{\mathfrak{f}}(x).$$

Indeed, if $\mathfrak{f}$ is a facet, we can restrict the support of h so that it does not meet the other facets, etc.

Furthermore, as $\mu(\mathfrak{N}(\mathfrak{p},\mathfrak{f})_{[m]}(\frac{\partial}{\partial x})$ contains only derivatives which are orthogonal to $\mathfrak{f}$, it is uniquely determined. □

Remark 3 For simplicial polytopes, one can just use Lemma 2 instead of Theorem 4.

5.5 *Local Behavior*

In order to illustrate the local behavior of the asymptotic expansion of Theorem 5, we will compute the first terms for two triangles in the plane.

First we observe that the terms F_1 and F_2 are easily computed out of Theorem 5 for any lattice polyhedron $\mathfrak{p}$.

If $\mathfrak{p}$ is lattice, then $\mu(\mathfrak{N}(\mathfrak{p},\mathfrak{p})) = 1$, so the relation $m + \ell - \dim \mathfrak{f} = 1$ is obtained exactly when $\mathfrak{f}$ is a facet and $m = 0$. Now, for a facet, we have $\mu_{[0]}(\mathfrak{N}(\mathfrak{p},\mathfrak{f})) = -b_1 = \frac{1}{2}$. This proves the formula $\langle F_1, h\rangle = \frac{1}{2}\int_{\partial\mathfrak{c}} h$.

If $\mathfrak{p}$ is a Delzant polyhedron, the next term $\langle F_2, h\rangle$ is also easy to compute with Theorem 5. For a facet $\mathfrak{f}$, let $u_\mathfrak{f}$ be the primitive generator of the transverse cone. For a face $\mathfrak{f}$ of codimension 2, let $(u_\mathfrak{f}^1, u_\mathfrak{f}^2)$ be the primitive edge generators of the transverse cone and let $C_\mathfrak{f} = \frac{1}{4} + Q(u_\mathfrak{f}^1, u_\mathfrak{f}^2)(\frac{1}{\|u_\mathfrak{f}^1\|^2} + \frac{1}{\|u_\mathfrak{f}^2\|^2})$. Then, if $\mathfrak{p}$ is Delzant,

$$\langle F_2, h\rangle = -\frac{1}{12}\sum_{\mathfrak{f},\mathrm{codim}\mathfrak{f}=1}\int_\mathfrak{f} \partial_{u_\mathfrak{f}}\cdot h + \sum_{\mathfrak{f},\mathrm{codim}\,\mathfrak{f}=2}\int_\mathfrak{f} C_\mathfrak{f} h.$$

If $\mathfrak{p}$ is not Delzant, the two-dimensional transverse cones are still simplicial, though maybe not unimodular, so the term F_2 involves arithmetic expressions (see [3]).

Example 9 We write these formulas for the following two triangles $\mathfrak{p}$ and $\mathfrak{p}'$.

$\mathfrak{p}$ is the triangle with vertices $(0,0)$, $(1,0)$,$(0,1)$. Let $\mathfrak{f}_1$ be its horizontal edge, $\mathfrak{f}_2$ the vertical one and $\mathfrak{f}_3$ the diagonal one. Then

$$\frac{1}{t^2}\sum_{(x_1,x_2)\in t\mathfrak{p}\cap\mathbb{Z}^2} h\left(\frac{x_1}{t},\frac{x_2}{t}\right) \sim \int_\mathfrak{p} h + \frac{1}{2t}\int_{\partial(\mathfrak{p})} h +$$
$$\frac{1}{t^2}\left(-\frac{1}{12}\left(\int_{\mathfrak{f}_1}\partial_{x_2}h + \int_{\mathfrak{f}_2}\partial_{x_1}h - \int_{\mathfrak{f}_3}\left(\frac{\partial_{x_1}+\partial_{x_2}}{2}\right)\cdot h\right) + \right.$$
$$\left.\frac{1}{4}h(0,0) + \frac{3}{8}h(1,0) + \frac{3}{8}h(0,1)\right) + \cdots .$$

$\mathfrak{p}'$ is the triangle with vertices $(0,0)$, $(2,0)$,$(0,3)$, horizontal edge $\mathfrak{f}_1'$, vertical edge $\mathfrak{f}_2'$, diagonal edge $\mathfrak{f}_3'$. Then

$$\frac{1}{t^2}\sum_{(x_1,x_2)\in t\mathfrak{p}'\cap\mathbb{Z}^2} h\left(\frac{x_1}{t},\frac{x_2}{t}\right) \sim \int_{\mathfrak{p}'} h + \frac{1}{2t}\int_{\partial(\mathfrak{p}')} h +$$
$$\frac{1}{t^2}\left(-\frac{1}{12}\left(\int_{\mathfrak{f}_1'}\partial_{x_2}h + \int_{\mathfrak{f}_2'}\partial_{x_1}h - \int_{\mathfrak{f}_3'}\left(\frac{3\partial_{x_1}+2\partial_{x_2}}{13}\right)\cdot h\right) + \right.$$
$$\left.\frac{1}{4}h(0,0) + \frac{19}{52}h(2,0) + \frac{5}{13}h(0,3)\right) + \cdots .$$

Indeed, the two expansions coincide if the test function h is supported near $(0,0)$.

5.6 The Remainder?

For the remainder at order k of the asymptotic expansion, a closed formula (in terms of derivatives of locally polynomial measures supported on faces) looks elusive.

As an example, we compute the remainder at order $k = 1$ (Riemann sum minus integral) for the triangle $\mathfrak{p}$ of Example 9.

Example 10 Let $D_n = \frac{1}{2}(\partial_{x_1} + \partial_{x_2})$ be the normal derivative to the diagonal edge $\mathfrak{f}_3$ and let $D_t = \frac{1}{2}(\partial_{x_1} - \partial_{x_2})$ the tangential derivative. For $t \in \mathbb{N}$, we have

$$\frac{1}{t^2} \sum_{(x_1,x_2)\in t\mathfrak{p}\cap\mathbb{Z}^2} h\left(\frac{x_1}{t}, \frac{x_2}{t}\right) - \int_{\mathfrak{p}} h(x_1,x_2)dx_1dx_2 =$$

$$\int_{\mathfrak{p}} A(t)dx_1dx_2 + \int_{\mathfrak{f}_3} B(t)dm_{\mathfrak{f}_3} + \frac{1}{t^2}h(1,0),$$

where

$$A(t) = \frac{1}{t}((\{tx_1\} - 1)\partial_{x_1}h + (\{tx_2\} - 1)\partial_{x_2}h) +$$

$$\frac{1}{t^2}(\{tx_1\} - 1)(\{tx_2\} - 1)\partial^2_{x_1x_2}h,$$

$$B(t) = \frac{3}{2t}h + \frac{1}{t^2}\{tx_1\}(\{tx_1\} - 1)D_n \cdot h + \frac{1}{2t^2}(\{tx_1\} - 1)D_t \cdot h.$$

6 Riemann Sums with Real Scaling Parameter over Semi-Rational Polyhedra

6.1 Asymptotic Expansions with Step-Polynomial Coefficients

We recall now the Euler-Maclaurin summation formula on the half-line $[s, \infty[$, for any real s, for a test function $h(x)$, [5, Theorem 9.2.2].

$$\sum_{x\in\mathbb{Z},x\geq s} h(x) = \int_s^\infty h(x)dx - \sum_{k=1}^{n} \frac{B_k(\{-s\})}{k!}h^{(k-1)}(s)$$

$$- \int_s^\infty \frac{B_n(\{-x\})}{n!}h^{(n)}(x)\,dx. \quad (30)$$

For any real $t > 0$, let

$$h_t(x) = \frac{1}{t}h\left(\frac{x}{t}\right).$$

Substituting h_t for h and ts for s in this formula, and changing variables in the integrals on the right-hand-side, we obtain

$$\frac{1}{t}\sum_{x\in\mathbb{Z},x\geq ts} h\left(\frac{x}{t}\right) = \int_s^\infty h(x)dx - \sum_{k=1}^{n-1}\frac{1}{t^k}\frac{B_k(\{-ts\})}{k!}h^{(k-1)}(s)$$
$$-\frac{1}{t^n}\left(\frac{B_n(\{-ts\})}{n!}h^{(n-1)}(s) + \int_s^\infty \frac{B_n(\{-tx\})}{n!}h^{(n)}(x)dx\right). \quad (31)$$

This formula can be considered as an asymptotic expansion (with a closed expression for the remainder) when $t \to +\infty$, $t \in \mathbb{R}$, where we allow the coefficient of $\frac{1}{t^k}$ to be a polynomial function of the fractional part $\{-ts\}$, that is, a step-polynomial function in the sense of the following definition.

Definition 3 A *step-polynomial* function $F(t)$ on $\mathbb{R}$ is an element of the algebra generated by the functions $t \mapsto \{\gamma t\}$, with $\gamma \in \mathbb{R}$.

If $F(t)$ is a step-polynomial, there exists

(1) A finite set Γ of real numbers linearly independent over $\mathbb{Q}$.
(2) For each $\gamma \in \Gamma$ a finite set of non zero integers $n_{\gamma,k}$.
(3) A polynomial in several variables, $P(X_{\gamma,k})$,

such that

$$F(t) = P(\{n_{\gamma,k}\gamma t\}). \quad (32)$$

An example of step-polynomial function with irrational γ's is displayed in Fig. 1.

A step polynomial function is defined for all $t \in \mathbb{R}$, and has a discrete set of discontinuities.

One must be aware that this set of data is not uniquely defined by $F(t)$. A simple example is

$$1 - \{t\} - \{-t\} = (1 - \{2t\} - \{-2t\})(1 - \{3t\} - \{-3t\}).$$

Definition 4 A function $\phi(t)$, defined for $t > 0$, has an *asymptotic expansion with step-polynomial coefficients* $F_k(t)$, when $t \to +\infty$, $t \in \mathbb{R}$, written as

$$\phi(t) \sim F_0(t) + \frac{1}{t}F_1(t) + \cdots + \frac{1}{t^n}F_n(t) + \cdots, \quad (33)$$

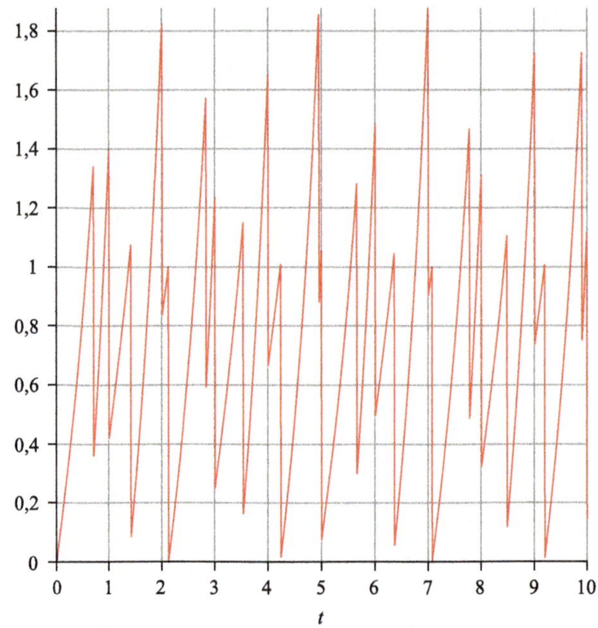

Fig. 1 Graph of $\{\sqrt{2}t\} + \{t\}^3$

if $\phi(t) = F_0(t) + \frac{1}{t}F_1(t) + \cdots + \frac{1}{t^n}F_k(t) + O(\frac{1}{t^{n+1}})$ when $t \to +\infty$, for every $n \geq 1$.

For lack of a reference, we give a proof for the following result.

Proposition 3 *If such an asymptotic expansion exists, it is unique.*

Proof First we prove that if $F(t)$ is a step-polynomial function and $F(t) = O(\frac{1}{t})$ when $t \to +\infty$, then $F(t) = 0$ for all $t \in \mathbb{R}$.

If $F(t)$ has an expression with a single γ, then $F(t)$ is periodic, so it must be identically 0.

Let us consider the case where Γ consists of two elements, 1 and $\gamma \notin \mathbb{Q}$. For instance, assume that

$$F(t) = P(\{t\}, \{-5t\}, \{\gamma t\}, \{2\gamma t\}) = O\left(\frac{1}{t}\right).$$

For $t = N + a$ with $N \in \mathbb{N}$ and a fixed, we have

$$F(N + a) = P(\{a\}, \{-5a\}, \{\gamma N + \gamma a\}, \{2(\gamma N + \gamma a)\}) = O\left(\frac{1}{N}\right).$$

Fix b such that $0 < b < 1$. By Kronecker theorem in one dimension, [9, Theorem 438] there exists a strictly increasing sequence $N_m \in \mathbb{N}$ such that

$\lim_{m\to\infty}\{\gamma N_m+\gamma a\} = b$. As $\{2x\} = 2x$ if $x \in [0, \frac{1}{2}[$ and $\{2x\} = 2x-1$ if $x \in [\frac{1}{2}, 1[$, we conclude that $P(\{a\}, \{-5a\}, b, 2b) = 0$ for any a and $b \in]0, \frac{1}{2}[$, hence for any b since P is a polynomial, and similarly $P(\{a\}, \{-5a\}, b, 2b-1) = 0$ for any b. Therefore, $P(\{a\}, \{-5a\}, \{b\}, \{2b\}) = 0$ for any a and b, hence $F(t)$ is identically 0.

A similar argument works in the general case. If $\Gamma = (\gamma_0, \gamma_1, \dots, \gamma_p)$, we can assume that $\gamma_0 = 1$. By Kronecker theorem, [9, Theorem 442], the set of points $(\{N\gamma_1 + u_1\}, \dots, \{N\gamma_p + u_p\}), N \in \mathbb{N}$ is dense in $[0, 1[^p$ for any $(u_1, \dots, u_p)$.

Thus we have proved that $F(t) = 0$. The proposition follows by induction on k as usual. □

6.2 Step-Polynomials on *V*. μ Function of a Semi-Rational Affine Cone

We fix a Euclidean scalar product Q on V. We recall now the properties of the holomorphic function $\mu_Q(\mathfrak{a})$ when $\mathfrak{a}$ is a semi-rational affine cone (a rational cone shifted by a real vertex, not just a rational one). We begin with some examples.

Example 11 In dimension one, $V = \mathbb{R}$ and $\Lambda = \mathbb{Z}$, for the cone $s + \mathbb{R}_{\geq 0}$, where $s \in \mathbb{R}$, we have

$$\mu(s + \mathbb{R}_{\geq 0})(\xi) = \frac{e^{\{-s\}\xi}}{1 - e^{\xi}} + \frac{1}{\xi} = -\sum_{m\geq 0} B_{m+1}(\{-s\})\xi^m. \tag{34}$$

Example 12 In $\mathbb{R}^d$ with standard lattice and Euclidean scalar product, let $\mathfrak{c}$ be the one-dimensional cone $\mathfrak{c} = \mathbb{R}_{\geq 0}e_1$, let $s = (s_1, \dots, s_d)$.

Then

$$\begin{aligned}
&\mu(s + \mathbb{R}_{\geq 0}e_1)(\xi)\\
&= \frac{e^{\{-s_1\}\xi_1}}{1 - e^{\xi_1}} + \frac{1}{\xi_1} = -\sum_{m\geq 0} B_{m+1}(\{-s_1\})\xi_1^m, \text{ if } (s_2, \dots, s_d) \in \mathbb{Z}^{d-1},\\
&= 0 \text{ otherwise .}
\end{aligned}$$

Therefore

$$\mu(s + \mathbb{R}_{\geq 0}e_1)(\xi) = -\prod_{j=2}^{d}(1 - \{s_j\} - \{-s_j\}) \sum_{m\geq 0} B_{m+1}(\{-s_1\})\xi_1^m. \tag{35}$$

We see that these functions involve step-polynomials in several variables.

Definition 5 A step-polynomial function on V is an element of the algebra generated by the functions $s \mapsto \{\langle\gamma, s\rangle\}$, with $\gamma \in V^*$. A *rational step-polynomial* function is an element of the algebra generated by the functions $s \mapsto \{\langle\gamma, s\rangle\}$, with rational γ's.

Step-polynomials occur when we consider the generating function of an affine cone with real vertex [1, Theorem 2.22]. We recall the result.

Let $\mathfrak{c} \subset V$ be a pointed cone and $s \in V$. Consider the *shifted generating function*

$$M(s, \mathfrak{c})(\xi) = \mathrm{e}^{-\langle\xi, s\rangle} S(s + \mathfrak{c})(\xi). \tag{36}$$

Lemma 4 *Let $\mathfrak{c} \subset V$ be a pointed cone. Let $v_1, \dots, v_n$ be the set of its lattice edge generators.*

The homogeneous components of $M(s, \mathfrak{c})(\xi)$ are of the form $M(s, \mathfrak{c})_{[m]}(\xi) = \frac{P_{m+n}(s,\xi)}{\prod_j \langle\xi, v_j\rangle}$ where $P_{m+n}(s, \xi)$ is a polynomial of degree $m + n$ in ξ with coefficients which are rational step-polynomial functions of s.

Let us now state the properties of the μ-function of a pointed affine cone.

Proposition 4 *Let $\mathfrak{a} \subset V$ be a semi-rational affine cone.*

- *(i) For any $v \in \Lambda$, we have $\mu(v + \mathfrak{a}) = \mu(\mathfrak{a})$.*
- *(ii) Let s vary in V. For every integer $m \geq 0$, the homogeneous component $\mu(s + \mathfrak{a})_{[m]}(\xi)$ is a polynomial function of $\xi \in V^*$ with coefficients which are step-polynomial functions of s.*
- *(iii) For every integer $m \geq 0$, the homogeneous component $\mu_{[m]}(t\mathfrak{a})(\xi)$ is a polynomial function of $\xi \in V^*$ with coefficients step-polynomial functions of $t \in \mathbb{R}$. If the scalar product Q is rational, the coefficients are rational step-polynomials.*

Proof We deduce properties (i) and (ii) from the definition of $\mu(\mathfrak{a})$ by renormalization 10. (iii) follows immediately from (ii)

We have $S(v + \mathfrak{a}) = \mathrm{e}^{\langle\xi, v\rangle} S(\mathfrak{a})$ if $v \in \Lambda$, hence (i).

(ii) is a consequence of Lemma 4: the renormalization $R_Q\left(\frac{P_{m+n}(s,\xi)}{\prod_j \langle\xi, v_j\rangle}\right)$ of $M(s, \mathfrak{c})_{[m]}(\xi)$ is also a polynomial in ξ with coefficients which are step-polynomial functions of s. □

6.3 *Local Euler-Maclaurin Asymptotic Expansion for a Semi-Rational Polyhedron*

We can now state the local Euler-Maclaurin asymptotic expansion for Riemann sums over a semi-rational polyhedron when the scaling parameter is real, not only integral. Given the above discussion on asymptotic expansions with step-polynomial coefficients, the proof is parallel to the case of a lattice polyhedron and an integral scaling parameter in Sect. 5. So we will leave the details to the reader.

The main step is the following analogue of Theorem 1 about Riemann sums over a cone, in the case of a semi-rational affine cone and a real scaling parameter.

Theorem 6 *Let $\mathfrak{c} \subset V$ be a pointed cone of dimension ℓ. Let $s \in V$. Let $\lambda \in V^*$ be such that $-\lambda$ lies in the dual cone of $\mathfrak{c}$. Consider the distribution on V given by*

$$\langle R_t(s + \mathfrak{c}), h\rangle = \frac{1}{t^\ell} \sum_{x \in (ts+\mathfrak{c}) \cap \Lambda} h(\frac{x}{t}). \tag{37}$$

It has an asymptotic expansion with step-polynomial coefficients when $t \to \infty$, t real, the Fourier transform of which is given by

$$\begin{aligned} \mathcal{F}(R_t(s + \mathfrak{c}))(\xi) &= \frac{1}{t^\ell} \lim_\lambda \left(S(ts + \mathfrak{c}) \left(-i\frac{\xi}{t} \right) \right) \qquad (38) \\ &= \frac{1}{t^\ell} \mathrm{e}^{-i\langle \xi, s\rangle} \lim_\lambda \left(M(ts, \mathfrak{c}) \left(-i\frac{\xi}{t} \right) \right) \\ &\sim \mathrm{e}^{-i\langle \xi, s\rangle} \sum_{k=0}^{\infty} \frac{1}{t^k} \lim_\lambda (M(ts, \mathfrak{c})_{[k-\ell]}(-i\xi)). \end{aligned}$$

Proof If $\mathfrak{c} = \mathbb{R}_{\geq 0}$ and s is any real number, (38) follows from the classical dimension one Euler-Maclaurin formula (30) for a half-line. The general case is reduced to dimension one by subdivision of cones, as in the proof of Theorem 1.

□

From there, we deduce the case of a polyhedron in a similar manner to Sect. 5.

Theorem 7 *Let V be a vector space with lattice Λ. Fix a Euclidean scalar product Q on V. Let $\mathfrak{p} \subseteq V$ be a semi-rational polyhedron of dimension ℓ. For any test function $h(x)$ on V, the following asymptotic expansion with step-polynomial coefficients holds when $t \to +\infty$ ($t \in \mathbb{R}$),*

$$\frac{1}{t^\ell} \sum_{x \in t\mathfrak{p} \cap \Lambda} h\left(\frac{x}{t}\right) \sim$$
$$\int_{\mathfrak{p}} h(x)dx + \sum_{k \geq 1} \frac{1}{t^k} \sum_{\substack{m \geq 0, \mathfrak{f} \in \mathcal{F}(\mathfrak{p}) \\ m+\ell-\dim \mathfrak{f} = k}} \int_{\mathfrak{f}} \left(\left(\mu(t\mathfrak{N}(\mathfrak{p}, \mathfrak{f}))_{[m]} \left(\frac{\partial}{\partial x} \right) \right) \cdot h \right) (x) dm_{\mathfrak{f}}(x). \tag{39}$$

Moreover, the differential operators which appear in (39) are unique in the sense that, if $\rho_m(\mathfrak{p}, \mathfrak{f}, t)(\xi)$ is a family of polynomials on V^* with step-polynomial coefficients of $t \in \mathbb{R}$, such that (39) holds for any test function $h(x)$ and such that $\rho_m(\mathfrak{p}, \mathfrak{f}, t)(\xi)$ is homogeneous of degree m and depends only on the Q-projection of

ξ *on* $(\mathrm{lin}(\mathfrak{f}))^{\perp}$, *then*

$$\rho_m(\mathfrak{p}, \mathfrak{f}, t)(\xi) = \mu_{[m]}(t\mathfrak{N}(\mathfrak{p}, \mathfrak{f}))(\xi).$$

Example 13 For the triangle $\mathfrak{p}$ with vertices $(0, 0)$, $(1, 0)$,$(0, 1)$, the asymptotic expansion with step-polynomial coefficients, for $t \in \mathbb{R}$, is

$$\frac{1}{t^2} \sum_{(x_1,x_2)\in t\mathfrak{p}\cap\mathbb{Z}^2} h\left(\frac{x_1}{t}, \frac{x_2}{t}\right) \sim \int_{\mathfrak{p}} h + \frac{1}{t} \int_{\partial(\mathfrak{p})} \left(\frac{1}{2} - \{t\}\right) h +$$

$$\frac{1}{t^2} \left(-\frac{1}{12} \left(\int_{\mathfrak{f}_1} \partial_{x_2} h + \int_{\mathfrak{f}_2} \partial_{x_1} h \right) + \left(-\frac{1}{12} + \frac{\{t\}}{2} - \frac{\{t\}^2}{2} \right) \int_{\mathfrak{f}_3} \left(-\frac{\partial_{x_1} + \partial_{x_2}}{2} \right) \cdot h \right)$$

$$+ \frac{1}{4} h(0, 0) + \frac{3}{8} h(1, 0) + \frac{3}{8} h(0, 1) + \cdots$$

When we restrict this to $t \in \mathbb{N}$, we recover the formula with constant coefficients of Example 9.

Appendix: Renormalization

Decomposition of the Space of Rational Functions with Poles in a Finite Set of Hyperplanes

In this section which is elementary linear algebra, the base field need not be $\mathbb{R}$.

Let $\Delta = \{v_1, \ldots, v_N\}$ be a finite set of non zero elements of V. We assume that Δ does not contain collinear vectors. We denote by $\mathcal{R}_\Delta$ the algebra of rational functions on V^* of the form

$$\frac{P(\xi)}{\prod_j \langle v_j, \xi\rangle^{n_j}}$$

where $P(\xi)$ is a polynomial function on V^* and $n_j \geq 0$. We will denote this fraction simply by $\frac{P}{\prod_j v_j^{n_j}}$. Its poles are contained in the union of the hyperplanes $\langle v_j, \xi\rangle = 0$.

All statements and proofs of this appendix are quite easy when the v_j's are linearly independent.

Definition 6 If L is a subspace of V, we denote by $G_{\Delta\cap L}$ the subspace of $\mathcal{R}_\Delta$ spanned by the fractions

$$\frac{1}{\prod_{j\in J} v_j^{n_j}}$$

with $n_j > 0$, $v_j \in L$, and $\{v_j, j \in J\}$ span L.

We denote by $B_{\Delta\cap L}$ the subspace of $G_{\Delta\cap L}$ spanned by the fractions $\frac{1}{\prod_{j\in J} v_j}$ where $(v_j, j \in J)$ is a basis of L. Thus $B_{\Delta\cap L}$ is the homogeneous summand of degree $-\dim L$ of $G_{\Delta\cap L}$. For $L = V$, we denote these spaces simply by G_Δ and B_Δ.

Definition 7 A *complement map* is a map $L \mapsto C(L)$ which associates to any subspace $L \subset V$ a complementary subspace $C(L)$, so that $V = L \oplus C(L)$.

For example, a scalar product Q defines a complement map.

We recall that the symmetric algebra $\mathrm{Sym}(V)$ is canonically identified with the algebra of polynomial functions on V^* and the symmetric algebra of the dual, $\mathrm{Sym}(V^*)$, is canonically identified with the algebra of constant coefficients differential operators on V^* (for clarity, we will denote by ∂_γ the differentiation with respect to $\gamma \in V^*$). Given a complement map, if L is a subspace of V, we can consider $\mathrm{Sym}(C(L))$ as a subspace of $\mathrm{Sym}(V)$.

Theorem 8 *Let C be a complement map on the set of subspaces of V. We have a direct sum decomposition*

$$\mathcal{R}_\Delta = \oplus_{L\subseteq V} \mathrm{Sym}(C(L)) \otimes G_{\Delta\cap L}.$$

Of course, the sum runs only over the set of subspaces which are spanned by elements of Δ. The summand corresponding to $L = \{0\}$ is the space of polynomials $\mathrm{Sym}(V)$.

Remark 4 If the base field is $\mathbb{R}$, and Δ is contained in an open half-space $\lambda > 0$, (this is realized by replacing v_j by $-v_j$ if necessary), then for $R \in \mathcal{R}_\Delta$, the algebraic decomposition of Theorem 8 can be translated into a decomposition of the inverse Fourier transform $\mathcal{F}^{-1}(\lim_\lambda R)$ into a sum of distributions each supported on a subspace L. Thus the uniqueness part of the theorem is easy to prove, as in the proof of Theorem 5.

Example 14 If $\dim V = 1$, then Δ consists of just one element say $\langle v_1, \xi\rangle = \xi$, there are two subspaces $L = V$ and $L = \{0\}$. Then $\mathcal{R}_\Delta$ consists of fractions $\frac{f(\xi)}{\xi^n}$. Such a fraction can be written in a unique way as $\frac{a_n}{\xi^n} + \cdots + \frac{a_1}{\xi} + g(\xi)$, where a_j are constants and $g(\xi)$ is a polynomial.

Proof The result is certainly not new. It is implicit in [6] and [4]. For completeness, we will give a full proof, to begin with, the following elementary lemma.

Lemma 5 *$G_{\Delta\cap L}$ is spanned by fractions of the form*

$$\frac{1}{\prod_{j\in J} v_j^{n_j}} \text{ where } (v_j, j \in J) \text{ is a basis of } L. \tag{40}$$

For example, $\frac{1}{(v_1+v_2)v_1v_2} = \frac{1}{(v_1+v_2)^2 v_2} + \frac{1}{(v_1+v_2)^2 v_1}$.

Proof It is enough to prove that the space spanned by fractions of the form 40 is stable by multiplication by $\frac{1}{v^n}$ where $v \in \Delta \cap L$. Theorem 2

We decompose $v = \sum_{j\in J} c_j v_j$ in the basis $(v_j, j \in J)$ of L. Then

$$\frac{1}{v^n \prod_{j\in J} v_j^{n_j}} = \frac{v}{v^{n+1} \prod_{j\in J} v_j^{n_j}} = \sum_{j\in J} c_j \frac{1}{v^{n+1} v_j^{n_j-1} \prod_{k\in J, k\neq j} v_k^{n_k}}.$$

Pick j such that $c_j \neq 0$. If $n_j = 1$, the j term is of the form (40), with v_j replaced by v in the basis $(v_j, j \in J)$ of L. If $n_j > 1$, we repeat the procedure on the j term, with v^n replaced by v^{n+1}, so the assertion follows by induction on $\sum_j n_j$. □

The equality $\mathcal{R}_\Delta = \sum_{L\subseteq V} \mathrm{Sym}(C(L)) \otimes G_{\Delta\cap L}$ follows easily from this lemma, by using the decomposition $\mathrm{Sym}(V) = \mathrm{Sym}(C(L)) \otimes S(L)$. There remains to prove that the sum is direct. The main step is [4], Theorem 1, which we recall in the following lemma.

We introduce the subspace $N_\Delta \subset \mathcal{R}_\Delta$ spanned by fractions $\frac{P}{\prod_{j\in J} v_j^{n_j}}$ where $(v_j, j \in J)$ *do not* span V. Thus

$$N_\Delta = \sum_{L\subsetneq V} \mathrm{Sym}(C(L)) \otimes G_{\Delta\cap L}.$$

Lemma 6 *Assume that Δ spans V. Then (i) $\mathcal{R}_\Delta = G_\Delta \oplus N_\Delta$.*

(ii) N_Δ is the torsion $\mathrm{Sym}(V^*)$ *submodule of $\mathcal{R}_\Delta$.*
(iii) G_Δ is a free $\mathrm{Sym}(V^*)$ *module.*

Proof Let us prove that N_Δ is a torsion $\mathrm{Sym}(V^*)$-module. Let $\phi = \frac{P}{\prod_{j\in J} v_j^{n_j}}$, where $(v_j, j \in J)$ **do not** span V.

There exists $\gamma \neq 0$ such that $\langle \gamma, v_j \rangle = 0$ for every $j \in J$. For N large enough, $\partial_\gamma^N \cdot \phi = \frac{\partial_\gamma^N \cdot P}{\prod_{j\in J} v_j^{n_j}} = 0$.

We prove the lemma by induction over $\dim V$.

We consider the subspace B_Δ of G_Δ spanned by fractions of the form

$$\frac{1}{\prod_{j\in B} v_j}$$

where $(v_j, j \in B)$ is a basis of V. It follows from Lemma 5 that G_Δ is generated as $\mathrm{Sym}(V^*)$-module by B_Δ.

Let us first prove that $B_\Delta \cap N_\Delta = 0$. We use the notion of partial residue. If $\phi \in B_\Delta$, the poles of ϕ are simple and contained in Δ. Let us fix an element of Δ, say v_1. Let $V_0 = V/\mathbb{R}v_1$, thus $V_0^* = v_1^\perp$. Let Δ_0 be the set of projections of the other elements of Δ on V_0. Thus $\mathcal{R}_{\Delta_0}$ consists of functions on V_0^*.

Definition 8 If v_1 is a simple pole of $\phi \in \mathcal{R}_\Delta$, the residue $\mathrm{Res}_{v_1}\phi$ is the element of $\mathcal{R}_{\Delta_0}$ defined by

$$\mathrm{Res}_{v_1}\phi = (v_1\phi)|_{v_1^\perp}.$$

Let $\phi \in B_\Delta \cap N_\Delta$. It is clear that $\mathrm{Res}_{v_1}\phi \in B_{\Delta_0}$. Let us show that $\mathrm{Res}_{v_1}\phi = 0$. We write $\phi = \sum_L g_L$ with $g_L \in \mathrm{Sym}(C(L)) \otimes G_{\Delta\cap L}$. Let $(L_a, a \in A)$ be the subspaces which contains v_1. For each $a \in A$, we pick $\gamma_a \in L_a^\perp$, $\gamma_a \neq 0$. For N large enough, $\partial_{\gamma_a}^N \cdot g_{L_a} = 0$. Let $D = \prod_{a\in A}\partial_{\gamma_a}^N$. Then $D \cdot g_{L_a} = 0$ for every $a \in A$.

If $v_1 \notin L$, then v_1 is not a pole of g_L, hence not a pole of $D \cdot g_L$ either. Therefore v_1 is not a pole of $D \cdot \phi$, so $\mathrm{Res}_{v_1}(D \cdot \phi) = 0$. The elements γ_a are in $v_1^\perp = V_0^*$, so that $\prod_a {\gamma_a}^N \in \mathrm{Sym}(V_0^*)$, and

$$D \cdot \mathrm{Res}_{v_1}\phi = \mathrm{Res}_{v_1}(D \cdot \phi) = 0.$$

Thus $\mathrm{Res}_{v_1}\phi$ lies in the $\mathrm{Sym}(V_0^*)$-torsion module of $\mathcal{R}_{\Delta_0}$. As $\mathrm{Res}_{v_1}\phi \in B_{\Delta_0}$, we have $\mathrm{Res}_{v_1}\phi = 0$ by the induction hypothesis. Thus v_1 is not a pole of ϕ. Since v_1 was any element of Δ, ϕ has no poles, so $\phi = 0$.

Next, let us prove $G_\Delta \cap N_\Delta = \{0\}$. Let $\phi_1, \ldots, \phi_s$ be a basis of B_Δ (over the base field). Then any fraction $\phi \in G_\Delta$ can be written as

$$\phi = \sum_{j=1}^{s} D_j \cdot \phi_j$$

for some D_j's in $\mathrm{Sym}(V^*)$. Assume $\phi \in N_\Delta$. We want to prove that $D_j = 0$ for every j. We can assume that ϕ is homogeneous, so D_j's are homogeneous with the same degree $k > 0$ (otherwise, $\phi \in B_\Delta \cap N_\Delta$, so $\phi = 0$). Assume $D_1 \neq 0$. There exists $v \in V$ such that the Lie bracket $[v, D_1] = v \circ D_1 - D_1 \circ v$ is not zero. The brackets $[v, D_j]$ are elements of $\mathrm{Sym}(V^*)$ of degree $k - 1$.

We have $v\phi \in N_\Delta$ and $\sum_{j\in J} D_j \cdot v\phi_j \in N_\Delta$, hence

$$\sum_{j\in J}[v, D_j] \cdot \phi_j \in G_\Delta \cap N_\Delta.$$

Iterating, we obtain a linear relation $\sum_{j\in J} c_j\phi_j \in N_\Delta$ with $c_1 \neq 0$, which contradicts the fact that $B_\Delta \cap N_\Delta = 0$.

We have proved $G_\Delta \cap N_\Delta = \{0\}$, hence (i). We have also proved that $(\phi_1, \ldots, \phi_s)$ is a basis of G_Δ as $\mathrm{Sym}(V^*)$-module, hence (iii), hence also (ii). □

We will now prove that $\mathrm{Sym}(C(L)) \otimes G_{\Delta\cap L}$ is a direct summand by induction on $\mathrm{codim}\,L$. The case $\mathrm{codim}\,L = 0$ is Lemma 6, (i). So, let us assume that $\mathrm{Sym}(C(L)) \otimes G_{\Delta\cap L}$ is a direct summand for all subspaces L (spanned by elements of Δ) which have codimension $\leq k$. Let $L_0, L_1, \ldots, L_s$ be the list of subspaces (spanned by elements of Δ) which have codimension $k+1$. If $s=0$, then we must have $k+1=\dim V$, hence $L_0=\{0\}$, so we are done. So we assume that $s\geq 1$.

Assume that we have a family $f_L \in \mathrm{Sym}(C(L))\otimes G_{\Delta\cap L}$ such that

$$\sum_{L,\operatorname{codim} L\leq k+1} f_L = 0.$$

We are going to prove that $f_{L_0}=0$. For $j=1,\dots,s$, we choose $\gamma_j\in L_j^{\perp}\subset V^*$ such that $\gamma_j\notin L_0^{\perp}$. For N large enough, $D=\prod_{j=1}^{s}\partial_{\gamma_j}^{N}$ kills all terms f_L if L is contained in some L_j with $j=1,\dots,s$. So $D\cdot f_{L_0}=0$ as well. We decompose

$$V^* = C(L_0)^{\perp}\oplus L_0^{\perp}.$$

Then

$$\mathrm{Sym}(V^*)=\mathrm{Sym}(C(L_0))^{\perp})\otimes \mathrm{Sym}(L_0^{\perp})=\oplus_n S_n(C(L_0)^{\perp})\otimes \mathrm{Sym}(L_0^{\perp}).$$

Let $D=D_0+D_{>0}$ with respect to the degree on $\mathrm{Sym}(C(L_0))^{\perp})$. Let us show that $D_0\neq 0$.

Let $\gamma_j=\alpha_j+\beta_j$ in the decomposition $V^*=C(L_0)^{\perp}\oplus L_0^{\perp}$. Then $\alpha_j\neq 0$, and $D_0=\prod_{j=1}^{s}\alpha_j^N$ is not zero.

We decompose $f_{L_0}\in \mathrm{Sym}(C(L_0)))\otimes G_{L_0\cap\Delta}=\oplus_n S_n(C(L_0))\otimes G_{L_0\cap\Delta}$ with respect to the degree on $\mathrm{Sym}(C(L_0))$,

$$f_{L_0}=f_n+f_{n-1}+\cdots.$$

We observe that $D_{>0}$ lowers strictly the degree with respect to $\mathrm{Sym}(C(L_0))$, while D_0 acts only on the $G_{L_0\cap\Delta}$ factor. Therefore we obtain

$$D_0\cdot f_n=0.$$

We deduce $f_n=0$ by applying Lemma 6, (iii), with V replaced by the subspace L_0, and L_0^* identified with $C(L_0)^{\perp}$. Thus, we have proved $f_{L_0}=0$. □

Renormalization of Meromorphic Functions with Hyperplane Singularities

From now on, we will assume that the complement map is given by a scalar product Q on V. We will denote it by C_Q.

Definition 9 The projection map on the summand corresponding to $L = \{0\}$ in Theorem 8 is called the renormalization map with respect to Q and is denoted by

$$R_Q : \mathcal{R}_\Delta \to \mathrm{Sym}(V).$$

Remark 5 The renormalization map extends into a map from the space of meromorphic functions with hyperplane singularities to the space of holomorphic functions near $\xi = 0$.

$$R_Q : \mathcal{M}_h(V^*) \to \mathcal{H}(V^*)$$

First, if f is a rational fraction with hyperplane singularities, we apply Theorem 8 with Δ any set containing the singular hyperplanes. For a meromorphic function f, we renormalize the Taylor expansion term by term.

Example 15 In dimension one, if $\phi(\xi)$ is holomorphic near 0, we have $\frac{\phi(\xi)}{\xi} = \frac{\phi(0)}{\xi} + \frac{\phi(\xi)-\phi(0)}{\xi}$, hence $R_Q(\frac{\phi(\xi)}{\xi}) = \frac{\phi(\xi)-\phi(0)}{\xi}$. For instance,

$$R_Q\left(\frac{1}{1-\mathrm{e}^{\xi}}\right) = \frac{1}{1-\mathrm{e}^{\xi}} + \frac{1}{\xi}.$$

In dimension 2, when Q is the standard scalar product, the decomposition of Theorem 8 is

$$\frac{\phi(\xi_1,\xi_2)}{\xi_1\xi_2} = \frac{\phi(0,0)}{\xi_1\xi_2} + \frac{1}{\xi_2}\left(\frac{\phi(\xi_1,0)-\phi(0,0)}{\xi_1}\right) + \frac{1}{\xi_1}\left(\frac{\phi(0,\xi_2)-\phi(0,0)}{\xi_2}\right) + \frac{\phi(\xi_1,\xi_2)-\phi(\xi_1,0)-\phi(0,\xi_2)+\phi(0,0)}{\xi_1\xi_2},$$

hence

$$R_Q\left(\frac{\phi(\xi_1,\xi_2)}{\xi_1\xi_2}\right) = \frac{\phi(\xi_1,\xi_2)-\phi(\xi_1,0)-\phi(0,\xi_2)+\phi(0,0)}{\xi_1\xi_2}.$$

For a function which has only simple poles which are linearly independent, renormalization provides a nice recursive formula for computing all the terms in the decomposition of Theorem 8. It is implemented in a Maple program [2].

Proposition 5 *Let $\Delta = \{v_1, \dots, v_s\}$ be a set of linearly independent vectors in V. For $J \subseteq \Delta$, let L_J be the subspace of V spanned by $(v_j, j \in J)$. Let $P \in \mathrm{Sym}(V)$ be a polynomial. Then the decomposition of $\frac{P}{v_1 \cdots v_s}$ in Theorem 8 is given by the following* *formula*

$$\frac{P}{v_1 \cdots v_s} = \sum_{J \subseteq \Delta} R_Q \left(\frac{P}{\prod_{k \notin J} v_k} |_{L_J^\perp} \right) \frac{1}{\prod_{j \in J} v_j}. \tag{41}$$

Here, $L_J^\perp \subseteq V^*$ *is identified with* $C_Q(L_J)^*$.

Example 15 illustrates the proposition.

Proof The proof is by induction on dim V. The decomposition of $\phi = \frac{P}{v_1 \cdots v_s}$ takes the form

$$\frac{P}{v_1 \cdots v_s} = \sum_{J \subseteq \Delta} \frac{P_J}{\prod_{j \in J} v_j}. \tag{42}$$

with $P_J \in \mathrm{Sym}(C_Q(L_J))$.

Let us prove that $P_J = R_Q \left(\frac{P}{\prod_{k \notin J} v_k} |_{L_J^\perp} \right)$ if $1 \in J$. We compute the residue Res_{v_1} of both sides.

We obtain

$$\frac{\bar{P}}{\bar{v}_2 \cdots \bar{v}_s} = \sum_{\{J, 1 \in J\}} \frac{\bar{P}_J}{\prod_{j \in J, j > 1} \bar{v}_j} \tag{43}$$

where $\bar{P}, \bar{P}_J, \bar{v}_j$ denote the restriction to $v_1^\perp$. When we identify $v_1^\perp$ with $C_Q(v_1)^*$, we have $\bar{v}_j \in C_Q(v_1)$, $\bar{P} \in \mathrm{Sym}(C_Q(v_1))$, and (43) is the decomposition of $\frac{\bar{P}}{\bar{v}_2 \cdots \bar{v}_s}$. For J such that $1 \in J$, we have $L_J^\perp \subseteq v_1^\perp$, so $\bar{P}_J = P_J$. So, by induction,

$$P_J = \bar{P}_J = R_Q \left(\frac{\bar{P}}{\prod_{k \notin J} \bar{v}_k} |_{\bar{L}_{\bar{J}}^\perp} \right),$$

where $\bar{J} = J \setminus \{1\}$, $\bar{L}_{\bar{J}} \subset C_Q(v_1)$ is the space spanned by $(\bar{v}_j, j \in \bar{J})$ and $\bar{L}_{\bar{J}}^\perp$ denotes the orthogonal of $\bar{L}_{\bar{J}}$ in $C_Q(v_1)$. Thus, $\frac{\bar{P}}{\prod_{k \notin J} \bar{v}_k} |_{\bar{L}_{\bar{J}}^\perp}$ is identified with $\frac{P}{\prod_{k \notin J} v_k} |_{L_J^\perp}$. □

Renormalization of the Generating Function of a Cone

In this section, V is a rational space. We will give an alternative definition of the μ-function of an affine (rational polyhedral) cone as the renormalization of the generating function of the cone, and derive the properties of the μ-function from this construction. Renormalization leads also a fast algorithm for computing the μ-function for a simplicial cone in fixed dimension, using Barvinok's decomposition into unimodular cones.

Definition 10 Fix a scalar product Q on V.

(1) Let $\mathfrak{c} \subset V$ be a cone and let $s \in V$. Consider the affine cone $s + \mathfrak{c}$. Define

$$\mu_Q(s + \mathfrak{c})(\xi) = R_Q(\mathrm{e}^{-\langle \xi, s \rangle} S(s + \mathfrak{c})(\xi)).$$

Then $\mu_Q(s + \mathfrak{c})(\xi)$ is a holomorphic function on V^* near $\xi = 0$.

(2) Let W be a rational quotient of V and let $\mathfrak{a} \subset W$ be an affine cone. Then W^* is a subspace of V^*.

The function $\mu_Q(\mathfrak{a})$ on W^* is extended to V^* by 0 on the orthogonal complement of W^* with respect to Q.

The valuation property of the μ-function follows immediately from this definition.

Proposition 6 *μ_Q is a valuation on the set of affine cones in V with a fixed vertex s.*

Proof The map $\mathfrak{c} \mapsto \mathrm{e}^{-\langle \xi, s \rangle} S(s + \mathfrak{c})(\xi)$ is a valuation on this set. □

We will now show that the local Euler-Maclaurin formula for generating functions of cones [3, Theorem 20] can be easily derived from this definition and the results on the poles and residues of $S(\mathfrak{a})$. (This result is also obtained in [8]).

Theorem 9 *Let V be a rational vector space. Fix a rational Euclidean scalar product on Q. Let $\mathfrak{a} \subset V$ be a rational affine cone. If $\mathfrak{f}$ is a face of $\mathfrak{a}$, let $I(\mathfrak{f})$ be the continuous generating function of $\mathfrak{f}$ (cf. Sect. 2.2) and let $\mathfrak{N}(\mathfrak{a}, \mathfrak{f}) \subset V/\operatorname{lin}(\mathfrak{f})$ denote the transversal cone to $\mathfrak{a}$ along $\mathfrak{f}$. Then*

$$S(\mathfrak{a}) = \sum_{\mathfrak{f}} \mu_Q(\mathfrak{N}(\mathfrak{a}, \mathfrak{f})) I(\mathfrak{f}). \tag{44}$$

The sum runs over the set of faces of $\mathfrak{a}$.

Proof The proof is by induction on $\dim \mathfrak{a}$ and relies on the determination of the residues of $S(\mathfrak{a})$ and $I(\mathfrak{a})$.

Lemma 7 *The poles of $S(\mathfrak{a})$ and $I(\mathfrak{a})$ are the edge generators of $\mathfrak{a}$. The residues $\operatorname{Res}_v S(\mathfrak{a}) = vS(\mathfrak{a})|_{v^\perp}$ and $\operatorname{Res}_v I(\mathfrak{a}) = vI(\mathfrak{a})|_{v^\perp}$ along an edge v are given by*

$$\operatorname{Res}_v S(\mathfrak{a}) = -S(\operatorname{proj}_{V/v} \mathfrak{a}), \tag{45}$$

$$\operatorname{Res}_v I(\mathfrak{a}) = -I(\operatorname{proj}_{V/v} \mathfrak{a}), \tag{46}$$

provided v is a primitive lattice vector.

The analogous formula for any *intermediate* generating function of $\mathfrak{a}$ is proven in [1, Proposition 3.5], using a Poisson summation formula. The following proof for the purely discrete and purely continuous generating functions is more elementary.

Proof Using a subdivision of $\mathfrak{a}$ into simplicial cones without adding edges, we can assume that $\mathfrak{a}$ is simplicial. Furthermore, we can also assume that $\mathfrak{a}$ is full dimensional. Let $(v_1, \ldots, v_d)$ be the primitive lattice edge generators of $\mathfrak{a}$, with $v = v_1$. Let s be the vertex of $\mathfrak{a}$. Let $\mathfrak{b} = \sum_j [0, 1[v_j$ and $S(s+\mathfrak{b})(\xi) = \sum_{x\in(s+\mathfrak{b})\cap\Lambda} e^{\langle\xi,x\rangle}$, a finite sum, so that $S(s+\mathfrak{b})(\xi)$ is analytic. We have

$$S(\mathfrak{a}) = S(s+\mathfrak{b})\frac{1}{\prod_{j=1}^{d} 1 - e^{v_j}}.$$

We have $\mathrm{Res}_{v_1} \frac{1}{1-e^{v_1}} = -1$, hence

$$\mathrm{Res}_{v_1} S(\mathfrak{a}) = -S(\mathrm{proj}_{V/v_1}(s+\mathfrak{b}))\frac{1}{\prod_{j=2}^{d} 1 - e^{\bar{v}_j}}$$

where $\bar{v}_j$ is the projection of v_j on V/v_1. The right-hand side is precisely $-S(\mathrm{proj}_{V/v_1}\mathfrak{a})$, even though the projected vectors $\bar{v}_j$ may not be primitive with respect to the projected lattice $\bar{\Lambda}$. So we have proved (45).

We have $I(\mathfrak{a}) = \frac{(-1)^d|\det(v_1,\ldots,v_d)|e^s}{\prod_{j=1}^{d} v_j}$, hence

$$\mathrm{Res}_{v_1} I(\mathfrak{a}) = \frac{(-1)^d|\det_\Lambda(v_1,\ldots,v_d)|e^{\bar{s}}}{\prod_{j=2}^{d}\bar{v}_j}.$$

Since v_1 is primitive, we have $|\det_\Lambda(v_1,\ldots,v_d)| = |\det_{\bar{\Lambda}}(\bar{v}_2,\ldots,\bar{v}_d)|$. Therefore the right-hand side of the residue above is equal to $-I(\mathrm{proj}_{V/v_1}\mathfrak{a})$. So we have proved (46). □

Let Δ be the set of primitive edge generators of $\mathfrak{a}$. We write the decomposition given by Theorem 8.

$$S(\mathfrak{a}) = \sum_L g_L. \tag{47}$$

First, we observe that $g_L \in \mathrm{Sym}(L^{\perp_Q}) \otimes B_L$, in other words, g_L has only simple poles. This is clear if $\mathfrak{a}$ is simplicial, so it is true also in the general case, using a subdivision of $\mathfrak{a}$ with no edges added.

Pick $v \in \Delta$. The set of faces of the projected cone $\mathrm{proj}_{V/v}\mathfrak{a}$ is in one-to-one correspondence with the set of faces of $\mathfrak{a}$ which have v as an edge. For such a face $\mathfrak{f}$, the transversal cone of $\mathrm{proj}_{V/v}\mathfrak{a}$ along its face $\mathrm{proj}_{V/v}\mathfrak{f}$ coincides with the transversal cone $\mathfrak{N}(\mathfrak{a},\mathfrak{f}) \subset V/\mathrm{lin}(\mathfrak{f})$. Moreover, if v is an edge of $\mathfrak{f}$, we have $I(\mathrm{proj}_{V/v}\mathfrak{f}) = -\mathrm{Res}_v I(\mathfrak{f})$ by Lemma 7. On the other hand, if v is not an edge of $\mathfrak{f}$, we have $\mathrm{Res}_v I(\mathfrak{f}) = 0$. So, by Lemma 7 and the induction hypothesis applied to the cone $\text{proj}_{V/v}\, \mathfrak{a}$, we have

$$\text{Res}_v\, S(\mathfrak{a}) = \sum_{\mathfrak{f}, v \in \mathfrak{f}} \mu(\mathfrak{N}(\mathfrak{a}, \mathfrak{f})) I(\text{proj}_{V/v}\, \mathfrak{f}) = \sum_{\mathfrak{f}} \text{Res}_v(\mu(\mathfrak{N}(\mathfrak{a}, \mathfrak{f})) I(\mathfrak{f})).$$

In conclusion, we have $\text{Res}_v\, g_L = 0$ if L is not a face and

$$\text{Res}_v\, g_{\text{lin}(\mathfrak{f})} = \text{Res}_v(\mu(\mathfrak{N}(\mathfrak{a}, \mathfrak{f})) I(\mathfrak{f}))$$

for any edge v of $\mathfrak{f}$. It is clear that an element of $\text{Sym}(L^{\perp_Q}) \otimes B_L$ is zero if all its residues are zero. Therefore, $g_L = 0$ if L is not a face and $g_{\text{lin}(\mathfrak{f})} = \mu(\mathfrak{N}(\mathfrak{a}, \mathfrak{f})) I(\mathfrak{f})$ for any face $\mathfrak{f}$ of $\mathfrak{a}$. □

References

1. V. Baldoni, N. Berline, J.A. De Loera, M.Köppe, M. Vergne, Intermediate sums on polyhedra II: bidegree and Poisson summation formula (2014), arXiv:1404.0065
2. V. Baldoni, N. Berline, M. Vergne, *General Mu*, Maple program, http://nicole.berline.perso.math.cnrs.fr/maple.html
3. N. Berline, M. Vergne, Local Euler-Maclaurin formula for polytopes. Mosc. Math. J. **7**(3), 355–386, 573 (2007)
4. M. Brion, M. Vergne, Arrangement of hyperplanes. I. Rational functions and Jeffrey-Kirwan residue. Ann. Sci. Éc. Norm. Sup. (4) **32**(5), 715–741 (1999)
5. H. Cohen, *Number Theory. Vol. II. Analytic and Modern Tools*. Graduate Texts in Mathematics, vol. 240 (Springer, New York, 2007)
6. C. De Concini, C. Procesi, *Topics in Hyperplane Arrangements, Polytopes and Box-Splines*. Universitext (Springer, New York, 2011)
7. V. Guillemin, S. Sternberg, Riemann sums over polytopes. Ann. Inst. Fourier (Grenoble) **57**(7), 2183–2195 (2007), Festival Yves Colin de Verdière.
8. L. Guo, S. Paycha, B. Zhang, Renormalization and the Euler-Maclaurin formula on cones (2015), arXiv:1306.3420v2
9. G.H. Hardy, E.M. Wright, *An Introduction to the Theory of Numbers*, 6th edn. (Oxford University Press, Oxford, 2008); Revised by D. R. Heath-Brown and J. H. Silverman, With a foreword by Andrew Wiles.
10. Y. Karshon, S. Sternberg, J. Weitsman, Euler-Maclaurin with remainder for a simple integral polytope. Duke Math. J. **130**(3), 401–434 (2005)
11. Y. Le Floch, A. Pelayo, Euler-Maclaurin formulas via differential operators (2013), arXiv:1312.5711v3
12. T. Tate, Asymptotic Euler-Maclaurin formula over lattice polytopes. J. Funct. Anal. **260**(2), 501–540 (2011)

Leray Spectral Sequence for Complements of Certain Arrangements of Smooth Submanifolds

Andrzej Weber

Abstract Let X be a complex algebraic manifold. Let U be the complement of a configuration of submanifolds. We study the Leray spectral sequence of the inclusion $U \hookrightarrow X$ computing the cohomology of U. Under some condition posed on the intersections of submanifolds we show that the Leray spectral sequence degenerates on E_3. This result generalizes well known properties of hyperplane arrangements. The main cause which rigidifies the spectral sequence is the weight filtration in cohomology.

Keywords Deligne spectral sequence • Hodge structure • Leray spectral sequence

Spectral sequence argument quite often appears in study the hyperplane arrangements or configuration spaces (see for example [2–4, 8, 11, 12, 17]). Very often the spectral sequence can be identified with the Leray spectral sequence of the embedding $j : U \hookrightarrow X$ of an open subset to a compact complex variety

$$E_2^{p,q} = H^p(X; R^q j_* \mathcal{L}) \;\Rightarrow\; H^{p+q}(U; \mathcal{L})\,. \tag{1}$$

When the coefficient system has geometric nature, then by Deligne the entries of the spectral sequence are equipped with the weight filtration in the sense of the mixed Hodge structure [5, 6] or in the sense of the Weil sheaves [1, 7]. For the constant coefficients $\mathbb{Q}_U$ it happens quite often that the entries of E_2 table are of pure weight. That is so for:

(1) Complements of the classical hyperplane arrangements in $\mathbb{P}^n$.
(2) Complements of elliptic hyperplane arrangements.
(3) In general: complements of hyperplane-like arrangements.

Purity implies the degeneration of the spectral sequence on E_3. The argument of purity can be applied in a much more general setup. We can add to the list above

A. Weber (✉)
Department of Mathematics, University of Warsaw, Banacha 2, 02-097 Warszawa, Poland
e-mail: aweber@mimuw.edu.pl

F. Callegaro et al. (eds.), *Configuration Spaces*, Springer INdAM Series 14,
DOI 10.1007/978-3-319-31580-5_4

(4) Configuration space of n points in a projective manifold Y considered as an open subset of $X = Y^n$

It was already noticed by Totaro [17] that if $X = Y^n$ for a smooth compact algebraic variety Y and U, the set of tuples consisting of pairwise distinct points, i.e., when U is the configuration space of n points in Y, then the Leray spectral sequence has only one nontrivial differential.

Our goal is to add to the list of spaces with pure sheaves $R^q j_* \mathbb{Q}_U$ another class of "arrangements", which we call "admissible arrangements of submanifolds". The members of the arrangements are smooth subvarieties of X, but we allow some kind of singular intersections. For example we allow that two subvarieties are tangent along a smooth subvariety of certain codimension, see Definition 1. We add to the list of open sets for which Leray sequence degenerates another item

(5) Complements of admissible arrangements of submanifolds in a compact smooth algebraic variety.

The concept of weight is central for this paper. Talking about weight filtration in the cohomology of complex algebraic variety one obviously uses Deligne construction [5, 6]. While dealing with sheaves there are (at least) two approaches possible. It is natural to work in the category of mixed Hodge modules developed by Saito [14]. Nevertheless we feel more comfortable using the reduction to $\mathbb{Q}_\ell$-sheaves on a variety over a finite characteristic field, as in [1, 7]. If we say that a constructible sheaf of $\mathbb{Q}$-vector spaces is pure of certain weight we mean that after tensoring with $\mathbb{Q}_\ell$ and after a good reduction to a finite characteristic p with $(p, \ell) = 1$, the sheaf is pure in the sense of [1, §5.1] (i.e., in the sense of the action of the Frobenius automorphism on stalks and costalks).

Many results about arrangements of hyperplanes can be obtained purely topologically, one can say "elementarily". Sometimes the weight argument is hidden somewhere in the references. Our goal is to review some basic mechanism and extend a bit the situation when it can be applied. The weight argument works for cohomology with rational coefficients. Some of the results might hold for integer homology, but the weight argument is too weak to handle torsion.

For the twisted coefficients, i.e., when the local system $\mathcal{L}$ is not constant, the Leray spectral sequence allows to generalize proofs done in particular situations. The conditions on the monodromy eigenvalues **(Mon)** of Sect. 4 implies that the stalks $R^k j_* \mathcal{L}$ vanish at points $x \in X \setminus U$. Therefore

$$H^*(U; \mathcal{L}) = H^*_c(U; \mathcal{L}),$$

where $H^*_c(-)$ denotes cohomology with compact supports. To prove that no use of Hodge theory is needed. The situation when the stalks of $R^k j_* \mathcal{L}$ do not vanish is more interesting. Another condition, called $\mathbf{(Mon)}^*$ in the classical setup, guaranties that the local cohomology is generated by logarithmic forms. This condition was studied e.g., in [15, Theorem 4.1]. If the local system $\mathcal{L}$ is equipped with a pure Hodge structure it seems that the stalks should be pure. The problem lies in the choice of a convenient category in which purity is understood. A natural candidate would be the category of *mixed Hodge modules* of Saito [14]. Making this choice we would have to add highly nontrivial technical introduction. We limit ourself just to few formal remarks in the final section. The supporting evidence is that in the case of normal crossing divisor and unitary local system the Leray spectral sequence degenerates. This was proven by Timmerscheidt in [16].

This note was inspired by the Workshop on Configuration Spaces in Cortona. I'm grateful to Alex Suciu for conversations and encouragement.

1 The Results

We will consider the following situation. Let X be a smooth compact complex algebraic variety. Let r be a natural number. We define an admissible arrangement of codimension c submanifolds inductively.

Definition 1 An arrangement of distinct submanifolds $\mathcal{Z} = \{Z_i\}_{i=1,2,\ldots,s}$ is admissible of codimension c in X if

1. Each Z_i is smooth of codimension c.
2. For any set of indices $I \subset \{1, 2, \ldots, s\}$ and $j \notin I$ the set-theoretic intersection

$$\left(\bigcup_{i\in I} Z_i\right) \cap Z_j$$

is empty or it is the union of an admissible codimension c arrangement in each Z_j.

The definition of admissible arrangement of submanifolds is constructed in the way which allows to apply the argument of "deletion-restriction" of [4, Proposition 2.1] or [3, Lemma 2.1]. Clearly each arrangement of submanifolds satisfying the following condition is admissible:

★ *for any set of indices $I \subset \{1, 2, \ldots, s\}$ the set-theoretic intersection $\bigcap_{i\in I} Z_i$ is empty or each irreducible component of $\bigcap_{i\in I} Z_i$ is a smooth subvariety of codimension $c\ell$ for some $\ell \in \mathbb{N}$.*

To prove that ★ implies the condition of Definition 1 assume that Z_i's are connected and they do not repeat. Then the intersection $\left(\bigcup_{i\in I} Z_i\right) \cap Z_j = \bigcup_{i\in I}(Z_i \cap Z_j) \subset Z_j$ is the sum of smooth irreducible components. The members of the arrangement $\{Z_i \cap Z_j\}_{i\in I}$, if not empty, have codimension $\leq 2c$ in X and by the assumption ★ the codimension is a multiple of c. Therefore it has to be equal to $2c$. Hence $Z_i \cap Z_j$ has codimension c in Z_j or it is empty. Some members of this arrangement may repeat, but we replace the arrangement by the set of irreducible components of intersections $Z_i \cap Z_j$. It again satisfies the condition ★, therefore it satisfies the inductive step of Definition 1.

Remark 1 A condition related to ★ is considered in [8, §3.1]: the intersection $\bigcap_{i\in I} Z_i$ is assumed to be smooth. Then the topological components coincide with the irreducible components. On the other hand there is no assumption on the dimensions of intersections. The paper [8] is strictly related to ours.

In particular when $c = 1$ then ★ reduces to the condition

- $\{Z_i\}$ *is an arrangement of smooth hypersurfaces* Z_i*, such that for each set of indices* I *any irreducible component of the intersection* $\cap_{i\in I} Z_i$ *is smooth.*

Note that in [15, §3] there were considered hyperplane-like arrangements, i.e., arrangement of hypersurfaces which locally look like hyperplane arrangements. The hyperplane-like arrangements satisfy the Definition 1 with $c = 1$. Our definition allow more complicated examples: in the normal direction the intersection may look like the arrangement of three curves $x = 0$, $x = y^2$ and $y = 0$ on the plane. Another example of an admissible arrangement of hypersurfaces is the sum of two surfaces in $\mathbb{C}^3$ given by the equations $z = 0$ and $z = xy$. Again this is a picture in the normal slice, and the arrangement itself may be of high dimension.

Remark 2 Combinatorial nature of Leray spectral sequence for admissible arrangements of submanifolds has lead the authors of [8] to a definition of the combinatorial cover and a related spectral sequence. The relation of their construction with Leray spectral sequence is explained in [*loc.cit*, §2.2]. In [*loc.cit*, Theorem 3.1] a cover indexed by components of intersections $\bigcap_{i\in I} Z_i$ is constructed. The authors take sufficiently small tubular neighbourhoods of these components. This particular covering gives rise to a spectral sequence which is isomorphic to ours after some re-indexing of entries.

We will prove a theorem about degeneration of the spectral sequence for cohomology with trivial coefficients:

Theorem 1 *Let* $\mathcal{Z} = \{Z_i\}_{i=1,2,\dots,s}$ *be an admissible arrangement of codimension* c *submanifolds in* X*. Then the Leray spectral sequence of the inclusion*

$$j : U = X \setminus \bigcup_{i\in\{1,2,\dots s\}} Z_i \longrightarrow X$$

$$E_2^{p,q} = H^p(X; R^q j_* \mathbb{Q}_U) \Rightarrow H^{p+q}(U; \mathbb{Q})$$

has all differentials d_r*,* $r \geq 2$ *vanishing except from* d_{2c}*.*

The result follows from the local computation:

Theorem 2 *With the notation as above the sheaves* $R^k j_* \mathbb{Q}_U = 0$ *for* k *not divisible by* $2c - 1$ *and for* $k = (2c-1)\ell$ *the sheaf* $R^k j_* \mathbb{Q}_U$ *is pure of weight* $2c\ell = \frac{2c}{2c-1}k$.

After further work one can show:

Theorem 3 *With the notation as above the sheaf* $R^{(2c-1)\ell} j_* \mathbb{Q}_U$ *is isomorphic to a direct sum of constant sheaves supported by smooth subvarieties of codimensions* $c\ell$*.*

We would like to stress that our degeneration result, Theorem 1, specializes to many known cases mentioned in the introduction as the cases **(1)**–**(4)**.

2 Ideal Situation: Normal Crossing Divisor

Let us start with recalling the Deligne spectral sequence defining the weight filtration in the cohomology of the normal crossing divisor complement $X \setminus D$. The the sheaf of meromorphic forms on X with logarithmic poles along D admits the weight filtration defined by the number of factors in denominators. The iterated residue defines an isomorphism from the graded pieces of the filtration to the sheave of forms on the intersections of the divisors. One obtains a spectral sequence allowing to compute the cohomology of $X \setminus D$ from the cohomologies of the intersections. The main theorem says that the spectral sequence degenerates on E_2, hence the only thing one has to know is the first differential.

Let us be more precise. We fix the notation: Let X be a compact smooth complex algebraic variety. We assume that $D \subset X$ is a smooth divisor with normal crossings i.e., the irreducible components of $D = \bigcup_{i=1}^m D_i$ are smooth and locally D is given by the equation $z_1 z_2 \dots z_k = 0$ in a certain coordinate system. For a multindex $I = \{i_1, i_2, \dots, i_\ell\}$ define $X_I = D_{i_1} \cap D_{i_2} \cap \dots \cap D_{i_\ell}$ and $X^\ell = \bigsqcup_{|I|=\ell} X_I$ with $X^0 = X$. Deligne considers the spectral sequence associated to the weight filtration in the complex of logarithmic forms [5, (3.2.4.1)]. It converges to $H^*(X \setminus D)$. The first table of the weight spectral sequence lies in the second quarter and has the form

$$
{}_W E_1^{-p,q} = H^{q-2p}(X^p)\,,
$$

[5, (3.2.7)]. Here is the example for the surface case:

$$
\begin{array}{cccccccc}
0 & \to & H^0(X^2) & \to & H^2(X^1) & \to & H^4(X^0) & 4\\
 & & 0 & \to & H^1(X^1) & \to & H^3(X^0) & 3\\
 & & 0 & \to & H^0(X^1) & \to & H^2(X^0) & 2\\
 & & & & 0 & \to & H^1(X^0) & 1\\
 & & & & 0 & \to & H^0(X^0) & 0\\
-3 & & -2 & & -1 & & 0 &
\end{array}
$$

The differential ${}_W d_1$ is the alternating sum of Gysin maps

$$
H^*(X_I) \to H^{*+2}(X_{I\setminus\{i\}})
$$

induced by the inclusions $X_I \hookrightarrow X_{I\setminus\{i\}}$. Since we care about the Hodge structure we should rather write $H^*(X^I)(-1) \to H^{*+2}(X^{I\setminus\{i\}})$, where $V(-1)$ denotes the Tate twist rising the weight by 2. The differential

$$
{}_W d_1 : H^*(X^p)(-p) \to H^{*+2}(X^{p-1})(1-p)
$$

might be nontrivial. The weight principle says: the maps between Hodge structures of distinct weights vanish. Therefore the second differential of the spectral sequence is the zero map and the weight spectral sequence degenerates on ${}_W E_2$, [5, (3.2.10)].

The rows of the spectral sequence ${}_WE_2$ compute the graded pieces of the weight filtration in $H^*(U)$

$$Gr^W_{k+\ell}H^k(U) =$$
$$= H_\ell\big(0\leftarrow \underbrace{H^k(X^0)}_{0} \leftarrow \underbrace{H^{k-2}(X^1)(-1)}_{1} \leftarrow \underbrace{H^{k-4}(X^2)(-2)}_{2} \leftarrow \dots\big).$$

As a matter of fact the weight spectral sequence after a reorganization of indices coincides with the Leray spectral sequence of the inclusion $j : X \setminus D \hookrightarrow X$. We have

$$R^q j_* \mathbb{Q}_U = \bigoplus_{|I|=q} \mathbb{Q}_{X_I}(-q).$$

The Leray spectral sequence in this case has the following second table

$$E_2^{p,q} = H^p(X^q)(-q).$$

Here is the picture for the surface case

$$\begin{array}{cccccc}
2 & H^0(X^2)(-2) & 0 & 0 & 0 & 0 \\
1 & H^0(X^1)(-1) & H^1(X^1)(-1) & H^2(X^1)(-1) & 0 & 0 \\
0 & H^0(X^0) & H^1(X^0) & H^2(X^0) & H^3(X^0) & H^4(X^0) \\
 & 0 & 1 & 2 & 3 & 4
\end{array}.$$

The second differential

$$d_2 : E_2^{p,q} = H^p(X^q)(-q) \to E_2^{p+2,q-1} = H^{p+2}(X^{q-1})(1-q)$$

preserves the weight: both entries have weight $p + 2q$. Checking directly the definitions of both differentials we find, that $d_2 = {}_Wd_1$. According to the weight principle the Leray spectral sequence degenerates on E_3, [5, (3.2.13)].

Remark 3 Let $\mathcal{L}$ be an unitary local system on a normal crossing divisor complement. The degeneration of the Leray spectral sequence (1) was proven in [16].

The goal of this paper is to show that a similar degeneration happens for an admissible arrangement of submanifolds when cohomology are taken with constant coefficients, i.e., Theorem 1. In the final section we will briefly discuss a possible degeneration for nontrivial local systems, but we are bounded by the category in which our object live. We will assume that $\mathcal{L}$ is of geometric origin to be able to talk about purity.

3 Proofs

Proof of Theorem 2 The first part of our proof is essentially the "deletion-restriction" of [4, Proposition 2.1] or [3, Lemma 2.1] rewritten in the sheaf language. We will prove a statement a bit stronger: the sheaves $R^k j_*\mathbb{Q}_U$ are pure and pointwise pure (compare the definition of purity [1, 5.1.8]). We proceed inductively with respect to the number of elements of the arrangement $\mathcal{Z}$ and the dimension of X. If $\mathcal{Z} = \{Z\}$, then there is a long exact sequence

$$\to H^{k-1}(U) \to H^{k-2c}(Z)(-c) \to H^k(X) \to H^k(U) \to ,$$

which localizes to an exact sequence of sheaves

$$\to R^{k-1}j_*\mathbb{Q}_U \to \mathcal{H}^{k-2c}(\mathbb{Q}_Z)(-c) \to \mathcal{H}^k(\mathbb{Q}_X) \to R^k j_*\mathbb{Q}_U \to .$$

Therefore $R^0 j_*\mathbb{Q}_U = \mathbb{Q}_X$ and $R^{2c-1}j_*\mathbb{Q}_U = \mathbb{Q}_Z(-c)$. The derived images vanish in the remaining degrees. In particular the theorem holds when the dimension of Z_i is zero. Assume now that the theorem holds for all families in varieties of dimension smaller then $n = \dim(X)$. We also assume that the theorem holds for families consisting of $s-1$ elements. Consider the Mayer-Vietoris exact sequence for the pair $U_1 = X \setminus Z_1$, $U_2 = X \setminus \bigcup_{i>1} Z_i$:

$$\to H^{k-1}(U) \to H^k(U_1 \cup U_2) \to H^k(U_1) \oplus H^k(U_2) \to H^k(U) \to$$

which can be localized to the long exact sequence of sheaves

$$\to R^{k-1}j_*\mathbb{Q}_U \to R^k j'_*\mathbb{Q}_{U_1\cup U_2} \to R^k j_{1*}\mathbb{Q}_{U_1} \oplus R^k j_{2*}\mathbb{Q}_{U_2} \to R^k j_*\mathbb{Q}_U \to ,$$

where j_1, j_2, j' are inclusions of the sets U_1, U_2 and $U_1 \cup U_2$ into X. Note that

$$R^k j'_*\mathbb{Q}_{U_1\cup U_2} = \begin{cases} \mathbb{Q}_X & \text{for } k = 0\,, \\ 0 & \text{for } 0 < k \leq 2c\,, \\ R^{k-2c} j''_*\mathbb{Q}_{Z_1\cap U_2}(-c) & \text{for } k > 2c\,, \end{cases}$$

where $j'' : Z_1 \cap U_2 \to X$ is the inclusion. Topologically this formula can be justified by the following: locally, i.e., if X was a ball, then $U_1 \cup U_2$ would be the $2c$-fold suspension of $Z_1 \cap U_2$. To compute the weight we notice that for $k > 2c$, $x \in Y = Z_1 \cap \bigcup_{i>1} Z_i$ by Alexander duality in X

$$\left(R^k j'_*\mathbb{Q}_{U_1\cup U_2}\right)_x = \mathcal{H}_{2n-k-1}(Y)_x(n)\,,$$

where $\mathcal{H}_*$ is the sheaf of local homology.[1] Again, by Alexander duality in Z_1, which is of dimension $n-c$,

$$\mathcal{H}_{2n-k-1}(Y)_x(n) = \left(R^{k-2c} j''_* \mathbb{Q}_{Z_1 \cap U_2}\right)_x (-c) .$$

Hence

$$\left(R^k j'_* \mathbb{Q}_{U_1 \cup U_2}\right)_x \simeq \left(R^{k-2c} j''_* \mathbb{Q}_{Z_1 \cap U_2}\right)_x (-c) .$$

for $k > 2c$. This is a kind of suspension isomorphism. Using the inductive hypothesis we find that $R^k j'_* \mathbb{Q}_{U_1 \cup U_2}$ is nonzero only if $k = 0$ or $k-1$ is divisible by $2c-1$ and the weight of $R^{(2c-1)\ell+1} j'_* \mathbb{Q}_{U_1 \cup U_2}$ is equal to $2c\ell$. If $c > 1$ the map $R^k j'_* \mathbb{Q}_{U_1 \cup U_2} \to R^k j_{1*} \mathbb{Q}_{U_1} \oplus R^k j_{2*} \mathbb{Q}_{U_2}$ vanishes by dimensional reasons. For $c = 1$ we use the weight argument: the source is of weight $2k-1$, the target is of weight $2k$. We conclude that we obtain a short exact sequence of sheaves of weight $\frac{2c}{2c-1} k$,

$$0 \to R^k j_{1*} \mathbb{Q}_{U_1} \oplus R^k j_{2*} \mathbb{Q}_{U_2} \to R^k j_* \mathbb{Q}_U \to R^{k+1} j'_* \mathbb{Q}_{U_1 \cup U_2} \to 0 .$$

(For $k = 0$ the exact sequence is extended from the left by $R^0 j'_* \mathbb{Q}_{U_1 \cup U_2} \simeq \mathbb{Q}_X$, which is pure of weight 0.) By inductive assumption the edge sheaves are pure and pointwise pure. Therefore the middle one is pure and pointwise pure. For $k = 0$ the exact sequence is extended from the left by $R^0 j'_* \mathbb{Q}_{U_1 \cup U_2}$, which is pure of weight 0. □

Proof of Theorem 3 Again by induction we find that the sheaf $R^{(2c-1)\ell} j_* \mathbb{Q}_U$ fits to an exact sequence

$$0 \to \bigoplus_\alpha \mathbb{Q}_{A_\alpha}(c\ell) \to R^{(2c-1)\ell} j_* \mathbb{Q}_U \to \bigoplus_\beta \mathbb{Q}_{B_\beta}(c\ell) \to 0 .$$

Here A_α and B_β are families of smooth varieties of codimensions $c\ell$. By Beilinson et al. [1, 5.4.6] $R^{(2c-1)\ell} j_* \mathbb{Q}_U$ decomposes into the sum of constant sheaves. (In the

[1] Avoiding the sheaf language and forgetting Hodge structure we have

$$\left(R^k j'_* \mathbb{Q}_{U_1 \cup U_2}\right)_x \simeq H^k(B_x \cap (U_1 \cup U_2)) \simeq H^k(\partial B_x \cap (U_1 \cup U_2)) ,$$

where B_x is a small ball around x. The stalk of the sheaf of local cohomology $\mathcal{H}_\ell(Y)_x$ is isomorphic to

$$H_\ell(B_x \cap Y, \partial B_x \cap Y) \simeq \bar{H}_{\ell-1}(\partial B_x \cap Y) .$$

By Alexander duality in ∂B_x we have

$$\bar{H}^k(\partial B_x \cap (U_1 \cup U_2)) \simeq \bar{H}_{2n-k-2}(\partial B_x \cap Y) \simeq H_{2n-k-1}(B_x \cap Y, \partial B_x \cap Y) .$$

case of when the arrangement is homeomorphic to a vector space arrangement see [10, III §3.7] for a topological proof.) □

Proof of Theorem 1 We follow the Totaro argument [17, §4] (see also his remark at the beginning of p.1062). If a sheaf S is pure of weight w, the base space is compact, then the cohomology $H^p(X; S)$ is pure of weight $p + w$ [1, 5.1.13]. By Theorem 2 the entries of the Leray spectral sequence $E_2^{p,q} = H^p(X; R^q j_* \mathbb{Q}_U)$ are pure of weight $p + \frac{2c}{2c-1} q$ for q divisible by $2c - 1$. The remaining rows of the E_2 table are zero. The first possible differential is

$$d_{2c} : E_2^{p,(2c-1)\ell} \longrightarrow E_2^{p+2c,(2c-1)\ell-2c+1},$$

that is

$$d_{2c} : H^p(X; R^{(2c-1)\ell} j_* \mathbb{Q}_U) \longrightarrow H^{p+2c}(X; R^{(2c-1)(\ell-1)} j_* \mathbb{Q}_U).$$

The weights of both entries are equal to $p + 2c\ell = (p + 2c) + 2c(\ell - 1)$ and the differential d_{2c} can be nontrivial. Further differentials hit the entries $E_{2c+1}^{p+k,(2c-1)\ell-k+1}$ which are subquotients of $H^{p+k}(X; R^{(2c-1)\ell-k+1} j_* \mathbb{Q}_U)$ and have weights

$$(p + k) + \frac{2c}{2c - 1}((2c - 1)\ell - k + 1) = p + 2c\ell + \frac{2c - k}{2c - 1} < p + 2c\ell$$

for $k > 2c$. Therefore the higher differentials vanish. □

4 Twisted Coefficients

We would like to conclude with some remarks concerning the cohomology with coefficients in a local system. We restrict our attention to the arrangements of hypersurfaces ($c = 1$), when the local monodromy might be nontrivial. If $\mathcal{L}$ is a local system on $U = X \setminus D$ the Leray spectral sequence relates the local invariants of intersections of divisors with the global cohomology. Consider the case of a hyperplane-like arrangement, i.e., we assume as in [15] that in some local analytic coordinates the hypersurfaces are given by linear equation. Under the condition that:

- **(Mon)**: $\dim(\mathcal{L}) = 1$ *and for each* $x \in D$ *the product of the monodromies associated to the divisors passing through* x *is not equal to one.*

One obtains the vanishing of local cohomology

$$(R^k j_* \mathcal{L})_x = H^k(U \cap B_x; \mathcal{L}) = 0$$

for any k at any point belonging to D, see [3, Lemma 5.2]. (Here B_x is a sufficiently small ball centered in x.) For abelian local systems of higher rank the generalization of **(Mon)** would demand that the products of all possible eigenvalues are not equal to one. If **(Mon)** or an appropriate analogue for higher rank systems is satisfied then the full derived push forward reduces to the push forward with compact supports

$$Rj_*\mathcal{L} = j_!\mathcal{L} .$$

It follows that the Leray spectral sequence reduces to the bottom row, which is

$$E_2^{p,0} = H^p(X; j_!\mathcal{L}) = H^p(X, D; j_*\mathcal{L}) .$$

Hence

$$H^p(U; \mathcal{L}) = H^p(X, D; j_*\mathcal{L}) .$$

This is essentially the argument e.g., of [4, Theorem 5.1].

Now suppose that $\mathcal{L}$ is enriched, so that it belongs to the category studied in [1] (e.g., is of geometric origin, [*loc.cit.* 6.2.4]). Furthermore suppose that $\mathcal{L}$ is pure of weight w. Then [*loc.cit.* 5.1.13] $Rj_*\mathcal{L}$ is of weight $\geq w$ and $Rj_!\mathcal{L}$ is of weight $\leq w$. Therefore if the condition (**Mon**) is satisfied, then $Rj_*\mathcal{L} = j_!\mathcal{L}$ is pure. Under assumption that X is compact the cohomology group $H^*(X; Rj_*\mathcal{L})$ are pure. The admit a Hodge decomposition

$$H^k(U; \mathcal{L}) \simeq H^k(X; Rj_*\mathcal{L}) = \bigoplus_{p+q=k+w} H^{pq} .$$

The Hodge decomposition for arrangements of planes in $\mathbb{P}^3$ were studied in [13].

If a weaker condition is satisfied: (**Mon**)* of [15, Theorem 4.1] (see also [9]), then the local cohomology of $(R^k j_*\mathcal{L})_x = H^k(U \cap B_x \mathcal{L})$ is generated by logarithmic forms df_i/f_i, where f_i is an equation of a divisor component. We would like to ask:

Question 1 Suppose that $(R^k j_*\mathcal{L})_x$ is generated by logarithmic forms. Does it imply that $R^k j_*\mathcal{L}$ is pure?

The sheaf $R^k j_*\mathcal{L}$ is a sum of locally constant sheaves supported by the intersections of divisor components. Therefore pointwise purity implies that $R^k j_*\mathcal{L}$ is pure as a sheaf. We do not give the positive answer to that Question 1. The passage from the analytic category to arithmetic property is delicate. We would have to compare two notions of purity: one from the world of Weil sheaves, another—defined via differential forms, properly formulated in the sense variation of Hodge structure or mixed Hodge modules. Moreover, not every local system admits a reduction to a local system over a variety over a finite field. Work of Timmerscheidt [16] may be a hint that something can be done for unitary local systems. When we consider local systems of geometric origin in the sense of [1] we can freely use the notion of purity. Assuming purity we proceed as in the proof of Theorem 2. The Leray spectral sequence can have nontrivial second differential, but the higher differentials vanish. Therefore

Corollary 1 *Let $U \subset X$ be an open set in a compact algebraic variety and $\mathcal{L}$ be a locally constant sheaf on U. Suppose that $\mathcal{L}$ is of geometric origin. Further assume that the sheaves $R^k j_* \mathcal{L}$ are pure of weight $w + 2k$. The skew-rows joined by the second differential of the Leray spectral sequence compute the graded pieces of the weight filtration in $H^*(U; \mathcal{L})$*

$$Gr^W_{w+k+\ell} H^k(U; \mathcal{L}) =$$
$$= H_\ell\big(0 \leftarrow \underbrace{H^k(X; R^0 j_* \mathcal{L})}_{0} \leftarrow \underbrace{H^{k-2}(X; R^1 j_* \mathcal{L})}_{1} \leftarrow \underbrace{H^{k-4}(X; R^2 j_* \mathcal{L})}_{2} \leftarrow \dots\big).$$

Possibly using the methods of Timmerscheidt [16] one can prove degeneration of the Leray spectral sequence for the complement of hyperplane-like arrangement and a unitary local system satisfying **(Mon)***. We finish our discussion with the question:

Question 2 Does in general the condition **(Mon)*** of [15, Theorem 4.1] imply degeneration of the Leray spectral sequence $E_2^{p,q} = H^p(X; R^q j_* \mathcal{L}) \Rightarrow H^{p+q}(U; \mathcal{L})$ degenerate?

Acknowledgements This work was supported by NCN grant 2013/08/A/ST1/00804.

References

1. A.A. Beilinson, J. Bernstein, P. Deligne, Faisceaux pervers. *Analysis and topology on singular spaces, I (Luminy, 1981)*, Astérisque, vol. 100 (Société Mathématique de France, Paris, 1982), pp. 5–171
2. A. Björner, T. Ekedahl, Subspace arrangements over finite fields: cohomological and enumerative aspects. Adv. Math. **129**(2), 159–187 (1997)
3. M.W. Davis, T. Januszkiewicz, I.J. Leary, The ℓ^2-cohomology of hyperplanes complements. Groups Geom. Dyn. **1**, 301–309 (2007)
4. M.W. Davis, S. Settepanella, Vanishing results for the cohomology of complex toric hyperplane complements. Publ. Mat. **57**, 379–392 (2013)
5. P. Deligne, *Théorie de Hodge II*. Publications Mathématiques/Institut des Hautes Etudes Scientifiques, vol. 40 (Institut des Hautes études Scientifiques, Bures-sur-Yvette, 1972), pp. 5–57
6. P. Deligne, *Théorie de Hodge III*. Publications Mathématiques/Institut des Hautes Etudes Scientifiques, vol. 44 (Institut des Hautes études Scientifiques, Bures-sur-Yvette, 1974), pp. 5–77
7. P. Deligne, La conjecture de Weil. II. Inst. Hautes Études Sci. Publ. Math. **52**, 137–252 (1980)
8. G. Denham, A.I. Suciu, S. Yuzvinsky, Combinatorial covers and vanishing of cohomology. Selecta Math. (N.S.) **22**(2), 561–594 (2016)
9. H. Esnault, V. Schechtman, E. Viehweg, Cohomology of local systems on the complement of hyperplanes. Invent. math. **109**, 557–561 (1992); Erratum. 112 (1993)

10. M. Goresky, R. MacPherson, *Stratified Morse Theory* (Springer, Berlin, 1988)
11. K. Jewell, Complements of sphere and subspace arrangements. Topology Appl. **56**(3), 199–214 (1994)
12. K. Jewell, P. Orlik, B. Shapiro, On the complements of affine subspace arrangements. Topology Appl. **56**(3), 215–233 (1994)
13. Y. Kawahara, Hodge number of cohomology of local systems on the complement of hyperplanes in $\mathbb{P}^3$. J. Math. Sci. Univ. Tokyo **8**, 177–199 (2001)
14. M. Saito, Hodge modules. Publ. Res. Inst. Math. Sci. **26**(2), 221–333 (1990)
15. V. Schechtman, H. Terao, A. Varchenko, Cohomology of local systems and the Kac-Kazhdan condition for singular vectors. J. Pure Appl. Algebra **100**, 93–102 (1995)
16. K. Timmerscheidt, Mixed Hodge theory for unitary local systems. J. Reine Angew. Math. **379**, 152–171 (1987)
17. B. Totaro, Configuration spaces of algebraic varieties. Topology **35**(4), 1057–1067 (1996)

Higher Topological Complexity of Artin Type Groups

Sergey Yuzvinsky

Abstract We calculate the higher topological complexity TC_s for the complements of reflection arrangements, in other words for the pure Artin type groups of all finite complex reflection groups. In order to do that we introduce a simple combinatorial criterion for arrangements sufficient for the cohomological lower bound for TC_s to coincide with the dimensional upper bound.

Keywords Artin type groups • Schwarz genus

1 Introduction

Topological complexity of a topological space X, $TC(X)$, was defined by Farber in [6] as a specialization of the Schwarz genus [12]. Unlike the Schwarz genus in general, $TC(X)$ is an invariant of the homotopy type of X. Later Yu. Rudyak in [11] extended Farber's definition to higher topological complexity $TC_s(X)$ for $s = 2, 3, 4, \ldots$ such that $TC(X) = TC_2(X)$. One of the common features of these invariants is a lower bound determined by the ring structure of $H^*(X)$.

This paper is concerned with a special class of topological spaces—the complements of complex hyperplane arrangements. Previously TC_2 has been calculated for particular classes of arrangements such as Coxeter series in [7] and general position arrangements in [4, 13]. These examples prompted the Conjecture that for all arrangement complements TC_2 coincides with the cohomological lower bound. The only known results for arrangement complements and arbitrary s (besides the basic examples of the circle and tori in [2, 11]) is the calculation for the Coxeter series of type A in the recent preprint [8].

In the present paper we give a simple combinatorial condition sufficient for the cohomological lower bound to coincide with the dimensional upper bound. This allows us to compute TC_s for a wide class of arrangements (including all complex reflection arrangements, i.e., $K[\pi, 1]$'s for the pure Artin type groups). For all arrangements of this class the value of TC_s coincides with the cohomological low bound for this s.

S. Yuzvinsky (✉)
University of Oregon, Eugene, OR 97403, USA
e-mail: yuz@math.uoregon.edu

F. Callegaro et al. (eds.), *Configuration Spaces*, Springer INdAM Series 14,
DOI 10.1007/978-3-319-31580-5_5

The results of this paper constituted a talk given by the author at the conference "Configuration Spaces" in Cortona in September of 2014. The author is grateful to the organizers of the very useful and pleasant conference for inviting him.

2 Definition of TC_s and Main Properties

Definition 1 Let X be a path-connected topological space and s an integer at least 2. Then $\mathrm{TC}_s(X)$ is the Schwarz genus of the fibration

$$\phi_s : X^{[0,1]} \to X^s,$$

where

$$\phi_s(\gamma) = \left(\gamma(0), \gamma\left(\frac{1}{s-1}\right), \gamma\left(\frac{2}{s-1}\right), \ldots, \gamma\left(\frac{s-2}{s-1}\right), \gamma(1)\right).$$

In other words, it is the smallest number n such that X^s is partitioned into Euclidean neighborhood retracts W_i $(i = 0, 1, \ldots, n)$ and on each W_i there exists a section $f_i : W_i \to X^{[0,1]}$ of ϕ_s (i.e., $\phi_s \circ f_i = 1_{W_i}$). These data is called a *motion planning* (m.p.). We use the *reduced* (or normalized) version of TC such that $\mathrm{TC}_s(X) = 0$ for a contractible X and each s.

Note that ϕ_s is a fibrational substitute with the fiber $(\Omega X)^{s-1}$ for the diagonal imbedding $d_s : X \to X^s$.

The following properties can be found in [2, 11].

(1) $\mathrm{TC}(X)$ is an invariant of the homotopy type of X.
(2) $\mathrm{TC}_s(X) \leq s \cdot \mathrm{hdim}(X)$ where hdim is the homotopy dimension (*the dimensional upper bound)*.
(3) $\mathrm{TC}_s(X \times Y) \leq \mathrm{TC}_s(X) + \mathrm{TC}_s(Y)$ (*the product formula)*.
(4) The cohomological lower bound.

This is the only lower bound and it requires some definitions.

Definition 2 Let d_s be the diagonal embedding $X \to X^s$. Denote by $cl(X, s)$ the cup length in $\ker d_s^*$, i.e., the largest integer k for which there exist k elements $u_i \in H^*(X^s)$ such that $d_s^* u_i = 0$ for every i and $u_1 u_2 \cdots u_k \neq 0$.

We have the following *cohomological lower bound*:

$$\mathrm{TC}_s(X) \geq cl(X, s).$$

This inequality holds for cohomology with arbitrary coefficients, even for local coefficients. In the rest of the paper we will use cohomology with coefficients in $\mathbb{C}$ omitting coefficients from the notation.

Example $\mathrm{TC}_s(S^1) = s - 1$ for every s.

Indeed choose an orientation and denote by u the generator of $H^1(S^1)$. Then the elements

$$u^{(i)} = u \otimes 1 \otimes \cdots \otimes 1 - 1 \otimes \cdots \otimes 1 \otimes u \otimes 1 \otimes \cdots \otimes 1$$

where the second u is in the i-th position are in $\ker d_s^*$ and $\prod_{i=2}^{s} u^{(i)} \neq 0$. By the cohomological lower bound $\mathrm{TC}_s(S^1) \geq s - 1$.

For a m.p. one can use the covering of the torus $(S^1)^s$ by $D_k = \{(x_1, \ldots, x_s)\}$ $(k = 0, 1, \ldots, s-1)$ such that $x_j = x_{j+1}$ for precisely k indexes j. A path $\gamma_j(x)$ from x_j to x_{j+1} is constant if $x_j = x_{j+1}$ and the rotation along the fixed orientation of S^1 otherwise.

3 Complements of Hyperplane Arrangements

In this paper we will deal mostly with topological spaces that are hyperplane arrangement complements.

Definition 3 A (complex linear essential) hyperplane arrangement is a set $\mathcal{A}$ of n linear hyperplanes in $\mathbb{C}^r$ such that $\bigcap_{H \in \mathcal{A}} H = \{0\}$. The *arrangement complement* is the topological space $M = \mathbb{C}^r \setminus \bigcup_{H \in \mathcal{A}} H$.

Among the arrangement complements there are, for instance, $K[\pi, 1]$ where π is the pure Artin type group for an arbitrary finite complex reflection group. The most frequently used examples of that are the Braid arrangements.

Example Consider $n = \binom{\ell}{2}$ hyperplanes given by the equations $x_i = x_j$ for all $1 \leq i < j \leq \ell$. This arrangement is called Braid arrangement because M is $K[\pi, 1]$ where $\pi = \pi_1(M)$ is the pure Braid group on ℓ strings, that is the pure Artin group for Σ_ℓ.

For an arbitrary arrangement the algebra $A = H^*(M)$ is well-known from work of Arnold et al. [1, 3, 9].

For each hyperplane $H \in \mathcal{A}$ fix a linear form α_H with $\ker \alpha_H = H$. Then A can be identified with the subalgebra of the algebra of all the holomorphic differential forms on M generated by the logarithmic forms $\frac{d\alpha_H}{\alpha_H}$ $(H \in \mathcal{A})$. The classes e_H of these forms form a canonical basis of A^1 whence for every $x \in A^1$ we have $x = \sum_{H \in \mathcal{A}} x_H e_H$ for some $x_H \in \mathbb{C}$. Relations for the generators are explicitly described and can be found in [10]. These relations imply in particular that $H^p(M) = 0$ for $p > r$.

A stronger fact is that M has the homotopy type of a finite simplicial complex of dimension r (see [10]).

The relations imply also that A is determined by the combinatorics of $\mathcal{A}$, i.e., the collection of linearly independent subsets of $\mathcal{A}$ (called *simple matroid*). In particular the (square-free) monomials corresponding to dependent sets of hyperplanes vanish in A. In particular the (square-free) monomials corresponding to independent sets ("independent monomials") linearly generate A but they are not linearly independent (over $\mathbb{C}$) in general. The theory of Gröbner basis gives so called no-broken-circuit (**nbc**) monomials that do form a basis of A.

To define this basis we need to fix a linear order on $\mathcal{A}$ whence on $\{e_1, \ldots, e_n\}$ which gives the deg-lex order on the monomials. Then a *circuit* is a minimal dependent set of e_i and a *broken circuit* is a circuit with the smallest element (in the fixed order) omitted. Then an **nbc**-monomial is a monomial whose support does not contain any broken circuits. It is easy to see that the set of **nbc**-monomials forms the basis of A given by the Gröbner theory for the deglex monomial ordering.

Later in this paper we will use the following.

Property (*) of nbc Basis
Suppose an order is fixed on $\mathcal{A}$ and μ is a non-**nbc** monomial for this order. Then its representation as a linear combination of **nbc** monomials looks like

$$\mu = \sum_i \pm\mu_i$$

where for each **nbc**-monomial μ_i we have $\mu_i < \mu$ in the deglex order.

This follows immediately from relations of monomials corresponding to circuits.

4 Properties of $\mathrm{TC}_s(M)$

Recall that r is the rank of an arrangement.

1. The general upper bound for M can be made a little more tight. Namely

$$\mathrm{TC}_s(M) \leq sr - 1.$$

   Indeed, for a non-empty arrangement $M = \bar{M} \times \mathbb{C}^*$ where $\bar{M}$ is the projectivization of M that has the homotopy type of a CW-complex of dimension $r-1$. Hence by the product formula

$$\mathrm{TC}_s(M) \leq \mathrm{TC}_s(\bar{M}) + \mathrm{TC}_s(S^1) \leq s(r-1) + s - 1 = sr - 1.$$

2. To find a lower bound we need some preparation.

4.1 Products Over Subsets

Fix an integer $s \geq 2$ and for each generator $e_i \in H^1(M)$, and each j $(1 < j \leq s)$ put

$$e_i^{(j)} = e_i \otimes 1 \otimes \cdots \otimes 1 - 1 \otimes \cdots \otimes 1 \otimes e_i \otimes 1 \otimes \cdots \otimes 1$$

where e_i in the second summand is in the jth position. Clearly each $e_i^{(j)} \in \ker d_s^*$.

Then for every $I \subset \{e_i^{(j)}\}$ if the product of all elements from I does not vanish then $|I|$ is smaller than or equal to TC_s.

4.2 Products Over Pairs

In the rest of the paper we will identify subsets of $\bar{n}$ with the respective subsets of generators in A^1 and with subarrangements of $\mathcal{A}$. The rank $\operatorname{rk} S$ of a subset S is the rank of the respective subarrangement which is the cardinality of its base (i.e., a maximal independent set). The rank of $\bar{n}$ is r.

Let $Q = (B, C)$ be an ordered pair of disjoint subsets of $\bar{n}$. The *product over Q* is defined by the formula

$$\pi_Q = \pi_B \cdot \pi'_C$$

where

$$\pi_B = \prod_{i \in B} \prod_{j=2}^{s} e_i^{(j)}, \quad \pi'_C = \prod_{i \in C} e_i^{(2)}.$$

We put $\bar{Q} = B \cup C$.

4.3 Basic Pairs and Balanced Sets

Definition 4 A pair Q is *basic* if $|B| = r$ (i.e., B is a base) and B, C are **nbc** in $\bar{Q}$ for some linear order on it.

Remark 1 We can extend a linear order on $\bar{Q}$ to a linear order on $\mathcal{A}$ so that every element of $\bar{Q}$ are smaller than every element of $\mathcal{A} \setminus \bar{Q}$. Then any monomial with support in $\bar{Q}$ is **nbc** in $\bar{Q}$ if and only if it is **nbc** in the whole $\mathcal{A}$.

Definition 5 A subset $S \subset \bar{n}$ of full rank is *balanced* if for any of its (linearly) closed non-empty subsets S' we have $|S'| < 2\operatorname{rk}(S')$.

Theorem 1

(i) If a pair Q is basic then $\bar{Q}$ is balanced.
(ii) Every balanced set S is $\bar{Q}$ for some basic pair Q, i.e., it is the union of the elements of the pair.

Proof

(i) Suppose a pair $Q = (B, C)$ is basic and fix an order such that B and C are **nbc**. Also assume there exists a closed non-empty subset $D \subset \bar{Q}$ with $|D| \geq 2\,\mathrm{rk}(D)$. Since the sets $B \cap D$ and $C \cap D$ are independent they both are bases of D. Now if i is the least element of D then to whichever of two bases it belongs, it depends on the other base which contradicts to B and C being **nbc** in $\bar{Q}$.

(ii) Suppose a set S is balanced and choose a base B of S hence of $\mathcal{A}$. Put $C = S \setminus B$. Since S is balanced $s = |C| \leq r - 1$. Using that S is balanced again we can find $r - s + 1$ elements in B independent of C. Order them linearly from 1 to $r - s + 1$ and denote the set they form by B_1. Again by the same property there exist two elements in C independent of $B \setminus B_1$. Assign numbers $r - s + 2$ and $r - s + 3$ to these elements and denote the set of them by C_1. Notice that $|C \setminus C_1| = s - 2$ and $|B \setminus B_1| = s - 1$. Now we just repeat the reasoning. There exist two elements in $B \setminus B_1$ independent on C_1 and we can assign numbers $r - s + 4$ and $r - s + 5$ to them. Continuing this process we obtain at some step a linear ordering on S such that no element depends on the set of larger (in this ordering) elements. Thus B and C are **nbc** in S for this order whence the pair is basic. □

5 Calculation of a Lower Bound

Theorem 2 *Let $\mathcal{A}$ be a central arrangement. Then for every integer s, $s \geq 2$, and every basic pair $Q = (B, C)$ we have $\pi_Q \neq 0$. Hence $TC_s \geq (s - 1)r + |C|$.*

Proof By construction, π_Q is the sum of pure tensors with coefficients ± 1 among which there is $\mu = e_C \otimes e_B \cdots \otimes e_B$ where for every subset $S \subset [n]$ we put $e_S = \prod_{i \in S} e_i$. Since Q is basic all monomials in μ are **nbc** in some order on $\mathcal{A}$ that we fix. Thus it suffices to proof that no other simple tensor from π_Q contains μ in the decomposition of its monomials in the linear combinations of **nbc** monomials.

Suppose that $\nu = \nu_1 \otimes \nu_2 \otimes \cdots \otimes \nu_s$ is such a simple tensor. Since the monomials ν_j for $j > 2$ cannot have elements from C and have degree r then $\nu_i = e_B$ for $i > 2$. The first two monomials are e_{C_i} $(i = 1, 2)$ where (C_1, C_2) is a partition of $B \cup C$ (with $|C_2| = r$).

Using Property (*) of the **nbc** basis we obtain the following. If $e_{C_1} \otimes e_{C_2}$ contains $e_C \otimes e_B$ in the decomposition and at least one of e_{C_i} is not **nbc** then $e_{C_1} \geq e_C$ and $e_{C_2} \geq e_B$ where at least one of the inequalities is strict. This contradicts the fact that (C_1, C_2) and (C, B) are partitions of $C \cup B$. Thus $\nu = \mu$ whence μ cannot be cancelled. □

6 Large Arrangements

Definition 6 We call an arrangement *large* if there exists a basic pair (B, C) with $|C| = r - 1$.

Comparing this with the dimensional upper bound for M we obtain for large arrangements that

$$\mathrm{TC}_s(M) = sr - 1$$

for every s.

Large arrangements are easy to find due to the following sufficient condition.

Definition 7 A pair (B, C) is *well-balanced* in $\mathcal{A}$ if B is a base, $|C| = r - 1$, and no $b \in B$ depends on C. An arrangement is *well-balanced* if there is a well-balanced pair in it.

Theorem 3 *Every well-balanced pair is balanced.*

Proof Indeed suppose (B, C) is well-balanced but there is a non-empty $D \subset B \cup C$ with $|D| \geq 2\,\mathrm{rk}\,D$. Then $D \cap B$ and $D \cap C$ are independent whence both are bases of D. Hence every $b \in D \cap B$ depends of $D \cap C$ which contradicts the condition. □

Let $L(\mathcal{A})$ be the lattice of all intersections of hyperplanes from $\mathcal{A}$ ordered opposite to inclusion. For $X \in L(\mathcal{A})$ we put $\mathcal{A}_X = \{H \in \mathcal{A} | H \geq X\}$.

Definition 8 $L(\mathcal{A})$ is well-balanced if there exists $X \in L(\mathcal{A})$, $\mathrm{codim}\,X = r - 1$ such that for no $Y \in L(\mathcal{A}) \setminus \{0\}$ we have $\mathcal{A} = \mathcal{A}_X \cup \mathcal{A}_Y$.

Using rk instead of codim makes this definition applicable to general geometric lattice.

Theorem 4 *If $L(\mathcal{A})$ is well-balanced then there exists a well-balanced pair in $\mathcal{A}$.*

Proof Let C be a base of $\mathcal{A}_X$ from the definition. Then $|C| = r - 1$. Put $\mathcal{A}' = \mathcal{A} \setminus \mathcal{A}_X$. By definition $\mathrm{rk}\,\mathcal{A}' = r$. Let B be a base of $\mathcal{A}'$ whence also a base of $\mathcal{A}$. Since B is disjoint with $\mathcal{A}_X$ no $b \in B$ depends on C. □

Corollary 1 *Suppose for all $X \in L(\mathcal{A}) \setminus \{0\}$ we have*

$$|\mathcal{A}_X| < \frac{n}{2}. \tag{1}$$

Then $\mathcal{A}$ is large.

Clearly it suffices to check the inequality (1) for X of rank $r - 1$ only.

Example The arrangements of the following classes are clearly large.

(1) Generic arrangements with $|\mathcal{A}| \geq 2r - 1$.
(2) Every arrangement containing a large subarrangement of full rank.

7 Groups Generated by Reflections

Definition 9 Let V be a complex linear space of dimension r. A (complex) *reflection* is a finite order invertible linear transformation $\tau : V \to V$ whose fixed point set is a hyperplane (denoted H_τ). A finite subgroup of $GL(V)$ is a *reflection group* if it is generated by reflections.

For a reflection group W the set $\mathcal{A}_W = \{H_\tau\}$ is called the *reflection arrangement* of W.

A reflection group W is *irreducible* if its tautological representation to $GL(V)$ is irreducible. Then the rank of W is r.

Theorem 5 (See [1, 3, 5, 9]) *Let $M_W = V \setminus \bigcup_{H \in \mathcal{A}_W} H$ for an arbitrary reflection group W. Then M_W is a $K[\pi, 1]$.*

Example For $\ell > 1$ every hyperplane $H_{ij} \subset \mathbb{R}^\ell$ of the Braid arrangement is the collection of fixed points of a real reflection permuting x_i and x_j. Thus the (complexified) Braid arrangement is the reflection arrangement for the permutation group $W = \Sigma_\ell$. Here $\pi_1(M_W)$ is the pure Braid group on ℓ strings, that is the *pure Artin group of type* $A_{\ell-1}$.

Similarly, for any (complexified) finite Coxeter group W the group $\pi_1(M_W)$ is the pure Artin group of the respective type. Because of that $\pi_1(M_W)$ for an arbitrary finite complex reflection group W is called the *pure Artin type group for* W (or *the generalized pure Braid group associated to* W).

8 Calculation of $\mathrm{TC}_s(M_W)$

Here is the main theorem of the paper.

Theorem 6 *For every irreducible reflection group W of rank r and every $s > 1$ the arrangement $\mathcal{A}_W$ is well-balanced whence $TC_s(M_W) = sr - 1$.*

Proof Our proof consists of four parts.

1. If W has rank equal to 2 then the result is immediate since $r = 2$.
2. Infinite series.

For the infinite series well-balanced pairs can be exhibited explicitly. For that we give hyperplanes by their defining linear forms and $\mathcal{A}$ the product of them.

(a) **Full monomial types** $G(m, 1, r)$ (if $m = 1$ it is $\mathbf{A_r}$; if $m = 2$ it is $\mathbf{B_r}$) : $Q = \prod_{i=1}^{r} x_i \prod_{1\le i<j\le r}(x_i^m - x_j^m)$. Put $B = \{x_1, \dots, x_r\}$ and $C = \{x_1 - x_2, \dots, x_1 - x_r\}$.
(b) **Special monomial types** $G(m, m, r)$ $m \geq 2$: $Q = \prod_{1\le i<j\le r}(x_i^m - x_j^m)$ (if $m = 2$ it is $\mathbf{D_r}$). Put $B = \{x_1 - \zeta x_2, \dots, x_1 - \zeta x_r, x_2 - \zeta x_3\}$ and $C = \{x_1 - x_2, \dots, x_1 - x_r\}$ where ζ is a primitive root of 1 of order m.

In (a), the result is clear. In (b), B is independent since it generates the basis $\{x_1, \dots, x_r\}$ of V^*. Besides C lies in the kernel of the index (the linear map $ind : V^* \to \mathbb{C}$, $ind(x_i) = 1$) while no $b \in B$ does.

3. The exceptional groups different from Coxeter types $\mathbf{E}_m$

In this case, we check case-by-case that $L(\mathcal{A}_W)$ is well-balanced using Tables C.1, C.2, and C.23 from the book [10].

We use Corollary 1 from this book stating that $\mathcal{A}_X$ is the reflection arrangement for a reflection subgroup W_X of W. The numbers $n_X = |\mathcal{A}_X|$ can be found from Table 1 as the sums of covariants for W_X.

Table 1 is organized as follows. The first row consists of the Shephard-Todd classification numbers (23–34) of exceptional groups of ranks greater than 2 (no types $\mathbf{E}_m$). The second row consists of the cardinalities n of the respective arrangements. The third row consists of the maximal cardinalities of $\mathcal{A}_X$. It suffices to check inequality (1): $|\mathcal{A}_X| < \frac{n}{2}$.

4. Types $\mathbf{E}_m$.

For the types $\mathbf{E}_m$, the inequality (1) does not hold but it is easy to check that $L(\mathcal{A})$ is well-balanced by definition. The needed information is in the Table 2.

The second row has the same meaning as in the previous table. The last row consists of pairs combining the maximal cardinality of $\mathcal{A}_Y$ with $\operatorname{rk} Y = r - 1$ and the cardinality of another $\mathcal{A}_X$ also with $\operatorname{rk} X = r - 1$. One needs to check that the sum in each pair is less than the entry of the second row. This shows that $L(\mathcal{A}_W)$ is well-balanced and finishes the proof. □

Remark 2 Generic arrangements with $n < 2r-1$ are not-large. The general formula for generic arrangements is

$$\mathrm{TC}_s(M) = \min\{sr - 1, (s-1)n\}.$$

For instance, if $r = 3$, $n = 4$, $s = 2$ then $\mathrm{TC}_2(M) = 4 < 2r - 1$.

This result has been generalized and will be published in another paper.

Table 1

23	24	25	26	27	28	29	30	31	32	33	34
15	21	12	21	45	24	40	60	60	40	45	126
5	4	4	5	5	9	12	15	15	12	12	45

Table 2

E_6	E_7	E_8
36	63	120
(20,15)	(36,21)	(63,42)

Conjecture 1 For every complex hyperplane arrangement the topological complexity of its complement equals (for every s) the cohomological lower bound.

References

1. V.I. Arnol'd, The cohomology ring of the group of dyed braids. Math. Notes **5**, 138–140 (1969)
2. I. Basabe, J. Gonzalez, Y. Rudyak, D. Tamaki, Higher topological complexity and its symmetrization. Algebr. Geom. Topology **14**, 2103–2124 (2014)
3. D. Bessis, Finite complex reflection arrangements are $K(\pi, 1)$ (2006). arXiv:math.GT/0610777
4. D. Cohen, G. Pruidze, Motion planning in tori. Bull. Lond. Math. Soc. **40**, 249–262 (2008)
5. P. Deligne, Les immeubles des groupes de tresses generalises. Invent. Math. **17**, 273–302 (1972)
6. M. Farber, Topological complexity of motion planning. Discrete Comput. Geom. **29**, 211–221 (2003)
7. M. Farber, S. Yuzvinsky, Topological robotics: subspace arrangements and collision free motion planning. *Geometry, Topology, and Mathematical Physics*. American Mathematical Society Translations Series 2, vol. 212 (American Mathematical Society, Providence, 2000), pp. 145–156
8. J. Gonzalez, M. Grant, Sequential motion planning of non-colliding particles in Euclidean spaces (2013). arXiv:math1309.4346
9. P. Orlik, L. Solomon, Unitary reflection groups and cohomology. Invent. Math. **59**, 77–94 (1980)
10. P. Orlik, H. Terao, *Arrangements of Hyperplanes*. Grundlehren der Mathematischen Wissenschaften, vol. 300 (Springer, Berlin/Heidelberg/New York, 1992)
11. Y. Rudyak, On higher analogs of topological complexity. Topology Appl. **157**, 916–920 (2010)
12. A. Shwarz, The genus of a fiber space. AMS Transl. **23**(14), 5339–5354 (1995)
13. S. Yuzvinsky, Topological complexity of generic hyperplane complements. Contemp. Math. **438**, 115–119 (2007)

Random Simplicial Complexes

Armindo Costa and Michael Farber

Abstract In this paper we propose a model of random simplicial complexes with randomness in all dimensions. We start with a set of n vertices and retain each of them with probability p_0; on the next step we connect every pair of retained vertices by an edge with probability p_1, and then fill in every triangle in the obtained random graph with probability p_2, and so on. As the result we obtain a random simplicial complex depending on the set of probability parameters $(p_0, p_1, \dots, p_r)$, $0 \le p_i \le 1$. The multi-parameter random simplicial complex includes both Linial-Meshulam and random clique complexes as special cases. Topological and geometric properties of this random simplicial complex depend on the whole set of parameters and their thresholds can be understood as convex subsets and not as single numbers as in all the previously studied models. We mainly focus on foundations and on containment properties of our multi-parameter random simplicial complexes. One may associate to any finite simplicial complex S a reduced density domain $\tilde{\mu}(S) \subset \mathbf{R}^r$ (a convex domain) which fully controls information about the values of the multi-parameter for which the random complex contains S as a simplicial subcomplex. We also analyse balanced simplicial complexes and give positive and negative examples. We apply these results to describe dimension of a random simplicial complex.

Keywords Balanced simplicial complex • Density domains • Random simplicial complex • Reduced density domains

1 Introduction

The study of random topological objects (such as random simplicial complexes and random manifolds) is motivated by potential applications to modelling of large complex systems in various engineering and computer science applications. Random topological objects are also of interest from pure mathematical point view

A. Costa • M. Farber (✉)
School of Mathematical Sciences, Queen Mary University of London, London E1 4NS, UK
e-mail: a.costa@qmul.ac.uk; m.farber@qmul.ac.uk

F. Callegaro et al. (eds.), *Configuration Spaces*, Springer INdAM Series 14,
DOI 10.1007/978-3-319-31580-5_6

since they can be used for constructing curious examples of topological objects with rare combinations of topological properties.

Several models of random manifolds and random simplicial complexes were suggested and studied recently, see [13] for a survey. One may mention random surfaces [16], random 3-manifolds [7], random configuration spaces of linkages [9]. The present paper is was inspired by the model of random simplicial complexes developed by Linial, Meshulam and Wallach [14, 15]. In the first paper [14] the authors studied an analogue of the classical Erdös–Rényi [8] model of random graphs in the situation of 2-dimensional simplicial complexes. In the following paper [15] a more general model of d-dimensional random simplicial complexes was studied. The random simplicial complexes of [14, 15] have the complete $(d-1)$-skeleton and their "randomness is concentrated in the top dimension". More specifically, one starts with the full $(d-1)$-skeleton of an $(n-1)$-dimensional simplex and adds d-faces at random, independently of each other, with probability p.

A different model of random simplicial complexes was studied by Kahle [11, 12] and by some other authors. These are clique complexes of random Erdös–Rényi graphs; here one takes a random graph in the Erdös–Rényi model and declares as a simplex every subset of vertices which form a *clique*, i.e. such that every two vertices of the subset are connected by an edge. Compared with the Linial–Meshulam model, the clique complex has "randomness" in dimension one but it influences the structure in all the higher dimensions.

In this paper we propose a more general and more flexible model of random simplicial complexes with randomness in all dimensions. We start with a set of n vertices and retain each of them with probability p_0; on the next step we connect every pair of retained vertices by an edge with probability p_1, and then fill in every triangle in the obtained random graph with probability p_2, and so on. As the result we obtain a random simplicial complex depending on the set of probability parameters

$$(p_0, p_1, \ldots, p_r), \quad 0 \leq p_i \leq 1.$$

Our multi-parameter random simplicial complex includes both Linial–Meshulam and random clique complexes as special case. Topological and geometric properties of this random simplicial complex depend on the whole set of parameters and their thresholds can be understood as convex subsets and not as single numbers as in all the previously studied models. We mainly focus on foundations and on containment properties of our multi-parameter random simplicial complexes. One may associate to any finite simplicial complex S a reduced density domain $\tilde{\mu}(S) \subset \mathbf{R}^r$ (which is a convex domain) which fully controls information about the values of the multi-parameter for which the random complex contains S as a simplicial subcomplex. We also analyse balanced simplicial complexes and give positive and negative examples. We apply these results to describe dimension of a random simplicial complex.

In the following papers (see [3, 4]) we shall address other topological and geometric properties of random simplicial complexes depending on multiple parameters (such as their homology and the fundamental group).

The authors thank Thomas Kappeler for useful discussions.

2 The Definition and Basic Properties

2.1 The Model

Let Δ_n denote the simplex with the vertex set $\{1, 2, \ldots, n\}$. We view Δ_n as an abstract simplicial complex of dimension $n-1$. Given a simplicial subcomplex $Y \subset \Delta_n$, we denote by $f_i(Y)$ the number of *i-faces* of Y (i.e. i-dimensional simplexes of Δ_n contained in Y).

Definition 1 An external face of a subcomplex $Y \subset \Delta_n$ is a simplex $\sigma \subset \Delta_n$ such that $\sigma \not\subset Y$ but the boundary of σ is contained in Y, $\partial\sigma \subset Y$.

We shall denote by $e_i(Y)$ the number of i-dimensional external faces of Y. Note that for $i = 0$, we have $e_0(Y) + f_0(Y) = n$ and for $i > 0$,

$$f_i(Y) + e_i(Y) \leq \binom{n}{i+1}.$$

Fix an integer $r > 0$ and a sequence

$$\mathbf{p} = (p_0, p_1, \ldots, p_r)$$

of real numbers satisfying

$$0 \leq p_i \leq 1.$$

Denote

$$q_i = 1 - p_i.$$

We consider the probability space $Y_r(n; \mathbf{p})$ consisting of all subcomplexes

$$Y \subset \Delta_n^{(r)}$$

with $\dim Y \leq r$ where the probability function

$$\mathbb{P}_r : Y_r(n; \mathbf{p}) \to \mathbf{R}$$

is given by the formula

$$\mathbb{P}_r(Y) = \prod_{i=0}^{r} p_i^{f_i(Y)} \cdot \prod_{i=0}^{r} q_i^{e_i(Y)} \tag{1}$$

for $Y \in Y_r(n; \mathbf{p})$. We shall show below that $\mathbb{P}_r$ is indeed a probability function, i.e.

$$\sum_{Y \subset \Delta_n^{(r)}} \mathbb{P}_r(Y) = 1, \tag{2}$$

see Corollary 1.

If $p_i = 0$ for some i then according to (1) we shall have $\mathbb{P}_r(Y) = 0$ unless $f_i(Y) = 0$, i.e. if Y contains no simplexes of dimension i (in this case Y contains no simplexes of dimension $\geq i$). Thus, if $p_i = 0$ the probability measure $\mathbb{P}_r$ is concentrated on the set of subcomplexes of Δ_n of dimension $< i$.

In the special case when one of the probability parameters satisfies $p_i = 1$ one has $q_i = 0$ and from formula (1) we see $\mathbb{P}_r(Y) = 0$ unless $e_i(Y) = 0$, i.e. if the subcomplex $Y \subset \Delta_n^{(r)}$ has no external faces of dimension i. In other words, we may say that if $p_i = 1$ the measure $\mathbb{P}_r$ is concentrated on the set of complexes satisfying $e_i(Y) = 0$, i.e. such that any boundary of the i-simplex in Y is filled by an i-simplex.

Lemma 1 *Let*

$$A \subset B \subset \Delta_n^{(r)}$$

be two subcomplexes satisfying the following condition: the boundary of any external face of B of dimension $\leq r$ is contained in A. Then

$$\mathbb{P}_r(A \subset Y \subset B) = \prod_{i=0}^{r} p_i^{f_i(A)} \cdot \prod_{i=0}^{r} q_i^{e_i(B)}. \tag{3}$$

Proof We act by induction on r. For $r = 0$, $A \subset B$ are discrete sets of vertices and the condition of the Lemma is automatically satisfied (since the boundary of any 0-face is the empty set). A subcomplex $Y \subset \Delta_n^{(0)}$ satisfying $A \subset Y \subset B$ is determined by a choice of $f_0(Y) - f_0(A)$ vertices out of $f_0(B) - f_0(A)$ vertices. Hence using formula (1),

$$\begin{aligned}\mathbb{P}_0(A \subset Y \subset B) &= \sum_{k=0}^{f_0(B)-f_0(A)} \binom{f_0(B) - f_0(A)}{k} \cdot p_0^{f_0(A)+k} q_0^{n-f_0(A)-k} \\ &= p_0^{f_0(A)} \cdot q_0^{n-f_0(A)} \cdot \left(1 + \frac{p_0}{q_0}\right)^{f_0(B)-f_0(A)}\end{aligned}$$

$$= p_0^{f_0(A)} \cdot q_0^{n-f_0(B)}$$
$$= p_0^{f_0(A)} \cdot q_0^{e_0(B)},$$

as claimed.

Now suppose that formula (3) holds for $r-1$ and consider the case of r. Note the formula

$$\mathbb{P}_r(Y) = \mathbb{P}_{r-1}(Y^{r-1}) \cdot q_r^{g_r(Y)} \cdot \left(\frac{p_r}{q_r}\right)^{f_r(Y)} \tag{4}$$

where $g_r(Y) = e_r(Y) + f_r(Y)$ is the number of boundaries of r-simplexes contained in Y. Note that the first two factors in (4) depend only on the skeleton Y^{r-1}.

We denote by $g_r^B(Y)$ the number of r-simplexes of B such that their boundary $\partial\Delta^r$ lies in Y. Clearly the number $g_r^B(Y)$ depends only on the skeleton Y^{r-1}. Our assumption that the boundary of any external i-face of B is contained in A for $i \leq r$ implies that for any subcomplex $A \subset Y \subset B$ one has

$$g_r(Y) - g_r^B(Y) = e_r(B). \tag{5}$$

A complex Y is uniquely determined by its skeleton Y^{r-1} and by the set of its r-faces. Note that, given the skeleton Y^{r-1}, the number $f_r(Y)$ is arbitrary satisfying

$$f_r(A) \subset f_r(Y) \subset g_r^B(Y).$$

Thus using (4) we find that the probability

$$\mathbb{P}_r(A \subset Y \subset B) = \sum_{A \subset Y \subset B} \mathbb{P}_r(Y)$$

equals

$$\sum_{Y^{r-1}} \mathbb{P}_{r-1}(Y^{r-1}) \cdot q_r^{g_r(Y)} \cdot \sum_{k=0}^{g_r^B(Y)-f_r(A)} \binom{g_r^B(Y) - f_r(A)}{k} \cdot \left(\frac{p_r}{q_r}\right)^{f_r(A)+k}$$
$$= \sum_{Y^{r-1}} \mathbb{P}_{r-1}(Y^{r-1}) \cdot q_r^{g_r(Y)} \cdot \left(\frac{p_r}{q_r}\right)^{f_r(A)} \cdot \left(1 + \frac{p_r}{q_r}\right)^{g_r^B(Y)-f_r(A)}$$
$$= \sum_{Y^{r-1}} \mathbb{P}_{r-1}(Y^{r-1}) \cdot p_r^{f_r(A)} \cdot q_r^{g_r(Y)-g_r^B(Y)}$$
$$= p_r^{f_r(A)} \cdot q_r^{e_r(B))} \cdot \sum_{Y^{r-1}} \mathbb{P}_{r-1}(Y^{r-1}).$$

Here we used Eq. (5). Next we may combine the obtained equality with the inductive hypothesis

$$\mathbb{P}_{r-1}(A^{r-1} \subset Y^{r-1} \subset B^{r-1}) = \prod_{i=0}^{r-1} p_i^{f_i(A)} \cdot \prod_{i=0}^{r-1} q_i^{e_u(B)}$$

to obtain (3). □

Note that the assumption that the boundary of any external face of B is an external face of A is essential in Lemma 1; the lemma is false without this assumption.

Taking the special case $A = \emptyset$, $B = \Delta_n^{(r)}$ in (3) we obtain the following Corollary confirming the fact that $\mathbb{P}_r$ is a probability function.

Corollary 1

$$\sum_{Y \subset \Delta_n^{(r)}} \mathbb{P}_r(Y) = 1.$$

Example 1 The probability of the empty subcomplex $Y = \emptyset$ equals

$$\mathbb{P}(Y = \emptyset) = (1 - p_0)^n.$$

If $p_0 \to 0$ then $\mathbb{P}(Y = \emptyset) = (1 - p_0)^n \sim e^{-p_0 n}$. Hence, if $np_0 \to 0$ then $\mathbb{P}(Y = \emptyset) \to 1$, i.e. we may say that $Y = \emptyset$, a.a.s.

If $p_0 = c/n$ then

$$\mathbb{P}(Y = \emptyset) = (1 - c/n)^n \to e^{-c}$$

as $n \to \infty$. Thus, we see that for $p_0 = c/n$ the empty subset appears with probability e^{-c}, a.s.s.

Since we intend to study non-empty large random simplicial complexes, we shall always assume that $p_0 = \frac{\omega}{n}$ where ω tends to ∞.

2.2 The Number of Vertices of Y

Lemma 2 *Consider a random simplicial complex $Y \in Y_r(n, \mathbf{p})$, where $\mathbf{p} = (p_0, p_1, \ldots, p_r)$ is the probability multiparameter. Assume that $p_0 = \omega/n$ where $\omega \to \infty$. Then the number of vertices $f_0(Y)$ of Y is approximately ω; more precisely for any $0 < \epsilon < 1/2$ a.a.s. one has*

$$(1 - \delta)\omega \leq f_0(Y) \leq (1 + \delta)\omega, \tag{6}$$

where $\delta = \omega^{-1/2+\epsilon}$.

Proof For a vertex $i \in \{1, \ldots, n\}$, denote by $X_i : Y_r(n, \mathbf{p}) \to \mathbf{R}$ the random variable such that $X_i(Y) = 1$ if $i \in Y$ and $X_i(Y) = 0$ if $i \notin Y$. Then $f_0 = \sum_i X_i$ and by Lemma 1, $\mathbb{E}(X_i) = p_0$. Hence, $\mathbb{E}(f_0) = np_0 = \omega$.

The variance of f_0 equals $V(f_0) = \sum_{i,j} \mathbb{E}(X_iX_j) - \mathbb{E}(f_0)^2$. By Lemma 1, $\mathbb{E}(X_iX_j) = p_0^2$ for $i \neq j$ and $\mathbb{E}(X_iX_j) = p_0$ for $i = j$. Thus, the variance of f_0 equals

$$n(n-1)p_0^2 + np_0 - n^2p_0^2 = np_0(1-p_0) = \omega(1-p_0).$$

Applying the Chebychev inequality

$$\mathbb{P}\{|f_0 - \mathbb{E}(f_0)| \geq \epsilon\} \leq V(f_0)/\alpha^2$$

with $\alpha = \delta\omega$ we obtain

$$\mathbb{P}\{(1-\delta)\omega \leq f_0 \leq (1+\delta)\omega\} \geq 1 - \frac{(1-p_0)}{\delta^2 \cdot \omega} \geq 1 - \frac{1}{\omega^{2\epsilon}} \to 1.$$

□

A more precise estimate of the number of vertices is obtained in [3].

2.3 Special Cases

The models of random simplicial complexes which were studied previously contained randomness in a single dimension while our model allows various probabilistic regimes in different dimensions simultaneously. Thus we obtain more flexible constructions of random simplicial complexes.

The model we consider turns into some well known models in special cases:

When $r = 1$ and $\mathbf{p} = (1, p)$ we obtain the classical model of random graphs of Erdös and Rényi [8].

When $r = 2$ and $\mathbf{p} = (1, 1, p)$ we obtain the Linial–Meshulam model of random 2-complexes [14].

When r is arbitrary and fixed and $\mathbf{p} = (1, 1, \ldots, 1, p)$ we obtain the random simplicial complexes of Meshulam and Wallach [15].

For $r = n - 1$ and $\mathbf{p} = (1, p, 1, 1, \ldots, 1)$ one obtains the clique complexes of random graphs studied in [11].

2.4 Gibbs Formalism

In this subsection we briefly describe a more general class of models of random simplicial complexes which includes the model of Sect. 2.1 as a special case.

On the set of all subcomplexes $Y \subset \Delta_n^{(r)}$, one considers an energy function $H = H_{\beta,\gamma}$ having the form

$$H(Y) = H_{\beta,\gamma}(Y) = \sum_{i=0}^{r} [\beta_i f_i(Y) + \gamma_i e_i(Y)], \tag{7}$$

where β_i and γ_i are real parameters, $i = 0, 1, \ldots, r$. The partition function

$$Z = Z_{\beta,\gamma} = \sum_{Y \subset \Delta_n^{(r)}} e^{H_{\beta,\gamma}(Y)} \tag{8}$$

is a function of the parameters β_i, γ_i and n and

$$\mathbb{P}_{\beta,\gamma}(Y) = \frac{1}{Z_{\beta,\gamma}} e^{H_{\beta,\gamma}(Y)} \tag{9}$$

is a probability measure on the set of all subcomplexes $Y \subset \Delta_n^{(r)}$. Here the case $r = \infty$ is not excluded; then $Y \subset \Delta_n$ runs over all subcomplexes.

In the special case when the parameters β_i, γ_i satisfy

$$e^{\beta_i} + e^{\gamma_i} = 1, \quad i = 0, \ldots, r. \tag{10}$$

We may define the probability parameters p_i, q_i by

$$p_i = e^{\beta_i}, \quad q_i = e^{\gamma_i}. \tag{11}$$

One can easily check that under the assumptions (10) the probability measure $\mathbb{P}_{\beta,\gamma}$ coincides with the measure $\mathbb{P}_r$ given by (1). The relation (10) implies that the partition function $Z_{\beta,\gamma} = 1$ equals one, according to Corollary 1.

3 The Containment Problem

Consider a random simplicial complex $Y \in Y_r(n, \mathbf{p})$ where $\mathbf{p} = (p_0, p_1, \ldots, p_r)$. As in the classical containment problem for random graphs we ask under which conditions Y has a simplicial subcomplex isomorphic to a given r-dimensional finite simplicial complex S. The answer is slightly different from the random graph theory: we associate with S a convex set

$$\tilde{\mathcal{M}}(S) \subset \mathbf{R}^{r+1}$$

in the space of exponents of probability parameters which (as we show here) is fully responsible for the containment.

Write

$$p_i = n^{-\alpha_i}, \quad \text{where} \quad \alpha_i \geq 0, \quad i = 0, 1, \ldots, r.$$

For simplicity we shall assume in this paper that the numbers α_i are constant (do not depend on n). We shall use the following notation

$$\mathbf{p} = n^{-\alpha}, \quad \text{where} \quad \alpha = (\alpha_0, \ldots, \alpha_r). \tag{12}$$

Clearly, we must assume that

$$\alpha_0 < 1$$

since for $\alpha_0 \geq 1$ the complex Y is either empty or has one vertex, a.a.s. (see Example 1).

3.1 *The Density Invariants*

Let S be a fixed finite simplicial r-dimensional complex. As usual, $f_i(S)$ denotes the number of i-dimensional faces in S. Define the following ratios (density invariants):

$$\mu_i(S) = \frac{f_0(S)}{f_i(S)}, \quad i = 0, 1, \ldots, r.$$

We do not exclude the case when $f_j(S) = 0$; then $\mu_j(S) = \infty$. The 0-th number is always one, $\mu_0(S) = 1$. Compare [1], Definition 11.

Lemma 3 *If*

$$\sum_{i=0}^{r} \frac{\alpha_i}{\mu_i(S)} > 1, \tag{13}$$

then the probability that S is embeddable into $Y \in Y_r(n, \mathbf{p})$ tends to zero as $n \to \infty$.

Proof Let $F_0(S)$ denote the set of vertices of S. An embedding of S into $\Delta_n^{(r)}$ is determined by an embedding $J : F_0(S) \to [n]$. For any such embedding define a random variable $X_J : Y_r(n, \mathbf{p}) \to \mathbf{R}$ given by

$$X_J(Y) = \begin{cases} 1, \text{ if } & Y \supset J(S), \\ 0, \text{ otherwise.} \end{cases}$$

Then $X = \sum_J X_J$ is the random variable counting the number of isomorphic copies of S in Y. One has

$$\mathbb{E}(X_J) = \prod_{i=0}^{r} p_i^{f_i(S)}$$

by Lemma 1. Thus we have

$$\mathbb{E}(X) = \binom{n}{f_0(S)} \cdot f_0(S)! \cdot \prod_{i=0}^{r} p_i^{f_i(S)}$$

$$\sim n^{f_0(S) - \sum_{i=0}^{r} \alpha_i f_i(S)}$$

$$= \left(n^{1-\sum_{i=0}^{r} \frac{\alpha_i}{\mu_i(S)}}\right)^{f_0(S)}.$$

We see that (26) implies $\mathbb{E}(X) \to 0$. Hence

$$\mathbb{P}(X > 0) \leq \mathbb{E}(X)$$

also tends to zero. □

3.2 *The Density Domains*

Consider the Euclidean Space $\mathbf{R}^{r+1}$ with coordinates $(\alpha_0, \alpha_1, \ldots, \alpha_r)$.

Definition 2 For a finite simplicial complex S of dimension $\leq r$, we denote by

$$\mathcal{M}(S) \subset \mathbf{R}^{r+1}$$

the convex domain given by the following inequalities:

$$\alpha_0 + \frac{\alpha_1}{\mu_1(S)} + \frac{\alpha_2}{\mu_2(S)} + \cdots + \frac{\alpha_r}{\mu_r(S)} < 1,$$

$$\alpha_0 \geq 0, \quad \alpha_1 \geq 0, \quad \ldots, \quad \alpha_r > 0.$$

The domain $\mathcal{M}(S)$ is a simplex of dimension $r+1$ which has the origin as one of its vertices and the other vertices are of the form $\mu_i(S)e_i$ where $e_i = (0, \ldots, 0, 1, 0, \ldots, 0)$ with 1 standing on the i-th position, where $i = 0, 1, \ldots, r$.

We may restate Lemma 3 as follows:

Lemma 4 *If the vector of exponents* $\boldsymbol{\alpha} = (\alpha_0, \ldots, \alpha_r)$ *does not belong to the closure*

$$\boldsymbol{\alpha} \notin \overline{\mathcal{M}(S)}$$

then the complex S *is not embeddable a.a.s. into a random simplicial complex* $Y \in Y_r(n, \mathbf{p})$, *where* $\mathbf{p} = n^{-\boldsymbol{\alpha}}$, *i.e.* $\mathbf{p} = (n^{-\alpha_0}, n^{-\alpha_1}, \ldots, n^{-\alpha_r})$.

Next we define the domain $\tilde{\mathcal{M}}(S) \subset \mathbf{R}^{r+1}$ as the intersection

$$\tilde{\mathcal{M}}(S) = \bigcap_{T \subset S} \mathcal{M}(T). \tag{14}$$

Here T runs over all subcomplexes of S. Clearly, $\tilde{\mu}(S)$ is an $(r+1)$-dimensional convex polytope.

Lemma 5 *If*

$$\boldsymbol{\alpha} = (\alpha_0, \alpha_1, \ldots, \alpha_r) \in \tilde{\mathcal{M}}(S)$$

then S *is embeddable into a random complex* $Y \in Y_r(n, \mathbf{p})$ *where* $\mathbf{p} = n^{-\boldsymbol{\alpha}}$, *a.a.s.*

Proof As in the proof of Lemma 1, let $F_0(S)$ denote the set of vertices of S and let $J : F_0(S) \to [n]$ be an embedding. Any such embedding uniquely determines a simplicial embedding $J : S \to \Delta_n$. Consider a random variable $X_J : Y_r(n, \mathbf{p}) \to \mathbf{R}$ given by

$$X_J(Y) = \begin{cases} 1, \text{ if } & Y \supset J(S), \\ \\ 0, \text{ otherwise.} \end{cases}$$

Then $X = \sum_J X_J$ counts the number of isomorphic copies of S in Y and $\mathbb{E}(X_J) = \prod_{i=0}^{r} p_i^{f_i(S)}$ by Lemma 1. Hence,

$$\mathbb{E}(X) \sim n^{f_0(S)} \cdot \prod_{i \geq 0} p_i^{f_i(S)}.$$

We shall use the Chebyshev inequality

$$\mathbb{P}(X = 0) \leq \frac{\mathrm{Var(X)}}{\mathbb{E}(X)^2}. \tag{15}$$

One has

$$\mathrm{Var}(X) = \mathbb{E}(X^2) - \mathbb{E}(X)^2 = \sum_{J,J'} \left[\mathbb{E}(X_J X'_J) - \mathbb{E}(X_J)\mathbb{E}(X_{J'})\right]. \tag{16}$$

Given two simplicial embeddings $J, J' : S \to \Delta_n$, the product $X_J X_{J'}$ is a 0–1 random variable, it has value 1 on a subcomplex $Y \subset \Delta_n^{(2)}$ if and only if Y contains the union $J(S) \cup J'(S)$. Thus, by Lemma 1, we have

$$\mathbb{E}(X_J X_{J'}) = \prod_{i \geq 0} p_i^{2f_i(S) - f_i(T')},$$

where T' denotes the intersection $J(S) \cap J'(S)$. If $J(S)$ and $J'(S)$ are disjoint then $\mathbb{E}(X_J X_{J'})$ equals $\mathbb{E}(X_J) \cdot \mathbb{E}(X_{J'})$; thus in formula (16) we may assume that J and J' are such that the intersection $J(S) \cap J'(S) \neq \emptyset$

Denote by T by the subcomplex $T = J^{-1}(T') \subset S$. For a fixed subcomplex $T \subset S$ the number of pairs of embeddings $J, J' : S \to \Delta_n$ such that $J^{-1}(J(S) \cap J'(S)) = T$ is bounded above by

$$C_T n^{2f_0(S) - f_0(T)}$$

where C_T is the number of isomorphic copies of T in S.

Thus we obtain

$$\mathrm{Var}(X) \leq \sum_{T \subset S} C_T n^{2f_0(S) - f_0(T)} \prod_{i \geq 0} p_i^{2f_i(S) - f_i(T)}.$$

On the other hand,

$$\mathbb{E}(X) \geq \frac{1}{2} n^{f_0(S)} \prod_{i \geq 0} p_i^{f_i(S)}.$$

Therefore,

$$\begin{aligned}
\frac{\mathrm{Var(X)}}{\mathbb{E}(X)^2} &\leq 4 \cdot \sum_{T \subset S} C_T n^{-f_0(T)} \cdot \prod_{i \geq 0} p_i^{-f_i(T)} \\
&\leq 4 \cdot \sum_{T \subset S} C_T n^{-f_0(T) + \sum_{i \geq 0} \alpha_i f_i(T)} \\
&= 4 \cdot \sum_{T \subset S} C_T n^{\left[\sum_{i \geq 0} \frac{\alpha_i}{\mu_i(T)} - 1\right] f_0(T)}.
\end{aligned}$$

Here T runs over all nonempty subcomplexes of S. If $(\alpha_0, \dots, \alpha_r) \in \tilde{\mathcal{M}}(S)$ then for any $T \subset S$ we have

$$\sum_{i \geq 0} \frac{\alpha_i}{\mu_i(T)} < 1.$$

Thus we see that the ratio $\frac{\mathrm{Var}(X)}{\mathbb{E}(X)^2}$ tends to zero as $n \to \infty$. The result now follows from (15). □

We may summarise the obtained results as follows:

Lemma 6 *Let S be a fixed finite simplicial complex of dimension* $\dim S \leq r$.

1. *If $\alpha \in \tilde{\mathcal{M}}(S)$ then a random simplicial complex $Y \in Y_r(n, n^{-\alpha})$ contains S as a simplicial subcomplex, a.a.s.*
2. *If $\alpha \notin Closure(\tilde{\mathcal{M}}(S))$ then a random simplicial complex $Y \in Y_r(n, n^{-\alpha})$ does not contain S as a simplicial subcomplex, a.a.s.*

The first statement repeats Lemma 5. The second statement follows from Lemma 4 and from the equality

$$Closure(\tilde{\mathcal{M}}(S)) = \bigcap_{T \subset S} Closure(\mathcal{M}(T)).$$

3.3 The Reduced Density Domain

Since $\mu_0(S) = 1$ the simplex $\mathcal{M}(S) \subset \mathbf{R}^{r+1}$ contains the point $(1, 0, \dots, 0)$ as one of its vertices and $\mathcal{M}(S)$ is the cone with apex $(1, 0, \dots, 0) \in \mathbf{R}^{r+1}$ over the simplex

$$\mu(S) = \{(\alpha_1, \dots, \alpha_r); \sum_{i=1}^{r} \frac{\alpha_i}{\mu_i(S)} < 1, \quad \alpha_i \geq 0\} \subset \mathbf{R}^r. \tag{17}$$

Hence the convex domain

$$\tilde{\mathcal{M}}(S) \subset \mathbf{R}^r$$

is a cone with apex $(1, 0, \dots, 0) \in \mathbf{R}^{r+1}$ over the domain $\tilde{\mu}(S) \subset \mathbf{R}^r$ which is defined as the intersection

$$\tilde{\mu}(S) = \bigcap_{T \subset S} \mu(T). \tag{18}$$

We call $\tilde{\mu}(S) \subset \mathbf{R}^r$ *the reduced density domain* associated to S.

For a subset of vertices $W \subset V = V(S)$, denote by $S_W \subset S$ the simplicial complex induced on W. Then

$$\tilde{\mu}(S) = \bigcap_{W \subset V} \mu(S_W). \tag{19}$$

This follows from the observation that for a simplicial subcomplex $T \subset S$ one has

$$\mu_i(T) \geq \mu_i(S_W)$$

and therefore

$$\mu(T) \supset \mu(S_W),$$

where $W = V(T) \subset V(S)$ is the set of vertices of T.

3.4 *The Invariance Principle*

By Lemmas 4 and 5, the extended density domain $\tilde{\mathcal{M}}(S)$ controls embedability of S into a random simplicial complex $Y \in Y_r(n, n^{-\alpha})$ and since it is a cone with vertex $(1, 0, \ldots, 0)$ we obtain the following corollary:

Corollary 2 *If $\alpha = (\alpha_0, \alpha_1, \ldots, \alpha_r) \in \mathbf{R}^{r+1}$ and $\alpha' = (\alpha'_0, \alpha'_1, \ldots, \alpha'_r) \in \mathbf{R}^{r+1}$ lie on a line passing trough $(1, 0, \ldots, 0) \in \mathbf{R}^{r+1}$ then the probability spaces $Y_r(n, n^{-\alpha})$ and $Y_r(n, n^{-\alpha'})$ have identical embedability properties with respect to any fixed finite simplicial complex S, $\dim S \leq r$, i.e. S embeds into $Y \in Y_r(n, n^{-\alpha})$ a.a.s. if and only if S is embeds into $Y \in Y_r(n, n^{-\alpha'})$, a.a.s.*

Thus, instead of a vector of exponents $\alpha = (\alpha_0, \alpha_1, \ldots, \alpha_r)$ we may consider the vector

$$\alpha' = (0, \frac{\alpha_1}{1 - \alpha_0}, \frac{\alpha_2}{1 - \alpha_0}, \ldots, \frac{\alpha_r}{1 - \alpha_0})$$

which has the first coordinate 0, i.e. in this case $p_0 = 1$.

Conjecture 1 We conjecture that all geometric and topological properties of the random complex $Y_r(n, n^{-\alpha})$ remain invariant when the multi-exponent α moves along any line passing through the point $(1, 0, \ldots, 0)$.

3.5 Examples

Example 2 Let S be a closed triangulated surface, $r = 2$. Then the number of edges e and the number of faces f are related by $3f = 2e$. We obtain that

$$3\mu_1(S) = 2\mu_2(S)$$

i.e. the simplex $\mu(S) \subset \mathbf{R}^2$ has a fixed slope independent of the topology of the surface and of the details of a particular triangulation.

More specifically, if S is a closed surface then

$$\mu_1(S) = \frac{1}{3} + \frac{\chi(S)}{e(S)}, \quad \mu_2(S) = \frac{1}{2} + \frac{\chi(S)}{f(S)}, \tag{20}$$

where $\chi(S)$ is the Euler characteristic of S and $e(S), f(S)$ denote the numbers of edges and faces (i.e. 2-simplexes) in S. This follows from [1, p. 132] and [6, Sect. 2].

The following Figs. 1, 2, 3 show the density domains for closed surfaces Σ depending on whether $\chi(\Sigma)$ is positive, zero or negative.

Example 3 Let $Z = P^2 \cup D^2$ be the following 2-complex. Here P^2 is a triangulated real projective plane having a cycle C of length 5 representing the non-contractible loop. D^2 is a triangulated disc with boundary of length 5 which is identified with C (Fig. 4). To compute $\mu_i(Z)$ we shall use the formulae

$$\mu_1(Z) = \frac{1}{3} + \frac{\chi(Z) + L(Z)/3}{e(Z)}, \quad \mu_2(Z) = \frac{1}{2} + \frac{\chi(Z) + L(Z)/2}{f(Z)}, \tag{21}$$

where $e(Z)$ and $f(Z)$ denote the numbers of edges and faces of Z and $L(Z)$ denotes

$$\sum_e [2 - \deg_Z(e)].$$

Here e runs over the edges of Z and $\deg_Z(e)$ is the number of faces of Z containing e, see formula (2) in [2] and formula (8) in [6]. In our case, $L(Z) = -5$ and $\chi(Z) = 2$; therefore,

$$\mu_1(Z) = \frac{1}{3} + \frac{1}{3e(Z)}, \quad \mu_2(Z) = \frac{1}{2} - \frac{1}{2f(Z)}. \tag{22}$$

Example 4 For an integer t, let S_t be a 2-complex constructed as follows. The vertex set $F_0(S_k)$ is $\{1, 2, \ldots, t+1\}$, the set of 1-simplexes $F_1(S_t)$ is the set of all pairs (i, j) where $1 \leq i < j \leq t+1$ (i.e. the 1-skeleton of S_t is a complete graph on $t+1$ vertices), and the set of 2-simplexes $F_2(S_t)$ consists of triples (i, j, k) where $1 \leq i < j < k \leq t$. To describe the reduced density domain $\tilde{\mu}(S_t)$ we shall use the formula (2.4). Consider a subset $W \subset F_0(S_t) = \{1, \ldots, t+1\}$. If W does not contain

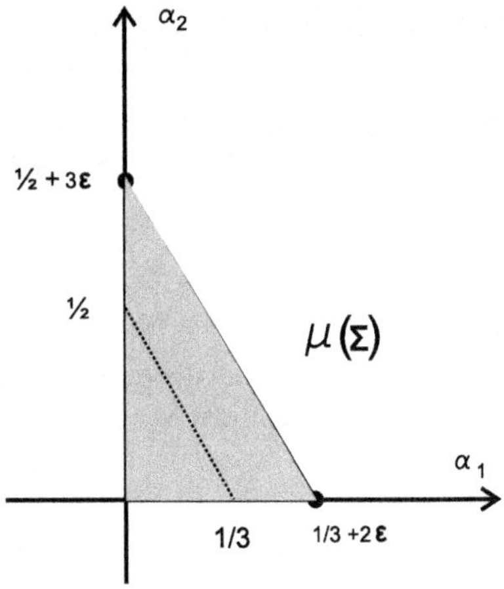

Fig. 1 The reduced density domain $\mu(\Sigma)$ for a triangulation of a closed surface with $\chi(\Sigma) > 0$. The number $\epsilon > 0$ depends on the genus and on the number of 2-simplexes in the triangulation

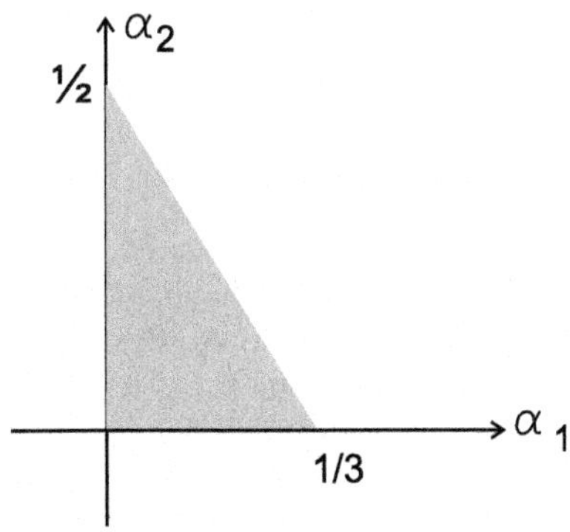

Fig. 2 The reduced density domain of a triangulated closed surface with $\chi(\Sigma) = 0$ (i.e. Σ is the torus or the Klein bottle)

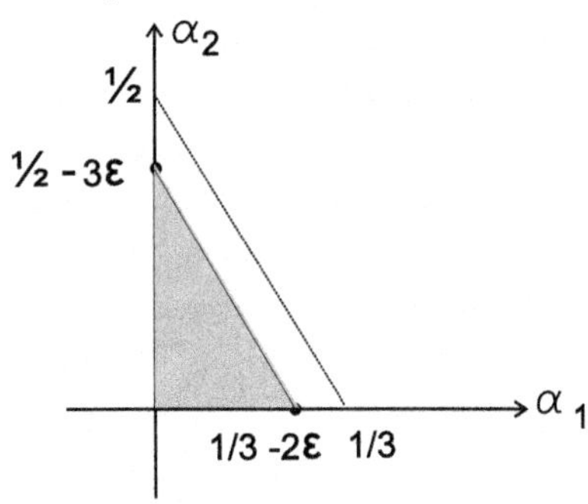

Fig. 3 The reduced density domain $\mu(\Sigma)$ of a triangulated closed surface with $\chi(\Sigma) < 0$. The number $\epsilon > 0$ depends on the genus and on the number of 2-simplexes in the triangulation

the vertex $t + 1$ then the induced complex S_W is the 2-skeleton of the τ-dimensional simplex where $|W| = \tau + 1$ and we have

$$\mu_1(S_W) = \frac{\tau + 1}{\binom{\tau+1}{2}} = \frac{2}{\tau}, \quad \mu_2(S_W) = \frac{\tau + 1}{\binom{\tau+1}{3}}. \tag{23}$$

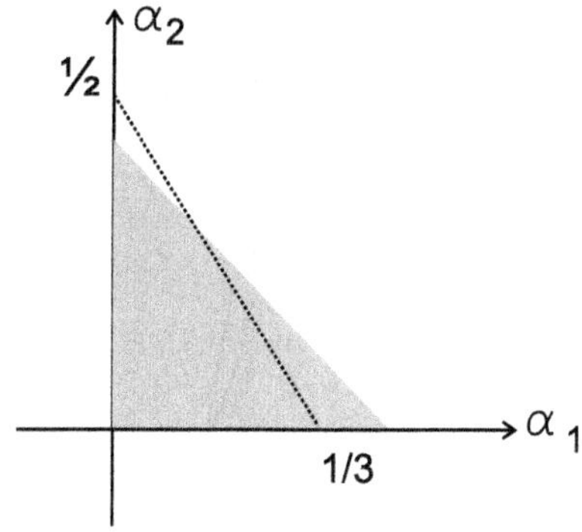

Fig. 4 The reduced density domain of the complex Z has a lower slope than in the case of surfaces

If W contains the last vertex $t+1$ then

$$\mu_1(S_W) = \frac{\tau+1}{\binom{\tau+1}{2}} = \frac{2}{\tau}, \quad \mu_2(S_W) = \frac{\tau+1}{\binom{\tau}{3}}. \tag{24}$$

In the first case, $\tau \leq t-1$ and in the second case $\tau \leq t$. We see that

$$\min_W \mu_1(S_W) = \frac{2}{t}$$

(is achieved for $W = W_1 = \{1, \ldots, t+1\}$) and

$$\min_W \mu_2(S_W) = \frac{t}{\binom{t}{3}}$$

(is achieved for $W = W_2 = \{1, \ldots, t\}$). The two lines given by the equations

$$\frac{\alpha_1}{\mu_1(S_{W_i})} + \frac{\alpha_2}{\mu_2(S_{W_i})} = 1, \quad i = 1, 2,$$

intersect at the point

$$(\alpha_1, \alpha_2) = \left(\frac{1}{t}, \frac{3(t+1)}{t(t-1)(t-2)}\right).$$

It is easy to check that this point (α_1, α_2) satisfies the inequality

$$\frac{\alpha_1}{\mu_1(S_W)} + \frac{\alpha_2}{\mu_2(S_W)} \leq 1,$$

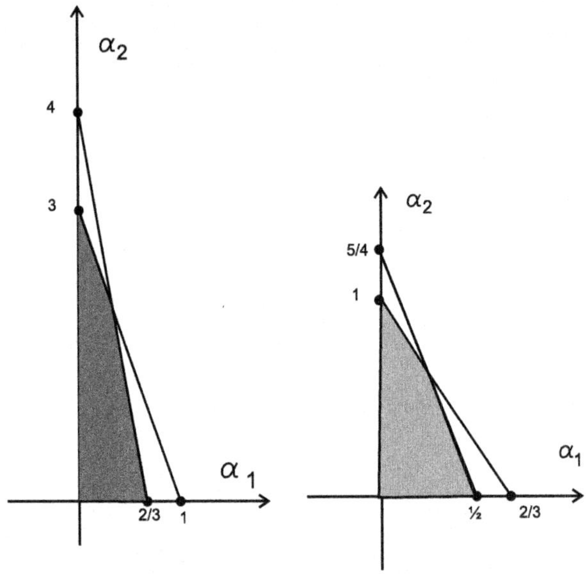

Fig. 5 The reduced density domain $\tilde{\mu}(S_t)$ of the complex S_t of Example 4 with $t = 3$ (*left*) and $t = 4$ (*right*)

for arbitrary subset $W \subset \{1, \dots, t+1\}$. This argument shows that in this example

$$\tilde{\mu}(S_t) = \mu(S_{W_1}) \cap \mu(S_{W_2})$$

and justifies our picture Fig. 5.

4 Balanced Simplicial Complexes

Definition 3 We shall say that an r-dimensional simplicial complex S is balanced if

$$\tilde{\mu}(S) = \mu(S) \subset \mathbf{R}^r.$$

The complex Z of Example 4 is unbalanced because its density domain $\tilde{\mu}(Z)$ is not a simplex, see Fig. 5.

Lemma 7 *The following properties are equivalent:*

a) S is balanced.
b) For any subcomplex $T \subset S$ one has $\mu(S) \subset \mu(T)$.
c) For any subcomplex $T \subset S$ and for any $i = 1, \ldots, r$ one has $\mu_i(T) \geq \mu_i(S)$.

The proof is obvious.

Example 5 Let S be a 2-dimensional triangulated disc having v vertices such that among them v_i are internal. It is easy to check (using the Euler–Poincaré formula) that $e = 2v + v_i - 3$ and $f = v + v_i - 2$. Hence,

$$\mu_1(S) = \frac{v}{2v + v_i - 3}, \quad \mu_2(S) = \frac{v}{v + v_i - 2}. \tag{25}$$

Let us assume that $v_i > 3$; then $\mu_1(S) < 1/2$ and $\mu_2(S) < 1$. Suppose that there exists a proper subdisc $T \subset S$ containing all the internal vertices. Then

$$\mu_1(T) = \frac{v'}{2v' + v_i - 3}, \quad \mu_2(S) = \frac{v'}{v' + v_i - 2}$$

where $v' < v$ and we see that

$$\mu_i(T) < \mu_i(S), \quad \text{for} \quad i = 1, 2.$$

This argument shows that there exist unbalanced triangulations of the disc.

Theorem 1 *Any triangulation of a closed surface S with $\chi(S) \geq 0$ is balanced.*

Proof Let S be a triangulated closed surface, $\chi(S) \geq 0$, and let $T \subset S$ be a proper subcomplex. We want to show that

$$\mu_i(T) \geq \mu_i(S), \quad \text{for} \quad i = 1, 2. \tag{26}$$

Using formulae (21) and our assumption $\chi(S) \geq 0$ we see that (26) would follow from the inequalities

$$L(T) \geq 3\chi(S, T), \quad L(T) \geq 2\chi(S, T), \tag{27}$$

since $e(T) \leq e(S)$ and $f(T) \leq f(S)$. Here $\chi(S, T) = \chi(S) - \chi(T)$. Clearly, every edge of T has degree 0, 1, or 2 and hence $L(T) \geq 0$; therefore, (27) follows automatically if $\chi(S, T) \leq 0$. In the case $\chi(S, T) > 0$ it is enough to show the left inequality in (27).

Let $N \supset T$ be a small tubular neighbourhood of T in S. We shall denote by $K = S - \text{Int(N)}$ the closure of the complement of N in S and apply Proposition 3.46

from [10]. This Proposition states that $H^i(K; \mathbf{Z}_2)$ is isomorphic to $H_{2-i}(S, S-K; \mathbf{Z}_2)$ and is valid without the assumption that S is orientable since we take $\mathbf{Z}_2$ coefficients. Thus,

$$\chi(K) = \chi(S, S-K) = \chi(S, T).$$

If we denote by $k = b_0(K) = b_0(S-T)$ the number of connected components of $S-T$ then

$$\chi(S, T) = \chi(T) \leq b_0(K) = k$$

and (26) would follow from

$$L(T) \geq 3k.$$

Note that

$$L(T) = e_1(T) + 2e_0(T)$$

where $e_i(T)$ denotes the number of edges of T which have degree $i = 0, 1$.

Consider the j-th connected component of the complement $S-T$; its set-theoretic boundary is a graph which is the image of a simplicial map $C_j \to S$ where C_j is a triangulation of the circle. Denoting by $|C_j|$ the number of edges we see that

$$\sum_{j=1}^{k} |C_j| = e_1(T) + 2e_0(T) = L(T).$$

Indeed, the image of the maps $\sqcup_j C_j \to S$ is the union of edges of T having degree 0 and 1 and each edge of degree 2 is covered twice. Clearly, $|C_j| \geq 3$ for each C_j and the inequality $L(T) \geq 3k$ follows. □

Remark 1 It is easy to show that the assumption $\chi(S) \geq 0$ of Theorem 1 is necessary. More specifically, *any closed surface with* $\chi(S) < 0$ *admits a non-balanced triangulation.* For the proof, see [6], Sect. 2.

Definition 4 We shall say that S is strictly balanced if for any proper subcomplex $T \subset S$ one has

$$\mu_i(T) \geq \mu_i(S) \quad \text{for} \quad i = 1, 2, \ldots, r$$

and at least one of these inequalities is strict.

Let S be a simplicial complex of dimension r. *The degree* of an i-dimensional simplex $\sigma \subset S$ is the number of $(i+1)$-dimensional simplexes containing σ; we denote this degree $\deg \sigma$.

Lemma 8 *Let S be a connected r-dimensional simplicial complex with the property that the degree of any i-dimensional simplex $\sigma \subset S$ depends only on i. Then S is strictly balanced.*

For the proof we need an expression of the density invariants $\mu_i(S)$ in terms of the average degree of simplexes which is described in the following lemma.

For a simplicial complex S we denote by $\bar{d}_i(S)$ the ratio

$$\frac{\sum\limits_{\dim \sigma = i} \deg \sigma}{f_i(S)}.$$

It has the meaning of the average degree of i-dimensional simplexes in S.

Lemma 9 *For an r-dimensional simplicial complex S one has*

$$\mu_i(S) = \frac{(i+1)!}{\bar{d}_0(S) \cdot \bar{d}_1(S) \cdots \bar{d}_{i-1}(S)} \quad \textit{where} \quad i = 1, 2, \ldots, r. \tag{28}$$

Proof We observe that

$$\sum_{\dim \sigma = i} \deg(\sigma) = (i+2) \cdot f_{i+1}(S)$$

and therefore

$$\bar{d}_i(S) = (i+2) \cdot \frac{f_{i+1}(S)}{f_i(S)}.$$

Multiplying these equalities we obtain

$$\bar{d}_0(S) \cdot \bar{d}_1(S) \cdots \bar{d}_{i-1}(S) = (i+1)! \cdot \frac{f_i(S)}{f_0(S)}.$$

The last equality is equivalent to the claim of the lemma. □

Proof (Proof of Lemma 8) By assumption, $\deg_i^S(\sigma) = d_i$ for any simplex $\sigma \subset S$, $\dim \sigma = i$. Let $T \subset S$ be a proper subcomplex. Then $\deg_i^T(\sigma) \leq d_i$ for any simplex $\sigma \subset T$, $\dim \sigma = i$. Moreover, $\deg_i^T(\sigma) < d_i$ for some σ and some i (for example, for the minimal i such that $f_{i+1}(T) < f_{i+1}(S)$). This shows that

$$d_i \geq \bar{d}_i(T), \quad i = 0, 1, \ldots, r-1.$$

and at least one of these inequalities is strict. Using the formula (28) we obtain that

$$\mu_i(S) \leq \mu_i(T), \quad i = 1, \ldots, r$$

and at least of these inequalities is strict. □

5 Dimension of a Random Simplicial Complex

Let S be the abstract simplex of dimension $s \leq r$. Then S is strictly balanced (by Lemma 8). The embeddability of S into a random complex $Y \in Y_r(n, \mathbf{p})$ means that $\dim Y \geq s$, hence we may use Lemma 6 to answer the question about the dimension of Y. We have

$$\mu_i(S) = \frac{s+1}{\binom{s+1}{i+1}}, \quad i = 0, 1, \ldots, s$$

and

$$\mu_i(S) = \infty \quad \text{for} \quad i = s+1, \ldots, r.$$

Therefore, applying Lemmas 8 and 6 we see that *the dimension of a random simplicial complex* $Y \in Y_r(n, n^{-\alpha})$ *satisfies* $\dim Y \geq s$ *if*

$$\sum_{i=0}^{s} \alpha_i \cdot \frac{\binom{s+1}{i+1}}{s+1} < 1. \tag{29}$$

Here $\alpha = (\alpha_0, \alpha_1, \ldots, \alpha_r)$, $p_i = n^{-\alpha_i}$. The inequality (29) can be rewritten as

$$\alpha_0 + \alpha_1 \cdot \frac{s}{2} + \alpha_2 \cdot \frac{s(s-1)}{6} + \cdots + \alpha_s \cdot \frac{1}{s+1} < 1.$$

For a vector $\alpha = (\alpha_0, \alpha_1, \ldots, \alpha_r)$ with $\alpha_i \geq 0$ define the quantities

$$\phi_s(\alpha) = \sum_{i=0}^{r} \frac{\binom{s+1}{i+1}}{s+1} \cdot \alpha_i = \sum_{i=0}^{r} \frac{\binom{s}{i}}{i+1} \cdot \alpha_i, \qquad s = 0, 1, \ldots, r. \tag{30}$$

Note that $\binom{s+1}{i+1} = 0$ for $i > s$. Clearly,

$$\frac{\binom{s}{i}}{i+1} \leq \frac{\binom{s+1}{i}}{i+1}$$

and hence one has

$$\phi_0(\alpha) \leq \phi_1(\alpha) \leq \phi_2(\alpha) \leq \cdots \leq \phi_r(\alpha). \tag{31}$$

Here

$$\phi_0(\alpha) = \alpha_0,$$
$$\phi_1(\alpha) = \alpha_0 + \frac{1}{2} \cdot \alpha_1,$$
$$\phi_2(\alpha) = \alpha_0 + \alpha_1 + \frac{1}{3} \cdot \alpha_2,$$
$$\phi_3(\alpha) = \alpha_0 + \frac{3}{2} \cdot \alpha_1 + \alpha_2 + \frac{1}{4} \cdot \alpha_3.$$

We obtain the following result:

Corollary 3 *Given a multi-index* $\alpha = (\alpha_0, \alpha_1, \ldots, \alpha_r)$, $\alpha_i \geq 0$ *and an integer* s *satisfying* $0 \leq s \leq r$.

(1) If $\phi_s(\alpha) < 1$ *then dimension of a random complex* $Y \in Y_r(n, n^{-\alpha})$ *satisfies* $\dim Y \geq s$*, a.a.s.*
(2) If $\phi_s(\alpha) > 1$ *then dimension of a random complex* $Y \in Y_r(n, n^{-\alpha})$ *satisfies* $\dim Y < s$*.*
(3) The convex domain given by the inequalities

$$\phi_s(\alpha) < 1 < \phi_{s+1}(\alpha), \quad \textit{and} \quad \alpha_0 \geq 0,\, \alpha_1 \geq 0,\, \ldots,\, \alpha_r \geq 0$$

describes the area of the multi-parameter $\alpha \in \mathbf{R}^{r+1}$ *such that dimension of a random complex* $Y \in Y_r(n, n^{-\alpha})$ *satisfies* $\dim Y = s$*, a.a.s.*

In particular we see that $\dim Y \geq 0$ if $\alpha_0 < 1$ in accordance with the result of Example 1.

As an illustration of Corollary 3 consider the special case when $\alpha_0 = 0$ and

$$\alpha_3 = \alpha_4 = \cdots = \alpha_r = 0,$$

i.e. we have only two nonzero parameters α_1 and α_2. Then Corollary 3 implies:

- $\dim Y = 1$ if $\alpha_1 < 2$ and $\alpha_1 + \frac{1}{3}\alpha_2 > 1$, a.a.s.
- $\dim Y = 2$ if $\alpha_1 + \frac{1}{3}\alpha_2 < 1$ and $\frac{3}{2} \cdot \alpha_1 + \alpha_2 > 1$, a.a.s.
- $\dim Y = 3$ if $\frac{3}{2} \cdot \alpha_1 + \alpha_2 < 1$ and $2\alpha_1 + 2\alpha_2 > 1$, a.a.s.
- $\dim Y = 4$ if $2\alpha_1 + 2\alpha_2 < 1$ and $\frac{5}{2} \cdot \alpha_1 + \frac{10}{3} \cdot \alpha_2 > 1$, a.a.s.,

and so on.

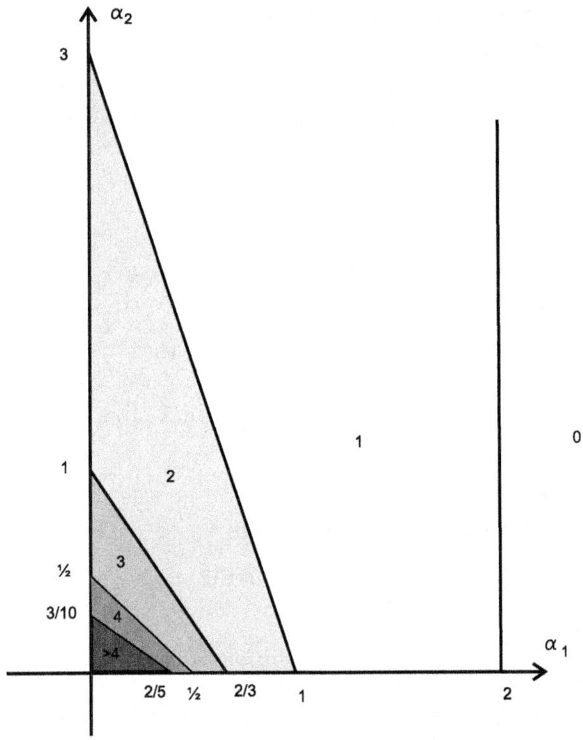

Fig. 6 Dimension of the random simplicial complex for various values of parameters α_1, α_2

Figure 6 depicts regions of the plane (α_1, α_2) where dimension is $1, 2, 3, 4$ and ≥ 5. Each of these regions is a polygonal convex domain with vertices in rational points.

Some further results are discussed in [5].

References

1. D. Cohen, A.E. Costa, M. Farber, T. Kappeler, Topology of random 2-complexes. J. Discrete Comput. Geom. **47**, 117–149 (2012)
2. A.E. Costa, M. Farber, Geometry and topology of random 2-complexes. Isr. J. Math. **209**, 883–927 (2015)
3. A.E. Costa, M. Farber, Large random simplicial complexes, I. J. Topol. Anal. (to appear, 2015). arXiv:1503.06285
4. A.E. Costa, M. Farber, Large random simplicial complexes, II; the fundamental group (2015). arXiv:1509.04837
5. A.E. Costa, M. Farber, Large Random Simplicial Complexes, III; The Critical Dimension. Preprint arXiv:1512.08714
6. A.E. Costa, M. Farber, D. Horak, Fundamental groups of clique complexes of random graphs. Trans. Lond. Math. Soc. **2**, 1–32 (2015)

7. N. Dunfield, W.P. Thurston, Finite covers of random 3-manifolds. Invent. Math. **166**(3), 457–521 (2006)
8. P. Erdős, A. Rényi, On the evolution of random graphs. Publ. Math. Inst. Hungar. Acad. Sci. **5**, 17–61 (1960)
9. M. Farber, Topology of random linkages. Algebr. Geom. Topol. **8**, 155–171 (2008)
10. A. Hatcher, *Algebraic Topology* (Cambridge University Press, Cambridge, 2002)
11. M. Kahle, Topology of random clique complexes. Discrete Math. **309**(6), 1658–1671 (2009)
12. M. Kahle, Sharp vanishing thresholds for cohomology of random flag complexes. Ann. Math. (2) **179**(3), 1085–1107 (2014)
13. M. Kahle, Topology of random simplicial complexes: a survey. Contemp. Math. **620**, 201–221 (2014)
14. N. Linial, R. Meshulam, Homological connectivity of random 2-complexes. Combinatorica **26**, 475–487 (2006)
15. R. Meshulam, N. Wallach, Homological connectivity of random k-complexes. Random Struct. Algorithm. **34**, 408–417 (2009)
16. N. Pippenger, K. Schleich, Topological characteristics of random triangulated surfaces. Random Struct. Algorithm. **28**(3), 247–288 (2006)

Combinatorial Topology of the Standard Chromatic Subdivision and Weak Symmetry Breaking for Six Processes

Dmitry N. Kozlov

Abstract In this paper we study a family of discrete configuration spaces, the so-called *protocol complexes*, which are of utmost importance in theoretical distributed computing. Specifically, we consider questions of the existance of compliant binary labelings on the vertices of iterated standard chromatic subdivisions of an n-simplex. The existance of such labelings is equivalent to the existance of distributed protocols solving Weak Symmetry Breaking task in the standard computational model.

As a part of our formal model, we introduce function $sb(n)$, defined for natural numbers n, called the *symmetry breaking* function. From the geometric point of view $sb(n)$ denotes the minimal number of iterations of the standard chromatic subdivision of an $(n-1)$-simplex, which is needed for the compliant binary labeling to exist. From the point of view of distributed computing, the function $sb(n)$ measures the minimal number of rounds in a protocol solving the Weak Symmetry Breaking task.

In addition to the development of combinatorial topology, which is applicable in a broader context, our main contribution is the proof of new bounds for the function $sb(n)$. Accordingly, the bulk of the paper is taken up by in-depth analysis of the structure of adjacency graph on the set of n-simplices in iterated standard chromatic subdivision of an n-simplex. On the algorithmic side, we provide the first distributed protocol solving Weak Symmetry Breaking task in the layered immediate snapshot computational model for some number of processes.

It is well known, that the smallest number of processes for which Weak Symmetry Breaking task is solvable is 6. Based on our analysis, we are able to find a very fast explicit protocol, solving the Weak Symmetry Breaking for six processes using only three rounds. Furthermore, we show that no protocol can solve Weak Symmetry Breaking in fewer than two rounds.

Keywords Protocol complexes • Standard chromatic subdivision • Weak symmetry breaking

D.N. Kozlov (✉)
Department of Mathematics, University of Bremen, 28334 Bremen, Germany
e-mail: dfk@math.uni-bremen.de

F. Callegaro et al. (eds.), *Configuration Spaces*, Springer INdAM Series 14,
DOI 10.1007/978-3-319-31580-5_7

1 Solvability of Weak Symmetry Breaking

1.1 Weak Symmetry Breaking as a Standard Distributed Task

Let n be an integer, such that $n \geq 2$. The Weak Symmetry Breaking task for n processes is an inputless task, where the possible outputs are 0 and 1. A distributed protocol is said to solve the Weak Symmetry Breaking task if in any execution without failed processes, there exists at least one process which has value 0 as well as at least one process which has value 1.

In the classical setting, the processes know their id's, and are allowed to compare them. It is however not allowed that any other information about id's is used. The protocols with this property are called *comparison-based*.[1] In practice this means that behavior of each process only depends on the relative position of its id among the id's of the processes it witnesses and not on its actual numerical value. As a special case, we note that each process must output the same value in case he does not witness other processes at all. Weak Symmetry breaking is a standard task in theoretical distributed computing, and its solvability in the standard computational models is a sophisticated question which has been extensively studied.

For the rest of this paper we shall fix the computational model to be the layered immediate snapshot model, see [9]. In this model the processes use two atomic operations being performed on shared memory. These operations are: *write* into the register assigned to that process, and *snapshot read*, which reads entire memory in one atomic step. Furthermore, it is assumed that the executions are well-structured in the sense that they must satisfy the two following conditions. First, it is only allowed that at each time a group of processes gets active, these processes perform a write operation together, and then they perform a snapshot read operation together; no other interleaving in time of the write and read operations is permitted. Such executions are called *immediate snapshot* executions. Second, each execution can be broken up in rounds, where in every round each non-faulty process gets activated precisely once.

Even though this computational model is seemingly quite restrictive, it has been proved, see, e.g., [9], that most of the commonly used shared memory computational models are equivalent, in the sense of which of the tasks are solvable, to this one. As we will see, this model has the major advantage that the protocol complexes have a comparatively simple topological structure.

1.2 Previous Work

Several groups of researchers have studied the solvability of the Weak Symmetry Breaking by means of comparison-based layered immediate snapshot protocols.

[1] Alternative terminology *rank-symmetric* is also used in the literature.

Due to primarily work of Castañeda and Rajsbaum, [6–8], it is known that the Weak Symmetry Breaking is solvable if and only if the number of processes is not a prime power; see also [4] for a counting-based argument for the impossibility part. This makes $n = 6$ the smallest number of processes for which this task is solvable.

The combinatorial structures arising in related questions on subdivisions of simplex paths have been studies in [2, 17]. The specific case $n = 6$ has been studied in [2], who has proved the existence of the distributed protocol which solves the Weak Symmetry Breaking task in 17 rounds.

We refer to the classical textbook [5], as well as the more recent monograph [3], for general background on theoretical distributed computing, and specifically on impossibility results. A general reference for topological methods in distributed computing is [9]. A general reference for combinatorial topology is [12]. Finally, we recommend the survey [11] for background on symmetry breaking tasks.

1.3 Our Results and Outline of the Paper

Our main mathematical result is the following theorem.

Theorem 1 *There exists a compliant binary labeling λ of the vertices of $\chi^3(\Delta^5)$, such that the restriction of λ to any 5-simplex of $\chi^3(\Delta^5)$ is surjective.*

This immediately implies the following theorem in theoretical distributed computing.

Theorem 2 *There exists a comparison-based layered immediate snapshot distributed protocol solving the Weak Symmetry Breaking task for six processes in three rounds.*

We also present the associated explicit distributed protocol. Because of the standard reduction to a mathematical question, see Theorem 3, we can derive Theorem 2 as a direct corollary of our main mathematical statement given in Theorem 1.

Our central method is the in-depth analysis of the simplicial structure of the second chromatic subdivision of a simplex. To this end, we need to develop the appropriate combinatorial language, in order to formalize and to work with standard chromatic subdivisions. This is done in Sects. 2 and 4, yielding as a byproduct the combinatorial framework which is applicable in a much bigger generality.

Once the combinatorics is clear, the plan of our proof is as follows. First, in Sect. 5, we define some specially designed binary labeling on the vertices of the second chromatic subdivision of Δ^5, calling this the *initial labeling*. This labeling is compliant, but it has many 1-monochromatic 5-simplices. In the next step, we find a perfect matching on the set of 1-monochromatic 5-simplices, such that whenever two 5-simplices are matched, they share a 4-simplex.

In Sect. 5 we describe the standard matching, which is easy to define, but which is not perfect. In Sect. 6 we modify this standard matching by connecting unmatched simplices by augmenting paths. This is a standard technique from matching theory, and it yields a perfect matching. Once we have a perfect matching, it is easy to use that to achieve a binary compliant labeling of the vertices of $\chi^3(\Delta^5)$ without any monochromatic simplices. This is done in the end of Sect. 6, and it proves our Main Theorem 1.

The remaining two sections are dealing with further specific issues. The explicit protocol for solving Weak Symmetry Breaking task for six processes in three rounds is given in Sect. 6. We finish with Sect. 7, where we give a very short argument for impossibility of solving Weak Symmetry Breaking task in one round for any number of processes.

While the focus of this paper has been on the case of six processes, it is possible to refine our analysis to a much more general setting, yielding new bounds for the symmetry breaking function, as well as producing distributed protocols for other numbers of processes. However, there is a number of technical issues which need to be resolved, and there is a number of new ideas which need to be introduced, since giving the simplex paths explicitly, as was done in Table 5, is prohibitive for higher values of n. For this reason, the extension of the techniques from this paper to higher values of n will appear in a separate article, see [18, 19].

2 Combinatorics of Iterated Chromatic Subdivisions

2.1 *Standard Chromatic Subdivision of a Simplex*

For an arbitrary nonnegative integer n, we set $[n] := \{0, \dots, n\}$. We let Δ^n denote the standard n-simplex. Note that Δ^n has $n + 1$ vertices, which we index by the set $[n]$. For brevity, we shall skip the curly brackets for the sets consisting of a single element, and write expressions like $A \cup x$ and $A \setminus x$, rather than $A \cup \{x\}$ and $A \setminus \{x\}$.

Definition 1 Given a set A and an element $x \in A$, we shall call the pair $\gamma = (A, x)$ a **node**. We shall say that x is the *color* of this node, and write $C(\gamma) = x$.

When W is any set of nodes, we set $C(W) := \{C(\gamma) \mid \gamma \in W\}$. The reasons for our terminology will become apparent soon.

Definition 2 An **ordered set partition** σ of a set A is an ordered tuple $(A_1, \dots, A_t)$ of nonempty subsets of A such that A is a disjoint union of $A_1, \dots, A_t$. We shall use the notation $\sigma = (A_1 \mid \dots \mid A_t)$.

Assume now that we are given an ordered set partition $\sigma = (A_1 \mid \dots \mid A_t)$ of a set A.

Definition 3 We shall call the set $V(\sigma) := \{(A_1 \cup \dots \cup A_{i(x)}, x) \mid x \in A\}$ the **set of nodes of** σ, where $i(x)$ denotes the unique index such that $x \in A_{i(x)}$.

Note that we have $V(\sigma) = \{(A_1 \cup \dots \cup A_k, x) \mid 1 \le k \le t,\ x \in A_k\}$.

Definition 4 If (B, x) is a node of an ordered set partition σ of a set A, such that $|B| \geq |A| - 1$, then we say that x is **almost maximal** with respect to σ.

Remark 1 For every ordered set partition σ of a set A, there exist at least two elements of A, which are almost maximal with respect to σ.

Indeed, if $\sigma = (A_1 \mid \ldots \mid A_t)$, then all elements of A_t are almost maximal. If $|A_t| = 1$, then also all elements of A_{t-1} are almost maximal.

Given a permutation $\pi = (\pi_0 \ldots \pi_n)$ of $[n]$, there is a natural ordered set partition of the set $[n]$ associated to it, namely $(\pi_0 \mid \ldots \mid \pi_n)$; we shall call it $\hat{\pi}$. This way, we obtain precisely all ordered set partitions consisting of singletons.

There is a standard mathematical reformulation of the solvability of the Weak Symmetry Breaking task which we now proceed to describe. The central role in this description is played by the following abstract simplicial complex, called *standard chromatic subdivision of a simplex.*

Definition 5 Let n be a nonnegative integer. The abstract simplicial complex $\chi(\Delta^n)$ is given as follows:

- The set of vertices $V(\chi(\Delta^n))$ consists of all nodes (V, x), such that $x \in V \subseteq [n]$.
- The maximal simplices of $\chi(\Delta^n)$ are indexed by ordered set partitions of $[n]$, where for each ordered set partition σ, its set of vertices is given by the corresponding set of nodes $V(\sigma)$.
- In general $S \subseteq V(\chi(\Delta^n))$ is a simplex if and only if there exists a maximal simplex σ such that $S \subseteq V(\sigma)$.

The color of the vertex of $\chi(\Delta^n)$ is the color of its indexing node, in particular, $C(-)$ can be applied to any subset of $V(\chi(\Delta^n))$. Furthermore, when σ is an n-simplex of $\chi(\Delta^n)$, we set $C(\sigma) := C(V(\sigma))$. Note that $\chi(\Delta^n)$ is a *pure* simplicial complex of dimension n, meaning that all of its maximal simplices have the same dimension.

The simplicial complex $\chi(\Delta^n)$ has been introduced by Herlihy and Shavit [10], see also a recent book [9], and is a widely used gadget in theoretical distributed computing. It has been proved in [13] that $\chi(\Delta^n)$ is a subdivision of an n-simplex. A wide generalization of this fact has been proved in [14, 15].

2.2 *Iterated Chromatic Subdivisions and the Weak Symmetry Breaking*

The construction from Definition 5 can be used to define chromatic subdivision of an arbitrary simplicial complex due to the following simple fact: when restricted to any of its boundary simplices τ, the standard chromatic subdivision $\chi(\Delta^n)$ is naturally isomorphic to $\chi(\Delta^m)$, where m is the dimension of τ. Indeed, given any simplicial complex K, we can simply replace each of its simplices with its standard chromatic subdivision, these will fit together nicely, and we can call the result the

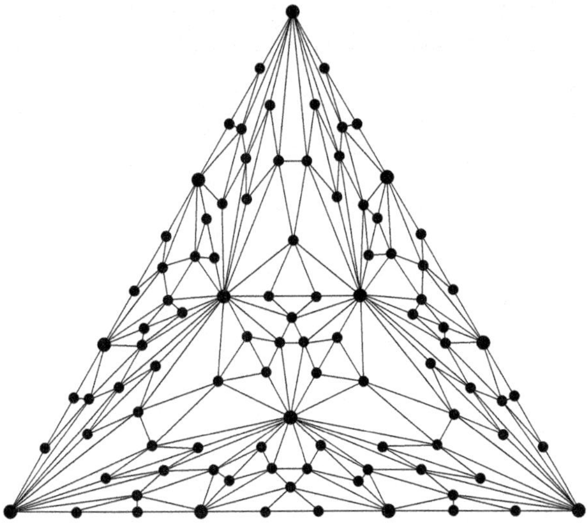

Fig. 1 The second iterated standard chromatic subdivision of a 2-simplex

standard chromatic subdivision of K. In particular, this means that we can define *iterated chromatic subdivisions* $\chi^k(\Delta^n)$. The simplicial complex $\chi^2(\Delta^2)$ is shown on Fig. 1.

For any finite set A we let Δ^A denote the standard simplex whose vertices are indexed by the elements of A. In this setting, our standard simplex Δ^n would be called $\Delta^{[n]}$. The boundary simplices are called Δ^I, for all subsets $I \subset [n]$.

Given two equicardinal subsets $I, J \subset [n]$, the corresponding boundary simplices Δ^I and Δ^J have the same dimension, and any bijection $f : I \to J$ induces a simplicial isomorphism from Δ^I to Δ^J. Clearly, the construction of the iterated standard chromatic subdivision will also induce a simplicial isomorphism from $\chi^d(\Delta^I)$ to $\chi^d(\Delta^J)$. We let $\varphi_{I,J}$ denote the simplicial isomorphism from $\chi^d(\Delta^I)$ to $\chi^d(\Delta^J)$ induced by the unique order-preserving bijection from I to J.

Definition 6 A binary labeling $\lambda : V(\chi^d(\Delta^n)) \to \{0, 1\}$ is called **compliant** if for all $I, J \subset [n]$, such that $|I| = |J|$, and all vertices $v \in V(\chi^d(\Delta^I))$, we have

$$\lambda(\varphi_{I,J}(v)) = \lambda(v).$$

In simple terms, this means that the restriction of the labeling to $\chi^d(\Delta^I)$ will only depend on the cardinality of the ordered set I.

The following well-known statement, see [2, 6–10, 17], is a useful reformulation of the solvability of the Weak Symmetry Breaking in the layered immediate snapshot model in purely mathematical terms.

Theorem 3 *In the layered immediate snapshot computational model, the Weak Symmetry Breaking task for n processes is solvable in d rounds if and only if there exists a compliant binary labeling* $\lambda : V(\chi^d(\Delta^{n-1})) \to \{0,1\}$*, such that for every* $(n-1)$*-simplex* $\sigma \in \chi^d(\Delta^{n-1})$*, the restricted map* $\lambda : V(\sigma) \to \{0,1\}$ *is surjective.*

Clearly, the reduction in Theorem 3 means that Theorem 1 immediately implies Theorem 2.

2.3 *The Symmetry Breaking Function and Its Estimates*

We are now ready to introduce the main function for our study.

Definition 7 Assume n is an arbitrary natural number. We let $sb(n)$ denote the minimal number d such that there exists a compliant binary labeling $\lambda : V(\chi^d(\Delta^{n-1})) \to \{0,1\}$, such that for every $(n-1)$-simplex $\sigma \in \chi^d(\Delta^{n-1})$, the restricted map $\lambda : V(\sigma) \to \{0,1\}$ is surjective. If no such d exists, we set $sb(n) := \infty$.

Note, that once such a labeling λ exists for some d, for any other $d' > d$ it is easy to extend it to a labeling $\lambda' : V(\chi^{d'}(\Delta^{n-1})) \to \{0,1\}$ satisfying the same conditions.

Furthermore, note that Theorem 3 implies the following remark.

Remark 2 The Weak Symmetry Breaking task for n processes is solvable in the layered immediate snapshot model if and only if $sb(n) \neq \infty$. The actual value $sb(n)$ is the minimal number of rounds needed for the distributed protocol to solve this task.

Table 1 summarizes our current knowledge of the function $sb(n)$.

Table 1 The known estimates of $sb(n)$

Estimate	Source
$sb(n) = \infty$ if and only if n is a prime power	[6–8]
$sb(n) = O(n^{q+3})$, if n is not a prime power and	[2]
q is the largest prime power in the prime factorization of n	
$sb(6) \leq 3$	Theorem 1
$sb(n) \geq 2$	Theorem 7

3 Combinatorial Description of the Simplicial Structure of Standard Chromatic Subdivisions

3.1 *Combinatorics of Partial Ordered Set Partitions and the Lower Simplices in the Standard Chromatic Subdivision*

Even though Definition 5 provides a well-defined simplicial complex, it is somewhat cumbersome to work with, mainly due to the fact that the lower-dimensional simplices lack a direct combinatorial description. We shall now mitigate this situation by giving an alternative combinatorial description of the simplicial structure of $\chi(\Delta^n)$, which has been first obtained in [16]. However, before we can do this, we need some additional terminology.

Definition 8 A **partial ordered set partition** of the set $[n]$ is a pair of ordered set partitions of nonempty subsets of $[n]$, $\sigma = ((A_1 \mid \dots \mid A_t), (B_1 \mid \dots \mid B_t))$, which have the same number of parts, such that for all $1 \leq i \leq t$, we have $B_i \subseteq A_i$. Given such a partial ordered set partition σ, we introduce the following terminology.

- The union $A_1 \cup \dots \cup A_t$ is called the **carrier set** of σ, and is denoted by carrier (σ).
- The union $B_1 \cup \dots \cup B_t$ is called the **color set** of σ, and is denoted by $C(\sigma)$.
- The **dimension** of σ is defined to be $|C(\sigma)| - 1$, and is denoted $\dim \sigma$.

When appropriate, we shall also write

$$\sigma = \begin{array}{|c|c|c|}\hline A_1 & \dots & A_t \\ \hline B_1 & \dots & B_t \\ \hline \end{array}, \tag{1}$$

which we shall call the *table form* of σ.

We note, that both nodes (A, x), for $A \subseteq [n]$, as well as ordered set partitions of $[n]$ are special cases of partial ordered set partitions of $[n]$. Indeed, a node (A, x), such that $A \subseteq [n]$ corresponds to the somewhat degenerate partial ordered set partition of $[n]$

$$\sigma = \begin{array}{|c|}\hline A \\ \hline x \\ \hline \end{array}.$$

Whereas an ordered set partition $(A_1 \mid \dots \mid A_t)$ corresponds to the partial ordered set partition of $[n]$

$$\sigma = \begin{array}{|c|c|c|}\hline A_1 & \dots & A_t \\ \hline A_1 & \dots & A_t \\ \hline \end{array},$$

i.e., a partial ordered set partition $((A_1 \mid \dots A_t), (B_1 \mid \dots \mid B_t))$, such that $A_i = B_i$ for all i, and $A_1 \cup \dots \cup A_t = [n]$.

Each partial ordered set partition has non-empty color set, which in turn is contained in its carrier set. The nodes correspond to the partial ordered set partitions with minimal color set, consisting of just one element, and ordered set partitions correspond to the partial ordered set partitions with maximal color set, namely the whole set $[n]$.

Definition 9 Assume we are given a partial ordered set partition $\sigma = ((A_1 \mid \dots \mid A_t), (B_1 \mid \dots \mid B_t))$ of the set $[n]$, such that $\dim \sigma \geq 1$, and we are also given an element $x \in C(\sigma)$, say $x \in B_k$, for some $1 \leq k \leq t$. To define the **deletion of x from σ** we consider three different cases.

Case 1. If $|B_k| \geq 2$, then the deletion of x from σ is set to be

$$((A_1 \mid \dots \mid A_t), (B_1 \mid \dots \mid B_{k-1} \mid B_k \setminus x \mid B_{k+1} \mid \dots \mid B_t)).$$

Case 2. If $|B_k| = 1$, and $k \leq t - 1$, then the deletion of x from σ is set to be

$$((A_1 \mid \dots \mid A_{k-1} \mid A_k \cup A_{k+1} \mid \dots \mid A_t), (B_1 \mid \dots \mid B_{k-1} \mid B_{k+1} \mid \dots \mid B_t)).$$

Case 3. If $|B_k| = 1$, and $k = t$, then the deletion of x from σ is set to be

$$((A_1 \mid \dots \mid A_{t-1}), (B_1 \mid \dots \mid B_{t-1})).$$

We denote the deletion of x from σ by $\mathrm{dl}\,(\sigma, x)$.

It is easy to see that $\mathrm{dl}\,(\sigma, x)$ is again a partial ordered set partition of $[n]$. We have $C(\mathrm{dl}\,(\sigma, x)) = C(\sigma) \setminus x$, $\dim \mathrm{dl}\,(\sigma, x) = \dim \sigma - 1$, and $\mathrm{carrier}\,(\mathrm{dl}\,(\sigma, x)) \subseteq \mathrm{carrier}\,(\sigma)$.

Definition 9 can be generalized as follows.

Definition 10 Let $\sigma = ((A_1 \mid \dots \mid A_t), \mathrm{ab}(B_1 \mid \dots \mid B_t))$ be a partial ordered set partition of the set $[n]$, and assume we are given a non-empty set $S \subset C(\sigma)$. Let $i_1 < \dots < i_m$ index all sets B_i, such that $B_i \not\subseteq S$, and set $i_0 := 0$. For all $1 \leq k \leq m$, we set

$$\tilde{A}_k := A_{i_{k-1}+1} \cup \dots \cup A_{i_k}, \text{ and } \tilde{B}_k := B_{i_k} \setminus S.$$

The obtained partial ordered set partition $((\tilde{A}_1 \mid \dots \mid \tilde{A}_m), (\tilde{B}_1 \mid \dots \mid \tilde{B}_m))$ is called **deletion of S from σ**, and is denoted $\mathrm{dl}\,(\sigma, S)$.

The properties of the deletion of a part of its color set from a partial ordered set partition are summarized in the following proposition.

Proposition 1 *Let* $\sigma = ((A_1 \mid \dots \mid A_t), \mathrm{ab}(B_1 \mid \dots \mid B_t))$ *be a partial ordered set partition of the set* $[n]$.

(1) *Assume we have a non-empty subset $S \subset C(\sigma)$. Then* $\mathrm{dl}\,(\sigma, S)$ *is again a partial ordered set partition of the set* $[n]$, *such that*

(a) $C(\mathrm{dl}\,(\sigma, S)) = C(\sigma) \setminus S$.
(b) $\dim \mathrm{dl}\,(\sigma, S) = \dim \sigma - |S|$.
(c) $\mathrm{carrier}\,(\mathrm{dl}\,(\sigma, S)) \subseteq \mathrm{carrier}\,(\sigma)$.

(2) *Assume S and T are disjoint non-empty subsets of* $\mathrm{supp}\,\sigma$, *such that $S \cup T \neq$* $\mathrm{supp}\,\sigma$. *Then, we have*

$$\mathrm{dl}\,(\mathrm{dl}\,(\sigma, S), T) = \mathrm{dl}\,(\sigma, S \cup T). \tag{2}$$

Proof All statements are immediate from Definition 10. □

Generalizing Definition 2 we can define the set of nodes of an arbitrary partial ordered set partition of the set $[n]$.

Definition 11 Let $\sigma = ((A_1 \mid \ldots \mid A_t), \mathrm{ab}(B_1 \mid \ldots \mid B_t))$ be a partial ordered set partition of the set $[n]$. We shall call the set $V(\sigma) = \{(A_1 \cup \cdots \cup A_{i(x)}, x) \mid x \in C(\sigma)\}$, the **set of nodes** of σ, where again $i(x)$ denotes the unique index such that $x \in B_{i(x)}$.

Note, that $V(\sigma) = \{(A_1 \cup \cdots \cup A_k, x) \mid 1 \leq k \leq t, x \in B_k\}$, $|V(\sigma)| = \dim \sigma + 1$, and $C(V(\sigma)) = C(\sigma)$. Comparing Definitions 10 and 11 yields the identity

$$V(\sigma) = \{\mathrm{dl}\,(\sigma, [n] \setminus \{x\}) \mid x \in C(\sigma)\}. \tag{3}$$

Furthermore, when an ordered set partition is viewed as a partial one, its set of nodes does not depend on which one of the Definitions 2 and 11 is used. Crucially, the set of nodes completely determines any partial ordered set partition of the set $[n]$.

Proposition 2 *Assume σ and τ are both partial ordered set partitions of the set $[n]$, such that $V(\sigma) = V(\tau)$, then $\sigma = \tau$.*

Proof Assume $\sigma \neq \tau$. Without loss of generality, we can write $\sigma = ((A_1 \mid \ldots \mid A_t), \mathrm{ab}(B_1 \mid \ldots \mid B_t))$ and $\tau = ((C_1 \mid \ldots \mid C_q), \mathrm{ab}(D_1 \mid \ldots \mid D_q))$, such that $t \leq q$. To start with, the sets of sets $\{A_1, A_1 \cup A_2, \ldots, A_1 \cup \cdots \cup A_t\}$ and $\{C_1, C_1 \cup C_2, \ldots, C_1 \cup \cdots \cup C_q\}$ must be equal. This immediately implies that $t = q$, and that $A_i = C_i$, for all $i = 1, \ldots, t$.

Since $C(\sigma) = C(\tau)$, we have $B_1 \cup \cdots \cup B_t = D_1 \cup \cdots \cup D_t$. Let k denote the minimal number such that $B_k \neq D_k$. Without loss of generality we can assume that we can find an element x, such that $x \in B_k$, $x \notin D_k$, and $x \in D_m$, for some $m > k$. Then, σ has a node $(A_1 \cup \cdots \cup A_k, x)$, whereas τ has a node $(A_1 \cup \cdots \cup A_m, x)$. Since $m \neq k$ we arrive at a contradiction. □

We are now ready to prove that partial ordered set partitions provide the right combinatorial language to describe the simplicial structure of $\chi(\Delta^n)$.

Proposition 3 *The nonempty simplices of $\chi(\Delta^n)$ can be indexed by all partial ordered set partitions of $[n]$. This indexing satisfies the following properties:*

(1) The dimension of the simplex indexed by $\sigma = ((A_1 \mid \dots \mid A_t), (B_1 \mid \dots \mid B_t))$ is equal to $\dim \sigma$.
(2) The vertices of the simplex σ indexed by $((A_1 \mid \dots \mid A_t), (B_1 \mid \dots \mid B_t))$ are indexed by $V(\sigma)$.
(3) In general, the set of subsimplices of the simplex σ indexed by $((A_1 \mid \dots \mid A_t), (B_1 \mid \dots \mid B_t))$ is precisely the set of simplices indexed by partial ordered set partitions from the set $\{\mathrm{dl}\,(\sigma, S) \mid S \subset C(\sigma)\}$.

Proof Assume $W \subseteq V(\chi(\Delta^n))$, such that W forms a simplex in $\chi(\Delta^n)$. By Definition 5 there exists an n-simplex σ of $\chi(\Delta^n)$, such that $W \subseteq V(\sigma)$. Furthermore, by (3) we get $W = \{\mathrm{dl}\,(\sigma, [n] \setminus x) \mid x \in C(W)\}$. We now set

$$\tilde{\sigma} := \mathrm{dl}\,(\sigma, [n] \setminus C(W)) \tag{4}$$

to be the partial ordered set partition of the set $[n]$ which indexes the simplex W. First, we note that by Proposition 1(1)(a) we have

$$C(\tilde{\sigma}) = [n] \setminus ([n] \setminus C(W)) = C(W). \tag{5}$$

Furthermore, we derive

$$\begin{aligned} V(\tilde{\sigma}) &= \{\mathrm{dl}\,(\tilde{\sigma}, C(\tilde{\sigma}) \setminus x) \mid x \in C(\tilde{\sigma})\} = \{\mathrm{dl}\,(\tilde{\sigma}, C(W) \setminus x) \mid x \in C(W)\} = \\ &= \{\mathrm{dl}\,(\mathrm{dl}\,(\sigma, [n]\setminus C(W)), C(W)\setminus x) \mid x \in C(W)\} = \{\mathrm{dl}\,(\sigma, [n]\setminus x) \mid x \in C(W)\} = W, \end{aligned} \tag{6}$$

where the penultimate equality is due to (2).

Assume τ is another n-simplex of $\chi(\Delta^n)$, such that $W \subseteq V(\tau)$. Set $\tilde{\tau} := \mathrm{dl}\,(\tau, [n] \setminus C(W))$. We have shown in (6) that $V(\tilde{\sigma}) = V(\tilde{\tau}) = W$. It follows from Proposition 2 that $\tilde{\sigma} = \tilde{\tau}$, and hence the partial ordered set partition of the set $[n]$, which indexes W does not depend on the choice of σ.

We can now show that the indexing given by (4) satisfies the 3 properties in the formulation of the proposition. To see property (1), let $W \subseteq V(\chi(\Delta^n))$ be the simplex of $\chi(\Delta^n)$ indexed by $\tilde{\sigma}$. On one hand, $\dim W = |W| - 1$. On the other hand, picking an n-simplex σ such that $W \subseteq V(\sigma)$, we get

$$\dim(\mathrm{dl}\,(\sigma, [n] \setminus C(W))) = \dim \sigma - (n+1-|C(W)|) = n-(n+1-|W|) = |W|-1,$$

hence $\dim W = \dim \tilde{\sigma}$.

Finally note, that property (2) has been proved by (6), whereas property (3) follows directly from (2). □

3.2 Combinatorial Language Encoding Simplices in Iterated Chromatic Subdivisions

The following combinatorial concept is the key to describing the iterated standard chromatic subdivisions.

Definition 12 Assume we are given a tuple $\sigma = (\sigma_1, \dots, \sigma_d)$ of partial ordered set partitions of the set A. We say that the tuple σ is **linked** if for all $1 \leq i \leq d-1$, we have $C(\sigma_i) = \text{carrier}\,(\sigma_{i+1})$.

Assume we have a partial ordered set partition $\sigma = ((A_1 \mid \dots \mid A_t), (B_1 \mid \dots \mid B_t))$, and $S \subset C(\sigma)$. Let i be the minimal index, such that $B_i \cup \dots \cup B_t \subseteq S$. We define $D(\sigma, S)$ to be $A_i \cup \dots \cup A_t$, if such an index i exists, and we let $D(\sigma, S)$ be empty otherwise. We can now generalize Proposition 3.

Proposition 4 *The simplices of $\chi^d(\Delta^n)$ can be indexed by linked d-tuples of partial ordered set partitions of the set $[n]$. This indexing satisfies following properties:*

(1) The dimension of the simplex indexed by $\sigma = (\sigma_1, \dots, \sigma_t)$ is equal to $\dim \sigma_t$.
(2) The subsimplices of $\sigma = (\sigma_1, \dots, \sigma_t)$ are all tuples $(\text{dl}\,(\sigma_1, S_1) \dots, \text{dl}\,(\sigma_t, S_t))$, *where $S_t \subset C(\sigma_t)$, and $S_i = D(\sigma_{i+1}, S_{i+1})$, for all $1 \leq i \leq t-1$.*

Proof Clearly, each simplex of $\chi^d(\Delta^n)$ can be obtained by first choosing a simplex of $\chi(\Delta^n)$, then viewing this simplex as Δ^k, for some $k \leq n$, then picking the next simplex in Δ^k, and so on, repeating d times in total. By Proposition 3 each next simplex can be indexed by a partial ordered set partition, and each time the color set of the previous simplex is the carrier of the next one. This is exactly the same as requiring for this tuple of simplices to be linked in the sense of Definition 12. Both (1) and (2) now follow immediately from Proposition 3 and the definition of the deletion operation. □

Generalizing (1), the index of a simplex in $\chi^d(\Delta^n)$ can be visualized as an array of subsets:

$$\begin{array}{|c|c|c||c||c|c|c|}\hline A_1^1 & \dots & A_{t_1}^1 & \dots & A_1^d & \dots & A_{t_d}^d \\ \hline B_1^1 & \dots & B_{t_1}^1 & \dots & B_1^d & \dots & B_{t_d}^d \\ \hline \end{array},$$

satisfying $B_1^k \cup \dots \cup B_{t_k}^k = A_1^{k+1} \cup \dots \cup A_{t_{k+1}}^{k+1}$, for all $1 \leq k \leq d-1$.

In particular, the top-dimensional simplices of $\chi^d(\Delta^n)$ are indexed by tuples $\alpha = ((A_1^1 \mid \dots \mid A_{t_1}^1), \dots, (A_1^d \mid \dots \mid A_{td}^d))$. We find it practical to use the following shorthand notation: $\alpha = (A_1^1 \mid \dots \mid A_{t_1}^1 \parallel \dots \parallel A_1^d \mid \dots \mid A_{td}^d)$.

3.3 *Combinatorics of the Pseudomanifold Structure of the Chromatic Subdivision of a Simplex*

One of the reasons, why the layered immediate snapshot computational model is amenable to detailed analysis is because the corresponding protocol complexes have a useful structure of a pseudomanifold.

Definition 13 A pure n-dimensional simplicial complex K is called a **pseudomanifold** of dimension n if the following two conditions are satisfied:

(1) Any $(n-1)$-simplex belongs to at most two n-simplices.
(2) Any two n-simplices σ and τ can be connected by a path of n-simplices $\sigma = \sigma_0, \sigma_1, \dots, \sigma_t = \tau$, such that for any $k = 1, \dots, t$, the n-simplices σ_{k-1} and σ_k share an $(n-1)$-simplex.

It is well known, see e.g., [9], that the iterated chromatic subdivision of an n-simplex is an n-pseudomanifold. We now provide the combinatorial language for describing how to move between n-simplices of this pseudomanifold.

To start with, consider an n-simplex σ of $\chi(\Delta^n)$, say $\sigma = (A_1 \,|\, \dots \,|\, A_t)$. We set

$$F(\sigma) := \begin{cases} [n] \setminus A_t, & \text{if } |A_t| = 1; \\ [n], & \text{otherwise.} \end{cases}$$

Definition 14 Given an n-simplex σ of the simplicial complex $\chi(\Delta^n)$, and $x \in F(\sigma)$, we let $\mathcal{F}(\sigma, x)$ denote the n-simplex of $\chi(\Delta^n)$ obtained in the following way. Let us say $\sigma = (A_1 \,|\, \dots \,|\, A_t)$, and $x \in A_k$.

Case 1. Assume $|A_k| \geq 2$, then we set

$$\mathcal{F}(\sigma, x) := (A_1 \,|\, \dots \,|\, A_{k-1} \,|\, x \,|\, A_k \setminus x \,|\, A_{k+1} \,|\, \dots \,|\, A_t).$$

Case 2. Assume $|A_k| = 1$ (that is $A_k = x$) and $k < t$. Then we set

$$\mathcal{F}(\sigma, x) := (A_1 \,|\, \dots \,|\, A_{k-1} \,|\, x \cup A_{k+1} \,|\, A_{k+2} \,|\, \dots \,|\, A_t).$$

Note that since $x \in F(\sigma)$, we cannot have the case $\sigma = (A_1 \,|\, \dots \,|\, A_{t-1} \,|\, x)$, hence $\mathcal{F}(\sigma, x)$ is well-defined. We say that $\mathcal{F}(\sigma, x)$ is obtained from σ by a *flip* of σ with respect to x. In this sense, $F(\sigma)$ is the set of all colors which can be flipped. Given any simplex σ we can always flip with respect to all colors except for at most one color. We can also flip back, so we have $\mathcal{F}(\mathcal{F}(\sigma, x), x) = \sigma$ for all σ, x.

Furthermore, we remark that for any $x \in F(\sigma)$, the n-simplices σ and $\mathcal{F}(\sigma, x)$ share an $(n-1)$-simplex τ given by $\tau = \mathrm{dl}\,(\sigma, x) = \mathrm{dl}\,(\mathcal{F}(\sigma, x), x)$.

Definition 14 can be generalized to iterated chromatic subdivisions. For an n-simplex $\sigma = (\sigma_1 \,\|\, \dots \,\|\, \sigma_d)$ of $\chi^d(\Delta^n)$ we set $F(\sigma) := F(\sigma_1) \cup \dots \cup F(\sigma_d)$. Effectively this means that $F(\sigma) = [n]$, unless $F(\sigma_1) = \dots = F(\sigma_d) = [n] \setminus p$, in which case we have $F(\sigma) = [n] \setminus p$.

Definition 15 Assume we are given an n-simplex $\sigma = (\sigma_1 \parallel \dots \parallel \sigma_d)$ of the simplicial complex $\chi^d(\Delta^n)$, and $x \in F(\sigma)$. Let k be the maximal index such that $x \in F(\sigma_k)$, by the definition of $F(\sigma)$, such k must exist. We define $\mathcal{F}(\sigma, x)$ to be the following d-tuple of ordered partitions:

$$\mathcal{F}(\sigma, x) := (\sigma_1 \parallel \dots \parallel \sigma_{k-1} \parallel \mathcal{F}(\sigma_k, x) \parallel \sigma_{k+1} \parallel \dots \parallel \sigma_n). \tag{7}$$

Again, it is easy to see that for any $x \in F(\sigma)$, the n-simplices σ and $\mathcal{F}(\sigma, x)$ will share an $(n-1)$-simplex, and that for all σ, x, we have the identity

$$\mathcal{F}(\mathcal{F}(\sigma, x), x) = \sigma. \tag{8}$$

3.4 *Standard Chromatic Subdivisions and Matchings*

Let us review some basic terminology of the graph theory, more specifically the matching theory. To start with, recall that a graph G is called *bipartite* if its set of vertices $V(G)$ can be represented as a disjoint union $A \cup B$, such that there are only edges of the type (v, w), with $v \in A$, $w \in B$. We shall call (A, B) a *bipartite decomposition*. Note, that such a decomposition need not be unique.

Classically, a *matching* in a graph G is a subset M of its set of edges $E(G)$, such that two different edges from M do not share vertices. We call edges which belong to M the *matching edges* and all other edges the *non-matching edges*. A vertex v is called *matched* if there exists an edge in M having v as an endpoint. We call the unmatched vertices *critical*, which is consistent with the terminology we used in the previous sections. A matching is called *perfect* if there are no critical vertices, otherwise we may call the matching *partial*. Clearly, an existence of a perfect matching implies that the sets A and B have the same cardinality. A matching is called *near-perfect* if there is exactly one critical vertex.

In this paper, the fundamental instance of a graph on which matchings are constructed is provided by Γ_n. This is the graph whose vertices are all n-simplices of $\chi(\Delta^n)$, and two vertices are connected by an edge if the corresponding n-simplices share an $(n-1)$-simplex. Sometimes, we shall abuse our language and call the vertices of Γ_n simplices. We color the edges of Γ_n as follows: the edge connecting σ with τ gets the color of the vertex of σ which does not belong to $\sigma \cap \tau$. This graph has also been studied in [1]; an example is shown on Fig. 2.

The graph Γ_n is bipartite and the bipartite decomposition is unique. Recall that n-simplices of $\chi(\Delta^n)$ are indexed by ordered set partitions $(A_1 \mid \dots \mid A_t)$ of the set $[n]$. The bipartite decomposition of Γ_n is then provided by sorting the n-simplices of $\chi(\Delta^n)$ into two groups according to the parity of the number t. For convenience of

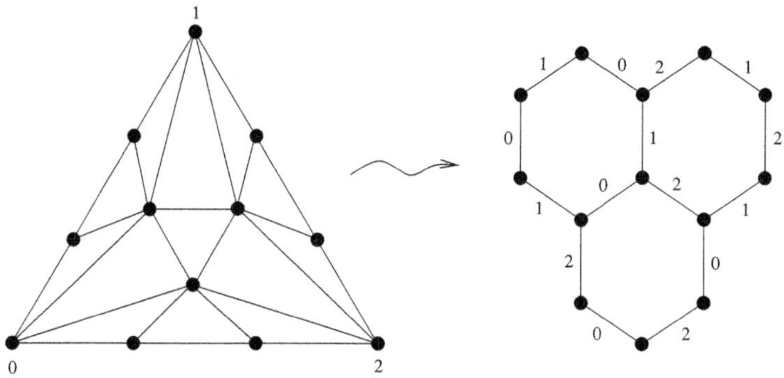

Fig. 2 The standard chromatic subdivision of a 2-simplex and the corresponding graph Γ_2, with labels showing the colors of the edges

notations we shall define a function $O : V(\Gamma_n) \to \{\pm 1\}$, which we call *orientation* of the simplex, as follows:

$$O(A_1 \mid \dots \mid A_t) := \begin{cases} 1, & \text{if } t \text{ is even;} \\ -1, & \text{if } t \text{ is odd.} \end{cases}$$

The matching itself in this context will mean to group n-simplices in pairs, so that in each pair the n-simplices share a boundary $(n-1)$-simplex. Accordingly, we can talk about critical n-simplices, etc.

In general, for any $d \geq 1$, we let Γ_n^d denote the graph, whose vertices are the n-simplices of $\chi^d(\Delta^n)$, and two vertices are connected by an edge if the corresponding n-simplices share an $(n-1)$-simplex. The edges of Γ_n^d are colored in the same way as those of Γ_n.

4 The Standard Matching for the Initial Labeling

4.1 Combinatorics of the Second Chromatic Subdivision of an n-Simplex

In this paper we will primarily need the combinatorial description of the simplicial structure of the second chromatic subdivision of an n-simplex. In this case, the Proposition 4 says that simplices of $\chi^2(\Delta^n)$ are indexed by pairs $(\sigma \parallel \tau)$ of partial ordered partitions of $[n]$, such that $C(\sigma) = \text{carrier}(\tau)$. The n-simplices of $\chi^2(\Delta^n)$

are simply pairs $(\sigma \parallel \tau)$ of n-simplices of $\chi(\Delta^n)$. The vertices of $\chi^2(\Delta^n)$ are indexed by pairs of partial ordered set partitions

$$v = \begin{array}{|c|c|c||c|} \hline A_1 & \dots & A_t & S \\ \hline B_1 & \dots & B_t & x \\ \hline \end{array}, \tag{9}$$

such that $S = B_1 \cup \dots \cup B_t$.

Assume we are given a simplex $\alpha = (\sigma \parallel \tau)$ of $\chi^2(\Delta^n)$, say $\sigma = ((A_1 \mid \dots \mid A_t), (B_1 \mid \dots \mid B_t))$, and $\tau = ((C_1 \mid \dots \mid C_q), (D_1 \mid \dots \mid D_q))$. Assume furthermore, we are given some subset $S \subset D_1 \cup \dots \cup D_q$. Let k be the minimal index such that $D_k \cup \dots \cup D_q \subseteq S$, then $\mathrm{dl}\,(\alpha, S) = (\mathrm{dl}\,(\sigma, C_k \cup \dots \cup C_q), \mathrm{dl}\,(\tau, S))$.

Finally, we remark that for an n-simplex $\alpha = (A_1 \mid \dots \mid A_t \parallel B_1 \mid \dots \mid B_q)$, and $x \in [n]$, the flip $\mathcal{F}(\alpha, x)$ is always defined unless $A_t = B_q = x$, and is explicitly given by (7).

4.2 *Description of the Initial Labeling*

For an arbitrary simplex $\sigma = (\sigma_1 \mid \dots \mid \sigma_d)$ of $\chi^d(\Delta^n)$ we set $\mathrm{supp}(\sigma) := \mathrm{carrier}\,(\sigma_1)$.

Definition 16 Let v be a vertex of $\chi^d(\Delta^n)$. We say that the vertex v is an **internal** vertex if $\mathrm{supp}(v) = [n]$, otherwise we say that v is a **boundary vertex**.

In the case $d = 2$, there is a handy criterion for deciding whether all vertices of an n-simplex are internal.

Proposition 5 *Let $\alpha = (A_1 \mid \dots \mid A_k \parallel B_1 \mid \dots \mid B_m)$ be an n-simplex of $\chi^2(\Delta^n)$. Then all vertices of α are internal if and only if $A_k \cap B_1 \neq \emptyset$.*

In general, assume that $A_k \cap B_1 = \emptyset$, let q be the largest index such that $A_k \cap B_i = \emptyset$ for all $1 \leq i \leq q$. The boundary vertices of α are precisely the vertices with colors from $B_1 \cup \dots \cup B_q$.

Proof Recall, that since α is an n-simplex, we have $A_1 \cup \dots \cup A_k = B_1 \cup \dots \cup B_m = [n]$. Pick $x \in [n]$, say $x \in B_l$. The vertex of α, which is colored by x has the index $(\tilde{\sigma} \parallel \tilde{\tau})$, where $\tilde{\sigma} = \mathrm{dl}\,(\sigma, B_{l+1} \cup \dots \cup B_m)$, and $\tilde{\tau} = (B_1 \cup \dots \cup B_l \mid x)$. This vertex is internal if and only if $\mathrm{carrier}\,(\tilde{\sigma}) = [n]$. By the definition of the deletion operation, this is the case if and only if $A_k \setminus (B_{l+1} \cup \dots \cup B_m) \neq \emptyset$, i.e.,

$$A_k \cap (B_1 \cup \dots \cup B_l) \neq \emptyset. \tag{10}$$

Clearly, the fact that (10) is true for all $l = 1, \dots, m$ is equivalent to the condition $A_k \cap B_1 \neq \emptyset$, and in general (10) yields the description of all boundary vertices of α. □

Table 2 The 136 exceptional boundary vertices of $\chi^2(\Delta^5)$

a	a					for all $0 \le a \le 5$;
a	a					
a,b	a	a,b	a,b			for all $0 \le a < b \le 5$;
a	a	a,b	b			
a,b,c	a	a,b,c	a,b	a,b,c	a,b	
a	a	a,b	a	a,b	b	
a,b,c	a,b,c	a,b,c	a,b,c			for all $0 \le a < b < c \le 5$.
a,b,c	b	a,b,c	c			

We are now in a position to describe the initial labeling of the vertices of $\chi^2(\Delta^5)$

$$I : V(\chi^2(\Delta^5)) \longrightarrow \{0, 1\}.$$

Before we do this, we would like to designate certain boundary vertices of $\chi^2(\Delta^5)$ as exceptional.

Definition 17 The set $\mathcal{V}$ consists of all vertices of $\chi^2(\Delta^5)$ listed in Table 2. The vertices in $\mathcal{V}$ are called **exceptional vertices**, while all other boundary vertices are called **regular vertices**.

Clearly, there are 136 exceptional vertices. Note, that $|\operatorname{supp}\sigma| \le 3$ whenever v is an exceptional boundary vertex.

The general rule for the labeling I is now as follows:

$$I(v) := \begin{cases} 1, & \text{if } v \text{ is an internal vertex;} \\ 1, & \text{if } v \text{ is an exceptional boundary vertex;} \\ 0, & \text{if } v \text{ is a regular boundary vertex.} \end{cases}$$

Clearly, this labeling is compliant.

We remark that since any 5-simplex contains an internal vertex, there are no 0-monochromatic 5-simplices in the labeling I. Of course there are quite many 1-monochromatic 5-simplices. We will eliminate them by passing to the third chromatic subdivision.

4.3 *The Matching Lemma*

Our next goal is to produce a perfect matching on the set of 1-monochromatic 5-simplices, so that each pair of matched 5-simplices shares a 4-simplex. We will start with producing a partial matching.

As a warm-up, let us as above consider the graph Γ_n, whose set of vertices is given by n-simplices of $\chi(\Delta^n)$. For $n \geq 1$, let f_n denote the number of vertices of Γ_n. We have $f_1 = 3, f_2 = 13$, etc. There is an easy recursion

$$f_n = \binom{n+1}{1} f_{n-1} + \binom{n+1}{2} f_{n-2} + \cdots + \binom{n+1}{n-1} f_1 + \binom{n+1}{n} + \binom{n+1}{n+1}, \tag{11}$$

see [14, Sect. 4.4] for a more general formula. One can see by induction that f_n is always odd. To do this, simply evaluate the right hand side of (11) modulo 2 and see that

$$f_n \equiv \binom{n+1}{1} + \binom{n+1}{2} + \cdots + \binom{n+1}{n-1} + \binom{n+1}{n} + \binom{n+1}{n+1} \equiv 2^{n+1} - 1 \equiv 1 \mod 2.$$

Since f_n is odd we cannot hope for a perfect matching. However, it is easy to produce a number of near-perfect matchings, each one matching all but one simplex. We need the following piece of terminology.

Definition 18 Let $\Sigma = (x_0, \ldots, x_k)$ be a fixed order on a nonempty subset of $[n]$, and let $\sigma = (A_1 \,|\, \ldots \,|\, A_t)$ be an n-simplex of $\chi(\Delta^n)$. We shall define the number $\mathrm{level}\,(\sigma, \Sigma) \in [n]$, called the **level of σ with respect to Σ** as follows. If $\sigma = (A_1 \,|\, \ldots \,|\, A_{t-k-1} \,|\, x_0 \,|\, \ldots \,|\, x_k)$, then $\mathrm{level}\,(\sigma, \Sigma)$ is not defined; else $\mathrm{level}\,(\sigma, \Sigma) := x_m$, where m is the highest number $0 \leq m \leq k$, such that $A_{t-k+m} \neq x_m$.

Definition 18 can be rephrased as follows. If the last set in the ordered set partition σ is not x_k, the $\mathrm{level}\,(\sigma, \Sigma) = x_k$, else we proceed with the next set in σ. If the last set in σ is x_k, but the penultimate one is not x_{k-1}, then $\mathrm{level}\,(\sigma, \Sigma) = x_{k-1}$, else we proceed with the next set in σ. One then repeats this step, moving down in the ordered set partition σ. The level is then not defined, if we went through whole Σ and did not stop. This will happen precisely when σ has the form $(A_1 \,|\, \ldots \,|\, A_{t-k-1} \,|\, x_0 \,|\, \ldots \,|\, x_k)$.

We are now ready to define a special kind of matching, which we shall call the *standard matching*.

Definition 19 Let $\Sigma = (x_0, \ldots, x_n)$ be any fixed order on the set $[n]$. Recall that $\widehat{\Sigma}$ is an ordered set partition of $[n]$ associated to Σ. The **standard matching M_Σ** associated to Σ is a bijection

$$M_\Sigma : \Gamma_n \setminus \widehat{\Sigma} \longrightarrow \Gamma_n \setminus \widehat{\Sigma},$$

defined by

$$\sigma \mapsto \mathcal{F}(\sigma, \mathrm{level}\,(\sigma, \Sigma)).$$

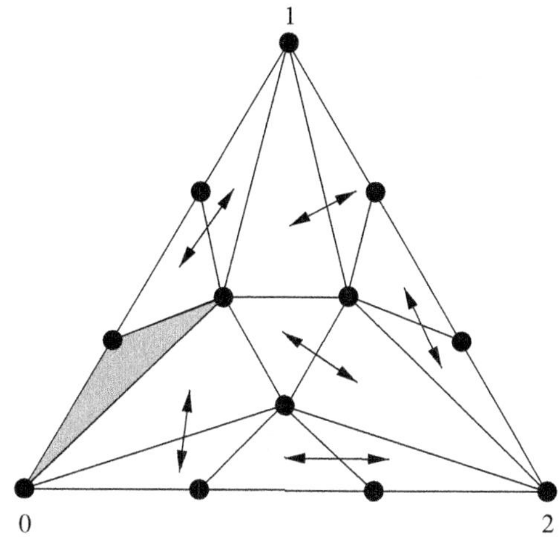

Fig. 3 The standard matching on $\chi(\Delta^2)$ for $\Sigma = (0, 1, 2)$

There is a simple verbal formulation for the standard matching. In order to match a simplex σ, simply flip with respect to the highest possible color, such that the flip preserves the level function, see also the proof of Proposition 6.

The next proposition says that in the standard matching on the set of all n-simplices of $\chi(\Delta^n)$ with respect to any permutation Σ, there is exactly one critical simplex, namely $\widehat{\Sigma}$. This is illustrated by Fig. 3.

Proposition 6 *The standard matching M_Σ described in Definition 19 is a well-defined near-perfect matching on Γ_n, with the only critical n-simplex being indexed by $\widehat{\Sigma}$.*

Proof Let $\sigma \in \Gamma_n \setminus \widehat{\Sigma}$. The crucial fact which we need is that

$$\text{level}\,(\sigma, \Sigma) = \text{level}\,(M_\Sigma(\sigma), \Sigma). \tag{12}$$

To see (12), assume $\sigma = (A_1 \mid \dots \mid A_m \mid x_{k+1} \mid \dots x_n)$, such that $A_m \neq x_k$, i.e., level $(\sigma, \Sigma) = k$. The ordered set partition $M_\Sigma(\sigma)$ is obtained from σ by either merging two sets, or by splitting one of the sets. If in the process the sets A_m, $x_{k+1}, \dots, x_n$ are untouched, then obviously level $(M_\Sigma(\sigma), \Sigma) = k$ as well. On the other hand, the only way the sets $x_{k+1}, \dots, x_n$ could be modified would be by merging them with a neighbor. However, this is impossible, since merges must always involve the set x_k, and we have assumed that $A_m \neq x_k$.

It remains to consider the case that the set A_m was altered. If it was split into x_k and $A_m \setminus x_k$, then again (12) is satisfied, since $A_m \setminus x_k \neq x_k$. Finally, if the set A_m was merged with the set A_{m-1}, then again $A_{m-1} \cup A_m \neq x_k$, and (12) is valid again.

Now it follows from (12) together with (8) that

$$M_\Sigma(M_\Sigma(\sigma)) = \mathcal{F}(\mathcal{F}(\sigma, \text{level}\,(\sigma, \Sigma)), \text{level}\,(\sigma, \Sigma)) = \sigma.$$

This means that we have a well-defined matching, and that all simplices σ for which level (σ, Σ) is defined are matched. We conclude that we have a near-perfect matching with the critical n-simplex being indexed by $\widehat{\Sigma}$. □

Note, that even if we have $\Sigma = (x_0, \dots, x_m)$, for $m < n$, the matching M_Σ is still defined for those simplices σ, for which the number level (σ, Σ) is defined. In this case, we will have a number of critical simplices, namely all those indexed by the tuples $(A_1 \mid \dots \mid A_k \mid x_0 \mid \dots \mid x_m)$, and the proof is simply the same as that of Proposition 6.

Definition 20 For a proper nonempty subset $V \subset [n]$ we let $\Gamma_n(V)$ denote the subgraph of Γ_n induced by set of all n-simplices $\sigma = (A_1 \mid \dots \mid A_t)$, such that $A_1 \not\subseteq V$.

Proposition 7 *Let $\Sigma = (x_0, \dots, x_k)$ be the set $[n] \setminus V$ taken in an arbitrary order. Then M_Σ provides a perfect matching of the simplices in $\Gamma_n(V)$.*

Proof The part of the proof of Proposition 6 showing (12) did not use that fact that Σ has cardinality $n + 1$, so we still have that identity. This means that M_Σ is still a well-defined matching on Γ_n, only now with many more critical simplices.

First, we note that level (σ, Σ) is defined for all $\sigma \in \Gamma_n(V)$. Indeed, of level (σ, Σ) is not defined, then $\sigma = (A_1 \mid \dots \mid A_k \mid x_0 \mid \dots \mid x_m)$, but then $A_1 \subseteq A_1 \cup \dots \cup A_k = [n] \setminus \{x_0, \dots, x_m\} = V$ yields a contradiction. Furthermore, we have the implication

$$\sigma \in \Gamma_n(V) \Rightarrow M_\Sigma(\sigma) \in \Gamma_n(V).$$

Indeed, assume $M_\Sigma(\sigma) = (B_1 \mid \dots \mid B_q)$, and $\sigma = (A_1 \mid \dots \mid A_t)$, with $A_1 \not\subseteq V$. Then, by definition of M_Σ, either we have $B_1 \supseteq A_1$, or $B_1 = x_k$, where $x_k =$ level $(\sigma, \Sigma) \in [n] \setminus V$. Either way, we get $B_1 \not\subseteq V$, so $M_\Sigma(\sigma) \in \Gamma_n(V)$. We now conclude that M_Σ is a perfect matching on $\Gamma_n(V)$. □

Definition 21 Given an n-simplex $\sigma = (A_1 \mid \dots \mid A_t)$, with $A_t = \{x_1, \dots, x_k\}$, such that $x_1 < \dots < x_k$, we set $\mathrm{St}(\sigma) := (x_1, \dots, x_k)$.

Definition 22 Given a subset $V \subset [n]$, we shall call a tuple $\mathcal{A} = (A_1 \mid \dots \mid A_k)$ of nonempty subsets of V a **prefix** in V, if $A_1, \dots, A_k$ are disjoint. We shall say that a prefix is **full** if $A_1 \cup \dots \cup A_k = V$. We also allow an empty prefix.

Furthermore, given a prefix $\mathcal{A} = (A_1 \mid \dots \mid A_k)$, we let $\Gamma_n(V, \mathcal{A})$ denote the subgraph of Γ_n induced by the set of all n-simplices $\sigma = (A_1 \mid \dots \mid A_k \mid A_{k+1} \mid \dots \mid A_t)$ of $\chi(\Delta^n)$, such that $A_{k+1} \not\subseteq V$.

In particular, the set $\Gamma_n(V, \emptyset)$ coincides with the previously defined set $\Gamma_n(V)$. We note, that for any $V \subset [n]$, the set of vertices of Γ_n, which is the set of all n-simplices of $\chi(\Delta^n)$, is a disjoint union of the sets of vertices of $\Gamma_n(V, \mathcal{A})$, where $\mathcal{A}$ runs over all prefixes in V. The following lemma, whose statement is illustrated by Fig. 4, provides a useful generalization of Proposition 7.

Fig. 4 The decomposition of Γ_2 for $V = \{0, 1\}$

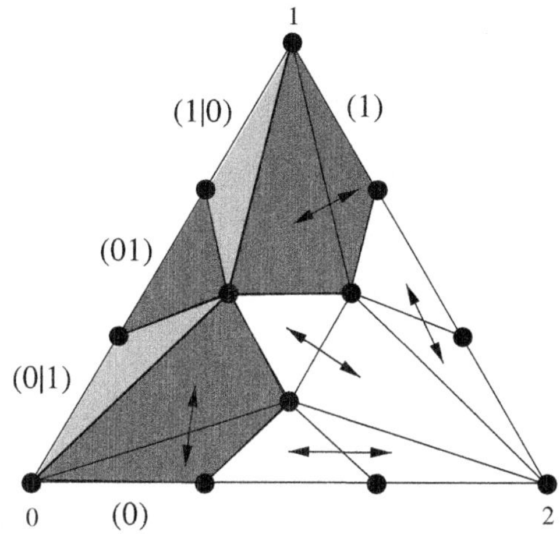

Lemma 1 *Assume* $V \subset [n]$*, and* $\Sigma = (x_0, \dots, x_m)$ *is an arbitrary permutation of the set* $[n] \setminus V$*. Let furthermore* $\mathcal{A} = (A_1 \mid \dots \mid A_k)$ *be an arbitrary prefix of* V*.*

(1) If the prefix $\mathcal{A}$ *is not full, then the standard matching* M_Σ *gives a perfect matching of simplices in* $\Gamma_n(V, \mathcal{A})$*.*
(2) If the prefix $\mathcal{A}$ *is full then the standard matching* M_Σ *of* $\Gamma_n(V, \mathcal{A})$ *has precisely one critical n-simplex, namely* $\sigma = (A_1 \mid \dots \mid A_k \mid x_0 \mid \dots \mid x_m)$*.*

Proof Our argument is very close to the proof of Proposition 7. Take $\sigma \in \Gamma_n(V, \mathcal{A})$. If level (σ, Σ) is not defined, then σ has the form $(B_1 \mid \dots \mid B_t \mid x_0 \mid \dots \mid x_m)$. Since $\sigma \in \Gamma_n(V, \mathcal{A})$, we see that $\sigma = (A_1 \mid \dots \mid A_k \mid x_0 \mid \dots \mid x_m)$, which is only possible if the prefix $\mathcal{A}$ is full. So, if $\mathcal{A}$ is full, we do have one critical simplex, otherwise we do not.

Assume now $\sigma = (A_1 \mid \dots \mid A_k \mid A_{k+1} \mid \dots \mid A_t)$, such that $\sigma \neq (A_1 \mid \dots \mid A_k \mid x_0 \mid \dots \mid x_m)$, and let $x_t = \text{level}(\sigma, \Sigma)$. We have $A_{k+1} \cap V \neq \emptyset$. The ordered set partition $M_\Sigma(\sigma)$ is obtained from σ by merging two sets or splitting a set. If the sets $A_1, \dots, A_{k+1}$ are untouched, then clearly $M_\Sigma(\sigma) \in \Gamma_n(V, \mathcal{A})$. In the other hand, the sets $A_1, \dots, A_k$ cannot be involved in any modification, so the only remaining case is that A_{k+1} has been modified. Either A_{k+1} has been replaced by x_t followed by $A_{k+1} \setminus x_t$, or $A_{k+1} = x_t$ and we replace A_{k+1}, A_{k+2} with $x_t \cup A_{k+2}$. In any case, $x_t \in B_{k+1}$, where $M_\Sigma(\sigma) = (A_1 \mid \dots \mid A_k \mid B_{k+1} \mid \dots)$, hence $V \cup B_{k+1} \neq \emptyset$. Thus, we derive the implication

$$\sigma \in \Gamma_n(V, \mathcal{A}) \Rightarrow M_\Sigma(\sigma) \in \Gamma_n(V, \mathcal{A}),$$

for non-critical σ. Hence M_Σ gives a perfect matching on the set of all n-simplices in $\Gamma_n(V, \mathcal{A})$, with the only possible exception being $\sigma = (A_1 \mid \dots \mid A_k \mid x_0 \mid \dots \mid x_m)$.

□

4.4 Using the Matching Lemma to Get a Partial Matching

Before we proceed with describing the partial matching on the set of all 1-monochromatic 5-simplices of the labeling I, let us give combinatorial description of all the elements in this set. To start with, all 5-simplices with only internal vertices are 1-monochromatic. These are described by a simple criterion in Proposition 5. The rest of 1-monochromatic 5-simplices are those which contain internal vertices and some exceptional boundary vertices. In Table 3 we list all exceptional boundary vertices and the 5-simplices which contain them.

Next we need to understand which of the 5-simplices from the right column of Table 3 contain *only* exceptional boundary vertices. Recall that a 5-simplex of $\chi^2(\Delta^5)$ is indexed by a pair of ordered set partitions of subsets of [5]. We shall describe the set of 5-simplices containing only exceptional boundary vertices by grouping them together according to the first ordered set partition.

Definition 23 For a 5-simplex σ of $\chi(\Delta^5)$, let $\mathcal{N}(\sigma)$ denote the subgraph of Γ_n induced by all 5-simplices τ of $\chi(\Delta^5)$ such that the 5-simplex $(\sigma \parallel \tau)$ of $\chi^2(\Delta^5)$ is 1-monochromatic.

The explicit description of the graph $\mathcal{N}(\sigma)$ is given in Proposition 8, using Table 4. We remark that the types of the 5-simplices listed in the left column of that table are meant to be mutually exclusive.

Proposition 8 *Let $\sigma = (S_1 \mid \dots \mid S_p)$ be a 5-simplex of $\chi(\Delta^5)$, and set $V := [5] \setminus S_p$. Then we have $\mathcal{N}(\sigma) = \Gamma_5(V) \cup \Lambda$, where the subset Λ is given as follows. If σ is of the type listed on the left column of Table 4, then*

$$\Lambda = \bigcup_{\mathcal{A}} \Gamma_5(V, \mathcal{A}), \tag{13}$$

where the union is taken over all prefixes $\mathcal{A}$ which are listed in the corresponding row of Table 4. We set Λ to be an empty set otherwise.

Proof To start with, we see that $V \subset [n]$. If V is empty, then we are looking at the central 5-simplex $\sigma = ([5])$, in which case all of the 5-simplices of $\mathcal{N}(\sigma)$ are internal.

Assume now V is non-empty. Let $\tau = (T_1 \mid \dots \mid T_q)$ be a 5-simplex of $\chi(\Delta^5)$. By Proposition 5, all the vertices of the 5-simplex $(\sigma \parallel \tau)$ of $\chi^2(\Delta^5)$ are internal if and only if $S_p \cap T_1 \neq \emptyset$, or equivalently $T_1 \not\subseteq V$. In other words, the 5-simplex $(\sigma \parallel \tau)$ is internal if and only if $\tau \in \Gamma_5(V)$. All internal 5-simplices are 1-monochromatic, and all further 1-monochromatic 5-simplices appear due to the exceptional vertices. So if did not have any, we would simply have $\mathcal{N}(\sigma) = \Gamma_5(V)$.

To take into account the influence of the exceptional vertices, we obtain the correction set Λ by a direct case-by-case analysis. Table 3 shows the cases which need to be considered. Table 4 describes in each one of these cases, which new 1-monochromatic simplices will arise. □

Table 3 Exceptional boundary vertices and 5-simplices containing them; here we assume that $a < b < c$

Exceptional vertices	5-Simplices with modified labels
$\begin{array}{c\|\|c} a & a \\ \hline a & a \end{array}$	$(a \mid S_1 \mid \dots \mid S_p \parallel a \mid T_1 \mid \dots \mid T_q)$
$\begin{array}{c\|\|c} a, b & a \\ \hline a & a \end{array}$	$(a, b \mid S_1 \mid \dots \mid S_p \parallel a \mid T_1 \mid \dots \mid T_q)$ $(b \mid a \mid S_1 \mid \dots \mid S_p \parallel a \mid T_1 \mid \dots \mid T_q)$
$\begin{array}{c\|\|c} a, b & a, b \\ \hline a, b & b \end{array}$	$(a, b \mid S_1 \mid \dots \mid S_p \parallel a, b \mid T_1 \mid \dots \mid T_q)$ $(a, b \mid S_1 \mid \dots \mid S_p \parallel a \mid b \mid T_1 \mid \dots \mid T_q)$
$\begin{array}{c\|\|c} a, b, c & a \\ \hline a & a \end{array}$	$(a, b, c \mid S_1 \mid \dots \mid S_p \parallel a \mid T_1 \mid \dots \mid T_q)$ $(c \mid a, b \mid S_1 \mid \dots \mid S_p \parallel a \mid T_1 \mid \dots \mid T_q)$ $(b \mid a, c \mid S_1 \mid \dots \mid S_p \parallel a \mid T_1 \mid \dots \mid T_q)$ $(b, c \mid a \mid S_1 \mid \dots \mid S_p \parallel a \mid T_1 \mid \dots \mid T_q)$ $(b \mid c \mid a \mid S_1 \mid \dots \mid S_p \parallel a \mid T_1 \mid \dots \mid T_q)$ $(c \mid b \mid a \mid S_1 \mid \dots \mid S_p \parallel a \mid T_1 \mid \dots \mid T_q)$
$\begin{array}{c\|\|c} a, b, c & a, b \\ \hline a, b & a \end{array}$	$(a, b, c \mid S_1 \mid \dots \mid S_p \parallel a, b \mid T_1 \mid \dots \mid T_q)$ $(c \mid a, b \mid S_1 \mid \dots \mid S_p \parallel a, b \mid T_1 \mid \dots \mid T_q)$ $(a, b, c \mid S_1 \mid \dots \mid S_p \parallel b \mid a \mid T_1 \mid \dots \mid T_q)$ $(c \mid a, b \mid S_1 \mid \dots \mid S_p \parallel b \mid a \mid T_1 \mid \dots \mid T_q)$
$\begin{array}{c\|\|c} a, b, c & a, b \\ \hline a, b & b \end{array}$	$(a, b, c \mid S_1 \mid \dots \mid S_p \parallel a, b \mid T_1 \mid \dots \mid T_q)$ $(c \mid a, b \mid S_1 \mid \dots \mid S_p \parallel a, b \mid T_1 \mid \dots \mid T_q)$ $(a, b, c \mid S_1 \mid \dots \mid S_p \parallel a \mid b \mid T_1 \mid \dots \mid T_q)$ $(c \mid a, b \mid S_1 \mid \dots \mid S_p \parallel a \mid b \mid T_1 \mid \dots \mid T_q)$
$\begin{array}{c\|\|c} a, b, c & a, b, c \\ \hline a, b, c & c \end{array}$	$(a, b, c \mid S_1 \mid \dots \mid S_p \parallel a, b, c \mid T_1 \mid \dots \mid T_q)$ $(a, b, c \mid S_1 \mid \dots \mid S_p \parallel a, b \mid c \mid T_1 \mid \dots \mid T_q)$ $(a, b, c \mid S_1 \mid \dots \mid S_p \parallel b \mid a \mid c \mid T_1 \mid \dots \mid T_q)$ $(a, b, c \mid S_1 \mid \dots \mid S_p \parallel a \mid b \mid c \mid T_1 \mid \dots \mid T_q)$ $(a, b, c \mid S_1 \mid \dots \mid S_p \parallel b \mid a, c \mid T_1 \mid \dots \mid T_q)$ $(a, b, c \mid S_1 \mid \dots \mid S_p \parallel a \mid b, c \mid T_1 \mid \dots \mid T_q)$
$\begin{array}{c\|\|c} a, b, c & a, b, c \\ \hline a, b, c & b \end{array}$	$(a, b, c \mid S_1 \mid \dots \mid S_p \parallel a, b, c \mid T_1 \mid \dots \mid T_q)$ $(a, b, c \mid S_1 \mid \dots \mid S_p \parallel a, c \mid b \mid T_1 \mid \dots \mid T_q)$ $(a, b, c \mid S_1 \mid \dots \mid S_p \parallel c \mid a \mid b \mid T_1 \mid \dots \mid T_q)$ $(a, b, c \mid S_1 \mid \dots \mid S_p \parallel a \mid c \mid b \mid T_1 \mid \dots \mid T_q)$ $(a, b, c \mid S_1 \mid \dots \mid S_p \parallel c \mid a, b \mid T_1 \mid \dots \mid T_q)$ $(a, b, c \mid S_1 \mid \dots \mid S_p \parallel a \mid b, c \mid T_1 \mid \dots \mid T_q)$

Table 4 New 1-monochromatic 5-simplices sorted by the supporting 5-simplex in $\chi(\Delta^5)$; here we assume $a < b < c$

Type of σ	Allowed prefixes
$(a \mid S_1 \mid \dots \mid S_p)$	(a)
$(ab \mid S_1 \mid \dots \mid S_p)$	(a), $(a \mid b)$
$(abc \mid S_1 \mid \dots \mid S_p)$	(a), $(a \mid b)$, (ab), $(ab \mid c)$, $(a \mid bc)$, $(a \mid b \mid c)$
$(a \mid b \mid S_1 \mid \dots \mid S_p)$	(a)
$(b \mid a \mid S_1 \mid \dots \mid S_p)$	(a), (b)
$(ab \mid c \mid S_1 \mid \dots \mid S_p)$	(a), $(a \mid b)$
$(ac \mid b \mid S_1 \mid \dots \mid S_p)$	(a), $(a \mid c)$
$(bc \mid a \mid S_1 \mid \dots \mid S_p)$	(a), (b), $(b \mid c)$
$(a \mid bc \mid S_1 \mid \dots \mid S_p)$	(a)
$(b \mid ac \mid S_1 \mid \dots \mid S_p)$	(a), (b)
$(c \mid ab \mid S_1 \mid \dots \mid S_p)$	(a), (c), (ab), $(a \mid b)$
$(a \mid b \mid c \mid S_1 \mid \dots \mid S_p)$	(a)
$(a \mid c \mid b \mid S_1 \mid \dots \mid S_p)$	(a)
$(b \mid a \mid c \mid S_1 \mid \dots \mid S_p)$	(a), (b)
$(b \mid c \mid a \mid S_1 \mid \dots \mid S_p)$	(a), (b)
$(c \mid a \mid b \mid S_1 \mid \dots \mid S_p)$	(a), (c)
$(c \mid b \mid a \mid S_1 \mid \dots \mid S_p)$	(a), (b), (c)

5 Modifying the Standard Matching

5.1 Conducting Graphs

In order to look at matchings in depth, we need certain standard tools from the matching theory. Given a partial matching, there is a classical way of modifying it, possibly enlarging, or even making it perfect. Following definition provides the key concepts.

Definition 24 Given a matching M on a graph G, assume γ is an edge path in G with endpoints v and w. We call the path γ

- **Alternating** if the edges of γ are alternatively matching and non-matching.
- **Properly alternating** if is alternating and those endpoints of the path which are not matched on the path are critical.
- **Augmenting** if it is properly alternating, starting and ending with non-matching edges.
- **Semi-augmenting** if it is properly alternating, starting with a matching edge and ending with a non-matching one.
- **Weakly semi-augmenting** if it is alternating, starting with a matching edge and ending with a non-matching one.
- **Non-augmenting** if it is alternating, starting and ending with matching edges.

Properly alternating non-self-intersecting edge paths allow us to modify matchings.

Definition 25 Assume we are given a matching M on a graph G, and a properly alternating non-self-intersecting edge path γ. We define $D(M, \gamma)$ be a new matching on G consisting of all edges from M which do not belong to γ together with all edges from γ which do not belong to M.

The next proposition shows that having a properly alternating edge path is enough to find a larger matching, and eventually to turn partial matchings into perfect ones.

Proposition 9 *Assume M is a matching on a bipartite graph G, and v, w are critical vertices of G with respect to M. Assume furthermore, that there exists a properly alternating edge path γ from v to w. Then, there exists a matching $\widetilde{M}$ on G, such that the set of critical vertices with respect to $\widetilde{M}$ is obtained from the set of critical vertices with respect to M by removing the vertices v and w.*

Proof If the path γ is non-self-intersecting, then the new matching $\widetilde{M}$ can be taken to be $D(M, \gamma)$. Assume γ is a self-intersecting path, and its vertices, listed in the path order, are $v = v_0, v_1, \dots, v_l = w$.

Since γ is self-intersecting, there exist $0 \leq k < m \leq l$, such that $v_k = v_m$, and there is no vertex duplication in the sequence $v_k, v_{k+1}, \dots, v_{m-1}$. Let $\tilde{\gamma}$ be obtained from γ by removing the loop consisting of the edges $(v_k, v_{k+1}), (v_{k+1}, v_{k+2}), \dots, v_{m-1}, v_m$. Since the graph G is bipartite, that loop must have an even length, hence the new path $\tilde{\gamma}$ is again properly alternating. Repeating this procedure we will eventually arrive at a properly alternating non-self-intersecting edge path connecting v with w. □

We note, that if the edge path γ in Proposition 9 is semi-augmenting, then the number of critical vertices does not change, whereas, if γ is non-augmenting, then the number of critical vertices increases by 2. Even though modifying a matching along a semi-augmenting path does not change the number of critical vertices, this modification is still useful as it could be thought of as transporting the critical vertex from one endpoint of the semi-augmenting path to the other.

Definition 26 Let G be a bipartite graph, with the corresponding bipartite decomposition (A, B). We say that G is **conducting** if one of the following situations occur:

(1) We have $|A| = |B| + 1$, and for any vertex $v \in A$, the graph G has a near-perfect matching, with the critical vertex v.
(2) We have $|A| = |B|$, and for any vertices $v \in A$, $w \in B$, the graph G has a matching, with precisely two critical vertices v and w.

More specifically, if a conducting graph G satisfies condition (1) we shall call it **conducting graph of the first type**, else we shall call it **conducting graph of the second type**.

One handy way of showing that a graph is conducting is by presenting a certain collection of paths, as the next proposition details.

Proposition 10 *Let G be a bipartite graph, with the corresponding bipartite decomposition (A, B).*

(1) Assume M is a near-perfect matching of G with critical vertex $v \in A$, such that for any other vertex $w \in A$ there exists a semi-augmenting path from v to w, then the graph G is conducting of the first type.
(2) Assume M is a perfect matching of G, such that for any vertices $v \in A$, $w \in B$ there exists a non-augmenting path from v to w, then the graph G is conducting of the second type.

Proof This is an immediate consequence of Proposition 9. □

Definition 27 Let Ω be a family of prefixes of a set $V \subset [n]$ containing the empty prefix. We say that Ω is **closed** if the following two conditions are satisfied:

(1) If $(A_1 \,|\, \dots \,|\, A_k \,|\, x) \in \Omega$, where $A_1, \dots, A_k \subseteq V$, $x \in V$, then $(A_1 \,|\, \dots \,|\, A_k) \in \Omega$.
(2) If $(A_1 \,|\, \dots \,|\, A_k) \in \Omega$, and $|A_m| \geq 2$, for some $1 \leq m \leq k$, then there exists $x \in A_m$, such that

$$(A_1 \,|\, \dots \,|\, A_{m-1} \,|\, x \,|\, A_m \setminus x \,|\, A_{m+1} \,|\, \dots \,|\, A_k) \in \Omega.$$

Furthermore, we let $\Gamma_n(V, \Omega)$ denote the subgraph of Γ_n induced by the vertices of $\bigcup_{\mathcal{A}\in\Omega} \Gamma_n(V, \mathcal{A})$.

Remark 3 By Lemma 1, the standard matching will induce a matching on $\Gamma_n(V, \Omega)$. However, this matching does not have to be perfect or near-perfect. As a matter of fact, the number of critical vertices will be equal to the number of full prefixes in Ω.

The next observation can be shown by direct inspection.

Remark 4 All families of prefixes Ω listed in Table 4 are closed.

We shall now proceed with analysing conductibility of graphs $\Gamma_n(V, \Omega)$. Let us fix a natural number $n \geq 3$. Let ρ denote the central simplex of $\chi(\Delta^n)$, indexed by $([n])$, and let ν denote the neighboring n-simplex indexed by $(n, [n-1])$. Consider the standard matching M_Σ, associated to $\Sigma = (0, 1, \dots, n)$. The only critical simplex of M_σ is $(0 \,|\, 1 \,|\, \dots \,|\, n)$, and clearly, M_σ matches ρ with ν. Our central tool will be the following technical lemma.

Lemma 2 *Let σ be a non-critical n-simplex in $\chi(\Delta^n)$, for $n \geq 3$. There exists a weakly semi-augmenting (with respect to M_Σ) path P starting at σ and ending either at ρ or at ν.*

Proof Since the endpoints of a weakly semi-augmenting path have the same orientation, we know that if such a path P exists, then its endpoint from the set $\{\rho, \nu\}$ is uniquely determined. For brevity, we shall say that P *goes from σ to $\{\rho, \nu\}$*. Taking the empty path we see that the claim is true for $\sigma = \rho$ and for $\sigma = \nu$. Assume that $\sigma \neq \rho, \sigma \neq \nu$. We shall use the notation $\sigma = (A_1 \,|\, \dots \,|\, A_t)$ throughout the proof.

When describing edges in a path, we shall usually not list entire combinatorial indices of the simplices, but rather the part of the ordered set partition where the actual change occurs. The path from σ to $\{\rho, \nu\}$ will be constructed by concatenating different pieces.

Let us briefly introduce some notations in which we encode our paths. Assume $\sigma = (A_1 \,|\, \dots \,|\, A_t)$ is an n-simplex of $\chi(\Delta^n)$, and set $k := \text{level}\,(\sigma, \Sigma)$.

- We let $\xrightarrow{k}$ denote the matching edge between σ and $M_\Sigma(\sigma) = \mathcal{F}(\sigma, k)$.
- Assume $A_i = \{x\}$, for some $1 \leq i \leq t-1$, and $x \neq k$. We let $\overset{x\cup}{\rightsquigarrow}$ denote the edge between σ and $(A_1 \,|\, \dots \,|\, A_{i_1} \,|\, x \cup A_{i+1} \,|\, \dots \,|\, A_t)$.
- Assume instead that $|A_i| \geq 2$, for some $1 \leq i \leq t$, $x \in A_i$, and $x \neq k$. We let $\overset{x|}{\rightsquigarrow}$ denote the edge between σ and $(A_1 \,|\, \dots \,|\, A_{i_1} \,|\, x \,|\, A_i \setminus x \,|\, \dots \,|\, A_t)$.
- If $|A_i| \geq 2$, for some $1 \leq i \leq t$, and level $(\sigma, \Sigma) \notin A_i$, we use the notation $\overset{A_i}{\rightsquigarrow}$ to denote the set of edges $\overset{x|}{\rightsquigarrow}$, for all $x \in A_i$.

We shall call the move $\overset{A}{\rightsquigarrow}$ the *generic split*. When a generic split is used in our notation of the path, it means that it does not matter at this point which of the elements of A we split off, and any will do. Formally, this means that we output a set of paths, rather than one path. Later, we will need to choose paths satisfying certain additional conditions. This can be done from this set of paths, by substituting specific splits instead of generic ones. For example, condition (2) of Definition 27 only guarantees the existance of *some* element x, which we are allowed to split. If at this point we used generic split of A_m in our path, then we will be able to pick a suitable path from our set.

We break our argument into considering three different cases.

Case 1. Assume that level $(\sigma, \Sigma) = n$. Pick i such that $n \in A_i$. Our assumption means that either $A_i = n$, and $i \leq t-1$, or $|A_i| \geq 2$.

Our first goal, is to reduce this to the case $i = 1$. Assume that $i \geq 2$, and $|A_{i-1}| = 1$, say $A_{i-1} = a$. We can use the moves (14) and (15) to lower the index i by 1.

$$(\ldots \,|\, a \,|\, n \cup B \,|\, \ldots) \xrightarrow{n} (\ldots \,|\, a \,|\, n \,|\, B \,|\, \ldots) \overset{a\cup}{\rightsquigarrow} (\ldots \,|\, \{n, a\} \,|\, B \,|\, \ldots). \tag{14}$$

$$(\ldots \,|\, a \,|\, n \,|\, B \,|\, \ldots) \xrightarrow{n} (\ldots \,|\, a \,|\, n \cup B \,|\, \ldots) \overset{a\cup}{\rightsquigarrow} (\ldots \,|\, \{n, a\} \cup B \,|\, \ldots). \tag{15}$$

Assume now $i \geq 2$, and $|A_{i-1}| \geq 2$. We can use the moves (16) and (17) to lower the cardinality $|A_{i-1}|$ by 1, eventually reducing this to the case above.

$$(\ldots \,|\, A_{i-1} \,|\, n \cup B \,|\, \ldots) \xrightarrow{n} (\ldots \,|\, A_{i-1} \,|\, n \,|\, B \,|\, \ldots) \overset{A_{i-1}}{\rightsquigarrow} (\ldots \,|\, x \,|\, A_{i-1} \setminus x \,|\, n \,|\, B \,|\, \ldots) \tag{16}$$

$$(\ldots \,|\, A_{i-1} \,|\, n \,|\, B \,|\, \ldots) \xrightarrow{n} (\ldots \,|\, A_{i-1} \,|\, n \cup B \,|\, \ldots) \overset{A_{i-1}}{\rightsquigarrow} (\ldots \,|\, x \,|\, A_{i-1} \setminus x \,|\, n \cup B \,|\, \ldots) \tag{17}$$

In either case, we are able to lower the index i by 1, and in the end to reach the case $n \in A_1$.

Next, we want to achieve $|A_1| = 2$, or $|A_1| = |A_2| = 1$. Assume first $|A_1| \geq 3$, then use the move (18) to achieve $|A_1| = |A_2| = 1$.

$$(n \cup A \,|\, \ldots) \xrightarrow{n} (n \,|\, A \,|\, \ldots) \overset{A}{\rightsquigarrow} (n \,|\, x \,|\, A \setminus x \,|\, \ldots) \tag{18}$$

Assume now $|A_1| = 1$, but $|A_2| \geq 2$. Since $\sigma \not\to \nu$, we must have $|A_3| \geq 1$ If $|A_3| \geq 2$, then use the move (19) to reduce to the case $|A_1| \geq 3$, which we just dealt with.

$$(n \,|\, A_2 \,|\, A_3 \,|\, \ldots) \xrightarrow{n} (n \cup A_2 \,|\, A_3 \,|\, \ldots) \overset{A_3}{\rightsquigarrow} (n \cup A_2 \,|\, x \,|\, A_3 \setminus x \,|\, \ldots). \tag{19}$$

If $|A_3| = 1$ and $A_4 \neq \emptyset$, then use the move (20) to reduce to the case $|A_1| \geq 3$ again.

$$(n \,|\, A_2 \,|\, b \,|\, A_4 \,|\, \ldots) \xrightarrow{n} (n \cup A_2 \,|\, b \,|\, A_4 \,|\, \ldots) \overset{b\cup}{\rightsquigarrow} (n \cup A_2 \,|\, b \cup A_4 \,|\, \ldots). \tag{20}$$

Finally, assume $|A_3| = 1$, $A_4 = \emptyset$. Then, use the move (21), picking $a \in \{n-1, n-2\} \cap A_2$, to reduce to the case $|A_1| = 2$.

$$(n \,|\, A_2 \,|\, b) \xrightarrow{n} (n \cup A_2 \,|\, b) \overset{a|}{\rightsquigarrow} (a \,|\, (n \cup A_2) \setminus a \,|\, b)$$
$$\xrightarrow{n} (a \,|\, n \,|\, A_2 \setminus a \,|\, b) \overset{a\cup}{\rightsquigarrow} (\{n, a\} \,|\, A_2 \setminus a \,|\, b). \tag{21}$$

So at this point we have $n \in A_1$, and $|A_1| = 2$, or $|A_1| = |A_2| = 1$.

Using one of the moves (22) and (23), whenever $|A_k| \geq 2$, we will eventually arrive at the simplex which is either encoded by $(n \,|\, a_1 \,|\, \ldots \,|\, a_n)$ or by $(\{n, a_1\} \,|\, a_2 \,|\, \ldots \,|\, a_n)$.

$$(n \,|\, a_1 \,|\, \ldots \,|\, A_k \,|\, \ldots) \xrightarrow{n} (\{n, a_1\} \,|\, \ldots \,|\, A_k \,|\, \ldots) \overset{A_k}{\rightsquigarrow} (\{n, a_1\} \,|\, \ldots \,|\, x \,|\, A_k \setminus x \,|\, \ldots) \tag{22}$$

$$(\{n, a_1\} \,|\, \ldots \,|\, A_k \,|\, \ldots) \xrightarrow{n} (n \,|\, a_1 \,|\, \ldots \,|\, A_k \,|\, \ldots) \overset{A_k}{\rightsquigarrow} (n \,|\, a_1 \,|\, \ldots \,|\, x \,|\, A_k \setminus x \,|\, \ldots). \tag{23}$$

We now continue with moves (24) and (25) reducing our simplex either to $(\{n, a\} \,|\, A)$ or to $(n \,|\, a \,|\, A)$.

$$(n \,|\, a_1 \,|\, \ldots \,|\, a_t \,|\, A) \xrightarrow{n} (\{n, a_1\} \,|\, a_2 \,|\, \ldots \,|\, a_t \,|\, A) \overset{a_t\cup}{\rightsquigarrow} (\{n, a_1\} \,|\, a_2 \,|\, \ldots \,|\, a_{t-1} \,|\, \{a_t\} \cup A) \tag{24}$$

$$(\{n, a_1\} \,|\, \ldots \,|\, a_t \,|\, A) \xrightarrow{n} (n \,|\, a_1 \,|\, a_2 \,|\, \ldots \,|\, a_t \,|\, A) \overset{a_t\cup}{\rightsquigarrow} (n \,|\, a_1 \,|\, a_2 \,|\, \ldots \,|\, a_{t-1} \,|\, \{a_t\} \cup A). \tag{25}$$

If $\sigma = (\{n, a\} \,|\, A)$, we use the move (26) to arrive at ν.

$$(\{n, a\} \,|\, A) \xrightarrow{n} (n \,|\, a \,|\, A) \overset{a\cup}{\rightsquigarrow} (n \,|\, [n-1]). \tag{26}$$

So we can assume that $\sigma = (n \,|\, a \,|\, A)$. If $a = n-1$, we use the move (27) to arrive at ρ.

$$(n \,|\, n-1 \,|\, A) \xrightarrow{n} (\{n-1, n\} \,|\, A) \overset{n-1|}{\rightsquigarrow} (n-1 \,|\, n \,|\, A) \xrightarrow{n} (n-1 \,|\, n \cup A) \overset{n-1\cup}{\rightsquigarrow} ([n]). \tag{27}$$

Else write $A = n-1 \cup B$ and use the move (28) to arrive at the considered case $a = n-1$.

$$
\begin{aligned}
(n\,|\,a\,|\,A) &\xrightarrow{n} (\{n,a\}\,|\,A) \overset{n-1|}{\rightsquigarrow} (\{n,a\}\,|\,n-1\,|\,B) \xrightarrow{n} (n\,|\,a\,|\,n-1\,|\,B) \\
&\overset{a\cup}{\rightsquigarrow} (n\,|\,\{n-1,a\}\,|\,B) \xrightarrow{n} (\{n,n-1,a\}\,|\,B) \overset{n-1|}{\rightsquigarrow} (n-1\,|\,\{n,a\}\,|\,B) \xrightarrow{n} (n-1\,|\,n\,|\,a\,|\,B) \\
&\overset{n-1\cup}{\rightsquigarrow} (\{n-1,n\}\,|\,a\,|\,B) \xrightarrow{n} (n\,|\,n-1\,|\,a\,|\,B) \overset{a\cup}{\rightsquigarrow} (n\,|\,n-1\,|\,a \cup B).
\end{aligned} \tag{28}
$$

Case 2. Assume level $(\sigma, \Sigma) \leq n-2$. In this case we have $A_{t-1} = n-1$ and $A_t = n$. We use the move (29) to reduce it to Case 1, where $k = \text{level}\,(\sigma, \Sigma)$.

$$
(\beta\,|\,n-1\,|\,n) \xrightarrow{k} (\tilde{\beta}\,|\,n-1\,|\,n) \overset{n-1\cup}{\rightsquigarrow} (\tilde{\beta}\,|\,\{n-1,n\}) \tag{29}
$$

Case 3. Assume level $(\sigma, \Sigma) = n-1$. If $\sigma = (\ldots\,|\,n-1 \cup B\,|\,\ldots\,|\,A\,|\,n)$, iteratively use the moves (30) and (31) to reduce it to Case 1.

$$
\begin{aligned}
(\ldots\,|\,n-1 \cup B\,|\,\ldots\,|\,A\,|\,n) &\xrightarrow{n-1} (\ldots\,|\,n-1\,|\,B\,|\,\ldots\,|\,A\,|\,n) \\
&\overset{A}{\rightsquigarrow} (\ldots\,|\,n-1\,|\,B\,|\,\ldots\,|\,x\,|\,A \setminus x\,|\,n), \text{ where } |A| \geq 2.
\end{aligned} \tag{30}
$$

$$
\begin{aligned}
(\ldots\,|\,n-1 \cup B\,|\,\ldots\,|\,a\,|\,n) &\xrightarrow{n-1} (\ldots\,|\,n-1\,|\,B\,|\,\ldots\,|\,a\,|\,n) \\
&\overset{a\cup}{\rightsquigarrow} (\ldots\,|\,n-1\,|\,B\,|\,\ldots\,|\,\{n,a\}).
\end{aligned} \tag{31}
$$

If $\sigma = (\ldots\,|\,n-1\,|\,B\,|\,\ldots\,|\,A\,|\,n)$, iteratively use the moves (32) and (33) to reduce it to Case 1.

$$
\begin{aligned}
(\ldots\,|\,n-1\,|\,B\,|\,\ldots\,|\,A\,|\,n) &\xrightarrow{n-1} (\ldots\,|\,n-1 \cup B\,|\,\ldots\,|\,A\,|\,n) \\
&\overset{A}{\rightsquigarrow} (\ldots\,|\,n-1 \cup B\,|\,\ldots\,|\,x\,|\,A \setminus x\,|\,n), \text{ where } |A| \geq 2.
\end{aligned} \tag{32}
$$

$$
\begin{aligned}
(\ldots\,|\,n-1\,|\,B\,|\,\ldots\,|\,a\,|\,n) &\xrightarrow{n-1} (\ldots\,|\,n-1 \cup B\,|\,\ldots\,|\,a\,|\,n) \\
&\overset{a\cup}{\rightsquigarrow} (\ldots\,|\,n-1 \cup B\,|\,\ldots\,|\,\{n,a\}).
\end{aligned} \tag{33}
$$

Assume now $\sigma = (\dots \mid n-1 \cup A \mid n)$. Use the moves (34) and (35) to reduce it to the previously considered cases.

$$(\dots \mid n-1 \cup A \mid n) \xrightarrow{n-1} (\dots \mid n-1 \mid A \mid n) \overset{A}{\rightsquigarrow} (\dots \mid n-1 \mid x \mid A \setminus x \mid n), \text{ where } |A| \geq 2. \quad (34)$$

$$(\dots \mid \{n-1, a\} \mid n) \xrightarrow{n-1} (\dots \mid n-1 \mid a \mid n) \overset{a\cup}{\rightsquigarrow} (\dots \mid n-1 \mid \{n, a\}). \quad (35)$$

Finally assume $\sigma = (\dots \mid B \mid n-1 \mid A \mid n)$. We consider three cases. If $|B| \geq 2$, we use the move (36); if $|B| = 1$, we use the move (37); if $B = \emptyset$ use the move (38), which is possible since $n \geq 3$.

$$(\dots \mid B \mid n-1 \mid A \mid n) \xrightarrow{n-1} (\dots \mid B \mid n-1 \cup A \mid n) \overset{B}{\rightsquigarrow} (\dots \mid x \mid B \setminus x \mid n-1 \cup A \mid n) \quad (36)$$

$$(\dots \mid b \mid n-1 \mid A \mid n) \xrightarrow{n-1} (\dots \mid b \mid n-1 \cup A \mid n) \overset{b\cup}{\rightsquigarrow} (\dots \mid \{n-1, b\} \cup A \mid n) \quad (37)$$

$$(n-1 \mid [n-2] \mid n) \xrightarrow{n-1} ([n-1] \mid n) \overset{n-2|}{\rightsquigarrow} (n-2 \mid [n-3] \cup n-1 \mid n). \quad (38)$$

In each case we reduce to the already considered case $\sigma = (\dots \mid A \cup n-1 \mid n)$. □

Remark 5 Curiously, Lemma 2 is not true when $n = 2$ for $\sigma = (1 \mid 0 \mid 2)$. It still remains true for all other simplices σ, and it is also trivially true for $n = 1$. All of this can be seen by direct inspection.

Theorem 4 *For any natural number n, the graph Γ_n is conducting of the first type.*

Proof The cases $n = 1$ and $n = 2$ can be verified directly, so we assume $n \geq 3$. Recall now, that by Proposition 6 the standard matching M_Σ is a near-perfect matching with a critical n-simplex indexed by Σ. Let τ be any other n-simplex such that $O(\tau) = O(\Sigma)$. By Proposition 10(1) we need to show that there exists a weakly semi-augmenting path between τ and Σ. Let σ be any of the neighboring n-simplices of Σ. Since both σ and τ are non-critical, Lemma 2 implies that there exist a weakly semi-augmenting path Q_1 between τ and $\{\rho, \nu\}$, and a weakly semi-augmenting path Q_2 between σ and $\{\rho, \nu\}$. Since $O(\sigma) \neq O(\tau)$, the paths Q_1 and Q_2 will link them to the different vertices in $\{\rho, \nu\}$. Let Q be the concatenation of Q_1, followed by the edge (ρ, ν), then by Q_2, and finally by the edge $(\sigma.\Sigma)$. Clearly, Q is a semi-augmenting path from τ to Σ. □

Theorem 5 *Assume $[n] \supset V \neq \emptyset$, such that $|V| \leq n-2$, then $\Gamma_n(V)$ is conducting of the second type.*

Proof Note, that the assumption $n \geq 3$ is automatic here, since $1 \leq |V| \leq n-2$. Without loss of generality, we can assume that $V = [k]$, with $0 \leq k \leq n-3$, and set $\Sigma := \{k+1, \dots, n\}$. By Proposition 7, the standard matching M_Σ is a perfect matching on the graph $\Gamma_n(V)$. Let ρ and ν be as above, we still have $\rho, \nu \in \Gamma_n(V)$ and $\nu := M_\Sigma(\rho)$. Pick an arbitrary $\sigma \in \Gamma_n(V)$, which is non-critical with respect to M_Σ. By Lemma 2 there exists a weakly semi-augmenting path Q from σ to $\{\rho, \nu\}$. All we need to do is to see that this path lies entirely within $\Gamma_n(V)$.

Recall, that there are 4 types of edges used in the construction of our path: $\stackrel{k}{\longrightarrow}$, $\stackrel{x\cup}{\leadsto}$, $\stackrel{x|}{\leadsto}$, and $\stackrel{A_i}{\leadsto}$. Clearly, if $\sigma \in \Gamma_n(V)$, then any of the edges $\stackrel{k}{\longrightarrow}$, $\stackrel{x\cup}{\leadsto}$, $\stackrel{A_i}{\leadsto}$ will lead to a vertex in $\Gamma_n(V)$ as well. The edges $\stackrel{x|}{\leadsto}$ occur only in paths (21), (27), (28), and (38). In all these cases we have $x \in \{n-2, n-1\}$, in particular $x \notin V$; hence also here the edge $\stackrel{x|}{\leadsto}$ leads to a vertex in $\Gamma_n(V)$.

We conclude that the entire path stays within $\Gamma_n(V)$. Now, we have seen in the proof of (1) that the existence of such paths means that any two vertices σ, τ of $\Gamma_n(V)$, such that $O(\sigma) = O(\tau)$, are connected by an augmenting path. By Proposition 10(2) this implies that $\Gamma_n(V)$ is conducting of the second type. □

Theorem 6 *The graphs $\Gamma_5(V, \Omega)$ are conducting for all families of prefixes Ω listed in Table 4, under the assumption $|S_p| \geq 3$.*

Proof Again without loss of generality we can assume that $n-2, n-1, n \in S_p$, all other cases can be reduced to this one by appropriate renaming. Choose Σ to be any order on $[n] \setminus V$ with $n-2, n-1, n$ coming last, i.e., $\Sigma = (\dots, n-2, n-1, n)$. By Remark 3 the standard matching M_Σ is a well-defined matching on the graph $\Gamma_n(V, \Omega)$ with one critical simplex for each full prefix in Ω. As above, we still have $\rho, \nu \in \Gamma_n(V)$ and $\nu := M_\Sigma(\rho)$.

Let σ be a non-critical simplex in $\Gamma_n(V, \Omega)$. By Lemma 2 there exists a weakly semi-augmenting path Q from σ to $\{\rho, \nu\}$. This time, what we need to do is to see that this path lies entirely within $\Gamma_n(V, \Omega)$. Again we need to perform a detailed analysis of moves (14) through (38).

Take any edge $e = (\sigma, \tau)$ on our path. Assume $\sigma \in \Gamma_n(V, \Omega)$, say $\sigma = (A_1, \dots, A_k, A_{k+1}, \dots, A_t)$, where $A_1, \dots, A_k \subseteq V$, and $A_{k+1} \nsubseteq V$. Of course we have $(A_1 | \dots | A_k) \in \Omega$. We need to show that $\tau \in \Gamma_n(V, \Omega)$. We consider separately each of the 4 possible types of edge e.

Case 1. If e is of the type $\stackrel{k}{\longrightarrow}$, then the τ starts with the same prefix as σ, so the claim follows.

Case 2. Assume e is of the type $\stackrel{A_i}{\leadsto}$. If $1 \leq i \leq k$, then, since Ω is a closed prefix, by Definition 27(2) there exists $x \in A_i$, such that the edge $\stackrel{x|}{\leadsto}$ will lead to $\tau \in \Gamma_n(V, \Omega)$. If $k+2 \leq i \leq t$, then τ starts with the same prefix as σ, so again $\tau \in \Gamma_n(V, \Omega)$. Finally, the case $i = k+1$ never occurs on our path.

Case 3. Assume e is of the type $\stackrel{x|}{\leadsto}$. As mentioned in the proof of Theorem 5, in all of these cases we have $x \in \{n-2, n-1\}$. This means that $x \notin V$, and τ starts with the same prefix as σ, so $\tau \in \Gamma_n(V, \Omega)$.

Case 4. Assume finally e is of the type $\stackrel{x\cup}{\leadsto}$. We must have $\{x\} = A_i$, for some $1 \leq i \leq t$. If $i \geq k+1$, then τ starts with the same prefix as σ, so $\tau \in \Gamma_n(V, \Omega)$.

If $i = k$, then τ starts with the prefix $(A_1 \mid \dots \mid A_{k-1})$. Since Ω is a closed prefix, Definition 27(1) implies that $(A_1 \mid \dots \mid A_{k-1}) \in \Omega$, hence $\tau \in \Gamma_n(V, \Omega)$. Finally, the case $i \leq k-1$ never occurs on our path.

Thus we have shown that $\tau \in \Gamma_n(V, \Omega)$ in all 4 cases. If Ω has no full prefixes, then by Lemma 1(1) $M(\Sigma)$ is a perfect matching and $\Gamma_n(V, \Omega)$ is conducting. If Ω has a single full prefix, then by Lemma 1(2) it has a near-perfect matching. Let γ denote the critical n-simplex. By an argument identical to the one used in the proof of Lemma 2, we see that there is a weakly semi-augmenting path from γ to any n-simplex σ of $\Gamma_n(V, \Omega)$, such that $O(\gamma) = O(\sigma)$. Using Proposition 10(1), we then conclude that $\Gamma_n(V, \Omega)$ is conducting.

Finally, when considering the row $(\{a, b, c\}, \{d, e, f\})$, with $a < b < c$, $d < e < f$, we get a family of prefixes with 3 full prefixes, hence we get 3 critical simplices. These simplices are $\tau_1 = (\{a, b\} \mid c \mid d \mid e \mid f)$, $\tau_2 = (a \mid b \mid c \mid d \mid e \mid f)$, $\tau_3 = (a \mid \{b, c\} \mid d \mid e \mid f)$. We now extend our matching M_Σ by matching τ_2 with τ_3, obviously these share an $(n-1)$-simplex, so the new matching is well-defined. All that remains is to find semi-augmenting paths from τ_2 and τ_3 to $\{\rho, \nu\}$. To find such a path starting from τ_2 we use the move (39), while to find such a path starting from τ_3 we use the move (40).

$$a \mid b \mid c \mid d \mid e \mid f \longrightarrow a \mid \{b, c\} \mid d \mid e \mid f \overset{eU}{\rightsquigarrow} a \mid \{b, c\} \mid d \mid \{e, f\} \tag{39}$$

$$a \mid \{b, c\} \mid d \mid e \mid f \longrightarrow a \mid b \mid c \mid d \mid e \mid f \overset{eU}{\rightsquigarrow} a \mid b \mid c \mid d \mid \{e, f\}. \tag{40}$$

In each case, we arrive at a non-critical simplex, from which we have already found a weakly semi-augmenting path to $\{\rho, \nu\}$. □

5.2 *Explicit Description of 21 Paths*

Let us now return to the labeling I, where we want to find a perfect matching on the set of all 1-monochromatic 5-simplices. Using the notation of Sect. 4.4, this set is a disjoint union of the sets $\mathcal{N}(\sigma)$, where σ ranges over all 5-simplices of $\chi(\Delta^5)$. The complete description of $\mathcal{N}(\sigma)$ has been obtained in Proposition 8.

At this point we would like to distinguish some of the 5-simplices of Γ_5. We set

$$\begin{aligned}
\mathcal{W}_1 &:= \{(a \mid [5] \setminus a) \mid a \in [5]\} \\
\mathcal{W}_2 &:= \{(\{a, b\} \mid [5] \setminus \{a, b\}) \mid a, b \in [5],\ a < b\} \\
\mathcal{W}_3 &:= \{(\{a, b, c\} \mid [5] \setminus \{a, b, c\}) \mid a, b, c \in [5],\ a < b < c\} \\
\mathcal{W} &:= \mathcal{W}_1 \cup \mathcal{W}_2 \cup \mathcal{W}_3 \cup ([5]).
\end{aligned}$$

Clearly, $|\mathcal{W}| = 42$, and we shall call the 5-simplices in $\mathcal{W}$ *exceptional simplices*.

For every $\sigma \in \Gamma_5$, $\sigma = (A_1 \mid \dots \mid A_t)$, we now fix a matching on $\mathcal{N}(\sigma)$. Specifically, we set $\Sigma := \mathrm{St}(\sigma)$, see Definition 21, and let $\widetilde{\mathcal{M}}_\sigma$ to be the standard matching $\mathcal{M}_\Sigma$, for all $\sigma \notin \mathcal{W}_3$. For $\sigma \in \mathcal{W}_3$, say $\sigma = (\{a,b,c\},\{d,e,f\})$, with $a < b < c$, $d < e < f$, we obtain $\widetilde{\mathcal{M}}_\sigma$ from the standard matching $\mathcal{M}_\Sigma$, by extending it by one more edge, matching $(a \mid b \mid c \mid d \mid e \mid f)$ with $(a \mid \{b,c\} \mid d \mid e \mid f)$, as in the proof of Theorem 6. We summarize what we know about these matchings in the following proposition.

Proposition 11 *The matching $\widetilde{\mathcal{M}}_\sigma$ is perfect for all $\sigma \notin \mathcal{W}$, and near-perfect for all $\sigma \in \mathcal{W}$.*

Proof If $\sigma = ([5])$, Proposition 6 implies that $\widetilde{\mathcal{M}}_\sigma$ is near-perfect. Assume now $\sigma \neq ([5])$. By Proposition 7 the matching $\widetilde{\mathcal{M}}_\sigma$ is perfect for all σ which do not appear in Table 4. Assume finally that σ does appear in Table 4. A line-by-line analysis shows that the corresponding prefix family Ω has full prefixes if and only if $\sigma \in \mathcal{W}$. It has one full prefix if $\sigma \in \mathcal{W}_1 \cup \mathcal{W}_2$, and it has three full prefixes if $\sigma \in \mathcal{W}_3$. The result now follows from Lemma 1, and the fact that two out of three critical (with respect to the standard matching) simplices for $\sigma \in \mathcal{W}_3$ have been matched in the extended matching $\widetilde{\mathcal{M}}_\sigma$. □

We let $\widetilde{\mathcal{M}}$ denote the matching on Γ_5 obtained as a union of all matchings $\widetilde{\mathcal{M}}_\sigma$. By Proposition 11 this matching has 42 critical simplices. We connect these 42 critical simplices by 21 non-intersecting augmenting paths; Table 5 contains the explicit description of the 21 paths. This will allow to use the construction from Definition 25 and to deform the matching $\widetilde{\mathcal{M}}$ to a perfect matching $\mathcal{M}$.

We define $R : \mathcal{W} \to \{\pm 1\}$, by setting $R(\sigma) := O(\tau)$, where $\tau \in \Gamma_5^2$ is the unique critical 5-simplex in $\mathcal{N}(\sigma)$. Note that $R([5]) = -1$, $R(\sigma) = 1$ for $\sigma \in \mathcal{W}_1 \cup \mathcal{W}_2$, and $R(\sigma) = -1$ for $\sigma \in \mathcal{W}_3$. We now need the following result.

Lemma 3 *Assume that we have a path Q in Γ_5 connecting two exceptional 5-simplices σ and τ, such that*

(1) $R(\sigma) = -R(\tau)$.
(2) Any 5-simplex γ on the path Q other than σ and τ is non-exceptional, and the corresponding graph $\mathcal{N}(\gamma)$ is conducting of the second type.

Then there exists an augmenting path T in Γ_5^2 such that

(1) T connects the critical simplex in $\mathcal{N}(\sigma)$ with the critical simplex in $\mathcal{N}(\tau)$.
(2) T lies entirely in $\bigcup_{\gamma \in Q} \mathcal{N}(\gamma)$.

Proof Assume $Q = (\gamma_1, \dots, \gamma_t)$, with $\gamma_1 = \sigma$, $\gamma_t = \tau$. For all $1 \leq i \leq t-1$, choose 5-simplices $\varphi_i \in \mathcal{N}(\sigma_i)$, $\psi_i \in \mathcal{N}(\sigma_{i+1})$ such that

(1) φ_i and ψ_i are adjacent in Γ_5^2.
(2) $O(\varphi_i) = R(\sigma)$.

To see that such simplices can be chosen let us pick α and β to be two adjacent 5-simplices in Γ_5, say $\alpha = (A_1 \mid \dots \mid A_p)$, $\beta = (B_1 \mid \dots \mid B_q)$, and $\beta = \mathcal{F}(\alpha, x)$. Without loss of generality we can assume that $A_p \subseteq B_q$; in fact, we could always

Table 5 Explicit description of the 21 paths

012	→	0 | 12	→	0 | 1 | 2	→	0 | 1	→	01
013	→	0 | 13	→	0 | 3 | 1	→	0 | 3	→	03
014	→	0 | 14	→	0 | 4 | 1	→	0 | 4	→	04
015	→	0 | 15	→	0 | 5 | 1	→	0 | 5	→	05
023	→	0 | 23	→	0 | 2 | 3	→	0 | 2	→	02
024	→	4 | 02	→	4 | 2 | 0	→	4 | 2	→	24
025	→	2 | 05	→	2 | 5 | 0	→	2 | 5	→	25
034	→	3 | 04	→	3 | 4 | 0	→	3 | 4	→	34
035	→	3 | 05	→	3 | 5 | 0	→	3 | 5	→	35
045	→	4 | 05	→	4 | 5 | 0	→	4 | 5	→	45
123	→	1 | 23	→	1 | 2 | 3	→	1 | 2	→	12
124	→	1 | 24	→	1 | 4 | 2	→	1 | 4	→	14
125	→	1 | 25	→	1 | 5 | 2	→	1 | 5	→	15
134	→	3 | 14	→	3 | 1 | 4	→	3 | 1	→	13
234	→	2 | 34	→	2 | 3 | 4	→	2 | 3	→	23
135	→	1 | 35	→	1 | 3 | 5	→	1 | 3	→	1
145	→	4 | 15	→	4 | 1 | 5	→	4 | 1	→	4
235	→	3 | 25	→	3 | 2 | 5	→	3 | 2	→	3
245	→	2 | 45	→	2 | 4 | 5	→	2 | 4	→	2
345	→	5 | 34	→	5 | 3 | 4	→	5 | 3	→	5
		0	→	012345				

assume that either $A_p = B_q$, or $A_p \cup x = B_q$, but we do not need such detail. Pick any $y \in A_p$. Then any 5-simplex $(\alpha \parallel C_1 \mid \dots \mid C_r \mid x)$ in Γ_5^2, such that $y \in C_1$ belongs to $\mathcal{N}(\alpha)$. Furthermore, it is adjacent to the 5-simplex $(\beta \parallel C_1 \mid \dots \mid C_r \mid x)$, which in turn belongs to $\mathcal{N}(\beta)$. There are 154 such 5-simplices, exactly half of which have orientation 1, so there is plenty of choice for the simplices φ_i and ψ_i.

Note that $O(\psi_i) = R(\tau)$ follows automatically from $O(\varphi_i) = -O(\psi_i)$ and $R(\sigma) = -R(\tau)$. Set ψ_0 to be the critical 5-simplex in $\mathcal{N}(\sigma)$, and set φ_t to be the critical 5-simplex in $\mathcal{N}(\tau)$. By Theorem 6 there exist semi-augmenting paths Q_i, for $1 \leq i \leq t$, such that each Q_i connects ψ_{i-1} with φ_i. Concatenating these paths will yield the desired path T. □

Since the paths presented in Table 5 are non-intersecting, it follows from Lemma 3 that the corresponding 21 augmenting paths in Γ_5^2 are non-intersecting as well. As we said above, this means that we can deform the matching $\widetilde{\mathcal{M}}$ to a perfect matching $\mathcal{M}$, which leads to the proof of our main theorem. Before we proceed with the proof, we need one last piece of terminology.

Definition 28 Assume $v = (\sigma_1 \parallel \dots \parallel \sigma_d)$ is a vertex of $\chi^d(\Delta^n)$. The **predecessor** of v is the vertex of the simplex $\tau = (\sigma_1 \parallel \dots \parallel \sigma_{d-1})$ of $\chi^{d-1}(\Delta^n)$, which has the same color as v; we shall denote it by Pred (v).

Proof (of Theorem 1) Let us describe an assignment L of values 0 and 1 to vertices of $\chi^3(\Delta^5)$. Let $w = (\sigma_1 \parallel \sigma_2 \parallel \sigma_3)$ be such a vertex, and set $\tau := (\sigma_1 \parallel \sigma_2)$, τ is a simplex of $\chi^2(\Delta^5)$. If $\dim \tau \leq 3$, then we set $L(w) := I(\text{Pred}\,(w))$. It is a compliant labeling, since the initial labeling I was chosen to be compliant. We shall call such the assignment $I(\text{Pred}\,(-))$ the *default value*.

Assume now $\dim \tau = 4$. If there is only one 5-simplex of $\chi^2(\Delta^5)$ containing τ, or if the two 5-simplices containing τ are not matched by $\mathcal{M}$, then we let $L(w)$ be the default value. Otherwise, we set $L(w) := 0$.

Finally, consider the case $\dim \tau = 5$. If τ is not a 1-monochromatic 5-simplex in I, then we let $L(w)$ be the default value. Otherwise, let $\mathcal{F}(\tau, c)$ be the 5-simplex which is matched to τ by $\mathcal{M}$. We set

$$L(w) := \begin{cases} 1, & \text{if } c = C(\sigma_3); \\ 0, & \text{otherwise.} \end{cases} \tag{41}$$

We shall now see that the obtained assignment L has no monochromatic 5-simplices. Let $\sigma = (\sigma_1 \parallel \sigma_2 \parallel \sigma_3)$ be a 5-simplex of $\chi^3(\Delta^5)$, and let $\tau = (\sigma_1 \parallel \sigma_2)$ be the corresponding 5-simplex of $\chi^2(\Delta^5)$ containing σ. If τ is not 1-monochromatic in I, then by our construction, all values of L on vertices of τ are the default ones, which means that the vertices of σ have the same values under L as vertices of τ of the same color under I. In particular, σ is not monochromatic.

Assume now τ is 1-monochromatic under I, and assume τ is matched to $\mathcal{F}(\tau, c)$ under $\mathcal{M}$. By our construction the vertex of σ which has color c is assigned the value 1, hence σ cannot be 0-monochromatic. Let $d \in [n]$, such that $c \neq d$, and d is almost maximal with respect to σ_3; such an element exists due to Remark 1. Let w be the color of σ with the color d. By our construction $L(w) = 0$, hence σ is not 1-monochromatic either. □

6 The Distributed Protocol Solving Weak Symmetry Breaking for Six Processes in Three Rounds

In this section we describe the application of our main theorem in theoretical distributed computing and present an explicite distributed protocol solving Weak Symmetry Breaking for six processes in three rounds. To start with, we record the information obtained in Sect. 5 in a table, which we call **ExSimp**. Namely, for each path from Table 5 use Lemma 3 to produce an augmenting path Q in Γ_5^2. For every simplex $(\sigma \parallel \tau)$ from Q we add an entry (σ, τ, c) to **ExSimp**, where c is the color of the edge from Q which is adjacent to $(\sigma \parallel \tau)$, and which does not belong to original matching $\widetilde{\mathcal{M}}$ (it does belong to the final matching $\mathcal{M}$).

The formal protocol is given in Fig. 5. The verbal description is as follows. Let p be the id of the process running the protocol.

```
protocol WSB6 (input: `$p$`)
`{\bf write}`; `{\bf read}`; `{\bf write}`; `{\bf read}`;
assume the current view is `$(\sigma'\ds\tau')$`
if `$\car(\sigma')\neq[n]$` and `$(\sigma'\ds\tau')$` is
not in Table `\ref{table:exv2}` then `{\bf decide 0}`
`{\bf write}`; `{\bf read}`;
assume the current view is `$(\sigma\ds\tau\ds\gamma)$`
if `$|\car(\gamma)|\leq n-1$` then `{\bf decide~1}`
if `$\car(\gamma)=[n]$` then `$\xi:=\tau$`
else assume `$\tau=(A_1\s\dots\s A_t\ds B_1\s\dots\s B_t)$`,
set `$q:=[n]\setminus(B_1\cup\dots\cup B_t)$`
        if `$q\notin A_1\cup\dots\cup A_t$` then `{\bf decide~1}`
        else pick `$1\leq k\leq t$` such that `$q\in A_k$`
        set `$\xi:=(A_1\s\dots\s A_{k-1}\s q\s A_k\sm q\s
        A_{k+1}\s\dots\s A_t)$`
if `$p=\match(\sigma,\xi)$` then `{\bf decide~1}`
else `{\bf decide~0}`

function `\match` (input: `$\xi,\sigma$`)
if `$(\xi,\sigma,c)$` is in `\exsimp` then return `$c$`
else return `$\lev(\xi,\st(\sigma))$`
```

Fig. 5 Protocol solving Weak Symmetry Breaking for six processes in three rounds

Step 1. Execute two rounds of write-read sequence. Assume that the view of the process p after two rounds is $v = (\sigma \parallel \tau)$. If carrier $(\sigma) = [n]$, then proceed to Step 2. Else we have carrier $(\sigma) \neq [n]$, in which case we check if v is one of the exceptional second round views from Definition 17, see Table 2. If it is an exceptional second round view from that table then proceed to Step 2, else **decide 0** and stop.

Step 2. Execute one more round of write-read. Assume that the view of the process p after three rounds is $(\sigma \parallel \tau \parallel \gamma)$.

- If $|\text{carrier}(\gamma)| \leq n-1$, then **decide 1** and stop.
- If carrier $(\gamma) = [n]$, set $\xi := \tau$ and proceed to Step 3.
- If $|\text{carrier}(\gamma)| = n$, assume $\tau = (A_1 \mid \dots \mid A_t \parallel B_1 \mid \dots \mid B_t)$. We have $|B_1 \cup \dots \cup B_t| = n$, so there exists $q \in [n]$, $q \neq p$, such that $B_1 \cup \dots \cup B_t = [n] \setminus q$. If $A_1 \cup \dots \cup A_t = [n] \setminus q$, then **decide 1** and stop. Else, there exists $1 \leq k \leq t$, such that $q \in A_k$. Now, set

$$\xi := (A_1 \mid \dots \mid A_{k-1} \mid q \mid A_k \setminus q \mid A_{k+1} \mid \dots \mid A_t).$$

Step 3. If $p = \textbf{Match}(\sigma, \xi)$, then **decide 1**, else **decide 0** and stop.

Here is the procedure **Match**, which has two ordered set partitions σ, ξ of $[n]$ as input and returns a value from $[n]$.
`Procedure` **Match**.

If (σ, ξ, c) is in **ExSimp**, set output $:= c$, else set

$$\text{output} := \text{level}\,(\xi, \text{St}\,(\sigma)).$$

We note that the views from Table 2 can also be described verbally. For example, the view

a	a
a	a

means that process a has only seen itself both in the first and in the second round. The view

a, b, c	a, b
a, b	a

has a more lengthy interpretation. It means, that the process a has seen one other process b in the second round, whose id was larger than that of a. Furthermore, a has seen three processes in the first round, including itself and b. The third process it has seen has id larger than that of both a and b. Finally, the views of b and a in the first round were the same (a knows that after the second round). We leave it to the reader to provide similar interpretations for other views listed in Table 2.

7 Weak Symmetry Breaking Cannot be Solved in One Round

Currently, there are no lower bounds for the number of rounds needed to solve the Weak Symmetry Breaking task. Here we give the first such lower bound, stating that if Weak Symmetry Breaking can be solved at all, one would need at least two rounds to do that.

Theorem 7 *We have $sb(n) \geq 2$, for all $n \geq 2$. In other words, the Weak Symmetry Breaking task cannot be solved in one round for any value of $n \geq 2$.*

Proof Let $\mathcal{V}^n$ denote the set of pairs of integers $\{(r, p) \mid 1 \leq p \leq r \leq n\}$. Let us assume that we have a distributed protocol running exactly one round and having output in the set $\{0, 1\}$. After the execution of this protocol each process ρ has seen a certain number of other processes and has determined the relative position of its own id among the id's of the processes it has seen. Such an information can be encoded by a pair of integers (r, p), such that $1 \leq r \leq n$, this is the number of processes, including itself, which ρ has seen, and $1 \leq p \leq r$, which is the relative position of the id of ρ. Hence, the decision function for such a protocol is precisely a function $\delta : \mathcal{V}^n \to \{0, 1\}$.

Assume now we are given such a decision function δ solving the Weak Symmetry Breaking. Let us break up $\mathcal{V}^n = \mathcal{V}_1^n \cup \cdots \cup \mathcal{V}_n^n$, where $\mathcal{V}_i^n = \{(i,p) \mid 1 \leq p \leq i\}$. Let k be the maximum index such that $\delta_{\mathcal{V}_i^n}$ is not surjective. Since $\delta_{\mathcal{V}_1^n}$ is not surjective, the number k is well-defined. Without loss of generality we can assume that $\delta(k,1) = \delta(k,2) = \cdots = \delta(k,k) = 0$. For all $k+1 \leq i \leq n$ we let α_i be any number such that $\delta(i,\alpha_i) = 0$. Such a number exists for all $k+1 \leq i \leq n$, since by our construction $\delta_{\mathcal{V}_i}$ is surjective for these values of i.

We shall now construct an execution E of the protocol after which the decision function will assign the value 0 to all processes, which of course contradicts to the claim that the protocol solves Weak Symmetry Breaking. The execution E will start with a simultaneous activation of k processes, after which the rest of the processes will activate one at a time. In total the execution E has $n-k+1$ rounds.

More specifically, we can find numbers $\xi(1),\ldots,\xi(n)$, such that the execution E is described as follows:

(1) The id's of the processes participating in the first round are $\xi(1),\ldots,\xi(k)$.
(2) For each $r = 2,\ldots,n-k+1$, the id of the process participating in the round r is $\xi(k+r-1)$.

These numbers can be determined constructively as follows. To start with, the numbers $\xi(1),\ldots,\xi(n)$ are initialized to be 0. We then proceed in $n-k+1$ steps. In the first step, we set $\xi(i) := i$ for all $1 \leq i \leq k$. Next, for i running from $k+1$ to n, repeat the following: set $\xi(i) := \alpha_i$, and increase by 1 each $\xi(j)$, such that $j < i$ and $\xi(j) \geq \alpha_i$. The result after $n-k+1$ steps is then taken as the final output.

It easy to see that each process activated in round r, for $2 \leq r \leq n-k+1$, will see $k+r-1$ processes, including itself, and that the relative position of his id in what he sees will be α_{k+r-1}. □

Acknowledgements We would like to thank Maurice Herlihy for useful discussions as well as the anonymous referee for multiple suggestions which have improved the presentation.

References

1. H. Attiya, A. Castañeda, A non-topological proof for the impossibility of k-set agreement, in *Stabilization, Safety, and Security of Distributed Systems*. Lecture Notes in Computer Science, vol. 6976 (2011), pp. 108–119
2. H. Attiya, A. Castañeda, M. Herlihy, A. Paz, Upper bound on the complexity of solving hard renaming, in *PODC* (2013), pp. 190–199
3. H. Attiya, F. Ellen, *Impossibility Results for Distributed Computing*. Synthesis Lectures on Distributed Computing Theory (Morgan & Claypool, San Rafael, 2014)
4. H. Attiya, A. Paz, Counting-based impossibility proofs for renaming and set agreement, in *26th International Symposium, DISC 2012, Salvador*. Lecture Notes in Computer Science, vol. 7611 (2012), pp. 356–370
5. H. Attiya, J. Welch, *Distributed Computing: Fundamentals, Simulations, and Advanced Topics*. Wiley Series on Parallel and Distributed Computing, 2nd edn. (Springer, Heidelberg, 2004), 432 pp.

6. A. Castañeda, S. Rajsbaum, New combinatorial topology upper and lower bounds for renaming, in *Proceeding of the 27th Annual ACM Symposium on Principles of Distributed Computing* (ACM, New York, 2008), pp. 295–304
7. A. Castañeda, S. Rajsbaum, New combinatorial topology bounds for renaming: the lower bound. Distrib. Comput. **25**(5), 287–301 (2010)
8. A. Castañeda, S. Rajsbaum, New combinatorial topology bounds for renaming: the upper bound. J. ACM **59**(1), Article No. 3, 49 pp. (2012)
9. M. Herlihy, D.N. Kozlov, S. Rajsbaum, *Distributed Computing Through Combinatorial Topology* (Elsevier/Morgan Kaufmann, Waltham, 2014), 336 pp.
10. M. Herlihy, N. Shavit, The topological structure of asynchronous computability, J. ACM **46**(6), 858–923 (1999)
11. D. Imbs, S. Rajsbaum, M. Raynal, The universe of symmetry breaking tasks. *Structural Information and Communication Complexity*. Lecture Notes in Computer Science, vol. 6796 (Springer, Heidelberg, 2011), pp. 66–77
12. D.N. Kozlov, *Combinatorial Algebraic Topology*. Algorithms and Computation in Mathematics, vol. 21 (Springer, Berlin/Heidelberg, 2008), pp. XX, 390, 115 illus.
13. D.N. Kozlov, Chromatic subdivision of a simplicial complex. Homology Homotopy Appl. **14**(2), 197–209 (2012)
14. D.N. Kozlov, *Witness structures and immediate snapshot complexes*, 26 pp. (preprint, 2014). arXiv:1404.4250
15. D.N. Kozlov, Topology of immediate snapshot complexes. Topol. Appl. **178**, 160–184 (2014)
16. D.N. Kozlov, Topology of the view complex. Homology Homotopy Appl. **17**(1), 307–319 (2015). doi:http://dx.doi.org/10.4310/HHA.2015.v17.n1.a15
17. D.N. Kozlov, *Weak symmetry breaking and abstract simplex paths*. Math. Struct. Comput. Sci. **25**(6), 1432–1462 (2015)
18. D.N. Kozlov, *Structure Theory of Flip Graphs with Applications to Weak Symmetry Breaking*, 42 pp. Preprint, arXiv:1511.00457
19. D.N. Kozlov, *All Binomial Identities Are Orderable*, 5 pp. Preprint, arXiv:1603.08743

Cryptomorphisms for Abstract Rigidity Matroids

Emanuele Delucchi and Tim Lindemann

Abstract This note contributes to the structure theory of abstract rigidity matroids in general dimension. In the spirit of classical matroid theory, we prove several cryptomorphic characterizations of abstract rigidity matroids (in terms of circuits, cocircuits, bases, hyperplanes). Moreover, the study of hyperplanes in abstract rigidity matroids leads us to state (and support with significant evidence) a conjecture about characterizing the class of abstract rigidity matroids by means of certain "prescribed substructures". We then prove a recursive version of this conjecture. (This is an extended version of the second author's bachelor thesis at University of Bremen.).

Keywords Abstract rigidity matroids • Cryptomorphisms • Matroids • Rigidity

1 Introduction

Rigidity matroids are combinatorial structures introduced by Graver in [2] which, roughly speaking, model some aspects of the theory of rigid bar-link framework (whose combinatorial study goes back to work of Laman [5]).

More precisely, a *framework* is a (finite) graph (V, E) together with a straight-line embedding into $\mathbb{R}^d$, and it is called *rigid* if the only continuous motions of the embedding of the vertices which fix the distance of adjacent vertices are composites of rotations and translations. Testing rigidity of a framework involves thus checking for nontrivial solutions of a system of linear equations, every equation corresponding to an edge of the graph. The rows of this matrix define a *matroid* on the ground set E (see Definition 1). This matroid depends of course on the chosen embedding, but it always satisfies some abstract properties (usually defined in terms of the closure operator, see Definition 5). An *abstract rigidity matroid* is any matroid

E. Delucchi (✉)
Department of Mathematics, University of Fribourg, 1700 Fribourg, Switzerland
e-mail: emanuele.delucchi@unifr.ch

T. Lindemann
Department of Mathematics, University of Bremen, 28334 Bremen, Germany
e-mail: s_hzvazy@uni-bremen.de

F. Callegaro et al. (eds.), *Configuration Spaces*, Springer INdAM Series 14,
DOI 10.1007/978-3-319-31580-5_8

defined on the set of edges of a graph which satisfies these additional properties. For example, the 'usual' cycle matroid of the given graph is in fact an abstract rigidity matroid, but abstract rigidity matroids are a much bigger class. For a comprehensive introduction to the combinatorial study of rigidity of frameworks we point to the book of Graver et al. [3].

Here, we approach the subject from the point of view of pure matroid theory, which we briefly introduce in Sect. 1, and focus on two main subjects.

First, the fact that classical matroid theory features a web of equivalent definitions—so-called *cryptomorphic* approaches, see Remark 1—has been recognized as one of its main theoretical strength since at least the seminal work of Crapo and Rota [1]. In Sect. 2, we develop the theory of abstract rigidity matroids in a corresponding way and derive different, equivalent characterizations, each reflecting one of the classical approaches to matroid theory.

Second, we focus on structural aspects and ask whether being an abstract rigidity matroid can be identified as a "structural" matroid property, e.g., whether the class of abstract rigidity matroids can be characterized through some "prescribed substructures". This is a quite standard organizational process in graph theory and matroid theory: in particular, prominent open questions and important results about characterizing certain classes of matroids through "excluded minors" abound. In Sect. 3, we prove a characterization of abstract rigidity matroids in terms of "prescribed classes of hyperplanes" in all restrictions (Theorem 5) and conjecture a nonrecursive characterization of abstract rigidity matroids via the requirement of the existence of a prescribed class of hyperplanes. We offer some stringent evidence towards this conjecture (e.g., Propositions 3 and 4) as well as in support of the general significance of the suggested substructures (e.g. through Proposition 5).

2 Review

In this section we will review some basic definitions and results about abstract rigidity matroids. In particular, we will state Viet-Hang Nguyens's combinatorial characterization of abstract rigidity matroids (Proposition 1). We will assume the reader familiar with the basics of matroid theory, and suggest Oxley's textbook [7] for an introduction to the subject.

2.1 Graphs

We consider a graph to be a pair (V, E) consisting of a finite set V and a set E of two-element subsets of V (thus, our graphs will not have loops nor parallel edges). Given two vertices $u, v \in V$ we will often use uv as a shorthand for $\{u, v\}$. For any finite set W we will let $K(W) := \{uv \mid u, v \in W, u \neq v\}$ so that $(W, K(W))$ is the complete graph on the vertex set W. Given a natural number n we will use the

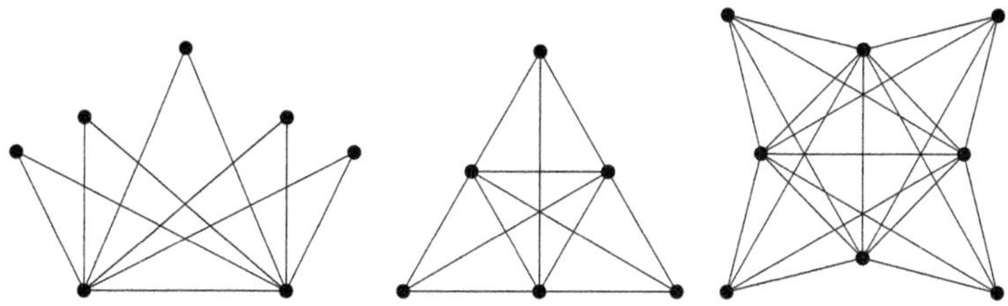

Fig. 1 Examples of some graphs of the form bigstar(V'), where $m := |V'|$

notation K_n to refer to any (and thus every) $K(W)$ with $|W| = n$. For any $E \subseteq K(V)$ let $V(E) := \{u \in V \mid uw \in E \text{ for some } w \in V\}$; $V(E)$ is called the *support* of the edge set E.

In the following, we will often consider graphs on a fixed vertex set and will then, if no confusion can occur, refer to edge-sets as "graphs".

- For $v \in V$ let

$$\text{star}(v) := \{vu \in K(V) \mid u \neq v\}.$$

  Sets of the form star(v) with $m - 1$ arbitrary edges deleted are simply called "vertex stars minus $m - 1$ edges".
- For a set $V' \subseteq V$ let

$$\text{bigstar}(V') := K(V) \backslash K(V \backslash V').$$

  Thus, the set bigstar(V') is the edge-set of the graph where every vertex v_0 not in V' is attached to $K(V')$ by the family of edges $\{v_0 v' \mid v' \in V'\}$ (one edge to every vertex in V'). See Fig. 1 for some illustrations.

2.2 *Matroids*

Matroid theory finds its origins in the attempt, by Hassler Whitney, to define combinatorial structures abstracting some properties of linear independency in vectorspaces. For instance, it is an easy check that the set of bases of a vectorspace (say, over a finite field) satisfies Definition 1 below.

Our goal in this introductory paragraph is to define matroids and some of the related terminology, and to explain what matroid theorists mean by *cryptomorphism* (see Remark 1). Indeed the word may sound unusual, but the concept is one that is both useful in applications and—most importantly for us here—as a theoretical feature which was singled out as one of the main aspects of interest of matroid theory ever since at least Crapo and Rota's seminal treatise [1].

Definition 1 Let S be a finite set and let $\mathcal{B}$ be a collection of subsets of S which fulfills

(i) $\mathcal{B} \neq \emptyset$.
(ii) If $B_1, B_2 \in \mathcal{B}$, then $|B_1| = |B_2|$.
(iii) For all $B_1, B_2 \in \mathcal{B}$, $x \in B_1 \backslash B_2$, there exists $y \in B_2 \backslash B_1$, such that $(B_1 \backslash \{x\}) \cup \{y\} \in \mathcal{B}$.

Then the pair $\mathcal{M} = (S, \mathcal{B})$ is called a *matroid* and $\mathcal{B}$ is called the collection of *bases* of $\mathcal{M}$. A set $I \subseteq S$ is called *independent* if there is a basis $B \in \mathcal{B}$ such that $I \subseteq B$, otherwise *dependent*. An inclusion-minimal dependent subset of S is called a *circuit* of $\mathcal{M}$. A maximal set which does not contain a basis is called a *hyperplane*.

The *closure* of a set $A \subseteq S$, denoted by $\sigma_{\mathcal{B}}(A)$, is the intersection of all hyperplanes containing A (if no such hyperplane exist, the closure is defined as S).

A direct consequence of the basis axioms is that the set of complements of bases does also fulfill these axioms; the matroid $\mathcal{M}^* := (S, \mathcal{B}^*)$ is called *dual* to $\mathcal{M}$, and $(\mathcal{M}^*)^* = \mathcal{M}$. If a subset of S is a circuit (or a hyperplane, or a basis etc) of $\mathcal{M}^*$ one says that it is a *cocircuit* (resp. cohyperplane, cobasis) of $\mathcal{M}$.

We will have use for the following standard material, which can be found, e.g., in [7].

Lemma 1 (Prop. 2.1.11, 2.1.21, 1.4.11 of [7]) *Let $\mathcal{M}$ be a matroid on a finite set S.*

(i) If C is a circuit and C' is cocircuit, then $|C \cap C'| \neq 1$.
(ii) H is a hyperplane if and only if $S \backslash H$ is a cocircuit.
(iii) $F \subseteq S$ is closed if and only if, for every circuit C, $|C \backslash F| \leq 1$ implies $C \subseteq F$.

An easy check shows that any function of the form $\sigma_{\mathcal{B}}(\cdot)$ satisfies the following definition.

Definition 2 Let S be a finite set and let $\sigma : 2^S \to 2^S$ be a function such that for all $A, B \subseteq S$:

(i) $A \subseteq \sigma(A)$.
(ii) $A \subset B$ implies $\sigma(A) \subseteq \sigma(B)$.
(iii) $\sigma(\sigma(A)) = \sigma(A)$.
(iv) If $x, y \in S$ and $x \in \sigma(A \cup \{y\})$, then $y \in \sigma(A \cup \{x\})$.

Then $\sigma(\cdot)$ is called a *matroid closure operator*, and any set $A \subseteq S$ is called *closed* (or *flat*) if $\sigma(A) = A$.

Theorem 1 (See Section 1.4 of [7]) *The function $\mathcal{B} \mapsto \sigma_{\mathcal{B}}(\cdot)$ is a bijection between the set of families $\mathcal{B}$ satisfying Definition 1 and the set of functions $\sigma(\cdot)$ satisfying Definition 2.*

Remark 1 (Cryptomorphisms) In the parlance of matroid theory, Theorem 1 is referred to as a *cryptomorphism* between the definition of matroids in terms of bases and the definition in term of closure operator. In fact, just like in Definition 2, one can isolate some distinguishing properties of the family of circuits (or of cocircuits, or hyperplanes, etc.) of a matroid and prove that any family of sets satisfying those formal properties can be obtained as the set of circuits (or… etc.) of a matroid defined e.g. as in Definition 1.

We close this short presentation of matroids by defining two more concepts we will have use for later.

Definition 3 Let $\mathcal{M}$ be a matroid on the set S and consider $T \subseteq S$. The *restriction* of $\mathcal{M}$ to T, written $\mathcal{M}[T]$, is the matroid with ground set T and closure operator defined for each $X \subseteq T$ as $\sigma(X) := \sigma_{\mathcal{M}}(X) \cap T$.

Definition 4 Let $\mathcal{M} = (S, \mathcal{B})$ be a matroid. The *rank* of any $X \subseteq S$ is

$$\mathrm{rk}(X) := \max |\{B \cap X \mid B \in \mathcal{B}\}|,$$

i.e., the size of the biggest independent set contained in X.

Lemma 2 (See Proposition 1.7.8 of [7]) *Let $\mathcal{M} = (S, \sigma)$ be a matroid, and $T \subseteq S$ be a closed set of $\mathcal{M}$. Consider any unrefinable chain $T = T_0 \subsetneq T_1 \subsetneq \ldots \subsetneq Tj = S$ of closed sets. Then*

$$\mathrm{rk}(S) - \mathrm{rk}(T) = j.$$

2.3 Abstract Rigidity Matroids

We now are ready to introduce the main character of this note. As these structures are less classical than graphs or matroids, we will go into some more detail. Notice that the ground set of an abstract rigidity matroid is the set of edges of a complete graph. These edges can be thought of as bars joined at the graph's vertices by means of freely movable joints – in fact, although perhaps not evident from our abstract point of view, 'rigidity' of a set of edges is meant to be related to (and indeed comes from) the concept of rigidity of such a bar-and-joints framework in m-space. In this intuition, taking the closure of a certain set of elements of an abstract rigidity matroid (i.e., a set of edges of the underlying graph) corresponds to increasing the given set by all edges whose presence would not change the degree of rigidity (i.e., the degrees of freedom) of the associated framework.

Definition 5 Let V be a finite set and let $\mathcal{A} = (K(V), \sigma)$ be a matroid on $K(V)$ with closure operator σ. A set $E \subseteq K(V)$ is then called **rigid** (with respect to $\mathcal{A}$) if $\sigma(E) = K(V(E))$. Let $m \in \mathbb{N}_{>0}$; the matroid $\mathcal{A}$ is called a **m-dimensional abstract rigidity matroid** when

C1. if $E, F \subseteq K(V)$ and $|V(E) \cap V(F)| < m$, then $\sigma(E \cup F) \subseteq K(V(E)) \cup K(V(F))$
C2. if $E, F \subseteq K(V)$ are rigid and $|V(E) \cap V(F)| \geq m$, then $E \cup F$ is rigid.

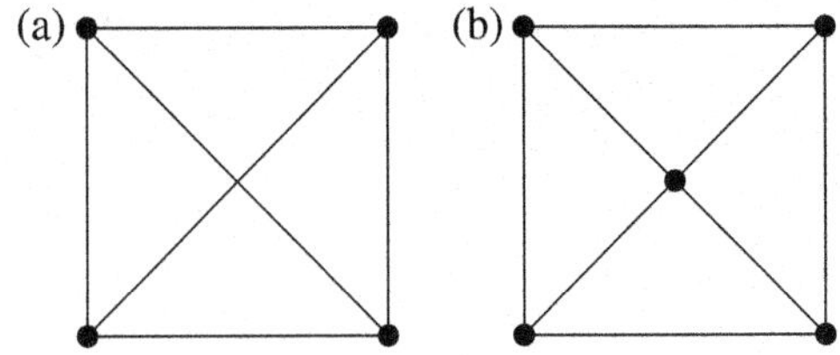

Fig. 2 (**a**), (**b**) Circuits of a 2-dimensional abstract rigidity matroid on five vertices

Roughly speaking condition (C1) says that edge-sets which do not share enough common vertices cannot unite to a rigid set, while (C2) says that rigid sets which are connected through enough vertices form a rigid union. Note that (C1) also states that $\sigma(E) \subseteq K(V(E))$ for all $E \subseteq K(V)$. As another consequence of (C1), we have that an edge set E is independent whenever $|V(E)| \leq m + 1$.

Example 1 Let $V := \{1, 2, 3, 4, 5\}$. Then, the set $K(V)$ of our rigidity matroid has $\binom{5}{2}$ elements, which we visualize as the set of edges of a complete graph K_5. Consider now any[1] 2-dimensional abstract rigidity matroid on V. By the last observation, we see immediately that every triangle (i.e., any set of edges forming a 3-cycle) in K_5 must be rigid and independent, and that every 'triangle with a dangling edge' is also independent as well as, by (C1), closed. However, by (C2), the union of two triangles along an edge is rigid, and thus its closure contains the last edge of the K_4 it spans. We conclude that every K_4 must be a circuit (see Fig. 2a).

To see that the graph in Fig. 2b – which we call $G_{(b)}$ – is also a circuit start again from the rigid set given by two triangles sharing an edge and attach a third triangle along another edge: again by (C2) we must have a rigid set spanning the whole K_5, thus the set of edges of the depicted graph is dependent. We leave to the reader the check that removing any edge from $G_{(b)}$ yields an independent set.

We close this example by noticing that there cannot be any other circuits. In fact, we know that every dependent set E must have $|V(E)| \geq 4$. Because K_4s are circuits, they are the only closed sets on four vertices. Closed sets on five vertices which do not contain $G_{(b)}$ as a subgraph must then arise by deleting an edge from $G_{(b)}$ and adding (at least) another. But if we delete one of the edges incident with the vertex of degree four then there are only two ways to add a new edge: one creates a K_4, the other a new subgraph of the form $G_{(b)}$—thus no new circuits can be created this way. Similarly, removing an edge between trivalent vertices allows (up to symmetry) for only one possible addition of another edge, which will create a K_4 subgraph. Thus, no new circuits on five vertices this way either.

We conclude that the set of circuits of any 2-dimensional abstract rigidity matroid on five vertices must be the family of all edge-sets of the form given in Fig. 2. Since this family satisfies circuit elimination we see that it *is* the set of circuits of *the unique* well-defined abstract rigidity matroid of dimension 2 on five vertices.

[1]In fact, there exists only one abstract 2-dimensional rigidity matroid on five vertices.

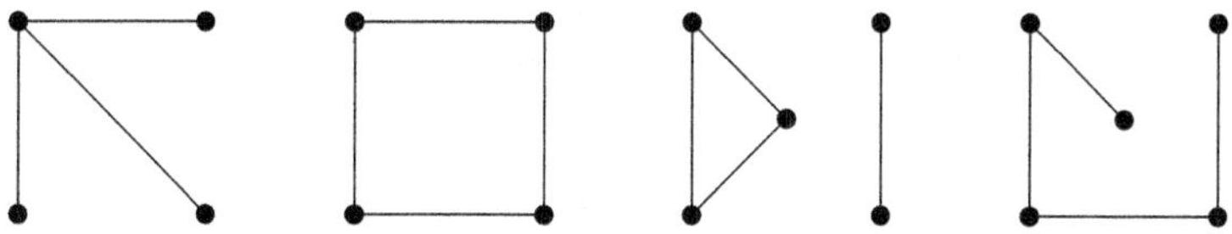

Fig. 3 The set of cocircuits of the 2-dimensional abstract rigidity matroid on five vertices

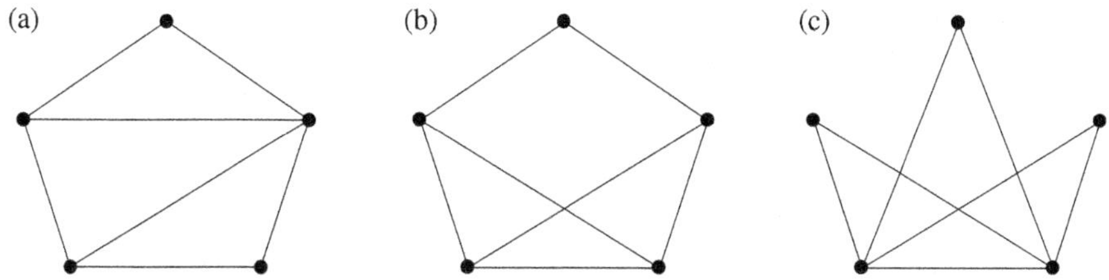

Fig. 4 (**a**), (**b**), (**c**) Bases of the 2-dimensional abstract rigidity matroid on five elements

Remark 2 The reasoning done with the graph of Fig. 2 can be generalized to a general m-dimensional abstract rigidity matroid $\mathcal{A}$ over a set V, to prove that every $K(V')$ with $|V'| = m + 2$ is a circuit. Moreover, with the notation of Sect. 2.1 by a similar argument we have that every vertex star minus $m - 1$ edges is a cocircuit of $\mathcal{A}$.

Example 2 The cocircuits of the matroid defined in Example 1 are depicted in Fig. 3: we see that the 'vertex stars minus 1 edge' are cocircuits.

Remark 3 For the sake of simplicity, if not otherwise specified, throughout the text we will consider m-dimensional rigidity matroids on $K(V)$ where $m \in \mathbb{N}_{>0}$ and $|V| \geq m + 1$ be fixed. Indeed one easily checks that every abstract rigidity matroid on $K(V)$ with $|V| \leq m$ is trivial in the sense that every edge-set would be an independent set in such a matroid.

Example 3 In the situation of Example 1 one easily checks, using (C1), that the edge-sets of the graphs depicted in Fig. 4 are independent. Moreover, they can be seen to be rigid: this is a direct application of (C2) for Fig. 4a and Fig. 4c, while for Fig. 4b remember that two triangles sharing an edge define a rigid set, thus the closure of the edge-set of Fig. 4b contains that of Fig. 4a, which is rigid. These three types of edge-sets are then bases of our 2-dimensional abstract rigidity matroid. One can check that there are no other bases.

Definition 6 An edge-set is said to fulfill *Laman's condition* in dimension m, if for all $F \subseteq E$ with $|V(F)| \geq m$, we have $|F| \leq m|V(F)| - \binom{m+1}{2}$.

In fact every independent set in a m-dimensional abstract rigidity matroid fulfills Laman's condition.

Example 4 We can easily check that the bases in Fig. 4 fulfill (sharply!) Laman's condition.

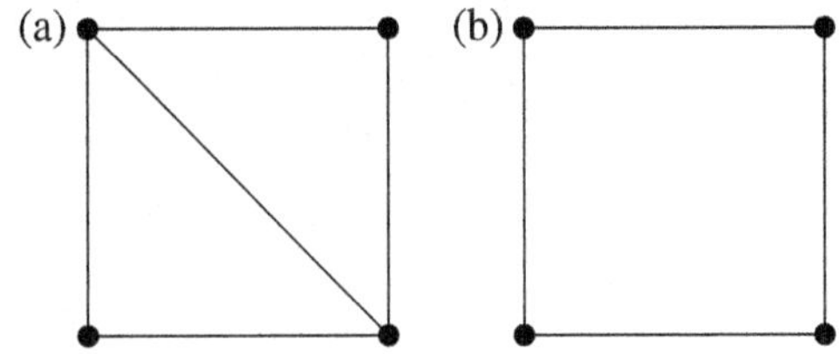

Fig. 5 (**a**) Independent and rigid. (**b**) Independent, not rigid

Lemma 3 ([3, Lemma 2.5.6.]) *Let $U \subseteq V$ and let $\mathcal{A}$ be a m-dimensional abstract rigidity matroid on $K(V)$. Then*

$$r(K(U)) = \begin{cases} \binom{|U|}{2} & \text{if } |U| \leq m+1 \\ m|U| - \binom{m+1}{2} & \text{if } |U| \geq m+1. \end{cases}$$

Therefore every m-dimensional abstract rigidity matroid on $K(V)$ is of rank $m|V| - \binom{m+1}{2}$, and rigid edge-sets on at least $m+1$ vertices have at least $m|V| - \binom{m+1}{2}$ edges.

Example 5 As an illustration of the fact that independent sets can be both rigid or nonrigid consider, in the situation of Example 1, the edge-sets in Fig. 5. Both edge-sets are independent because they do not contain any circuit. The graph in (a) is rigid, since it spans a K_4-circuit. The graph in (b) is not rigid because it is closed (one checks that it does not contain any set of the form 'circuit minus one element').

Definition 7 An edge-set which is both rigid and independent in an m-dimensional rigidity matroid is called *m-isostatic*.

A result in rigidity theory states that 2-isostatic sets are at least 2-vertex connected [3, Exercise 4.7]. Our Proposition 5 will prove that in fact every rigid set in a m-dimensional rigidity matroid is m-vertex connected.

Note that every rigid edge-set must have an isostatic subset, so rigidity of a graph can be seen as related to connectivity. This is emphasized by the following result.

Lemma 4 (Theorem 3.11.8. of [3]) *There is only one* 1*-dimensional abstract rigidity matroid on $K(V)$, the cycle matroid on $K(V)$.*

However, for $m > 1$, there can be many different m-dimensional abstract rigidity matroids on $K(V)$. This makes the theory interesting, and is investigated in more detail for example in [3, 8].

Here we will only briefly recall some of the structural aspects of abstract rigidity matroids in general dimension, especially as related to the problem of characterizations alternative to Definition 5 (e.g., towards "cryptomorphisms"—cf. Remark 1)

Definition 8 Let $E \subseteq K(V)$ be an edge-set, let $v_1, \dots v_k \in V(E)$ and $w \in V \backslash V(E)$. Then the set $E \cup \{v_1 w, \dots v_k w\}$ is called a ***k*-valent 0-extension** of E.

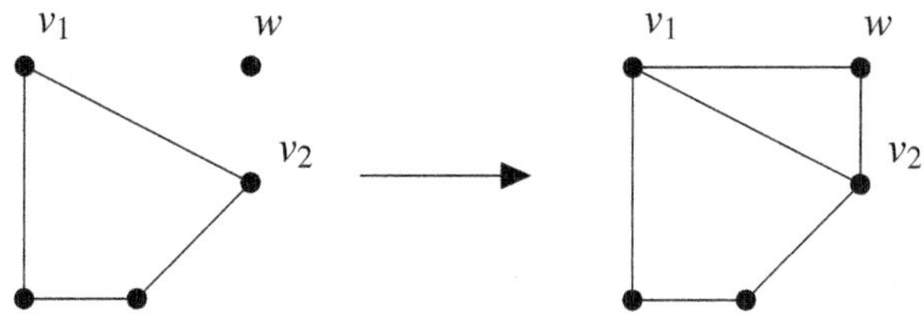

Fig. 6 An example of a 2-valent-0-extension

Intuitively, we are here attaching a vertex to a given edgeset by means of exactly k new edges (see Fig. 6 for an example).

Lemma 5 (Lemma 0.1 of [4]) *Let $\mathcal{A}$ be a matroid on $K(V)$ such that all vertex stars minus $m-1$ edges are cocircuits. Then:*

(i) No circuit of $\mathcal{A}$ contains a vertex of valence less than $m+1$.
(ii) For $k \leq m$, every k-valent 0-extension of an independent edge-set is independent.
(iii) For $k \geq m$, every k-valent 0-extension of a rigid edge-set is rigid if (C2) holds in $\mathcal{A}$ (with respect to m).

Corollary 1 *For any $V' \subseteq V$ with $|V'| = m$,* bigstar(V') *is a basis in any abstract rigidity matroid of dimension m.*

Proof For any $V' \subseteq V$ with $|V'| = m$, bigstar(V') is the set $K(V')$ with every vertex of $V \setminus V'$ attached to it via m-valent 0-extensions. Note that $K(V')$ is independent and rigid, because it is a complete graph on fewer than $m+2$ vertices. Hence Lemma 5 implies that bigstar(V') is a basis. ■

The problem of finding alternative characterizations of abstract rigidity matroids has been raised in the literature, and we conclude this short review with the 2010 result of Nguyen which gave an answer to this problem. This will be the starting point for our considerations.

Proposition 1 ([6, Theorem 2.2.] [4, Theorem 0.2]) *Let $m \in \mathbb{N}$ and let V be a finite set with $|V| \geq m+1$. Furthermore, let $\mathcal{A}$ be a matroid on $K(V)$.*

Then, $\mathcal{A}$ is a m-dimensional abstract rigidity matroid if and only if any two of the following conditions hold.

(i) All vertex stars minus $m-1$ edges are cocircuits of $\mathcal{A}$.
(ii) All K_{m+2} are circuits of $\mathcal{A}$.
(iii) $r(K(V)) = m|V| - \binom{m+1}{2}$.

In particular, any two of the conditions imply the third.

3 Axiomatizations of Rigidity Matroids

In this section we will derive several cryptomorphic definitions of abstract rigidity matroids. Our starting point will be Proposition 1, which states conditions on the circuits, cocircuits and on the rank function. We strive for characterizations fitting into the classical reformulations of matroid theory—in particular: cocircuits, hyperplanes, bases, circuits.

We start by removing from Proposition 1 the reference to rank and circuits.

Proposition 2 *Let $\mathcal{A}$ be a matroid on $K(V)$. Then $\mathcal{A}$ is a m-dimensional abstract rigidity matroid if and only if*

D1. *all cocircuits have strictly more than $|V| - m$ vertices and*
D2. *all vertex stars minus $m - 1$ edges are cocircuits.*

Proof First of all let $\mathcal{A}$ be an m-dimensional abstract rigidity matroid. We have already seen in Remark 2 that all vertex stars minus $m - 1$ edges are cocircuits in abstract rigidity matroids, so (D2) holds. We prove (D1) by contraposition, and so let $C \subseteq K(V)$ be a set with $|V(C)| \leq |V| - m$. If C is empty, then it is independent in any matroid and thus not a cocircuit. Let then $C \neq \emptyset$. Thus, C contains an edge with endpoints, say, $v_1, v_2 \in V(C)$. Because of the constraint on the cardinality of $V(C)$, we can choose

$$v_3, \ldots, v_{m+2} \notin V(C).$$

By Proposition 1.(ii), $K(\{v_1, \ldots, v_{m+2}\})$ is a circuit in $\mathcal{A}$ which intersects the set C in the single edge $v_1 v_2$. Therefore, C cannot be a cocircuit by Proposition 1.

Now let $\mathcal{A}$ be a matroid satisfying (D1) and (D2). By Proposition 1, to prove that $\mathcal{A}$ is an abstract rigidity matroid, we only need to verify that $r(\mathcal{A}) = m|V| - \binom{m+1}{2}$. So choose an m-element subset V' of V. Since, by (D2), all vertex stars minus $m - 1$ edges are cocircuits, bigstar(V') is independent by Lemma 5. Moreover, its complement is $K(V \setminus V')$—a complete graph on $|V| - m$ vertices—and thus coindependent. So bigstar(V') is also a spanning set, hence a basis, and we can use it to show that the rank of $\mathcal{A}$ has the correct value, as follows.

$$\begin{aligned} r(\mathcal{A}) &= r(\mathrm{bigstar}(V')) = |K(V')| + m(|V| - m) \\ &= \binom{m}{2} + m(|V| - m) = m|V| + \left(\binom{m}{2} - m^2\right) = m|V| - \binom{m+1}{2}. \end{aligned}$$

■

We now state hyperplane axioms for abstract rigidity matroids, as we will be concerned extensively with hyperplanes in the next section. They are easily derived from the cocircuit axioms by complementation.

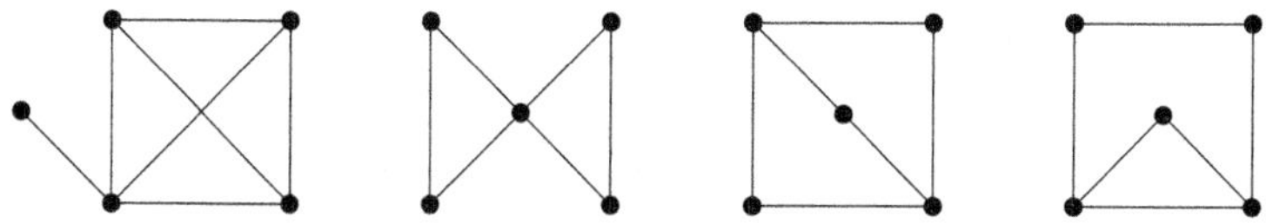

Fig. 7 The set of hyperplanes of the 2-dimensional abstract rigidity matroid on five vertices

Definition 9 For $v \in V$ and $A \subseteq V \backslash \{v\}$ let

$$\Delta_v^A := K(\{v\} \cup A) \cup K(V \backslash \{v\}).$$

Theorem 2 *Let $\mathcal{A}$ be a matroid on $K(V)$ with hyperplanes $\mathcal{H}$. Then $\mathcal{A}$ is an m-dimensional abstract rigidity matroid if and only if $\mathcal{H}$ fulfills:*

H1. if $H \in \mathcal{H}$ then at most $m-1$ vertices in $V(H)$ have valence $|V|-1$.
H2. for all $v \in V$, the sets of the form Δ_v^A where $|A| = m-1$ (i.e., $(m-1)$-valent 0-extensions of $K(V \backslash v)$) are hyperplanes.

Example 6 The set of hyperplanes of the abstract 2-dimensional rigidity matroid on five vertices are those corresponding to subgraphs of K_5 of the forms listed in Fig. 7. The first picture on the left corresponds to the type Δ_v^A with $|A| = m-1$.

We now move to a characterization of abstract rigidity matroids in terms of bases.

Theorem 3 *Let $\mathcal{A}$ be a matroid on $K(V)$ with bases $\mathcal{B}$. Then $\mathcal{A}$ is a m-dimensional abstract rigidity matroid if and only if $\mathcal{B}$ fulfills:*

B1. if $B \in \mathcal{B}$ then $V(B) = V$ and every vertex in B has at least valence m,
B2. for all $V' \subseteq V$ with $|V'| = m$, the set bigstar(V') *is a basis.*

Proof Suppose $\mathcal{A}$ is a m-dimensional abstract rigidity matroid and there is a basis B of $\mathcal{A}$ with a vertex v of degree less than m. Let $N(v)$ denote the set of neighbors of v in the graph induced by B, and choose some $A \subseteq V \backslash \{v\}$ with $|A| = m-1$ and $N(v) \subseteq A$. Then, $B \subseteq \Delta_v^A$, which is a hyperplane by (H2) and should then not have maximal rank. Thus no such basis exists and (B1) holds.

To prove (B2) notice that, whenever $|V'| = m$, bigstar(V') is a m-valent-0-extensions of some K_m, thus independent. Moreover, such a bigstar(V') cannot be contained in any hyperplane, because m of its vertices have valence $|V|-1$, thus it violates (H2).

We now turn to the reverse direction and let $\mathcal{B}$ fulfill (B1) and (B2). We will prove that, then, the set of hyperplanes of the given matroid satisfies (H1) and (H2). First, by (B2) every set of edges inducing a graph with m or more vertices of valence $|V|-1$ contains a basis: thus no hyperplane has more than $m-1$ vertices of valence $|V|-1$ and (H1) holds. To prove (H2) consider any $v \in V$ and any $A \subseteq V \backslash \{v\}$ with $|A| = m-1$, and let $H := \Delta_v^A$. Note that, by (B1), H cannot contain a basis because v has degree $m-1$ in H. Moreover, any edge e not in H must have v as an endpoint. So, by (B2) $H \cup \{e\}$ contains a basis. We conclude that H is a maximal set which does not contain a basis—i.e., a hyperplane—and (H2) holds. ■

We close this section with circuit axioms.

Theorem 4 *Let $\mathcal{A}$ be a matroid on $K(V)$ with set of circuits C. Then $\mathcal{A}$ is a m-dimensional abstract rigidity matroid if and only if C fulfills:*

Z1. *if $C \in C$, no vertex of $V(C)$ has valence less than $m + 1$.*
Z2. *all K_{m+2} are circuits.*

Proof Let $\mathcal{A}$ be a m-dimensional abstract rigidity matroid. Then, Proposition 1.(ii) gives directly (Z2). Proposition 1.(i) says that all vertex stars minus $m-1$ edges are cocircuits of $\mathcal{A}$ and so, by Lemma 5.(i), (Z1) holds.

For the reverse implication, let C satisfy (Z1) and (Z2), and let $\mathcal{B}$ denote the set of bases of $\mathcal{A}$. To prove that (B2) holds, choose then any subset $V' \subseteq V$ with $|V'| = m$, and consider the set bigstar(V'). Every edge xy not in bigstar(V') is contained in the circuit $K(V' \cup \{x, y\})$ and thus the closure of bigstar(V') is $K(V)$. Also, bigstar(V') does not contain any circuit: by (Z1), all vertices of $V \backslash V'$ have valence m, and so any circuit in bigstar(V') must be a subset of $K(V')$, which itself is independent because it only contains vertices of valence $m - 1$. Summarizing, there is no circuit in bigstar(V') and so it is independent, thus a basis and (B2) holds. In order to prove (B1), suppose by way of contradiction that there is a basis B with a vertex v of valence less than m. Then, there is a vertex $w \in V$ such that $vw \notin B$. But, by Remark 3, $B \cup \{vw\}$ must be dependent: thus it contains a circuit C which must contain the edge vw. In particular, v has at most valence m in C, contradicting (Z1). Hence (B1) holds. ■

4 Hyperplanes of Abstract Rigidity Matroids

All axiomatizations given in Sect. 3 share a similar structure: they require some "prescribed substructure" (e.g., conditions (H2), (B2), (Z2), (D2)) and impose some condition on the valency of vertices (e.g., (H1), (B1), (Z1), (D1)). From a structural point of view one can ask whether it is possible to characterize abstract rigidity matroids from a "prescribed substructure" point of view alone.

This question turns out to be intriguing and not trivial. We will investigate it from the vantage point of hyperplanes, where the "prescribed substructures" are hyperplanes of the form Δ_v^A for $|A| = m - 1$ (see Definition 9). We suggest to look at a bigger family of "prescribed substructures", namely the following.

Definition 10 Given a vertex set V and any $m \geq 1$, define

$$\mathcal{H}_m := \{H = K(V_1) \cup K(V_2) \mid V_i \nsubseteq V_j \text{ for } i \neq j, V_1 \cup V_2 = V, |V_1 \cap V_2| = m - 1\}.$$

The family $\mathcal{H}_m$ consists of sets of edges of complete graphs which intersect in a complete graph on $m - 1$ vertices. Figure 8 shows how these sets may look like.

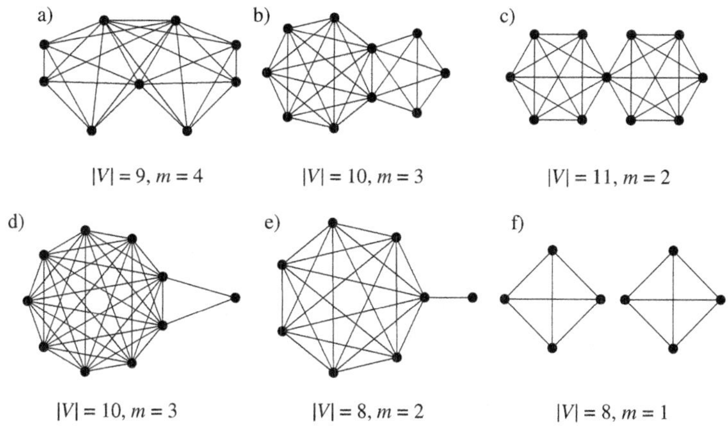

Fig. 8 Examples of elements of $\mathcal{H}_m$. (**a**) $|V| = 9$, $m = 4$, (**b**) $|V| = 10$, $m = 3$, (**c**) $|V| = 11$, $m = 2$, (**d**) $|V| = 10$, $m = 3$, (**e**) $|V| = 8$, $m = 2$, (**f**) $|V| = 8$, $m = 1$

Question 1 Is every matroid $\mathcal{A}$ on the ground set $K(V)$ which satisfies $\mathcal{H}_m \subseteq \mathcal{H}(\mathcal{A})$ an m-dimensional abstract rigidity matroid?

Evidence towards an affirmative answer to this question is given in the next two propositions.

Proposition 3 *Let $\mathcal{A}_m$ be an m-dimensional abstract rigidity matroid on $K(V)$ with hyperplanes $\mathcal{H}$. Then $\mathcal{H}_m \subseteq \mathcal{H}$.*

Proof Let $H \in \mathcal{H}_m$ such that $H = K(V_1) \cup K(V_2)$ with $V_1, V_2 \subseteq V$, $V_1 \cup V_2 = V$, $|V_1 \cap V_2| = m - 1$ and $V_i \nsubseteq V_j$ for $i \neq j$. We have:

$$\begin{aligned}\sigma(H) = \sigma(K(V_1) \cup K(V_2)) &\overset{(C1)}{\subseteq} K(V(K(V_1))) \cup K(V(K(V_2))) \\ &= K(V_1) \cup K(V_2) = H \subseteq \sigma(H),\end{aligned}$$

thus $\sigma(H) = H \neq K(V)$.
Let $V' := V_1 \cap V_2$ and let $e \in K(V)\backslash H$, $e = v_1 v_2$. Then without loss of generality $v_i \in V_i$ $(i = 1, 2)$, $v_1, v_2 \notin V'$.
Note that $K(V' \cup \{v_1, v_2\})$ is rigid, as is $K(V_1)$. In addition, $|V(K(V' \cup \{v_1, v_2\})) \cap V_1| = |V' \cup \{v_1\}| = m$, so by (C2), also their union $K(V' \cup \{v_1, v_2\}) \cup K(V_1)$ is rigid, as is $K(V' \cup \{v_1, v_2\}) \cup K(V_2)$. The union of these sets is $H \cup e$ and

$$\begin{aligned}\sigma(H \cup \{e\}) &= \sigma((K(V' \cup \{v_1, v_2\}) \cup K(V_1)) \cup K((V' \cup \{v_1, v_2\}) \cup K(V_2))) \\ &\overset{(C2)}{=} K(V).\end{aligned}$$

■

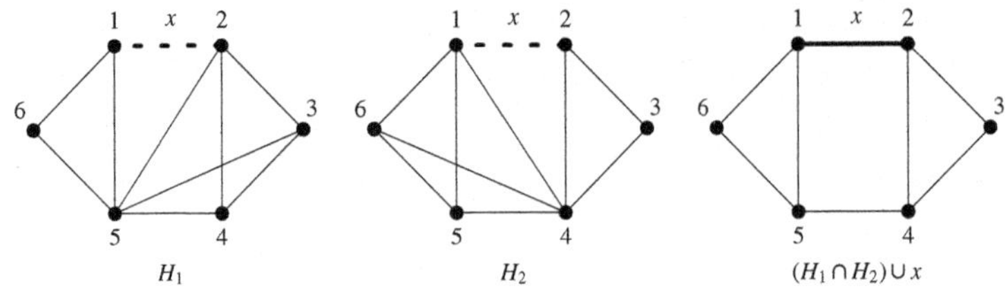

Fig. 9 An example where $|V| = 6$ and $m = 2$

Remark 4 Notice that, in general, the inclusion $\mathcal{H}_m \subseteq \mathcal{H}(\mathcal{A})$ is strict. Figure 9 illustrates this.

The edge-sets on the left and in the middle are elements of $\mathcal{H}_2(V)$ and x is not in their union. By the hyperplane axioms for matroids, the set on the right has to be subset of another hyperplane which is at least 2-vertex-connected, hence not an element of $\mathcal{H}_2(V)$. For a concrete instance of this see also Fig. 7.

Proposition 4 *Let $\mathcal{A}$ be matroid on $K(V)$ with hyperplanes $\mathcal{H}$ and $\mathcal{H}_m \subseteq \mathcal{H}$. Then its closure-operator fulfills (C1).*

Proof Let $E, F \subseteq K(V)$ such that $|V(E) \cap V(F)| < m$. For given $v \in V$, the complement of star(v) minus $m-1$ arbitrary edges is a hyperplane. So we are able to find hyperplanes which intersect in $K(V \backslash \{v\})$. This means that for all $v \in V$, the set $K(V \backslash \{v\})$ is closed in $\mathcal{A}$ and thus is every complete edgeset. Hence $\sigma(E) \subseteq K(V(E))$ and $\sigma(F) \subseteq K(V(F))$, so we may assume without loss of generality that $V(E) \nsubseteq V(F)$ and vice versa. Now we are able to choose $V_1, V_2 \subseteq V$ such that $K(V_1) \cup K(V_2)$ is an element of $\mathcal{H}_m$—a hyperplane—and

$$(K(V_1) \cup K(V_2)) \cap (K(V(E \cup F))) = K(V(E)) \cup K(V(F)).$$

Because the sets which intersect on the left-hand side are closed in $\mathcal{A}$, $K(V(E)) \cup K(V(F))$ must also be closed. It contains $E \cup F$, so $\sigma(E \cup F) \subseteq K(V(E)) \cup K(V(F))$. ∎

We give one more token of the advantage of considering the set $\mathcal{H}_m$: the following proposition gives an immediate proof, for all dimension, of a strengthening of the problem posed in [3, Exercise 4.7] for dimension 2.

Proposition 5 *Let $\mathcal{A}$ be a matroid on $K(V)$, with hyperplanes $\mathcal{H}$ and $\mathcal{H}_m \subseteq \mathcal{H}$. Furthermore let E be a rigid edgeset (i.e. $\sigma(E) = K(V(E))$). Then $(V(E), E)$ is at least m-vertex-connected.*

Proof Otherwise choose a vertex-set $U \subseteq V(E)$ such that $|U| < m$ and deleting U out of $(V(E), E)$ would leave a disconnected graph. Like in the proof before, we are able to find $V_1, V_2 \subseteq V$ such that $K(V_1) \cup K(V_2)$ is a hyperplane in $\mathcal{H}_m$, $E \subseteq K(V_1) \cup K(V_2)$ and $U \subseteq V_1 \cap V_2$. We can construct V_1, V_2 in such a way, that at least one vertex v_1 of one component of the graph after deleting U lies in $V_1 \backslash V_2$, while another vertex v_2 of another such component lies in $V_2 \setminus V_1$. This implies that $v_1, v_2 \in V(E)$ but $v_1v_2 \notin K(V_1) \cup K(V_2)$. Therefore $v_1v_2 \notin \sigma(E) = K(V(E))$. This is a contradiction. ■

4.1 Inductive Characterization

The last part of our study moves from the observation that certain restrictions of rigidity matroids are again rigidity matroids.

Definition 11 Let $\mathcal{A}$ be an m-dimensional abstract rigidity matroid on the set $K(V)$. Given $X \subseteq V$, let $\mathcal{A}[X]$ denote the restriction of $\mathcal{A}$ to the set $K(X)$.

Lemma 6 *Let $\mathcal{A}$ be an m-dimensional abstract rigidity matroid on V. For every $X \subseteq V$, $|X| \geq m + 1$, $\mathcal{A}[X]$ is an m-dimensional rigidity matroid.*

Proof We check condition (ii) and (iii) of Proposition 1. Condition (ii) is clearly inherited from $\mathcal{A}$ (and possibly empty) and Lemma 3 implies (iii). ■

The following definition identifies the "prescribed substructures" through which we will be able to give an intrinsic characterization of abstract rigidity matroids.

Definition 12 Recall Definition 10 and let

$$\mathcal{H}_m^{(1)} := \{K(V_1) \cup K(V_2) \in \mathcal{H}_m \mid |V_1| = m\} = \{\Delta_v^A \mid |A| = m - 1\}.$$

The fact that the subfamily $\mathcal{H}_m^{(1)}$ enforces nontrivial structure on any matroid with $\mathcal{H}_m^{(1)} \subseteq \mathcal{H}(\mathcal{A})$ can be already seen from Lemma 5, whose premise can be rephrased as "$\mathcal{H}_m^{(1)} \subseteq \mathcal{H}(\mathcal{A})$". Moreover, the same requirement is equivalent to condition (i) in Proposition 1—hence, answering our question requires a detailed study of the rank function of matroids on $K(V)$ for which $\mathcal{H}_m^{(1)}$ is a subset of the hyperplane set.

Our main result in this section is the following.

Theorem 5 *Let $|V| \geq m + 1$ and $\mathcal{A}$ a matroid on $K(V)$. The matroid $\mathcal{A}$ is an m-dimensional abstract rigidity matroid if and only if, for all $X \subseteq V$,*

$$\mathcal{H}_m^{(1)}(K(X)) \subseteq \mathcal{H}(\mathcal{A}[X]).$$

Remark 5 The proof will actually work also with milder assumptions: it is enough to assume that there exists an enumeration $v_1, \ldots, v_k$ of the vertices such that $\mathcal{H}_2^{(1)}(K(\{v_1, \ldots, v_j\})) \subseteq \mathcal{H}(\mathcal{A}[\{v_1, \ldots, v_j\}])$ for all $j = 2, \ldots k$.

The proof of the theorem relies on the following two general lemmas.

Lemma 7 *Let $m \geq 2$, $|V| \geq m+1$ and let $\mathcal{A}$ be a matroid on the ground set $K(V)$. Choose any $v_0 \in V$.*

- *If all sets in $\mathcal{H}_m^{(1)}$ are closed in $\mathcal{A}$, $K(V \backslash v_0)$ is closed in $\mathcal{A}$.*
- *If $\mathcal{H}_m^{(1)} \subseteq \mathcal{H}(\mathcal{A})$, the rank of the flat $K(V \backslash v_0)$ in $\mathcal{A}$ is $r(K(V)) - m$.*

Proof Choose $v_0 \in V$ and let $A := V \backslash v_0$. Choose $a_1, \ldots, a_m \in A$ and let $A_0 := \{a_1, \ldots, a_m)\}$. Moreover, set $A_j := \{v_0\} \cup A_0 \backslash a_j$.

For $j = 1, \ldots m-1$ let

$$H_j := \Delta_{v_0}^{A_j} \in \mathcal{H}_m^{(1)}$$

and

$$F_j := \bigcap_{i \leq j} H_i.$$

Moreover, set $F_0 := K(V)$ and let $r_0 = r_{\mathcal{A}}(F_0)$.

As intersections of hyperplanes, all F_j are flats, so we have a chain

$$F_m \subsetneq F_{m-1} \subsetneq \ldots \subsetneq F_0$$

in the lattice of flats of $\mathcal{A}$. Since $K(V \backslash v_0) = F_m$, we have proved the first claim. For the second it will be enough to prove that $r_{\mathcal{A}}(F_j) = r_0 - j$. First notice that, for every $j > 1$, F_j covers F_{j-1}. In fact, by definition F_0 covers F_1 and, for $j > 0$,

$$F_j = K(A) \cup K(\{v_0, a_{j+1}, \ldots, a_m\})$$

so we see that the cardinality $|F_j \backslash F_{j-1}| = 1$. Therefore clearly there can be no G with $F_j \subsetneq G \subsetneq F_{j+1}$. The chain $F_m \subsetneq \ldots \subsetneq F_0$ is therefore saturated, and its length thus equals the difference between the rank of F_0 and the rank of F_m. ∎

Lemma 8 *Let $m \geq 2$, $|V| \geq m+1$ and let $\mathcal{A}$ be a matroid on the ground set $K(V)$. If all sets in $\mathcal{H}_m^{(1)}$ are closed in $\mathcal{A}$, $K(V')$ is independent in $\mathcal{A}$ for every $V' \subseteq V$ with $|V'| = m$.*

Proof Suppose by way of contradiction that $K(V')$ contains a circuit C, let uv be an element of this circuit and $A := (V' \backslash \{v, u\}) \cup v_0$, where v_0 is any element of $V \backslash V'$ (nonempty for cardinality reasons). Then, Δ_v^A is closed in $\mathcal{A}$ and contains all of $K(V')$ except uv, a contradiction to the fact that uv is in a circuit of $K(V')$. ∎

Proof (of Theorem 5) One direction ("left to right") is proved by Lemma 6 and Proposition 3. For the other direction let $V_1 \subseteq V_2 \subseteq \ldots \subseteq V_n$ be a filtration of V such that $|V_i| = i$ for all i. Notice that the hereditarity assumption allows us to apply Lemma 7 recursively to get that

$$r(K(V_i)) = r(K(V)) - m(|V| - |V_i|)$$

for all $i \geq m$, and with Lemma 8 we have $r(K(V_m) = \binom{m}{2}$. Thus we can write

$$\binom{m}{2} = r(K(V)) - m(|V| - m)$$

hence

$$r(K(V)) = m|V| - m^2 + (m^2 - m)/2 = m|V| - \binom{m+1}{2}$$

and we conclude with Proposition 1. ■

Corollary 2 *Question 1 is now equivalent to the following.*

Question 2 *Let* $m \in \mathbb{N}_{>0}$ *and* V *a set with* $|V| \geq m + 1$. *Is it true that, for any matroid* $\mathcal{A}$ *on the vertex set* $K(V)$, $\mathcal{H}_m \subseteq \mathcal{H}(\mathcal{A})$ *implies* $\mathcal{H}_m^{(1)} \subseteq \mathcal{H}(\mathcal{A}[X])$ *for all* X *with* $|X| \geq m + 1$?

Acknowledgements Emanuele Delucchi acknowledges partial support from SNSF-Professorship grant PP00P2_150552/1. Tim Lindemann acknowledges partial support from Swiss European Mobility Programme at University of Fribourg.

References

1. H.H. Crapo, G.-C. Rota, *On the Foundations of Combinatorial Theory: Combinatorial Geometries*, preliminary edition (The M.I.T. Press, Cambridge, MA/London, 1970)
2. J.E. Graver, Rigidity matroids. SIAM J. Discrete Math. **4**(3), 355–368 (1991)
3. J. Graver, B. Servatius, H. Servatius, *Combinatorial Rigidity*. Graduate Studies in Mathematics, vol. 2 (American Mathematical Society, Providence, RI, 1993)
4. J.E. Graver, B. Servatius, H. Servatius, Abstract rigidity in m-space, in *Jerusalem Combinatorics '93*. Contemporary Mathematics, vol. 178 (American Mathematical Society, Providence, RI, 1994), pp. 145–151
5. G. Laman, On graphs and rigidity of plane skeletal structures. J. Eng. Math. **4**, 331–340 (1970)
6. V.-H. Nguyen, On abstract rigidity matroids. SIAM J. Discrete Math. **24**(2), 363–369 (2010)
7. J. Oxley, *Matroid Theory*. Oxford Graduate Texts in Mathematics, vol. 21, 2nd edn. (Oxford University Press, Oxford, 2011)
8. S. Patkar, B. Servatius, K.V. Subrahmanyam, Abstract and generic rigidity in the plane. J. Combin. Theory Ser. B **62**(1), 107–113 (1994)

Induced and Complete Multinets

Jeremiah Bartz

Abstract Multinets are certain configurations of lines and points with multiplicities in the complex projective plane $\mathbb{P}^2$. They appear in the study of resonance and characteristic varieties of complex hyperplane arrangement complements and cohomology of Milnor fibers. In this paper, two properties of multinets, inducibility and completeness, and the relationship between them are explored with several examples presented. Specializations of multinets plays an integral role in our findings. The main result is the classification of complete 3-nets.

Keywords Complete multinet • Free arrangement • Hyperplane arrangement • Induced multinet • $K(\pi$ • 1) arrangement • Multinet • Net

1 Introduction

Multinets are certain configurations of lines and points with multiplicities in the complex projective plane $\mathbb{P}^2$. More specifically they are multi-arrangements of projective lines partitioned into three or four blocks with some additional combinatorial properties (see Sect. 2). They originally arose in the study of resonance and characteristic varieties of the complement of a complex hyperplane arrangement in [5, 10]. Multinets have also appeared while studying the cohomology of Milnor fibers in [4].

Very few examples of multinets with non-trivial multiplicities were known initially. It was observed in [5] that several of the earliest known examples satisfied an extra property which implied the underlying arrangements were $K(\pi, 1)$-arrangements. These multinets are referred to as *complete multinets*. More recently a systematic method of constructing multinets was introduced in [2] and produced a variety of new examples known as *induced multinets*. Not all multinets are induced. In fact two line arrangements can support the same multinet structure while not being lattice equivalent. Such arrangements are referred to as *specializations* of a given multinet. In the paper we recall some definitions and known results, give

J. Bartz (✉)
Francis Marion University, 4822 E Palmetto St, Florence, SC 29506, USA
e-mail: jbartz@fmarion.edu

F. Callegaro et al. (eds.), *Configuration Spaces*, Springer INdAM Series 14,
DOI 10.1007/978-3-319-31580-5_9

examples of induced and non-induced multinets, and investigate the completeness of the induced multinets presented in [2]. Specializations of multinets plays an integral role in our findings. The main result is the classification of complete 3-nets.

The paper is organized as follows. In Sect. 2 we recall basic definitions and relevant properties of multinets. In Sect. 3 we exhibit examples of induced and non-induced multinets while exploring the notion of specializations of a multinet. Section 4 discusses completeness of multinets and contains the main result of the paper, the classification of complete 3-nets. Finally some open problems are listed in Sect. 5.

2 Preliminaries

We recall basic definitions and relevant properties of multinets. Several well-known examples are collected. A description of monomial groups and their arrangements is given. These arrangements were used in [1, 2] to systematically construct examples of multinets known as *induced multinets*. A synopsis is given of this construction and illustrated by obtaining a $(3,4)$-net realizing $\mathbb{Z}/2\mathbb{Z} \times \mathbb{Z}/2\mathbb{Z}$ as an induced multinet.

2.1 *Pencils of Curves and Multinets*

There are several equivalent ways to define multinets. Here we present them using pencils of plane curves. A *pencil of plane curves* is a line in the projective space of homogeneous polynomials from $\mathbb{C}[x_1, x_2, x_3]$ of some fixed degree d. Any two distinct curves of the same degree generate a pencil, and conversely a pencil is determined by any two of its curves C_1, C_2. An arbitrary curve C in the pencil (called a *fiber*) is $C = aC_1 + bC_2$ where $[a : b] \in \mathbb{P}^1$. Every two fibers in a pencil intersect in the same set of points $\mathcal{X} = C_1 \cap C_2$, called the *base* of the pencil. If fibers do not have a common component (called a *fixed component*), then the base is a finite set of points.

A curve of the form $\prod_{i=1}^{q} \alpha_i^{m_i}$, where α_i are distinct linear forms and $m_i \in \mathbb{Z}_{>0}$ for $1 \le i \le q$, is called *completely reducible*. Such a curve is called *reduced* if $m_i = 1$ for each i. We are interested in connected pencils of plane curves without fixed components and at least three completely reducible fibers. By connectivity we mean the nonexistence of a completely reducible fiber whose distinct components intersect only at $\mathcal{X}$. For conciseness we refer to such a pencil as a *Ceva pencil*.

Definition 1 The union of all completely reducible fibers (with a fixed partition into fibers, also called *blocks*) of a Ceva pencil of degree d is called a *(k, d)-multinet* where k is the number of the blocks. The base $\mathcal{X}$ of the pencil is determined by the multinet structure and called the *base* of the multinet.

If the intersection of each two fibers is transversal, i.e. $|\mathcal{X}| = d^2$ and hence all blocks are reduced, then the multinet is called a *net*. If $|\mathcal{X}| < d^2$ we call the multinet *proper*. If all blocks are reduced the multinet is said to be *light*. If there are non-reduced blocks we call the multinet *heavy*. A block of a multinet is said to be a *pencil* if all of its lines intersect at a common point.

From a projective geometry perspective, a (k, d)-multinet is a multi-arrangement $\mathcal{A}$ of lines in $\mathbb{P}^2$ provided with multiplicities $m(\ell) \in \mathbb{Z}_{>0}$ $(\ell \in \mathcal{A})$ and partitioned into k blocks $\mathcal{A}_1, \ldots, \mathcal{A}_k$ $(k \geq 3)$ subject to the following two conditions.

(i) Let $\mathcal{X}$ be the set of the intersections of lines from different blocks. For each point $p \in \mathcal{X}$, the number

$$n_p = \sum_{\ell \in \mathcal{A}_i, p \in \ell} m(\ell)$$

is independent on i. This number is called the *multiplicity* of p.

(ii) For every two lines ℓ and ℓ' from the same block, there exists a sequence of lines from that block $\ell = \ell_0, \ell_1, \ldots, \ell_r = \ell'$ such that $\ell_{i-1} \cap \ell_i \notin \mathcal{X}$ for $1 \leq i \leq r$. This is the connectivity condition.

Multinets can be defined purely combinatorially using incidence relations. Note that the multiplicity $m(\ell)$ for each $\ell \in \mathcal{A}$ equals the multiplicity of its corresponding linear factor in the completely reducible fibers of the Ceva pencil. From this viewpoint a net is a multinet with $m(\ell) = n_p = 1$ for all $\ell \in \mathcal{A}$ and $p \in \mathcal{X}$.

From a combinatorial viewpoint (k, d)-nets are the realization of $k - 2$ pairwise orthogonal Latin squares of size d (after identifying all blocks). If $k = 3$, the Latin square gives a multiplication table of a quasi-group G and the associated net is said to *realize* G. The classification of groups which can be realized by nets has been completed in [8, 9, 16].

2.2 Properties of Multinets and Examples

Several important properties of multinets are listed below which have been collected from [5, 14, 17].

Proposition 1 *Let $\mathcal{A}$ be a (k, d)-multinet. Then:*

1. *$\sum_{\ell \in \mathcal{A}_i} m(\ell) = d$, independent of i.*
2. *$\sum_{\ell \in \mathcal{A}} m(\ell) = dk$.*
3. *$\sum_{p \in \mathcal{X}} n_p^2 = d^2$ (Bézout's theorem).*
4. *$\sum_{p \in \mathcal{X} \cap \ell} n_p = d$ for every $\ell \in \mathcal{A}$.*
5. *There are no multinets with $k \geq 5$.*
6. *All multinets with $k = 4$ are nets.*

Example 1 An arrangement comprised of k lines which intersect at a common point supports a $(k, 1)$-net where each block consists of one line. This case corresponds to a so-called local resonance component. It is considered to be trivial and we will often tacitly assume that $d > 1$.

Example 2 For each $n \geq 2$, there is a Ceva pencil generated by $x^n - y^n$ and $y^n - z^n$ with third completely reducible fiber given by $x^n - z^n$. It is commonly referred to as the Fermat pencil. The corresponding arrangement supports a $(3, n)$-net realizing $\mathbb{Z}/n\mathbb{Z}$ with each block being a pencil. The common intersection point of each block lies outside of the base of the net. For $n = 3$, this is one of the specializations of a Pappus arrangement (cf. Sect. 3.2). Yuzvinsky showed in Proposition 3.3 of [16] that a $(3, n)$-net with all blocks being pencils is projectively equivalent to the arrangement defined by $Q = [x^n - y^n][x^n - z^n][y^n - z^n]$.

Example 3 For each $n \geq 1$, a $(3, 2n)$-multinet is given by the pencil generated by polynomials $x^n(y^n - z^n)$ and $y^n(x^n - z^n)$ with the third completely reducible fiber being $z^n(x^n - y^n)$. These are the projectivizations of the reflection arrangements for the full monomial groups $G(n, 1, 3)$ (see Sect. 2.3). For $n = 1$, it gives the only (up to projective isomorphism) $(3, 2)$-net of Coxeter type A_3; for $n = 2$, it is the $(3, 4)$-multinet of Coxeter type B_3. These multinets are heavy when $n > 1$.

Example 4 The cubics xyz and $x^3 + y^3 + z^3$ generate a Ceva pencil with 4 completely reducible fibers. They give a $(4, 3)$-net known as the Hesse configuration. This is the only currently known multinet with 4 blocks. A long-standing conjecture posed by Yuzvinsky is that the Hesse configuration is the unique 4-net up to projective isomorphism. Dunn et al. proved in [3] that the Hesse configuration is the unique (4,3)-net in $\mathbb{P}^2$ and no $(4, d)$-nets exists in $\mathbb{P}^2$ for $d = 4, 5, 6$. An alternate proof of the non-existence of (4,4)-nets in $\mathbb{P}^2$ using tropical geometry was given by Güntürkün and Kişisel in [7].

2.3 Monomial Groups and Their Arrangements

Examples of multinets can be derived from certain reflection arrangements of finite complex monomial groups. Historically monomial groups played a useful role in the representation theory of groups (see [11]). Reflection arrangements of finite complex monomial groups and their subarrangements have been studied extensively, especially regarding their connections with free arrangements (see [12]). Below we summarize the description of finite complex monomial groups and their arrangements given in [12].

Let $V = \mathbb{C}^\ell$ with $\ell > 1$ and choose a basis $\{e_1, \ldots, e_\ell\}$ of V. For any integer $n > 1$, let $\mathbb{Z}/n\mathbb{Z}$ denote the cyclic group of order n generated by $\theta = \exp(2\pi i/n)$. Put $I = \{1, 2, \ldots, \ell\}$ and let $\epsilon : I \to \mathbb{Z}/n\mathbb{Z}$ be any function. The *full monomial group*, denoted $G(n, 1, \ell)$, is the subgroup of $GL(V)$ consisting of all transformations

$$g(\sigma, \epsilon)e_i = \varepsilon(i)e_{\sigma(i)}$$

where $\sigma \in \mathrm{Sym}(\ell)$, the symmetric group on ℓ symbols. Equivalently, the full monomial group is the wreath product of $\mathbb{Z}/n\mathbb{Z}$ and $\mathrm{Sym}(\ell)$, consisting of all $\ell \times \ell$ monomial matrices with entries in $\mathbb{Z}/n\mathbb{Z}$. Its reflection arrangement is said to be *of type* $G(n, 1, \ell)$ and given by

$$Q = x_1 \ldots x_\ell \prod_{1 \le i < j \le \ell} (x_i^n - x_j^n).$$

Another family of reflection arrangements can be defined using certain irreducible subgroups of the full monomial group $G(n, 1, \ell)$. Let p be divisor of n and $G(n, p, \ell)$ be the subgroup of the full monomial group consisting of all $g(\sigma, \epsilon)$ where $\prod \epsilon(i)$ is a power of θ^p. These groups are generated by reflections and irreducible since $n > 1$. If $p < n$, the subgroup $G(n, p, \ell)$ contains the reflections $e_i \mapsto \theta^p e_i$ and $e_i \mapsto e_j$ for $i \neq j$. In this case, the corresponding reflection arrangement is the same as the reflection arrangement of the full monomial group $G(n, 1, \ell)$. On the other hand, if $p = n$, the reflection arrangement of the subgroup $G(n, n, \ell)$ is defined by

$$Q = \prod_{1 \le i < j \le \ell} (x_i^n - x_j^n)$$

and said to be of *type* $G(n, n, \ell)$. For $n = 2$ the full monomial group $G(2, 1, \ell)$ is the Coxeter group of type B_ℓ and $G(2, 2, \ell)$ is the Coxeter group of type D_ℓ.

2.4 *Induced Multinets*

Few examples of multinets with non-trivial multiplicities were known initially. Then a systematic method of constructing multinets was introduced in [1, 2] and produced a variety of new examples known as *induced multinets* which we define below. We briefly describe the method of producing induced multinets and give a summary of their combinatorial properties.

The notion of multinets can be generalized to $\mathbb{P}^r$ ($r > 2$) by using pencils of homogeneous polynomials of $r + 1$ variables. Presently the only known multinets in $\mathbb{P}^r$ for $r > 2$ are the $(3, 2n)$-nets in $\mathbb{P}^3$ given for each $n \in \mathbb{Z}_{>0}$ by the defining polynomial

$$Q_n = [(x_0^n - x_1^n)(x_2^n - x_3^n)][(x_0^n - x_2^n)(x_1^n - x_3^n)][(x_0^n - x_3^n)(x_1^n - x_2^n)]$$

where the brackets determine the blocks. This arrangement is the collection of all (projectivizations of) reflection hyperplanes of the finite complex monomial group $G(n, n, 4)$ (see Sect. 2.3). For $n = 2$ it is the Coxeter group of type D_4.

Each block of Q_n is partitioned in two *half-blocks* (determined by parentheses) of degree n each. Notice that all the planes of a half-block intersect at one line, called the *base of the half-block*. For instance the base of the leftmost half-block is given by the system $x_0 = 0, x_1 = 0$.

Multinets can be constructed as follows. Intersect Q_n with a plane H that does not belong to Q_n. The resulting multi-arrangement in H is denoted by $\mathcal{A}^H$ and referred to as the *arrangement induced by* Q_n. The pencil in $\mathbb{P}^3$ corresponding to Q_n induces a pencil in $\mathbb{P}^2$ with 3 completely reducible fibers. It may happen that the pencil has a fixed component. In this case, we cancel the fixed components obtaining a smaller arrangement $\mathcal{A}_0^H$ with a multinet structure. Abusing the notation slightly we will call $\mathcal{A}^H$ (if there is no fixed component) or $\mathcal{A}_0^H$, provided with the partitions into fibers of the induced pencil, the *induced multinet*.

A systematic study of the possible combinatorics of induced multinets obtained from Q_n was performed in [1, 2]. Induced multinets from Q_1 are either $(3, 2)$-nets realizing $\mathbb{Z}/2\mathbb{Z}$ or trivial (cf. Example 1). The following theorem from [2] gives a summary of the possibilities for $n > 1$. The first five cases are heavy multinets whereas the last five cases are light multinets.

Theorem 1 *There are 10 possibilities for the combinatorics for induced multinets from Q_n. Each possibility exists and is described below.*

1. *If $n > 1$, a heavy $(3, 2n)$-multinet can have three lines of multiplicity n and remaining lines of multiplicity 1. This is projectively equivalent to the multinets realizing $G(n, 1, 3)$ discussed in Example 3. For $n = 2$ this is the $(3, 4)$-multinet of Coxeter type B_3.*
2. *If $n > 1$, a heavy $(3, 2n)$-multinet can have a unique line of multiplicity n and all other lines of multiplicity 1. The base $\mathcal{X}$ consists of two points of multiplicity n and all remaining points with multiplicity 1.*
3. *If $n > 1$ is even, a heavy $(3, 2n)$-multinet can have three lines of multiplicity 2 and remaining lines of multiplicity 1. The base $\mathcal{X}$ consists of $3n - 3$ points of multiplicity 2 and all remaining points of multiplicity 1. For $n = 2$ this is the $(3, 4)$-multinet of Coxeter type B_3.*
4. *If $n > 1$ is odd, a heavy $(3, 2n)$-multinet can have two lines of multiplicity 2 and remaining lines of multiplicity 1. The base $\mathcal{X}$ consists of $2n - 1$ points of multiplicity 2 and all remaining points of multiplicity 1.*
5. *If $n > 1$, a heavy $(3, 2n)$-multinet can have an unique line of multiplicity 2 and all other lines of multiplicity 1. The base $\mathcal{X}$ consists of n points of multiplicity 2 and all other points of multiplicity 1.*
6. *A light $(3, 2n)$-multinet can have a unique point in $\mathcal{X}$ of multiplicity n. All other points in $\mathcal{X}$ have multiplicity 1. For $n = 2$ this is the $(3, 2)$-net of Coxeter type A_3.*

7. *If $n > 1$, a light $(3, 2n-1)$-multinet can have a unique point in $\mathcal{X}$ of multiplicity $n-1$. All other points in $\mathcal{X}$ have multiplicity 1. For $n = 2$ this gives a $(3, 3)$-net realizing $\mathbb{Z}/3\mathbb{Z}$ with each block in general position.*
8. *If $n > 2$, a light $(3, 2n-2)$-multinet can have a unique point in $\mathcal{X}$ of multiplicity $n-2$. All other points in the base locus have multiplicity 1. For $n = 3$ this gives a $(3, 4)$-net realizing $\mathbb{Z}/2\mathbb{Z} \times \mathbb{Z}/2\mathbb{Z}$ with each block having exactly three concurrent lines and fourth line in general position (cf. Example 5).*
9. *If $n > 1$, a light $(3, 2n)$-multinet can have several points of multiplicity 2 if it does not have points of multiplicity greater than 2. The number of these points is bounded independently of n by 2^{96}.*
10. *A light $(3, 2n)$-multinet can be a net which realizes the dihedral group of order $2n$.*

Example 5 Intersect Q_3 by the hyperplane H defined by $x_0 = (\xi + 1)x_1 - \xi x_2$ where ξ is a primitive 3rd root of unity. Then $\mathcal{A}^H$ has two common factors $x_1 - x_2$ and $x_1 - \xi x_2$. Canceling results in a $(3, 4)$-multinet realizing $\mathbb{Z}/2\mathbb{Z} \times \mathbb{Z}/2\mathbb{Z}$. Using a convenient choice of labels for the lines, the three blocks are $\mathcal{A}_1 = \{\ell_{11}, \ell_{12}, \ell_{13}, \ell_{14}\}$, $\mathcal{A}_2 = \{\ell_{21}, \ell_{22}, \ell_{23}, \ell_{24}\}$, and $\mathcal{A}_3 = \{\ell_{31}, \ell_{32}, \ell_{33}, \ell_{34}\}$ where the equations for the lines in $(\mathbb{P}^2)^*$ are

$$\begin{array}{lll} \ell_{11} = [0:1:-1] & \ell_{21} = [1:0:-1] & \ell_{31} = [1:\xi^2:\xi] \\ \ell_{12} = [2\xi:1:0] & \ell_{22} = [1:0:-\xi] & \ell_{32} = [1:\xi^2:1] \\ \ell_{13} = [0:1:-\xi] & \ell_{23} = [1:0:-\xi^2] & \ell_{33} = [1:-\xi^2:0] \\ \ell_{14} = [0:1:-\xi^2] & \ell_{24} = [\xi:2:0] & \ell_{34} = [\xi:1:1]. \end{array}$$

The base locus consists of sixteen points of multiplicity 1, namely

$$\begin{array}{ll} \ell_{11} \cap \ell_{21} \cap \ell_{31} = [1:1:1] & \ell_{13} \cap \ell_{21} \cap \ell_{33} = [1:\xi:1] \\ \ell_{11} \cap \ell_{22} \cap \ell_{32} = [\xi:1:1] & \ell_{13} \cap \ell_{22} \cap \ell_{34} = [\xi:\xi:1] \\ \ell_{11} \cap \ell_{23} \cap \ell_{33} = [\xi^2:1:1] & \ell_{13} \cap \ell_{23} \cap \ell_{31} = [\xi^2:\xi:1] \\ \ell_{11} \cap \ell_{24} \cap \ell_{34} = [-2\xi^2:1:1] & \ell_{13} \cap \ell_{24} \cap \ell_{32} = [-2:\xi:1] \\ \ell_{12} \cap \ell_{21} \cap \ell_{32} = [1:-2\xi:1] & \ell_{14} \cap \ell_{21} \cap \ell_{34} = [1:\xi^2:1] \\ \ell_{12} \cap \ell_{22} \cap \ell_{31} = [\xi:-2\xi^2:1] & \ell_{14} \cap \ell_{22} \cap \ell_{33} = [\xi:\xi^2:1] \\ \ell_{12} \cap \ell_{23} \cap \ell_{34} = [\xi^2:-2:1] & \ell_{14} \cap \ell_{23} \cap \ell_{32} = [\xi^2:\xi^2:1] \\ \ell_{12} \cap \ell_{24} \cap \ell_{33} = [0:0:1] & \ell_{14} \cap \ell_{24} \cap \ell_{31} = [-2\xi^2:\xi^2:1]. \end{array}$$

Using this choice of labels, we can see that this net realizes $\mathbb{Z}/2\mathbb{Z} \times \mathbb{Z}/2\mathbb{Z}$ by observing that the point $\ell_{1i} \cap \ell_{2j} \cap \ell_{3k} \in \mathcal{X}$ appears in the associated Latin square

$$\begin{bmatrix} 1 & 2 & 3 & 4 \\ 2 & 1 & 4 & 3 \\ 3 & 4 & 1 & 2 \\ 4 & 3 & 2 & 1 \end{bmatrix}$$

as k in the (i, j)-th position. Note that each block has exactly three concurrent lines and a fourth line in general position.

3 Specializations of Multinets and Inducibility from Q_n

Although induced multinets from Q_n provide a wealth of examples of multinets, it is known that not all multinets can be obtained in this manner. For instance, every $(3, 2n+1)$-net for $n \in \mathbb{Z}_{>1}$ is not an induced multinet. An example of a proper multinet which is not induced is given in Problem 4 of [2]. Unlike the examples for nets, there is a specialization of this latter example which is an induced multinet from Q_n. To be precise, a *specialization* of a multinet is any line arrangement in $\mathbb{P}^2$ which satisfies the incidence relations of the given multinet. A multinet may have more than one specialization. That is, there may be line arrangements with nonisomorphic intersection lattices which satisfy the same multinet incidence relations.

In [13] Stipins constructs (k, d)-nets from $k-2$ mutually orthogonal Latin squares of order d. These Latin squares contain the combinatorial data regarding the incidence relations of the associated nets and are used to generate a parameterized family of explicit defining equations. In this section, we explore the possible specializations of several multinets using the results of Stipins and incidence relations of the associated multinets.

3.1 (3,2)-Nets

Any $(3, 2)$-net in $\mathbb{P}^2$ is projectively equivalent to the arrangement with defining polynomial $[x(y-z)][y(x-z)][z(x-y)]$. In particular it consists of classes which are all in general position (which coincides with being pencil for $d = 2$) and is associated with the Latin square

$$\begin{bmatrix} 1 & 2 \\ 2 & 1 \end{bmatrix}.$$

Thus all $(3, 2)$-nets are lattice equivalent, realize $\mathbb{Z}/2\mathbb{Z}$, and have Coxeter type A_3. These nets can be induced from Q_1. Also, these arrangements are simplicial and denoted $\mathcal{A}(6, 1)$ in Grünbaum's catalogue of simplicial arrangements in [6].

3.2 (3,3)-Nets

Up to isotopy, there is a unique Latin square of order 3, namely

$$\begin{bmatrix} 1 & 2 & 3 \\ 2 & 3 & 1 \\ 3 & 1 & 2 \end{bmatrix}.$$

This is the multiplication table for $\mathbb{Z}/3\mathbb{Z}$. Assuming that the block $\mathcal{A}_1 = \{\ell_{11}, \ell_{12}, \ell_{13}\}$ is in general position, Stipins derived the family of line arrangements in $\mathbb{P}^2$ indexed by $[s_0 : s_1] \times [t_0 : t_1] \in \mathbb{P} \times \mathbb{P}$ given by

$$\begin{array}{lll} \ell_{11} = [1:0:0] & \ell_{21} = [1:1:1] & \ell_{31} = [s_0 : s_1 : s_1] \\ \ell_{12} = [0:1:0] & \ell_{22} = [s_0t_1 : s_1t_1 : s_1t_0] & \ell_{32} = [t_0 : t_1 : t_0] \\ \ell_{13} = [0:0:1] & \ell_{23} = [s_0t_0 : s_0t_1 : s_1t_0] & \ell_{33} = [s_0t_1 : s_0t_1 : s_1t_0]. \end{array}$$

This line arrangement is a $(3,3)$-net realizing $\mathbb{Z}/3\mathbb{Z}$ for generic indices with the other blocks given by $\mathcal{A}_2 = \{\ell_{21}, \ell_{22}, \ell_{23}\}$ and $\mathcal{A}_3 = \{\ell_{31}, \ell_{32}, \ell_{33}\}$. Note that the nine lines are distinct if and only if $s_0, s_1, t_0, t_1 \neq 0$, $s_0 \neq s_1$, $t_0 \neq t_1$, and $s_0/s_1 \neq t_0/t_1$. This family can be reindexed by the parameters λ and μ by normalizing the original indices. That is, put $[s_0 : s_1] = [1 : s_1/s_0] = [1 : \lambda]$ and $[t_0 : t_1] = [1 : t_1/t_0] = [1 : \mu]$ where $\lambda, \mu \neq 0, 1$ and $\lambda \neq \mu$. In terms of this reparameterization, the family of line arrangements can be written in $(\mathbb{P}^2)^*$ as

$$\begin{array}{lll} \ell_{11} = [1:0:0] & \ell_{21} = [1:1:1] & \ell_{31} = [1:\lambda:\lambda] \\ \ell_{12} = [0:1:0] & \ell_{22} = [\mu:\lambda\mu:\lambda] & \ell_{32} = [1:\mu:1] \\ \ell_{13} = [0:0:1] & \ell_{23} = [1:\mu:\lambda] & \ell_{33} = [\mu:\mu:\lambda]. \end{array}$$

Denoting points $\ell_{1i} \cap \ell_{2j} \cap \ell_{3k} \in \mathcal{X}$ as the triple (i,j,k), the nine points of $\mathcal{X}$ are

$$\begin{array}{lll} (1,1,1) = [0:1:-1] & (1,2,2) = [0:1:-\mu] & (1,3,3) = [0:\lambda:-\mu] \\ (2,1,2) = [1:0:-1] & (2,2,3) = [\lambda:0:-\mu] & (2,3,1) = [\lambda:0:-1] \\ (3,1,3) = [1:-1:0] & (3,2,1) = [\lambda:-1:0] & (3,3,2) = [\mu:-1:0]. \end{array}$$

The intersections points within the three blocks of this family are

$$\begin{aligned} \ell_{11} \cap \ell_{12} &= [0:0:1] \\ \ell_{11} \cap \ell_{13} &= [0:1:0] \\ \ell_{12} \cap \ell_{13} &= [1:0:0] \\ \ell_{21} \cap \ell_{22} &= [\lambda(1-\mu) : \mu - \lambda : \mu(\lambda - 1)] \\ \ell_{21} \cap \ell_{23} &= [\lambda - \mu : 1 - \lambda : \mu - 1)] \\ \ell_{22} \cap \ell_{23} &= [\lambda\mu(\lambda - 1) : \lambda(1-\mu) : \mu(\mu - \lambda)] \\ \ell_{31} \cap \ell_{32} &= [\lambda(1-\mu) : \lambda - 1 : \mu - \lambda] \\ \ell_{31} \cap \ell_{33} &= [\lambda(\lambda - \mu) : \lambda(\mu - 1) : \mu(1-\lambda)] \\ \ell_{32} \cap \ell_{33} &= [\mu(\lambda - 1) : \mu - \lambda : \mu(1-\mu)]. \end{aligned}$$

Each block consists of three lines which can either be a pencil or in general position. Recall it was assumed that the block $\mathcal{A}_1$ is in general position. Direct computations show that the block $\mathcal{A}_2$ is a pencil if and only if $\mu\lambda^2 - 3\mu\lambda + \lambda + \mu^2 = 0$. Similarly, $\mathcal{A}_3$ is a pencil if and only if $\lambda\mu^2 - 3\lambda\mu + \mu + \lambda^2 = 0$. If $\mathcal{A}_2$ and $\mathcal{A}_3$ are both pencils, solving the pair of equations for μ gives $\mu = \lambda$ or $\mu = 1$. However these values for

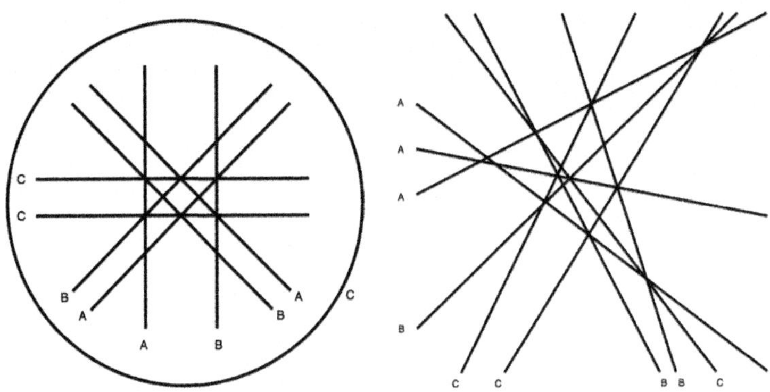

(a) Two blocks in general position (b) Three blocks in general position

Fig. 1 Equivalent (3, 3)-nets with different class structures. (**a**) Two blocks in general position. (**b**) Three blocks in general position

μ do not define a (3, 3)-net as noted above. On the other hand, given a generic value for λ, the equation associated with $\mathcal{A}_2$ can be solved for μ to find parameters so that $\mathcal{A}_2$ is a pencil and $\mathcal{A}_3$ is in general position (see Fig. 1a). For generic λ and μ, both equations will not be satisfied and yields a (3, 3)-net with all blocks in general position (see Fig. 1b).

The remaining possible specialization of the (3, 3)-net realizing $\mathbb{Z}/3\mathbb{Z}$ consists of all blocks being pencils. This configuration is possible and was discussed in Example 2. We summarize our findings in the following result.

Theorem 2 *Any* (3, 3)*-net realizes* $\mathbb{Z}/3\mathbb{Z}$ *and has exactly one of the following block structures:*

1. *Every block is in general position.*
2. *One block is a pencil, two blocks are in general position.*
3. *Every block is a pencil.*

In particular, there does not exist a (3, 3)-net in $\mathbb{P}^2$ with one block in general position and two blocks being pencils. The specialization with all blocks in general position can be induced from Q_2 using cancellation. The other cases are not inducible from Q_n. However, the specialization with all classes being pencils appears as a subarrangement of the (3, 6)-net realizing the dihedral group of order 6 induced from Q_3.

3.3 (3, 2n)-Multinets of Type G(n, 1, 3)

For each $n \geq 1$, the arrangement defined by $[x^n(y^n - z^n)][y^n(x^n - z^n)][z^n(x^n - y^n)]$ supports a $(3, 2n)$-multinet (see example 3). This multinet is the projectivization of the reflection arrangement of the monomial group of type $G(n, 1, 3)$ and thus referred to as being *of type* $G(n, 1, 3)$. We show below that any specialization of this multinet is projectively equivalent to the arrangement with defining polynomial $[x^n(y^n - z^n)][y^n(x^n - z^n)][z^n(x^n - y^n)]$. Thus any $(3, 2n)$-multinets of type $G(n, 1, 3)$ are lattice equivalent.

Any specialization of a $(3, 2n)$-multinet of type $G(n, 1, 3)$ consists of three lines of multiplicity n and $3n$ lines of multiplicity 1. The base $\mathcal{X}$ has three points of multiplicity n and n^2 points of multiplicity 1. It contains the $(3, n)$-net realizing $\mathbb{Z}/n\mathbb{Z}$ with all blocks being pencils as a subarrangement (see Example 2). By Proposition 3.3 of [16], this subarrangement is projectively equivalent to the arrangement defined by $[x^n - y^n][x^n - z^n][y^n - z^n]$. It remains to determine the possible ways to add the remaining three lines of multiplicity n to this subarrangement and obtain a multinet of type $G(n, 1, 3)$.

The three points of multiplicity n are the intersection of two lines of multiplicity n and n lines of multiplicity 1. Thus these three points are the common intersection point of each block, namely $[1 : 0 : 0]$, $[0 : 1 : 0]$ and $[0 : 0 : 1]$. The lines of multiplicity n are the three lines which pass through exactly two of these points, namely x, y, and z. Taking multiplicity into account, we can make the following conclusion.

Theorem 3 *Any* $(3, 2n)$*-multinet of type* $G(n, 1, 3)$ *is projectively equivalent to the arrangement with defining polynomial* $[x^n(y^n - z^n)][y^n(x^n - z^n)][z^n(x^n - y^n)]$.

These multinets are inducible from Q_n (see Sect. 2.4).

3.4 Light (3, 4)*-Multinet with Unique Double Point*

Next we turn our attention to specializations of the light $(3, 4)$-multinet with a unique point of multiplicity 2 and all other points of multiplicity 1 in the base $\mathcal{X}$. Two such examples with different block structures are presented in Fig. 2. The specialization in Fig. 2a is the simplicial arrangement known as $\mathcal{A}(12, 1)$ in Grünbaum's catalogue of simplicial arrangements in [6]. The specialization with two blocks in general position in Fig. 2b first appeared in [5].

Here each block consists of four lines. There are three possibilities for each block: (1) the four lines are a pencil; (2) the lines are in general position; or (3) exactly three of the four lines meet at a common point. A block in the latter configuration is said to be an *easel* due to its resemblance to an artists' easel.

We can eliminate the first possibility with the following lemma.

Lemma 1 *If a light and proper multinet has a block which is a pencil, then it is the trivial multinet with* $|\mathcal{X}| = 1$.

Proof Let p be the common intersection point from the block which is a pencil. Then $p \in \mathcal{X}$ because the multinet is light and proper. It follows that p lies on every line from the other two blocks, hence $\mathcal{X} = \{p\}$. This is the trivial multinet. □

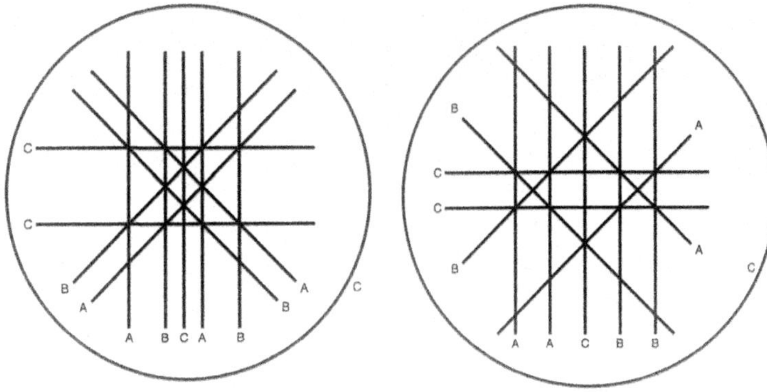

(a) No blocks in general position (b) Two blocks in general position

Fig. 2 Equivalent light and proper $(3, 4)$-multinets. (**a**) No blocks in general position. (**b**) Two blocks in general position

It follows that each block is either (1) in general position with six double points or (2) an easel with a unique triple point and 3 double points. To determine the specializations possible for this multinet, we use the multinet incidence structure to compute the realization space.

We may assume the unique point $p \in \mathcal{X}$ with multiplicity 2 has coordinates $[0 : 0 : 1]$ in $\mathbb{P}^2$. Since no block is a pencil, we may choose coordinates on $(\mathbb{P}^2)^*$ and labels for the lines in the each block, namely $\mathcal{A}_1 = \{\ell_{11}, \ell_{12}, \ell_{13}, \ell_{14}\}$, $\mathcal{A}_2 = \{\ell_{21}, \ell_{22}, \ell_{23}, \ell_{24}\}$, and $\mathcal{A}_3 = \{\ell_{31}, \ell_{32}, \ell_{33}, \ell_{34}\}$, so that

$$\begin{array}{lll} \ell_{11} = [1:1:1] & \ell_{21} = [1:0:0] & \ell_{31} = [s_0 : s_1 : s_2] \\ \ell_{12} = [0:1:\lambda] & \ell_{22} = [0:1:0] & \ell_{32} = [0:1:t] \\ \ell_{13} = [0:1:\mu] & \ell_{23} = [0:0:1] & \ell_{33} = [0:1:u] \\ \ell_{14} = [x_0 : x_1 : x_2] & \ell_{24} = [y_0 : y_1 : y_2] & \ell_{34} = [v_0 : v_1 : v_2]. \end{array}$$

Here we assumed that $p = \ell_{12} \cap \ell_{13} \cap \ell_{22} \cap \ell_{23} \cap \ell_{32} \cap \ell_{33}$. Using the incidence relations imposed by the multinet structure, we can express this family of line arrangements in terms of the two parameters λ and μ. The blocks $\mathcal{A}_1$ and $\mathcal{A}_2$ can be used to compute the coordinates for the twelve points of $\mathcal{X}$ with multiplicity 1 as

$$\begin{array}{ll} \ell_{11} \cap \ell_{21} = [0:1:-1] & \ell_{13} \cap \ell_{21} = [0:\mu:-1] \\ \ell_{11} \cap \ell_{22} = [1:0:-1] & \ell_{13} \cap \ell_{24} = [\mu y_1 - y_2 : -\mu y_0 : y_0] \\ \ell_{11} \cap \ell_{23} = [1:-1:0] & \ell_{14} \cap \ell_{21} = [0 : x_2 : -x_1] \\ \ell_{11} \cap \ell_{24} = [y_1 - y_2 : y_2 - y_0 : y_0 - y_1] & \ell_{14} \cap \ell_{22} = [x_2 : 0 : -x_0] \\ \ell_{12} \cap \ell_{21} = [0:\lambda:-1] & \ell_{14} \cap \ell_{23} = [x_1 : -x_0 : 0] \\ \ell_{12} \cap \ell_{24} = [\lambda y_1 - y_2 : -\lambda y_0 : y_0] & \ell_{14} \cap \ell_{24} = [z_0 : z_1 : z_2] \end{array}$$

where

$$\begin{aligned} z_0 &= x_2y_1 - x_1y_2 \\ z_1 &= x_0y_2 - x_2y_0 \\ z_2 &= x_1y_0 - x_0y_1. \end{aligned}$$

Consider the line ℓ_{32} and its intersections with lines from $\mathcal{A}_1$ and $\mathcal{A}_2$. Each intersection point lies in $\mathcal{X}$. The double point p is the intersection with $\ell_{12}, \ell_{13}, \ell_{22}$, and ℓ_{23}. It follows that ℓ_{32} passes through either $\ell_{11} \cap \ell_{21}$ and $\ell_{14} \cap \ell_{24}$, or $\ell_{11} \cap \ell_{24}$ and $\ell_{14} \cap \ell_{21}$. We may choose our labels so that $\ell_{11} \cap \ell_{21}$ and $\ell_{14} \cap \ell_{24}$ lie on ℓ_{32} which implies $t = 1$. It follows that $\ell_{11} \cap \ell_{24}$ and $\ell_{14} \cap \ell_{21}$ lie on ℓ_{33}.

Next consider the line ℓ_{22}. A similar argument shows ℓ_{22} passes through either $\ell_{11} \cap \ell_{31}$ and $\ell_{14} \cap \ell_{34}$, or $\ell_{11} \cap \ell_{34}$ and $\ell_{14} \cap \ell_{31}$. We may choose our labels so that $\ell_{11} \cap \ell_{31}$ and $\ell_{14} \cap \ell_{34}$ lie on ℓ_{22}. It follows that $\ell_{11} \cap \ell_{34}$ and $\ell_{14} \cap \ell_{31}$ lie on ℓ_{23}. The incidence relations for twelve points in $\mathcal{X}$ of multiplicity 1 based on these choice of labels are

$$\begin{array}{lll} \ell_{11} \cap \ell_{21} \cap \ell_{32} & \ell_{12} \cap \ell_{21} \cap \ell_{31} & \ell_{14} \cap \ell_{21} \cap \ell_{33} \\ \ell_{11} \cap \ell_{22} \cap \ell_{31} & \ell_{12} \cap \ell_{24} \cap \ell_{34} & \ell_{14} \cap \ell_{22} \cap \ell_{34} \\ \ell_{11} \cap \ell_{23} \cap \ell_{34} & \ell_{13} \cap \ell_{21} \cap \ell_{34} & \ell_{14} \cap \ell_{23} \cap \ell_{31} \\ \ell_{11} \cap \ell_{24} \cap \ell_{33} & \ell_{13} \cap \ell_{24} \cap \ell_{31} & \ell_{14} \cap \ell_{24} \cap \ell_{32}. \end{array}$$

It is straightforward to use these incidence relations to express this family in terms of the parameters λ and μ, namely

$$\begin{array}{lll} \ell_{11} = [1:1:1] & \ell_{21} = [1:0:0] & \ell_{31} = [\lambda:1:\lambda] \\ \ell_{12} = [0:1:\lambda] & \ell_{22} = [0:1:0] & \ell_{32} = [0:1:1] \\ \ell_{13} = [0:1:\mu] & \ell_{23} = [0:0:1] & \ell_{33} = [1:1:\lambda\mu] \\ \ell_{14} = [\lambda:1:\lambda\mu] & \ell_{24} = [\lambda:1+\lambda:\lambda(1+\mu)] & \ell_{34} = [1:1:\mu]. \end{array}$$

The base $\mathcal{X}$ consists of thirteen points. The twelve points of multiplicity 1 are

$$\begin{array}{ll} \ell_{11} \cap \ell_{21} \cap \ell_{32} = [0:1:-1] & \ell_{13} \cap \ell_{21} \cap \ell_{34} = [0:\mu:-1] \\ \ell_{11} \cap \ell_{22} \cap \ell_{31} = [1:0:-1] & \ell_{13} \cap \ell_{24} \cap \ell_{31} = [\lambda-\mu:\lambda\mu:-\lambda] \\ \ell_{11} \cap \ell_{23} \cap \ell_{34} = [1:-1:0] & \ell_{14} \cap \ell_{21} \cap \ell_{33} = [0:\lambda\mu:-1] \\ \ell_{11} \cap \ell_{24} \cap \ell_{33} = [1-\lambda\mu:\lambda\mu:-1] & \ell_{14} \cap \ell_{22} \cap \ell_{34} = [\mu:0:-1] \\ \ell_{12} \cap \ell_{21} \cap \ell_{31} = [0:\lambda:-1] & \ell_{14} \cap \ell_{23} \cap \ell_{31} = [1:-\lambda:0] \\ \ell_{12} \cap \ell_{24} \cap \ell_{34} = [\lambda-\mu:-\lambda:1] & \ell_{14} \cap \ell_{24} \cap \ell_{32} = [\lambda\mu-1:\lambda:-\mu] \end{array}$$

and the point of multiplicity 2 is

$$\ell_{12} \cap \ell_{13} \cap \ell_{22} \cap \ell_{23} \cap \ell_{32} \cap \ell_{33} = [1:0:0].$$

For the twelve lines and thirteen points of $\mathcal{X}$ to all be distinct, the parameters must satisfy the following conditions: $\lambda \neq \mu$, $\lambda\mu \neq 1$, and $\lambda, \mu \neq 0, 1$. Observe that the intersection points within each block are

$$\begin{array}{ll}
\ell_{11} \cap \ell_{12} = [1-\lambda : \lambda : -1] & \ell_{22} \cap \ell_{23} = [1 : 0 : 0] \\
\ell_{11} \cap \ell_{13} = [1-\mu : \mu : -1] & \ell_{22} \cap \ell_{24} = [\lambda + \lambda\mu : 0 : -\lambda] \\
\ell_{11} \cap \ell_{14} = [\lambda\mu - 1 : \lambda - \lambda\mu : 1-\lambda] & \ell_{23} \cap \ell_{24} = [1+\lambda : -\lambda : 0] \\
\ell_{12} \cap \ell_{13} = [1 : 0 : 0] & \ell_{31} \cap \ell_{32} = [\lambda - 1 : \lambda : -\lambda] \\
\ell_{12} \cap \ell_{14} = [1-\mu : -\lambda : 1] & \ell_{31} \cap \ell_{33} = [\mu - 1 : -\lambda\mu : 1] \\
\ell_{13} \cap \ell_{14} = [\mu - \lambda\mu : -\lambda\mu : \lambda] & \ell_{31} \cap \ell_{34} = [\lambda - \mu : \lambda\mu - \lambda : 1-\lambda] \\
\ell_{21} \cap \ell_{22} = [0 : 0 : 1] & \ell_{32} \cap \ell_{33} = [1 : 0 : 0] \\
\ell_{21} \cap \ell_{23} = [0 : 1 : 0] & \ell_{32} \cap \ell_{34} = [1-\mu : -1 : 1] \\
\ell_{21} \cap \ell_{24} = [0 : \lambda + \lambda\mu : -\lambda - 1] & \ell_{33} \cap \ell_{34} = [\lambda\mu - \mu : -\lambda\mu : 1].
\end{array}$$

Recall that each block is either (1) in general position with six double points or (2) an easel with a unique triple point and 3 double points. The block $\mathcal{A}_1$ is in general position except if $\mu = 2 - \lambda$ or $\mu = \lambda/(2\lambda - 1)$; $\mathcal{A}_2$ is in general position unless $\lambda = -1$ or $\mu = -1$; and $\mathcal{A}_3$ is in general position unless $\mu = 1/(2 - \lambda)$ or $\mu = (2\lambda - 1)/\lambda$. All three blocks are in general position for generic choice of λ and μ. Exactly two blocks are in general position by choosing $\lambda = -1$ and a generic value of μ (see Fig. 2b).

To resolve the remaining cases, choose $\lambda = -1$. Then $\mathcal{A}_1$ is in general position unless $\mu = 3$ or $\mu = 1/3$. Either choice of values for the parameters also satisfies the relation for $\mathcal{A}_3$, hence no blocks are in general position (see Fig. 2a). It is straightforward to verify that if any two blocks are easels, then the third block is also an easel. This gives us the following result.

Theorem 4 *Any light* (3, 4)*-multinet with base* $\mathcal{X}$ *consisting of a unique double point and all other points of multiplicity 1 has exactly one of the following blocks structures:*

1. *every block is in general position.*
2. *one block is an easel, two blocks are in general position.*
3. *every block is an easel.*

The specialization with all blocks in general position can be induced from Q_2. It is also possible to induce the specialization with every block being an easel from Q_3 using double cancellation. On the other hand, it is not possible to induce from Q_n the specialization with exactly two blocks in general position.

4 Complete Multinets

A Riemann-Hurwitz type formula was obtained for multinets in [5] by calculating the Euler characteristic of the blowup of $\mathbb{P}^2$ at the points of $\mathcal{X}$ using the Ceva pencil. This formula can be used to determine whether all singular fibers of a Ceva pencil associated to a multinet are completely reducible. In this case, the complement of the arrangement is aspherical and the multinet is referred to as a $K(\pi, 1)$-arrangement.

4.1 Classification of Complete 3-Nets

We recall the definition of complete multinets and results obtained in [5]. Then we present and establish our main result, namely the classification of complete 3-nets.

We begin by introducing some additional notation. Let $P_{\mathcal{A}}$ be the set of intersection points of $\mathcal{A}$. Let $\overline{\mathcal{X}}$ denote the set of intersection points of $\mathcal{A}$ not contained in $\mathcal{X}$. Thus $P_{\mathcal{A}} = \overline{\mathcal{X}} \sqcup \mathcal{X}$ and $\overline{\mathcal{X}} \cap P_{\mathcal{A}_i}$ is the set of intersection points of the block $\mathcal{A}_i$ not contained in $\mathcal{X}$. For $p \in \overline{\mathcal{X}}$, let m_p be the multiplicity of p in $\mathcal{A}$. The next two results and subsequent definition were introduced in [5].

Theorem 5 *Let $\mathcal{A}$ be a (k, d)-multinet, and let $\pi : \mathbb{P}^2 \to \mathbb{P}^1$ be the associated Ceva pencil. Then*

$$3 + |\mathcal{X}| \geq (2 - k)[3d - d^2 + \sum_{p \in \mathcal{X}} (n_p^2 - n_p)] + 2|\mathcal{A}| - \sum_{p \in \overline{\mathcal{X}}} (m_p - 1) \qquad (1)$$

with equality if and only if the blocks of $\mathcal{A}$ form the only singular fibers of π.

Corollary 1 *Equality holds in (1) if and only if the restriction of π to the complement $M = \mathbb{P}^2 - (\cup\mathcal{A})$ of $\mathcal{A}$ is a smooth bundle projection with base $B = \mathbb{P}^1 - (k$ points) and fiber a smooth surface with some points removed. In particular, $\mathcal{A}$ is a $K(\pi, 1)$-arrangement.*

Definition 2 A (k, d)-multinet $(d \geq 2)$ or its associated Ceva pencil is called *complete* if the equality holds in (1). When $k = 3$ this condition reduces to

$$\sum_{p \in \overline{\mathcal{X}}} (m_p - 1) \geq 2|\mathcal{A}| - |\mathcal{X}| - 3(d + 1) + \sum_{p \in \mathcal{X}} n_p. \qquad (2)$$

Thus the underlying arrangement of a complete multinet is a $K(\pi, 1)$-arrangement.

Falk and Yuzvinsky present several examples of complete multinets in [5], specifically the arrangements presented in Examples 2, 3, and 4. It follows that any arrangement which is lattice equivalent to the one of the arrangements defined by $[x^n - y^n][x^n - z^n][y^n - z^n]$, $[x^n(y^n - z^n)][y^n(x^n - z^n)][z^n(x^n - y^n)]$, or the Hesse configuration is complete. Currently, these are the only known examples of complete multinets. In fact, we show that the family of examples given by the Fermat pencil are the only complete 3-nets.

Theorem 6 *A complete* $(3, n)$*-net is projectively equivalent to the arrangement with defining polynomial* $[x^n - y^n][x^n - z^n][y^n - z^n]$.

Proof By Proposition 3.3 of [16], it suffices to show each block of a complete $(3, n)$-net is a pencil. Using Definition 2 and Proposition 1, the Riemann-Hurwitz type formula (2) for a complete $(3, n)$-net becomes

$$\begin{aligned}\sum_{p\in\overline{X}}(m_p - 1) &= 2|\mathcal{A}| - |X| - 3(n+1) + \sum_{p\in X} n_p \\ &= 2(3n) - |X| - 3(n+1) + |X| \\ &= 3(n-1).\end{aligned}$$

Consider the block $\mathcal{A}_i$. Note $n > 1$ and select a line $\ell_0 \in \mathcal{A}_i$. Since $\ell \cap \ell_0 \in \overline{X} \cap P_{\mathcal{A}_i}$ for each $\ell \in \mathcal{A}_i \setminus \{\ell_0\}$, we have

$$\begin{aligned}\sum_{p\in\overline{X}\cap\ell_0}(m_p - 1) &= \left(\sum_{p\in\overline{X}\cap\ell_0} m_p\right) - |\overline{X}\cap\ell_0| \\ &= [(n-1) + |\overline{X}\cap\ell_0|] - |\overline{X}\cap\ell_0| \\ &= n-1.\end{aligned}$$

Moreover, $\overline{X} \cap \ell_0 \subseteq \overline{X} \cap P_{\mathcal{A}_i}$ and $m_p \geq 2$ for each $p \in \overline{X} \cap P_{\mathcal{A}_i}$, hence

$$\sum_{p\in\overline{X}\cap P_{\mathcal{A}_i}}(m_p - 1) \geq \sum_{p\in\overline{X}\cap\ell_0}(m_p - 1) = n-1. \tag{3}$$

It follows that

$$\begin{aligned}\sum_{p\in\overline{X}}(m_p - 1) &\geq 3\cdot\min_i\left(\sum_{p\in\overline{X}\cap P_{\mathcal{A}_i}}(m_p - 1)\right) \\ &\geq 3(n-1).\end{aligned}$$

If $\overline{X} \cap \ell_0 = \overline{X} \cap P_{\mathcal{A}_i}$ for each i and every $\ell_0 \in \mathcal{A}_i$, or equivalently $\mathcal{A}_i$ is a pencil for each i, then equality holds and the 3-net is complete. Conversely, suppose the 3-net has another intersection point $p_0 \in (\overline{X} \cap P_{\mathcal{A}_i}) \setminus (\overline{X} \cap \ell_0)$ for some i. Then (3) is a strict inequality since the lefthand sum is larger by at least one and implies that the 3-net is not complete. The result now follows. □

Vallès shows in Theorem 2.7 of [15] that the union of all the singular fibers of a pencil of degree n plane curves whose base locus consisting of n^2 distinct points is a free divisor. Since all singular fibers of a Ceva pencil are completely reducible for a complete multinet, we obtain the following statement.

Theorem 7 *Complete nets are free.*

Alternatively, this can be established for 3-nets by combining Theorem 6 and Theorem 6.60 from [12]. Note that the Hesse configuration is the only currently known 4-net. It is complete (see [5]) and free (see Proposition 6.85 of [12]).

4.2 Completeness of Induced Multinets from Q_n

The known infinite families of complete 3-multinets are related to induced multinets from Q_n. The $(3, 2n)$-multinets of type $G(n, 1, 3)$ are induced multinets from Q_n by choosing H to be the plane $x_0 = 0$. Furthermore the $(3, n)$-nets from the Fermat pencil appear as their subarrangements. The latter family of nets are not inducible directly from Q_n with one exception, namely any $(3, 2)$-net is complete and can be induced from Q_1.

With induced multinets from Q_n providing numerous examples of multinets, we investigate this class of multinets for completeness. Let $\mathcal{A}_p = \{\ell \in \mathcal{A} : p \in \ell\}$ denote the lines of $\mathcal{A}$ passing through the point p. A useful tool is the following local test for completeness presented in [5].

Proposition 2 *Suppose $\mathcal{A}$ is a complete multinet. Then, for each $p \in \mathcal{X}$,*

$$2n_p - 2 = \sum_{\ell \in \mathcal{A}_p} (m(\ell) - 1). \tag{4}$$

In particular, if the multinet is light, then the multinet is complete only if it is a net.

Note that (4) holds for each $p \in \mathcal{X}$ of a net. Also since the only proper k-multinets occur when $k = 3$ (see Proposition 1), we focus our attention on this situation. Using combinatorial properties of multinets, Proposition 2 can be freshly reformulated to give a local test for completeness which is convenient to implement.

Corollary 2 *Suppose $\mathcal{A}$ is a complete 3-multinet. Then, for each $p \in \mathcal{X}$,*

$$|\mathcal{A}_p| = n_p + 2. \tag{5}$$

Proof For 3-multinets, observe

$$\sum_{\ell \in \mathcal{A}_p} (m(\ell) - 1) = 3n_p - |\mathcal{A}_p|.$$

Substituting into (4) and simplifying gives the statement. □

Theorem 8 *The only complete multinets induced from* Q_n *are the* $(3, 2)$*-net of Coxeter type* A_3 *and the* $(3, 2n)$*-multinets of type* $G(n, 1, 3)$.

Proof A complete description of multinets induced from Q_n was given in Theorem 1 and classified into ten types. We refer to a specific type based on the numbering conventions used there.

It follows from Proposition 2, Theorems 1, and 6 that any light induced multinet from Q_n is not complete with one exception. The $(3, 2)$-net of Coxeter type A_3 can be induced from Q_1. It remains to investigate the completeness of the heavy induced multinets.

Induced multinets of type 1 which realize $G(n, 1, 3)$ are complete. Also the induced multinet of type 3 from Q_2 realizes $G(2, 1, 3)$, hence is also complete. We use Corollary 2 to show the remaining types are not complete by exhibiting a point $p \in \mathcal{A}$ where (5) does not hold. For type 2, choose $p \in \mathcal{X}$ with multiplicity n and observe $|\mathcal{A}_p| = 2n + 1$. Next consider an induced multinet of type 3 (with $n > 2$), of type 4, or of type 5. Choose $p \in \mathcal{X}$ to be a double point which lies on exactly one of the lines of multiplicity 2. Then $n_p = 2$ and $|\mathcal{A}_p| = 5$. This completes the proof. □

5 Open Problems

We list several open problems.

Problem 1 Are there examples of complete multinets other than the ones exhibited in Examples 2, 3, and 4?

Problem 2 How many specializations are possible for the nets realizing Latin squares of small order such as $\mathbb{Z}/4\mathbb{Z}$, $\mathbb{Z}/2\mathbb{Z} \times \mathbb{Z}/2\mathbb{Z}$, $\mathbb{Z}/5\mathbb{Z}$, and Latin square of order 5 which is not isotopic to the multiplication table of a group (see [13])? These are nets constructed by Stipins.

Problem 3 Are there any general properties regarding the number of specializations of a given multinet?

Problem 4 (Yuzvinsky) Is the Hesse configuration the unique 4-net up to projective isomorphism?

Acknowledgements Thank you to Max Wakefield and the anonymous referee for their helpful comments for improving this paper, especially pointing out connections with the results of Vallès. Also, thank you to the organizers of *Configuration Spaces* and *Perspectives in Lie Theory*, Centro de Giorgi, Scuola Normale Superiore di Pisa, Istituto Nazionale di Alta Matematica, and many others involved for being wonderful hosts while working on this project during my visits to Italy.

References

1. J. Bartz, Multinets in $\mathbb{P}^2$ and $\mathbb{P}^3$. Ph.D. Thesis, University of Oregon, 2013
2. J. Bartz, S. Yuzvinsky, Multinets in $\mathbb{P}^2$, in *Bridging Algebra, Geometry, and Topology*, ed. by D. Ibadula, W. Veys. Springer Proceedings in Mathematics and Statistics, vol. 96 (Springer, Berlin, 2014)
3. C. Dunn, M. Miller, M. Wakefield, S. Zwicknagl, Equivalence classes of Latin squares and nets in $\mathbb{CP}^2$ (2008). arXiv:0703.0142v4
4. G. Denham, A. Suciu, Multinets, parallel connections, and Milnor fibrations of arrangements. Proc. Lond. Math. Soc. **108**(6), 1435–1470 (2014)
5. M. Falk, S. Yuzvinsky, Multinets, resonance varieties, and pencils of plane curves. Compos. Math. **143**, 1069–1088 (2007)
6. B. Grünbaum, A catalogue of simplicial arrangements in the real projective plane. ARS Math. Contemp. **2**, 1–25 (2009)
7. M.H. Güntürkün, A. Kişisel, Using tropical degenerations for proving the nonexistence of certain nets (2011). arXiv: 1107.5530v4
8. G. Korchmaros, G.P. Nagy, N. Pace, k-nets embedded in a projective plane over a field. Combinatorica **35**(1), 63–74 (2015)
9. G. Korchmaros, G. Nagy, N. Pace, 3-nets realizing a group in a projective plane. J. Algebr. Comb. **39**(4), 939–966 (2014)
10. A. Libgober, S. Yuzvinsky, Cohomology of the Orlik-Solomon algebras and local systems. Compos. Math. **121**, 337–361 (2000)
11. O. Ore, Theory of monomial groups. Trans. Am. Math. Soc. **51** (1942)
12. P. Orlik, H. Terao, *Arrangements of Hyperplanes* (Springer, Berlin, 1992)
13. J. Stipins, Old and new examples of k-nets in $\mathbb{P}^2$ (2007). arXiv:math/0701046
14. J. Stipins, On finite k-nets in the complex projective plane. Ph.D. Thesis, University of Michigan, 2007
15. J. Vallès, Free divisors in a pencil of curves. J. Singularities **11**, 190–197 (2015)
16. S. Yuzvinsky, Realization of finite Abelian groups by nets in $\mathbb{P}^2$. Compos. Math. **140**, 1614–1624 (2004)
17. S. Yuzvinsky, A new bound on the number of special fibers in a pencil of curves. Proc. Am. Math. Soc. **137**, 1641–1648 (2009)

Cohomology of the Milnor Fibre of a Hyperplane Arrangement with Symmetry

Alexandru Dimca and Gus Lehrer

Abstract We prove some general results concerning the cohomology of the Milnor fibre of a hyperplane arrangement, and apply them to the case when the arrangement has some symmetry properties, particularly the case of the set of reflecting hyperplanes of a unitary reflection group. We relate the isotypic components of the monodromy action on the cohomology to the cohomology degree and to the mixed Hodge structure of the cohomology. We also use monodromy eigenspaces to determine the spectrum in some cases, which in turn throws further light on the equivariant Hodge structure of the cohomology and on the determination of the equivariant Hodge-Deligne polynomials. When the arrangement is the set of reflecting hyperplanes of a unitary reflection group, then using eigenspace theory for reflection groups, we prove sum formulae for additive functions such as the equivariant weight polynomial and certain polynomials related to the Euler characteristic, such as the Hodge-Deligne polynomials. This leads to a case-free determination of the Euler characteristic in this case, answering a question of Denham-Lemire. We also give an alternative formula for the spectrum of an arrangement which permits its computation in low dimensions, and we provide several examples of such computations.

Keywords Equivariant Hodge-Deligne polynomial • Hyperplane arrangement • Milnor fibre • Monodromy

1 Introduction and Notation

Let $\mathcal{A}$ be a finite collection of hyperplanes in $V := \mathbb{C}^\ell$ [29]. For each hyperplane $H \in \mathcal{A}$, let $\ell_H \in V^*$ be a linear form such that $H = \mathrm{Ker}(\ell_H)$. Suppose $G \subset \mathrm{GL}(V)$ is a finite group which stabilises the arrangement $\mathcal{A}$, i.e. such that for each $H \in \mathcal{A}$

A. Dimca
University of Nice Sophia Antipolis, CNRS, LJAD, UMR 7351, 06100 Nice, France
e-mail: dimca@unice.fr

G. Lehrer (✉)
School of Mathematics and Statistics F07, University of Sydney, NSW 2006, Australia
e-mail: gustav.lehrer@sydney.edu.au

F. Callegaro et al. (eds.), *Configuration Spaces*, Springer INdAM Series 14,
DOI 10.1007/978-3-319-31580-5_10

and $g \in G$, $gH \in \mathcal{A}$; note that G may be trivial. For $g \in \mathrm{GL}(V)$, g acts on $f \in \mathbb{C}[V] = S(V^*)$ (S denoting the symmetric algebra) by $gf(v) = f(g^{-1}v)$. Define $Q_0 \in \mathbb{C}[V] = S(V^*)$ by $Q_0 = \prod_{H \in \mathcal{A}} \ell_H$, and write $d = \deg(Q_0) = |\mathcal{A}|$.

Clearly for each $g \in G$, $gQ_0 = \lambda(g)Q_0$ for some root of unity $\lambda(g) \in \mathbb{C}^*$. It follows that there are uniquely defined indices $e_H \in \mathbb{Z}_{>0}$ ($H \in \mathcal{A}$) such that $\widetilde{Q} = \prod_{H \in \mathcal{A}} \ell_H^{e_H}$ is invariant under G, i.e. $g\widetilde{Q} = \widetilde{Q}$ for all $g \in G$, and the e_H are minimal. This can be seen by looking at the G-orbits of hyperplanes. There is also evidently a minimal positive integer e such that $Q := Q_0^e$ is G-invariant. We shall work principally with Q and Q_0.

Definition 1

1. The *Milnor fibre* $F = F(\mathcal{A})$ of the arrangement $\mathcal{A}$ is the variety $F := Q^{-1}(1) \subset V$.
2. The *reduced Milnor fibre* $F_0 = F_0(\mathcal{A})$ of the arrangement $\mathcal{A}$ is the variety $F_0 := Q_0^{-1}(1) \subset V$, where $Q_0(z) := \prod_{H \in \mathcal{A}} \ell_H$.

Write $m := \deg(Q) = \sum_{H \in \mathcal{A}} e_H$, and define $\Gamma := G \times \mu_m$. Then $m = de$ and Γ acts on V as $(g, \zeta)v = \zeta^{-1}gv$ (for $g \in G$ and $\zeta \in \mu_m$), and hence on $\mathbb{C}[V]$ in the usual way ($\gamma f(v) = f(\gamma^{-1}(v))$ for $f \in \mathbb{C}[V]$ and $v \in V$). Moreover $Q \in \mathbb{C}[V]$ is invariant under Γ, whence the Milnor fibre F, and its cohomology, have a Γ-action, the former being an algebraic action, and the latter being the induced linear action.

Write $\Gamma_0 := \{\gamma \in \Gamma \mid \gamma(F_0) \subseteq F_0\}$. Then evidently Γ_0 acts on F_0 as a group of variety automorphisms, and hence acts on its cohomology.

The central problem to be discussed in this work is

1.1 Problem A

(i) Determine the equivariant Poincaré polynomial

$$P^{\Gamma}(F, t) := \sum_{i \geq 0} H^i(F, \mathbb{C})t^i, \tag{1}$$

as an element of $R(\Gamma)[t]$, where t is an indeterminate, and $R(\Gamma)$ is the Grothendieck ring of complex finite dimensional representations of Γ.

(ii) Determine the equivariant Poincaré polynomial

$$P^{\Gamma_0}(F_0, t) := \sum_{i \geq 0} H^i(F_0, \mathbb{C})t^i, \tag{2}$$

as an element of $R(\Gamma_0)[t]$, where t is an indeterminate.

Even when G is trivial, this problem is unsolved in many cases where a solution would have applications in several areas. A notable such example is the following. We shall refer to this example throughout this work to illustrate our results.

Example 1 Take coordinates $z_1, \dots, z_\ell$ in V and let $\mathcal{A}$ be the arrangement $\mathcal{A} = \{H_{ij} \mid 1 \leq i < j \leq \ell\}$, where H_{ij} is the hyperplane $z_i - z_j = 0$. Here we take $G = \mathrm{Sym}_\ell$, $e = 2$, $d = \frac{\ell(\ell-1)}{2}$ and $m = \ell(\ell - 1)$.

More generally, we would like to address the following problem.

1.2 Problem B

Determine each of the mixed Hodge components

$$H^{p,q}(H^j(F_0, \mathbb{C})) := Gr_F^p Gr_{p+q}^W H^j(F_0, \mathbb{C})$$

(see Sect. 6) of the reduced Milnor fibre. That is, for each j, p, q, determine (i) its dimension, and (ii) its structure as Γ_0-module.

We shall shortly see that Problems A(i) and (ii) are equivalent, and at various times we shall focus on one or the other. We shall start by proving some relations between the μ_d-action and cohomology degree, and later we shall relate the action to the mixed Hodge structure of the cohomology, and the determination of the spectrum. In Sect. 4 we shall treat the case when G is a reflection group. In particular, although by no means do we provide a complete solution of either problem, we obtain an improvement of the results of [12] on the equivariant Euler characteristic.

This work is organised as follows. In Sect. 2 we establish some basic algebraic and topological tools, and in Sect. 3 we use them to derive some relations between the order of the monodromy action and cohomology degree. In Sect. 4 we treat the case where G is a unitary reflection group. Our main results here are a general formula for additive functions, such as the equivariant weight polynomial, a closed formula for the equivariant Euler characteristic, and a complete solution of Problem A for the case where $G = \mathrm{Sym}_5$. In Sect. 5 we treat reflection groups of rank 2, and in Sect. 6 we prove a number of general Hodge-theoretic results about the Hodge-Deligne polynomials of the Milnor fibre. In particular, we give a formula (Proposition 15) for the spectrum of an arrangement in terms of its monodromy-equivariant Hodge structure. This permits the calculation of the spectrum of many arrangements in low dimensions, including that of $\mathcal{A}_4$, the set of reflecting hyperplanes of Sym_5.

Finally, we note that our results are far from the last word on these problems, which remain unsolved even in the ubiquitous and much studied case of Example 1, although many partial results are known.

2 Some General Algebraic and Topological Results

In this section, we prove some general results of relevance to our exposition. We begin with a result which is used to relate Problems A(i) and A(ii). The following notation will be used for the irreducible characters of any group μ_m of complex roots of unity. The character $\zeta \mapsto \zeta^i$ $(i = 0, 1, \ldots, m-1)$ is denoted by γ_i. We also denote by $I(H)$ the set of (equivalence classes of) irreducible complex representations of a finite group H.

Lemma 1 *Let $\Gamma = G \times \mu_m$ and $\Gamma_0 := \{\gamma \in \Gamma \mid \gamma Q_0 = Q_0\}$ be the groups defined above.*

(i) For any representation θ of Γ_0, there is a representation $\tilde{\theta}$ of Γ such that $\mathrm{Res}^{\Gamma}_{\Gamma_0}(\tilde{\theta}) = \theta$.
(ii) We have $\mathrm{Ind}^{\Gamma}_{\Gamma_0}(\theta) = \tilde{\theta} \otimes (1 \otimes \gamma_0 + \lambda \otimes \gamma_{\frac{m}{e}} + \lambda^2 \otimes \gamma_{\frac{2m}{e}} + \cdots + \lambda^{e-1} \otimes \gamma_{\frac{(e-1)m}{e}})$, where $\tilde{\theta}$ is as above.

Proof First we identify $\Gamma_0 \subseteq \Gamma$. Recall that for $g \in G$, $gQ_0 = \lambda(g)Q_0$ for some root of unity $\lambda(g)$, and the map $g \mapsto \lambda(g)$ is clearly a 1-dimensional representation of G. Moreover by choice of e as the minimal power of Q_0 which is G-invariant, the linear character λ has order e. Now $(g, \xi)Q_0(v) = \lambda(g)\xi^d Q_0(v)$. Hence $\Gamma_0 = \mathrm{Ker}(\lambda \otimes \gamma_d)$. Further, since there is an element $g \in G$ such that $\lambda(g)$ generates μ_e, we have $G\Gamma_0 = \Gamma$, and hence $G/(G \cap \Gamma_0) \cong \Gamma/\Gamma_0 \cong \mu_e$. Write $G_0 = G \cap \Gamma_0$.

Now to understand induction from Γ_0 to Γ, we use Clifford theory, applied to our simple case ([16, Theorem 1], [17]). The facts we require are as follows. If H is a finite group and K is a normal subgroup, we identify $I(H/K)$ with the subset of $I(H)$ consisting of representations in which K acts trivially. For $\beta \in I(H)$, define $F(\beta) = \{\xi \in I(H/K) \mid \beta\xi = \beta\}$. Then $\mathrm{Res}^H_K(\beta)$ is irreducible if and only if $|F(\beta)| = 1$.

In our present situation, taking $H = \Gamma$ and $K = \Gamma_0$, any element of $I(\Gamma)$ is of the form $\theta = \rho \otimes \gamma_i$, where $\rho \in I(G)$ and γ_i is as defined above. In particular, since $\Gamma_0 = \mathrm{Ker}(\lambda \otimes \gamma_d)$, $I(\Gamma/\Gamma_0)$ may be identified with $\{\lambda^j \otimes \gamma_{\frac{jm}{e}} = (\lambda \otimes \gamma_d)^j \mid j = 0, 1, \ldots, e-1\}$.

Now for any $\beta = \rho \otimes \gamma_i \in I(\Gamma)$ we have $\lambda^j \otimes \gamma_{\frac{jm}{e}} \beta = \lambda^j \rho \otimes \gamma_{\frac{jm}{e}+i} = \beta$ if and only if $j = 0$, since otherwise, the second factor is distinct from γ_i. Thus from Clifford theory we deduce that each element $\beta \in I(\Gamma)$ restricts to an irreducible representation of Γ_0. It follows easily by Frobenius reciprocity that for any element $\theta \in I(\Gamma_0)$, if $\tilde{\theta}$ is any irreducible component of $\mathrm{Ind}^{\Gamma}_{\Gamma_0}(\theta)$, then $\mathrm{Res}^{\Gamma}_{\Gamma_0}(\tilde{\theta}) = \theta$, and that $\mathrm{Ind}^{\Gamma}_{\Gamma_0}(\theta) = \sum_{j=0}^{e-1} (\lambda \otimes \gamma_d)^j \tilde{\theta}$, which proves both (i) and (ii). □

Let $R(H) = \sum_{\beta \in I(H)} \mathbb{Z}\beta$ be the Grothendieck ring of the finite group H, and let $R_+(H) = \sum_{\beta \in I(H)} \mathbb{Z}_{\geq 0}\beta$. If K is a subgroup of H, we have homomorphisms $\mathrm{Ind}^H_K : R(K) \to R(H)$ and $\mathrm{Res}^H_K : R(H) \to R(K)$, which both take R_+ to R_+, and which satisfy Frobenius reciprocity. These maps extend in the obvious way to maps between $R(H)[t]$ and $R(K)[t]$, where t is an indeterminate.

Corollary 1 *Maintain the notation of Lemma 1.*

(i) *For any element* $\theta(t) = \sum_i \theta_i t^i \in R_+(\Gamma_0)[t]$, *there is an element* $\tilde{\theta}(t) \in R_+(\Gamma)$ *such that* $\operatorname{Res}^{\Gamma}_{\Gamma_0} \tilde{\theta}(t) = \theta(t)$.
Moreover $\operatorname{Ind}^{\Gamma}_{\Gamma_0}(\theta(t)) = \sum_{j=0}^{e-1}(\lambda \otimes \gamma_d)^j \tilde{\theta}(t)$.

(ii) *In the notation of* (1) *and* (2), *we have* $P^{\Gamma}(F,t) = \sum_{j=0}^{e-1}(\lambda \otimes \gamma_d)^j \widetilde{P}^{\Gamma_0}(F_0,t)$, *where* $\widetilde{P}^{\Gamma_0}(F_0,t)$ *is an extension to* Γ *of* $\widetilde{P}^{\Gamma_0}(F_0,t)$, *as in (i). Similar statements hold for the equivariant Hodge-Deligne polynomials, to be defined later.*

Proof Part (i) is immediate from Lemma 1.

To see (ii), note that F has irreducible components $\{F_j \mid j = 0, \ldots, e-1\}$, where F_j is defined as the locus of $Q_0(v) = \exp\frac{2\pi\sqrt{-1}j}{e}$. These are permuted transitively by Γ, and by definition, the stabiliser of F_0 is Γ_0. It follows that for each degree p, as Γ-module, $H^p(F,\mathbb{C}) \cong \operatorname{Ind}^{\Gamma}_{\Gamma_0}(H^p(F_0,\mathbb{C})$. The statement is now clear from (i), as are the statements about the Hodge structure. □

It follows that knowledge of $P^{\Gamma}(F,t)$ is equivalent to knowledge of $P^{\Gamma_0}(F_0,t)$. It will sometimes be helpful below to restrict attention to the 'reduced case'.

Our next two results are topological.

Proposition 1 *Let* X *be a CW-complex with a free* G*-action, where* G *is a finite group. Assume that the quotient space* $Y = X/G$ *has the homotopy type of a finite CW-complex. Then* X *has the homotopy type of a finite CW-complex, and we have the following equation in the Grothendieck ring of* G.

$$\chi^G(X) = \chi(Y) \cdot \operatorname{Reg}_G,$$

where Reg_G *is the regular representation of* G *and* $\chi^G(X) = \sum_j (-1)^j H^j(X,\mathbb{C})$ *is the equivariant Euler characteristic of* X.

Proof It follows from [36] (see also [4, Sect. 3.13]) that the virtual representation $\chi^G(X)$ is an integer multiple $c \cdot \operatorname{Reg}_G$ of the regular representation of G. This may also be easily seen in our context because in view of the free nature of the action of G on X, the cohomology of X is the cohomology of a cochain complex such that G acts in each degree as a multiple of the regular representation. Thus this first assertion follows from the Hopf trace formula.

To determine c, take the inner product of $\chi^G(X) = c \cdot \operatorname{Reg}_G$ with 1_G. Since $\dim H^i(X/G) = (H^i(X), 1_G)_G$, we have $(\chi^G(X), 1_G)_G = \chi(X/G)$. But $(\operatorname{Reg}_G, 1_G)_G = 1$, whence $c = \chi(X/G)$. □

The following result, which goes back to Milnor [25, pp. 76–78] and Oka [26] may be found in [9, Proposition 3.1.21]. We provide a short proof below, which shows that it is a simple consequence of Proposition 1.

Proposition 2 *Let* $Q_0 \in \mathbb{C}[x_1, \ldots, x_n]$ *be homogeneous of degree* d. *Let* $F_0 = \{x = (x_1, \ldots, x_n) \in \mathbb{C}^n \mid Q_0(x_1, \ldots, x_n) = 1\}$ *and* $U = \{x \in \mathbb{P}^{n-1} \mid Q_0(x_1, \ldots, x_n) \neq 0\}$. *Let* $h : F_0 \to F_0$ *be given by* $h(x) = \zeta x$, *where* $\zeta \in \mathbb{C}^*$ *has order* d *(this is the* *'monodromy action'). Define the zeta function*

$$Z(t) = \prod_{p=0}^{n-1} \det\left(\mathrm{Id} - th^* | H^p(F_0, \mathbb{C})\right)^{(-1)^p},$$

where h^* is the action induced by h on cohomology. Then

$$Z(t) = (1 - t^d)^{\frac{\chi(F_0)}{d}} = (1 - t^d)^{\chi(U)},$$

where χ denotes Euler characteristic.

Proof Let $H = \langle h \rangle$. For any element $y \in H$, we have the evaluation homomorphism $\varepsilon_y : R(H) \to \mathbb{C}$, where $R(H)$ is the Grothendieck ring of H; this is defined by $\varepsilon_y(\theta) = \mathrm{Trace}(y, \theta)$, which we generally write simply as $\theta(y)$. This homomorphism extends in the obvious way to $R(H)[[t]]$, and we consider $Z(t)$ as the evaluation at $h \in H$ of the element $Z^H(t) := \prod_{p=0}^{n-1} \det(\mathrm{Id} - tH^p(F_0, \mathbb{C}))^{(-1)^p}$ of $R(H)[[t]]$.

Taking the formal logarithm of $Z^H(t)$, using the identity $\ln(1-x) = -(x + \frac{x^2}{2} + \frac{x^3}{3} + \frac{x^4}{4} + \ldots)$, we see that

$$-\ln(Z^H(t))(h) = \sum_{j=1}^{\infty} \left(\chi^H(h^j)\right) \frac{t^j}{j}.$$

But by Proposition 1, since H acts freely on F_0, the equivariant Euler characteristic $\chi^H(F_0) = \chi(U)\,\mathrm{Reg}_H$, whence it follows that the above expression simplifies to

$$-\ln(Z^H(t))(h) = \chi(U) \sum_{i=1}^{\infty} \frac{t^{id}}{i},$$

and this last expression is evidently equal to $-\ln\left((1 - t^d)^{\chi(U)}\right)$. □

3 Relations Between Monodromy and Cohomology Degree

In this section we prove some results which imply restrictions on the cohomology degree in which monodromy of a specified order may appear.

3.1 A Diagram of Spaces

Given a hyperplane arrangement in $V = \mathbb{C}^\ell$, write $M = V \setminus \cup_{H\in\mathcal{A}} H$ and $\widetilde{M} := \{(v,\xi) \in M \times \mathbb{C}^\times \mid Q(v) = \xi^m\}$. Then we have maps $p_1 : \widetilde{M} \to M$ and $\pi_1 : \widetilde{M} \to F$, where p_1 is the first projection, and $\pi_1(\xi, v) = \xi^{-1}v$. We also have $\pi : M \to U = \mathbb{P}(M)$ and $p : M \to U$ given by $\pi(v) = [v]$ and $p(v) = [v]$, the latter for $v \in F$. These maps fit together in a commutative diagram as follows.

$$\begin{array}{ccc} \widetilde{M} & \xrightarrow{p_1} & M \\ {\scriptstyle \pi_1}\downarrow & & \downarrow{\scriptstyle \pi} \\ F & \xrightarrow{p} & U = \mathbb{P}(M) \end{array} \tag{3}$$

The group Γ acts on $\widetilde{M}$ component-wise: $(g,\zeta)(v,\xi) := (gv, \zeta\xi)$ for $(g,\zeta) \in \Gamma$ and $(v,\xi) \in \widetilde{M}$. Since $\pi_1(v,\xi) = \xi^{-1}v$, it is easily checked that π_1 respects the Γ-action, and that the horizontal arrows are maps to the quotient by μ_m.

Note that the map $(v,\xi) \mapsto (\xi^{-1}v, \xi) : \widetilde{M} \to F \times \mathbb{C}^\times$ is an isomorphism of varieties, and the map $p_1 : F \times \mathbb{C}^\times \to M$ is given by $(v,\xi) \mapsto \xi v$. With this identification, the Γ-action on $F \times \mathbb{C}^\times$ is given by $(g,\zeta).(w,\xi) = (\zeta^{-1}gw, \zeta\xi)$, for $(g,\zeta) \in \Gamma$ and $(w,\xi) \in F \times \mathbb{C}^\times$. In particular, this realises M as the quotient of $F \times \mathbb{C}^\times$ by μ_m, the latter acting via $\zeta(w,\xi) = (\zeta^{-1}w, \zeta\xi)$.

Notice that the action of both G and of μ_m on F is free. We next record the result of applying Proposition 1 to these free actions. Observe that in the diagram (3) we could replace F by F_0; this will be implicit in the discussion below.

Proposition 3

1. *We have* $\chi^G(F) = \chi(F/G)\,\mathrm{Reg}_G$.
2. *We have* $\chi^{\mu_m}(F) = \chi(U)\,\mathrm{Reg}_{\mu_m}$.
   Now assume that G is a unitary reflection group in the sense of [23] and that $\mathcal{A} = \mathcal{A}_G$ *is the set of reflecting hyperplanes of G.*
3. *Let* $1 = m_1^* \le m_2^* \le \dots \le m_r^*$ *be the non-zero coexponents of G (see [23, Definition 10.27, p. 257]). Here* $r = \mathrm{codim}_V(\cap_{H\in\mathcal{A}_G} H)$ *is the rank of G. Then* $\chi(U) = (-1)^{r-1}(m_2^*-1)(m_3^*-1)\dots(m_r^*-1)$.
4. *We have* $\chi(F/G) = |G|^{-1}m(-1)^{r-1}(m_2^*-1)(m_3^*-1)\dots(m_r^*-1)$.

Proof The assertions (1) and (2) are immediate from Proposition 1. To see (3), note that by [23, p. 257], the Poincaré polynomial of M is $P_M(t) = \prod_{i=1}^r(1+m_i^*t)$, where the m_i^* are the coexponents of G. Moreover precisely r of the coexponents are non-zero, and the smallest of these is the degree of the Euler form, viz. $m_1^* = 1$. But from the diagram above, we see that $P_U(t) = \frac{P_M(t)}{(1+t)}$. This proves (3). Finally, it follows from (1) and (2) by evaluation at the identity element, that $\chi(F) = m\chi(U) = |G|\chi(F/G)$. Hence (4) is immediate from (3). □

3.2 Monodromy Order and Cohomology Degree

We will now apply Proposition 2 to obtain constraints upon the action of the monodromy on the cohomology of F_0 . Define $Y = \cup_{H\in\mathcal{A}}H$ and let $N = \mathbb{P}(Y)$ be the corresponding hypersurface in the projective space $\mathbb{P}(V)$. For any character $\gamma \in \hat{\mu}_d$, it is known that

$$\dim H^k(F_0)_\gamma = \dim H^k(U, L_\gamma)$$

where L_γ is the rank one local system on U with monodromy around each hyperplane in N given by multiplication by $\gamma(\zeta_d)$, see for instance [10, Proposition 6.4.6].

In order to prove vanishing results for the twisted cohomology groups $H^k(U, L_\gamma)$ we need a good compactification of U, which is the same as an embedded resolution of N. To describe this, we need the following basic notion about hyperplane arrangements. An *edge* X is an intersection of hyperplanes in $\mathcal{A}$. An edge X is *dense* if the subarrangement of hyperplanes $\mathcal{A}_X$ containing it is irreducible, i.e. the hyperplanes in $\mathcal{A}_X$ cannot be partitioned into two nonempty sets so that after a linear change of coordinates on V hyperplanes in different sets are given by linear equations in different, disjoint sets of coordinates. In particular, any hyperplane is a dense edge.

The condition of X being dense is a combinatorially determined condition which can be checked in a neighborhood of a given edge, see [32]. Let $D(\mathcal{A})$ denote the set of dense edges of the arrangement $\mathcal{A}$.

There is a canonical way to obtain an embedded resolution of the divisor N in $\mathbb{P}(V)$ (cf. [7]). First, blow up the points on $\mathbb{P}(V)$ which correspond to the dense 1-dimensional edges of $\mathcal{A}$ to obtain a map $p_1 : Z_1 \to \mathbb{P}(V)$. Then, blow up all the proper transforms under p_1 of projective lines on $\mathbb{P}(V)$ corresponding to the dense 2-dimensional edges in $D(\mathcal{A})$. Continuing in this way, we get a map $p = p_{\ell-2} : Z = Z_{\ell-2} \to \mathbb{P}(V)$ which is an embedded resolution of the divisor N in $\mathbb{P}(V)$, if $\ell = \dim V$. Then, $D = p^{-1}(N)$ is a normal crossing divisor in Z, with smooth irreducible components D_X corresponding to the edges $X \in D(\mathcal{A})$ with $\dim X > 0$. Furthermore, the map p induces an isomorphism $Z \setminus D = U$, see [30, 32] for details.

Let L be a rank one local system on $Z \setminus D = U$. Then the monodromy of L about an irreducible component D_X of the normal crossing divisor D is the nonzero complex number c_X obtained by taking the products of the complex numbers c_H which represent the monodromies of L about the hyperplanes $H \in \mathcal{A}$ such that $X \subset H$.

In particular, for $L = L_\gamma$ as above, we get $c_X = \gamma(\zeta_d)^{m_X}$, where m_X denotes the number of hyperplanes $H \in \mathcal{A}$ such that $X \subset H$.

The following vanishing result follows from [10, Theorem 6.4.18 and Remark 6.4.20]. See also [24] and [3].

Theorem 1 *Assume that for a given rank one local system L, there is a hyperplane $H \in \mathcal{A}$ such that $c_X \neq 1$ for any dense edge $X \in D(\mathcal{A})$ with $X \subseteq H$ and* $\operatorname{codim} X \leq c$. *Then $H^p(U, L) = 0$ for any p with $0 \leq p < c$.*

Taking into account standard affine Lefschetz theorems, this implies the next result which relates cohomology degree and monodromy order.

Proposition 4 *Let $\mathcal{A}$ be an arrangement of d hyperplanes in $\mathbb{C}^\ell$ and let $Q_0 = \prod_{H\in\mathcal{A}} \alpha_H$ where ℓ_H is a linear form corresponding to $H \in \mathcal{A}$. Let F_0 be the corresponding reduced Milnor fibre, and let γ be a character of μ_d which occurs with non-zero multiplicity in $H^s(F_0, \mathbb{C})$ for some $s > 0$. Then there is a dense edge X in $D(\mathcal{A})$ such that* $\operatorname{codim} X \leq s + 1$ *and the order of γ divides the multiplicity m_X.*

Recall that if G is a reflection group in V, then writing $E = \cap_{H\in\mathcal{A}} H$, the rank of the arrangement $\mathcal{A}$ is $r = \operatorname{codim} E$. The corresponding Milnor fibre F_0 is naturally isomorphic to a product $F_1 \times E$, with F_1 a hypersurface in an r-dimensional affine space. In particular $H^k(F_0) = H^k(F_1) = 0$ for $k \geq r$.

Moreover, for any $X \in D(\mathcal{A})$ with $\dim X > \dim E$ one has $m_X < d$. This yields the following statement.

Corollary 2 *In the notation above, $H^j(F_0, \mathbb{C}) = 0$ if $j \geq r$. Moreover if γ is a faithful character of μ_d, i.e., γ has order d, then γ occurs in $H^j(F_0)$ with non-zero multiplicity if and only if $j = r - 1$. The multiplicity of γ in $H^{r-1}(F_0)$ is equal to $|\chi(F_0)/d| = |\chi(U)|$.*

This yields the following result.

Corollary 3 *Let G be a reflection group in $V = \mathbb{C}^\ell$. Let r be the rank of the corresponding hyperplane arrangement, and let γ be a faithful character of $\mu_{\frac{m}{2}}$ where $m = \sum_{H\in\mathcal{A}_G} e_H$. Then γ occurs in $H^j(F_0, \mathbb{C})$ with non-zero multiplicity only if $j = r - 1$, and the multiplicity of γ in $H^{r-1}(F_0, \mathbb{C})$ is $|e(G)| = \prod_{i\geq 2}(m_i^* - 1)$.*

Proof Apply Corollary 2 to the polynomial Q_0 of Sect. 1. The first assertion is immediate, and in this case it follows from Proposition 3 that $\chi(F_0)/d = \chi(F)/m = \chi(F^{\mu_m}) = \chi(U) = (-1)^{r-1}(m_2^* - 1)(m_3^* - 1)\dots(m_{r-1}^* - 1)$. □

Remark 1

1. Note that Corollaries 2 and 3 are consistent with Proposition 3(2).
2. It may be easily shown, using [23, Corollary 10.4.1] that

$$e(G) = (-1)^\ell \sum_{g\in G} \det{}_V g^{-1} \dim(\operatorname{Fix} g).$$

Example 2 (Continuation of Example 1) We take $G = Sym_{\ell+1}$ acting by permutations on $\mathbb{C}^{\ell+1}$. In this case the rank of the corresponding braid arrangement $\mathcal{A}_\ell$ is $r = \ell$.

The edges $X \in L(\mathcal{A})$ are given, up to the induced action of G, by partitions $\kappa = (k_1 \geq k_2 \geq \dots k_s \geq 1)$ of $\ell + 1$. The dense edges correspond to partitions satisfying $k_2 = \ldots = k_s = 1$. If X is associated to such a partition, then X has codimension $\ell + 1 - s$ and multiplicity $m_X = \binom{\ell+2-s}{2}$.

This proves the following.

Corollary 4 *Let $G = Sym_{\ell+1}$ acting by permutations on $\mathbb{C}^{\ell+1}$, and as above, write $Q_0 = \prod_{H \in \mathcal{A}_G} \ell_H$. Let F_0 be the corresponding Milnor fibre given by $Q_0(x) = 1$. The group μ_d acts on $H^*(F_0, \mathbb{C})$ where $d = \frac{m}{2} = \binom{\ell+1}{2}$; suppose $\gamma \in \hat{\mu}_d$ appears with non-zero multiplicity in $H^i(F_0, \mathbb{C})$. Then*

1. *If $i = 0$ then $\gamma = \gamma_0 = 1_{\mu_d}$.*
2. *If $i = 1$ then $|\gamma|$ divides g.c.d.$(3, d)$.*
3. *If $i = 2$ then $|\gamma|$ divides g.c.d.$(6, d)$.*
4. *If $i = 3$ then $|\gamma|$ divides either g.c.d.$(6, d)$ or g.c.d.$(10, d)$.*
5. *If $i = 4$ then $|\gamma|$ divides at least one of the following integers: g.c.d.$(6, d)$, g.c.d.$(10, d)$ or g.c.d.$(15, d)$.*

Proof We apply Proposition 4 and Example 2. □

Example 3 To compare the above result with the computations in [34], we see that for $i = 2$, eigenvalues of order 6 occur if $\ell = 3$ (then $d = 6$), eigenvalues of order 2 occur for $\ell = 4$ (then $d = 10$), and eigenvalues of order 3 occur for $\ell = 5, 6$ (then $d = 15, 21$), i.e. in these cases our results are sharp.

However, it is known that for $\ell > 6$ only the trivial character γ_0 occurs in $H^2(F_0, \mathbb{C})$, [33], and this fact cannot be obtained with our techniques so far.

To apply the above results, the following lemma will be useful. In the statement, note that $H^i(U, \mathbb{C})$ is regarded as a G-module, and its structure is well known (cf. [18–20]).

Lemma 2 *Let Q etc. be as in Sect. 1. For elements $A, B \in R(\Gamma)[t]$ say that $A \geq B$ if $A - B \in R_+(\Gamma)[t]$.*

1. *We have $P^\Gamma(F, t) \geq (1 \otimes \gamma_0 + \varepsilon \otimes \gamma_{\frac{m}{2}}) \left(P^G(U, t) \otimes \gamma_0\right)$.*
2. *If only the trivial monodromy character $\gamma_0 \in \widehat{\mu_{\frac{m}{2}}}$ occurs in $H^i(F_0, \mathbb{C})$, then as Γ-module, we have*

$$H^i(F, \mathbb{C}) \simeq (1 \otimes \gamma_0 + \varepsilon \otimes \gamma_{\frac{m}{2}})(H^i(U, \mathbb{C}) \otimes \gamma_0).$$

   In this case, as Γ_0-module, $H^i(F_0, \mathbb{C}) \simeq H^i(U, \mathbb{C}) \otimes \gamma_0$.
3. *The multiplicity of $1_{\mu_{\frac{m}{2}}}$ in $H^i(F_0, \mathbb{C})$ is $b = \dim(H^i(U, \mathbb{C}))$.*

Proof We have $F/\mu_m \simeq F_0/\mu_{\frac{m}{2}} \simeq U$, whence $H^i(U, \mathbb{C})$ is isomorphic to the γ_0-isotypic component of $H^i(F_0, \mathbb{C})$ for each i. Further, using the fact that $H^i(F, \mathbb{C}) = \mathrm{Ind}_{\Gamma_0}^{\Gamma}(H^i(F_0, \mathbb{C})) = \mathrm{Ind}_{\Gamma_0}^{\Gamma}(\rho \otimes \gamma_0)$ (see Lemma 1) is easily shown that $H^i(F, \mathbb{C}) \geq (1 \otimes \gamma_0 + \varepsilon \otimes \gamma_{\frac{m}{2}})(\rho \otimes \gamma_0)$, where ρ is the γ_0-isotypic part of $H^i(F_0, \mathbb{C})$. Hence $\rho \simeq H^i(U, \mathbb{C})$, and the first statement follows.

Parts (2) and (3) follow easily. □

Note that as we have seen (Proposition 3), the Poincaré polynomial

$$P^G(U,t;1) = \sum_i \dim H^i(U,\mathbb{C})t^i$$

of U is known to be equal to $\prod_{i\geq 2}(1 + m_i^* t)$, where the m_i^* are the coexponents of G. Thus the multiplicities in part (3) of the Lemma are known.

4 The Case When G is a Reflection Group

We assume throughout this section that G is a unitary reflection group and that $\mathcal{A} = \mathcal{A}_G$ is the set of reflecting hyperplanes of G, cf. [29]. The notation of Sect. 1 remains in force for Q_0, Q, etc.

4.1 *Free Action and Eigenspaces*

Observe that if $v \in F$ and $(g,\xi) \in \Gamma$, then $(g,\xi)v = v \iff gv = \xi v$. Thus $v \in F$ is fixed by some element of Γ if and only if $v \in V(g,\xi)$, for some $g \in G, \xi \in \mathbb{C}^\times$, where $V(g,\xi)$ is the ξ-eigenspace of $g \in G$. Now every element $v \in F$ lies on no reflecting hyperplane of G, and is therefore regular. Conversely, every regular eigenspace intersects F, because F spans M. The following statement is immediate from this observation.

Lemma 3 *Let d be a regular number for G; that is, if $\zeta_d = \exp(\frac{2\pi i}{d})$, then any $\zeta_d(= \exp(\frac{2\pi\sqrt{-1}}{d}))$-eigenspace of $g \in G$ contains a regular vector (and is therefore maximal). Then d divides m, the degree of Q.*

Proof Let $E := V(g,\zeta_d)$ be a regular eigenspace. Then there is a vector $v \in E \cap F$. Since F is G-stable (since Q is G-invariant), it follows that $gv = \zeta_d v \in F$. Thus $Q(\zeta_d v) = \zeta_d^m Q(v) = 1 = Q(v)$, whence $\zeta_d^m = 1$, and so $d|m$. □

The basic facts concerning regular eigenspaces may be found in [23, Sect. 11.4]. The main facts we require are as follows. If $E := V(g,\zeta)$ contains a regular vector, then E is a maximal ζ-eigenspace, i.e. it is not properly contained in $V(x,\zeta)$ for $x \in G$; if the order of ζ is d, then g has order d and the centraliser $C_G(g)$ acts faithfully on E as a reflection group with invariant degrees $\{d_i \mid d \text{ divides } d_i\}$; if $V(g',\zeta)$ is another regular eigenspace, then g' is conjugate to g in G. For a regular element $g \in G$ of order d, we write $G(d) := C_G(g)$. This determines $G(d)$ up to conjugacy in G. If G is irreducible (as we have assumed) then $G(d)$ is irreducible for each regular d.

Now suppose that $\zeta = \zeta_d \in \mathbb{C}^\times$ is regular, as above. Then d divides $\bar{d} := \gcd\{d_i \mid d \text{ divides } d_i\}$. Moreover there is an element $\bar{g} \in G$ such that $V(\bar{g},\zeta_{\bar{d}}) \neq 0$, where

$\zeta_{\bar{d}} = \exp(\frac{2\pi i}{\bar{d}})$. Thus $V(\bar{g}, \zeta_{\bar{d}}) = V((\bar{g})^{\frac{\bar{d}}{d}}, \zeta_d)$, since both have the same dimension. Hence it suffices to consider regular numbers d such that $d = \bar{d}$. Let $\mathcal{P}$ denote the set of such integers. Note that $\bar{1} = \gcd\{d_1, \ldots, d_\ell\}$.

If $d, e \in \mathcal{P}$ and $d|e$, then for any regular (and hence maximal) ζ_e-eigenspace $E_e = V(g, \zeta_e)$, we have $E_e \subseteq V(g^{\frac{e}{d}}, \zeta_d) := E_d$. It follows that if $V(d)$ is the union of all ζ_d-eigenspaces for $d \in \mathcal{P}$ and $F(d) = V(d) \cap F$, then $d|e$ implies that $F(e) \subseteq F(d)$.

The following cyclic subgroups of Γ play an important role in the discussion below.

Definition 2 For $d \in \mathcal{P}$, let $(g_d, \zeta_d) \in \Gamma$ be such that $V(g_d, \zeta_d) \neq 0$. This determines (g_d, ζ_d) up to conjugacy in Γ. Define $R(d)$ to be the cyclic group $\langle (g_d, \zeta_d) \rangle$ of Γ.

Proposition 5 *We have*

1. $F = \cup_{d \in \mathcal{P}} F(d)$.
2. *If* $d|e$, $d, e \in \mathcal{P}$, *then* $F(d) \supseteq F(e)$.
3. *For any two integers* e_1, e_2, *we have* $F(e_1) \cap F(e_2) = F(\mathrm{lcm}(e_1, e_2))$.
4. *For each* $d \in \mathcal{P}$ *let* $F(d)^\circ = F(d) \setminus \cup_{d|e, e \neq d} F(e)$. *Then* $F = \bigsqcup_{d \in \mathcal{P}} F(d)^\circ$ *and for each* $d \in \mathcal{P}$, $\overline{F(d)^\circ} = F(d) = \bigsqcup_{d|e} F(e)^\circ$.
5. $\Gamma / R(\bar{1})$ *acts freely on* $F(\bar{1})^\circ := F^\circ$, *where* $R(\bar{1})$ *denotes the subgroup of* Γ *defined above.*

Proof All statements except (3) follow from the above discussion. As for (3), it is well known that if $P_1, \ldots, P_\ell$ is a set of basic invariants of the group G and P_i is homogeneous of degree d_i, then $V(d) = \cap_{i: d \nmid d_i} V(P_i)$ where $V(f)$ denotes the zero set of the polynomial f. It follows that $V(e_1) \cap V(e_2) = V(\mathrm{lcm}(e_1, e_2))$, and (3) is now evident. □

4.2 Additive Functions and Euler Characteristics

Proposition 5 shows that the closed subspaces $F(d)$ form an Eulerian collection in the sense of [12, Definition (2.1)]. Let $\mathcal{B}$ be the Boolean algebra of constructible subsets of F, and A be any abelian group. Recall [12, loc. cit.] that a function $\beta : \mathcal{B} \to A$ is additive if for any subsets $Y \supseteq Z$ in $\mathcal{B}$, we have $\beta(Y) = \beta(Z) + \beta(Y \setminus Z)$.

The decomposition of F as a disjoint union of locally closed Γ-invariant subvarieties in Proposition 5(4) implies that we have the following relation for any additive function (see [12, Proposition (2.2)]).

Lemma 4 *Let* β *be any additive function on the constructible subspaces of* F. *Then*

$$\beta(F) = \sum_{d \in \mathcal{P}} \beta(F(d)^\circ).$$

Lemma 4 applies to any additive function β, and in Sect. 6 below we shall explore examples of additive functions arising from the mixed Hodge structure on the cohomology. By [12, (2.6)], one such function is the compactly supported weight polynomial $W_c^\Gamma(F,t)$ (cf. [12, Definition (1.5)(ii)]. We therefore have

Corollary 5

$$W_c^\Gamma(F,t) = \sum_{d\in\mathcal{P}} W_c^\Gamma(F(d)^\circ, t).$$

Now the equivariant Euler characteristic is an additive function, so the next statement is immediate.

Corollary 6 *We have*

$$\chi^\Gamma(F) = \sum_{d\in\mathcal{P}} \chi^\Gamma(F(d)^\circ).$$

To prove a more explicit version of the last result, we shall require the following general fact.

The summands on the right of the expression in Lemma 6 may now be described more explicitly.

Proposition 6 *For each $d \in \mathcal{P}$ let $g_d \in G$ be a d-regular element, i.e. an element such that $E_d := V(g_d, \zeta_d) \neq 0$. Then writing $R(d)$ for the cyclic group $\langle (g_d, \zeta_d)\rangle$, we have*

$$\chi^\Gamma(F(d)^\circ) = c(d)\,\mathrm{Ind}_{R(d)}^\Gamma(1_{R(d)}),$$

where $c(d) = \chi(F(d)^\circ/(G(d)\times\mu_m))$.

Proof Note first that the intersections with F of the maximal ζ_d-eigenspaces of elements of G are disjoint. This is because if $E = V(g, \zeta_d)$ and $E' = V(g', \zeta_d)$ are two such eigenspaces, then for $v \in E \cap E' \cap F$, we have $gv = \zeta_d v = g'v$, whence by regularity, $g = g'$ and hence $E = E'$. It follows, since Γ acts transitively on the maximal ζ_d-eigenspaces, that if $E_d = V(g_d, \zeta_d)$ is one of these, and $\Gamma(d)$ is the stabiliser in Γ of $F(d)^\circ \cap E_d$, then

$$\chi^\Gamma(F(d)^\circ) = \mathrm{Ind}_{\Gamma(d)}^\Gamma(\chi^{\Gamma(d)}((F\cap E_d)^\circ)), \tag{4}$$

where $(F\cap E_d)^\circ = F(d)^\circ \cap E_d$. Next let us determine $\Gamma(d)$. If $v \in E_d \cap F$ and $(g,\xi) \in \Gamma$, then $(g,\xi)v = \xi^{-1}gv \in E_d$ implies that $g_d gv = \zeta_d gv = gg_d v$, whence by the regular nature of v, we have $g \in C_G(g_d) = G(d)$. The converse is easily checked, whence we see that $\Gamma(d) = G(d)\times\mu_m$.

Now $G(d)$ acts as a reflection group on E_d, and it is proved in [20] and in [5] that the reflecting hyperplanes of this action are the intersections with E_d of those of G. Suppose $(g,\xi) \in \Gamma(d)$ and $v \in (F\cap E_d)^\circ$ are such that $(g,\xi)v = v$. Then $gv = \xi v$.

It follows, since $v \in (F \cap E_d)^\circ$, that $\xi = \zeta_d^i$ for some i, and hence that $g_d^i v = gv$, whence $(g, \xi) = (g_d, \zeta_d)^i \in \langle (g_d, \zeta_d) \rangle$. Conversely, each element of the group $\langle (g_d, \zeta_d) \rangle$ fixes $F(d)^\circ \cap E_d$ pointwise. Hence the quotient group $\Gamma(d)/\langle (g_d, \zeta_d) \rangle = \Gamma(d)/R(d)$ acts freely on $(F \cap E_d)^\circ$, and it follows from Proposition 1 that

$$\chi^{\Gamma(d)}((F \cap E_d)^\circ) = c(d) \operatorname{Reg}_{\Gamma(d)/\langle (g_d, \zeta_d) \rangle} = c(d) \operatorname{Ind}_{R(d)}^{\Gamma(d)}(1_{R(d)}), \tag{5}$$

where $c(d) = \chi(F(d)^\circ/(G(d) \times \mu_m))$, and Reg_H denotes the regular representation of a finite group H.

Putting together (4) and (5) and using transitivity of induction, we obtain the statement of the Proposition. □

The above proof yields the more general statement below, which applies to most of the invariants we wish to investigate, such as the equivariant weight polynomials and the Hodge-Deligne polynomials.

Scholium 4.3 *Let β be an additive Γ-functor from constructible subspaces of F to the ring $S := A \otimes_{\mathbb{C}} R(\Gamma)$, where $R(\Gamma)$ is the complex Grothendieck ring of Γ and A is any commutative ring. Then S is naturally a Γ-module, and the statement that β is a Γ-functor means that β satisfies $\beta(\gamma(Y)) = \gamma(\beta(Y))$ for all $\gamma \in \Gamma$ and constructible $Y \subseteq X$. Then*

1. *$\beta(F) = \sum_{d \in \mathcal{P}} \operatorname{Ind}_{\Gamma(d)}^{\Gamma} \beta((F \cap E_d)^0)$, where $\Gamma(d) = G(d) \times \mu_m$ and E_d is any maximal ζ_d-eigenspace of V.*
2. *$\Gamma(d)/R(d)$ acts freely on $(F \cap E_d)^0$, where $R(d) = \langle (g_d, \zeta_d) \rangle$ is the cyclic group occurring in the proof of Proposition 6.*

Proof From Lemma 4 we have $\beta(F) = \sum_{d \in \mathcal{P}} \beta(F(d)^0)$. But by the argument at the beginning of the proof of Proposition 6, $F(d)^0$ is a disjoint union $F(d)^0 = \bigsqcup_E F(d)^0 \cap E$, where the sum is over the distinct (maximal) ζ_d-eigenspaces of G in V. It follows from the additivity property of β that $\beta(F(d)^0 = \sum_E \beta(F(d)^0 \cap E)$. Moreover, since the eigenspaces E are conjugate under G, and hence *a fortiori* under Γ, it follows from the defining property of β that the subspaces $\beta(F(d)^0 \cap E)$ of the $R(\Gamma)$-module $\beta(F(f)^0)$ are permuted transitively by Γ; further, again by the argument in the proof of Proposition 6, the stabilizer of one of these subspaces is $\Gamma(d)$. It follows from the functorial property and additivity that $\beta(F) = \operatorname{Ind}_{\Gamma(d)}^{\Gamma}(\beta(F(d) \cap E_d)^0)$, where E_d is a particular maximal ζ_d-eigenspace.

This proves (1), while (2) is proved in the proof above. □

Examples of functors β to which Scholium 4.3 may be applied include the Euler characteristic, the compactly supported weight polynomial, and the compactly supported Hodge-Deligne polynomial (see (31) in Sect. 6 below). The Euler characteristic case is the easiest to handle.

Let $U = \mathbb{P}(M) := M/\mathbb{C}^\times$ and consider the map $p : F \longrightarrow U$ given by $p(v) = [v]$. This is an unramified μ_m-covering, and U may be identified with F/μ_m (see 3 below). Let $U(d), U(d)^\circ$ be the analogues for the pair $E_d, G(d)$ of the spaces U, U°

for V, G. Thus in particular, $U(d)^\circ = p(E_d^\circ)$. Then evidently $F(d)^\circ/(G(d) \times \mu_m)$ may be identified with $U(d)^\circ/G(d)$.

Using this notation, Lemma 6 and Proposition 6 may be combined to yield the following statement.

Corollary 7 (cf. [6, Theorem 3.13]) *Let G be an irreducible unitary reflection group acting on $V = \mathbb{C}^\ell$. Let $M, Q, m, F, U, \mathcal{P}$ etc. be as in the discussion above, and for $d \in \mathcal{P}$ let $R(d)$ be the cyclic subgroup of $\Gamma = G \times \mu_m$ defined in Definition 2. Then*

$$\chi^\Gamma(F) = \sum_{d \in \mathcal{P}} \chi(U(d)^\circ/G(d)) \operatorname{Ind}_{R(d)}^\Gamma(1). \tag{6}$$

Remark 2 The value of $\chi(U^\circ/G)$ has been computed case by case in [6, Theorem 3.15]. We shall give an essentially case free proof below. The key result for the examples below is that for $G = G(r, 1, \ell)$ (an imprimitive reflection group) $\chi(U^\circ/G) = 0$ if $\ell > 2$ or if $\ell = 2$ and $r = 1$. If $\ell = 2$ and $r > 1$ then $\chi(U^\circ/G) = -1$; otherwise (if $\ell = 1$) $\chi(U^\circ/G) = 1$.

Remark 3 Since $\chi(F/G) \in \mathbb{Z}$, Proposition 3 implies the following divisibility result for any unitary reflection group G. Recall (cf. [21, (1.3)]) that $m = n_G + N_G$, where $n_G = \sum_i m_i = \sum_{H \in \mathcal{A}_H}(e_H - 1)$ and $N_G = \sum_i m_i^* = |\mathcal{A}_G|$, and the m_i and m_i^* are respectively the exponents and coexponents of G. Since $|G| = \prod_i (1 + m_i)$, it follows that

$$\prod_{i=1}^{\ell}(1 + m_i) \text{ divides } \left(\sum_{i=1}^{\ell}(m_i + m_i^*)\right) \prod_{i=2}^{\ell}(1 - m_i^*).$$

Example 4 Let G be the symmetric group $\mathrm{Sym}_{\ell+1}$ acting on $\mathbb{C}^{\ell+1}$ by permutation of coordinates. The non-zero coexponents in this case are $1, 2, \ldots, \ell$, and $m = \ell(\ell+1)$. Hence in this case we have $\chi(U) = (-1)^{\ell-1}(\ell - 1)!$, $\chi(F/G) = (-1)^{\ell+1}$ and $\chi(F) = (-1)^{\ell+1}(\ell + 1)!$.

We finish this section with a closed (but finitely recursive) formula for $\chi(U^0/G)$, which may be applied to give a case free proof of [6, Theorem 3.15].

Theorem 2 *Let G be a finite reflection group in the complex vector space V, M the corresponding hyperplane complement, $U = \mathbb{P}(M)$, etc, as above. Let $\mathcal{P}$ be the poset of integers d such that d is the* gcd *of the degrees of G which it divides, where $e \leq d$ in $\mathcal{P}$ if $d|e$. Then in the notation above,*

$$\chi(U^0/G) = |Z(G)| \sum_{d \in \mathcal{P}} \mu(d)|G(d)|^{-1} \prod_{2 \leq i \leq r(d)} (1 - m_i^*(d)),$$

where $\mu(d) = \mu(d, |Z(G)|)$ is the Möbius function of the poset $\mathcal{P}$, $r(d)$ is the rank of the reflection group $G(d)$, and the $m_i^*(d)$ are the coexponents of the reflection group $G(d)$ written so that $m_1^*(d) \le m_2^*(d) \le \cdots \le m_{r(d)}^*(d)$.

Proof It follows from Proposition 3 that the multiplicity of the trivial character γ_0 of the monodromy group μ_m in $\chi(F)$ is $\chi(U)$. Hence applying Corollary 7,

$$\chi(U) = \sum_{d \in \mathcal{P}} \chi(U(d)^\circ/G(d))(\mathrm{Ind}_{R(d)}^{\Gamma}(1), \gamma_0)_{\mu_m}, \tag{7}$$

where $(\, , \,)_{\mu_m}$ denotes the usual inner product of characters. But an easy application of Mackey's formula shows that $(\mathrm{Res}_{\mu_m}^{\Gamma} \mathrm{Ind}_{R(d)}^{\Gamma}(1), \gamma_0)_{\mu_m} = \frac{|G|}{d}$. It follows from (7) that

$$\chi(U) = (\chi^{\Gamma}(F), \gamma_0)_{\mu_m} = \sum_{d \in \mathcal{P}} \chi(U(d)^\circ/G(d))\frac{|G|}{d}. \tag{8}$$

If we define, for any reflection group G, $c(G) = \frac{\chi(U)}{|G|}$ and $c_0(G) = \chi(U^0/G)$, then (8) may be rearranged to read

$$c(G) = \sum_{d \in \mathcal{P}(G)} \frac{c_0(G(d))}{d}, \tag{9}$$

where $\mathcal{P}(G)$ indicates the poset $\mathcal{P}$ for the group G.

However if $e \le d$ in $\mathcal{P}(G)$, it is shown in [22] that

$$G(d)(e) = G(e). \tag{10}$$

Moreover using the fact that the degrees of $G(d)$ are precisely those degrees of G which are divisible by d, a short calculation verifies that

$$\mathcal{P}(G(d)) = \{e \in \mathcal{P}(G) \mid e \le d\}. \tag{11}$$

It follows from (10) and (11) that we may apply (8) to the group $G(d)$ to obtain, for any $d \in \mathcal{P}$

$$c(G(d)) = \sum_{e \in \mathcal{P}(G), e \le d} \frac{c_0(G(e))}{e}. \tag{12}$$

We may now invert the relation (12) using the Möbius function $\mu(e, d)$ of $\mathcal{P}$ to obtain for any $d \in \mathcal{P}$,

$$\frac{c_0(G(d))}{d} = \sum_{e \in \mathcal{P}(G), e \le d} \mu(e, d)c(G(e)). \tag{13}$$

Taking into account that the top element of $\mathcal{P}$ is $|Z(G)|$ (which is the gcd of the degrees of G), the stated relation is simply the case $d = |Z(G)|$ of (13), where we write $\mu(d) = \mu(d, |Z(G)|)$, and use the formula in Proposition 3(3) for $\chi(U)$. □

4.4 Quotient by G

It is easy to deduce from the formula (6) the μ_m-equivariant Euler characteristic of F/G. The result is as follows.

Proposition 7 *Let $\gamma_0, \gamma_1, \ldots, \gamma_{m-1}$ be the characters of μ_m, where $\gamma_i(\zeta_m) = \zeta_m^i$. Then*

$$\chi^{\mu_m}(F/G) = \sum_{d \in \mathcal{P}} \chi(U(d)^\circ/G(d)) \sum_{j=0}^{\frac{m}{d}-1} \gamma_{dj}. \tag{14}$$

Proof Clearly $\chi^{\mu_m}(F/G)$ is the 1_G-isotypic part of the virtual representation on the right side of (6). To compute this, note first that for any representation α of Γ, we have by Frobenius reciprocity

$$(\mathrm{Ind}_{R(d)}^{\Gamma}(1), \alpha)_\Gamma = (1_{R(d)}, \mathrm{Res}_{R(d)}^{\Gamma}(\alpha))_{R(d)}. \tag{15}$$

It is therefore evident that

$$(\mathrm{Ind}_{R(d)}^{\Gamma}(1), 1_G \otimes \gamma_i)_\Gamma = \begin{cases} 1 \text{ if } \gamma_i|_{\mu_d} = 1 \\ 0 \text{ otherwise.} \end{cases}$$

The stated formula follows easily. □

Remark 4 Let $f_1, \ldots, f_\ell$ be a set of basic invariants for G. The orbit map $V \longrightarrow V/G$ may be realised as $\pi : x \mapsto (f_1(x), \ldots, f_\ell(x))$; since Q is G-invariant, there is a unique polynomial $\Delta(y_1, \ldots, y_\ell)$ such that $Q(x) = \Delta(f_1(x), \ldots, f_\ell(x))$. The Milnor fibre F is mapped by π to $F/G = F_\Delta := \{y \mid \Delta(y) = 1\}$, and the monodromy μ_m acts on F_Δ via $y \mapsto (\zeta^{d_1} y_1, \ldots, \zeta^{d_\ell} y_\ell)$, $(\zeta \in \mu_m)$. For the symmetric group $S_{\ell+1}$, $\Delta(y)$ is just the usual discriminant of the polynomial $t^{\ell+1} - y_1 t^\ell + y_2 t^{\ell-1} - \cdots + (-1)^{\ell+1} y_{\ell+1}$. In view of the remark at the beginning of Example 5 below, one may take $y_1 = 0$ in this case, i.e. it suffices to consider the Milnor fibre of the polynomial $\Delta_0(y_2, \ldots, y_{\ell+1}) := \Delta(0, y_2, \ldots, y_{\ell+1})$.

Example 5 We take $G = \mathrm{Sym}_{\ell+1}$ acting by permutation of coordinates on $\mathbb{C}^{\ell+1}$. This action is irreducible on the hyperplane $x_1 + \cdots + x_{\ell+1} = 0$, and if F_0 is the Milnor fibre corresponding to this irreducible action, we have $F = F_0 \times \mathbb{C}$, so that the Γ-modules $H^*(F_0)$ and $H^*(F)$ are isomorphic. We consider here the graded space $H^*(F)$. The regular numbers are all integers d which divide ℓ or $\ell + 1$. The degrees of G are $1, 2, \ldots, \ell + 1$. For any regular d, since d is one of the degrees, $\bar{d} = d$, and so $\mathcal{P}$ coincides with the set of regular numbers.

Let $d \in \mathcal{P}$. It is easily verified that $G(d) \simeq G(d, 1, [\frac{\ell+1}{d}])$. Hence in the notation above, by Remark 2 $\chi(U(d)^\circ/G(d)) = 0$ unless $[\frac{\ell+1}{d}] \leq 2$, that is unless $d = \ell, \ell+1, \frac{\ell+1}{2}$ or $\frac{\ell}{2}$. The corresponding values of $\chi(U(d)^\circ/G(d))$ in these cases are respectively $1, 1, -1, -1$.

Thus the main results above may be written in this case as follows.

$$\chi^{\Gamma}(F) = \begin{cases} \mathrm{Ind}_{R(\ell)}^{\Gamma}(1) + \mathrm{Ind}_{R(\ell+1)}^{\Gamma}(1) - \mathrm{Ind}_{R(\frac{\ell}{2})}^{\Gamma}(1) \text{ if } \ell \text{ is even} \\ \mathrm{Ind}_{R(\ell)}^{\Gamma}(1) + \mathrm{Ind}_{R(\ell+1)}^{\Gamma}(1) - \mathrm{Ind}_{R(\frac{\ell+1}{2})}^{\Gamma}(1) \text{ if } \ell \text{ is odd} \end{cases} \tag{16}$$

where $R(d)$ is the cyclic group (of order d) generated by (g_d, ζ_d), where g_d is a product of $[\frac{\ell+1}{d}]$ d-cycles (see Definition 2).

For the $1_{\mathrm{Sym}_{\ell+1}}$-isotypic part, we have

$$\chi^{\mu_m}(F/G) = \begin{cases} \sum_{j=0}^{\ell} \gamma_{\ell j} + \sum_{j=0}^{\ell-1} \gamma_{(\ell+1)j} - \sum_{j=0}^{2\ell+1} \gamma_{\frac{\ell}{2}j} \text{ if } \ell \text{ is even} \\ \sum_{j=0}^{\ell} \gamma_{\ell j} + \sum_{j=0}^{\ell-1} \gamma_{(\ell+1)j} - \sum_{j=0}^{2\ell-1} \gamma_{\frac{(\ell+1)}{2}j} \text{ if } \ell \text{ is odd.} \end{cases} \tag{17}$$

These formulae may be written more transparently as

$$\chi^{\mu_m}(F/G) = \begin{cases} \sum_{j=0}^{\ell-1} \gamma_{(\ell+1)j} - \sum_{j=0}^{\ell} \gamma_{\frac{\ell}{2}(2j+1)} \text{ if } \ell \text{ is even} \\ \sum_{j=0}^{\ell} \gamma_{\ell j} - \sum_{j=0}^{\ell-1} \gamma_{\frac{(\ell+1)}{2}(2j+1)} \text{ if } \ell \text{ is odd.} \end{cases} \tag{18}$$

Let us check these formulae for compatibility with the results of [8]. It is shown in [2, Theorem 4] that if $B_{\ell+1}$ is the braid group on $\ell + 1$ strings, and $R = \mathbb{C}[q, q^{-1}]$ is the $B_{\ell+1}$-module upon which the generators of $B_{\ell+1}$ act as multiplication by $-q$, then for each degree k, we have

$$H^k(F/G, \mathbb{C}) \simeq H^{k+1}(B_{\ell+1}, R).$$

Given this, the main result [8, Theorem, p. 739], may be stated in the language of the above exposition as follows.

Proposition 8 *For any character $\gamma \in \widehat{\mu_m}$ write $|\gamma|$ for its order. For any integer $h \geq 2$ which divides ℓ or $\ell + 1$, we write $i(h) = \left[\frac{\ell+1}{h}\right]$, and $k(h) = i(h)(h-2)$. If $k \neq k(h)$ for some h, then $H^k(F/G, \mathbb{C}) = 0$. For the remaining cases we have the following equation in the Grothendieck ring of μ_m.*

$$H^{k(h)}(F/G, \mathbb{C}) = \begin{cases} \sum_{\gamma \in \widehat{\mu_m}, |\gamma| = 2h} \gamma \text{ if } h \text{ is odd,} \\ \sum_{\gamma \in \widehat{\mu_m}, |\gamma| = \frac{h}{2}} \gamma \text{ if } h \equiv 2 (mod\ 4), \\ \sum_{\gamma \in \widehat{\mu_m}, |\gamma| = h} \gamma \text{ if } h \equiv 0 (mod\ 4). \end{cases}$$

The above Proposition may be applied to deduce information about $H^*(F_0, \mathbb{C})$ as follows. Note that $H^*(F_0, \mathbb{C})$ is *a priori* a module for $\mu_{\frac{m}{2}}$. The next result provides some information concerning this action.

Corollary 8 *Let G etc. be as in Lemma 1 and take $e = 2$; thus we have Q_0, Γ_0, etc. Then $F/G \simeq F_0/G_0$, where $G_0 = \operatorname{Ker}(\varepsilon)$, and the $\mu_{\frac{m}{2}}$-action on $H^i(F_0/G_0, \mathbb{C})$ is just the restriction to $\mu_{\frac{m}{2}}$ of the μ_m-action on $H^i(F/G, \mathbb{C})$.*

Proof We have $F = F_0 \bigsqcup F_-$, and G permutes the two pieces; that is, for each $g \in G$, $gF_0 = F_0$ or $gF_0 = F_-$. Moreover there is an element of G which interchanges the two pieces. It follows that F/G may be identified with the equivalence classes of elements of F_0, equivalence being defined as $x \equiv y$ if $y = gx$ for some $g \in G$. But if $x, y \in F_0$, and $gx = y$, then g must be in G_0, which is the stabiliser of F_0, and so $F/G \simeq F_0/G_0$. □

Remark 5 Let $d = \frac{m}{2} = \deg Q_0$. Then $\mu_d = \Gamma_0 \cap \mu_m$, and μ_m is a normal subgroup of Γ, which is contained in the centre of Γ. Hence $\Gamma_0/\mu_d \cong \Gamma/\mu_m \cong G$. It follows that if $\gamma \in \widehat{\mu_d}$ and M is any Γ_0-module, then the γ-isotypic component M^γ of M is a module for $G = \Gamma_0/\mu_d$. We shall use this repeatedly below. Note that $\dim M^\gamma = (M, \gamma)_{\mu_d}$.

The next corollary of Proposition 8 provides a strengthening of the formula (18), in that it asserts that there is no cancellation in arriving at that formula. Note that in accord with our general practice, in the statement below, $H^{i,\mu_m}(F/G, \mathbb{C})$ denotes $H^i(F/G, \mathbb{C})$ as μ_m-module.

Corollary 9 *Let $G = \operatorname{Sym}_{\ell+1}$ as in example 5. Then*

$$\sum_{i \equiv \ell(\mod 2)} H^{i,\mu_m}(F/G, \mathbb{C}) = \begin{cases} \sum_{i=0}^{\ell-1} \gamma_{(\ell+1)i} \text{ if } \ell \equiv 0(\mod 2) \\ \sum_{i=0}^{\ell-1} \gamma_{\frac{\ell+1}{2}(2i+1)} \text{ if } \ell \equiv 1(\mod 2), \end{cases}$$

and

$$\sum_{i \equiv \ell+1(\mod 2)} H^{i,\mu_m}(F/G, \mathbb{C}) = \begin{cases} \sum_{i=0}^{\ell} \gamma_{\ell i} \text{ if } \ell+1 \equiv 0(\mod 2) \\ \sum_{i=0}^{\ell} \gamma_{\frac{\ell}{2}(2i+1)} \text{ if } \ell+1 \equiv 1(\mod 2). \end{cases}$$

Proof We use the notation of Proposition 8. First observe that a short computation shows that

$$\begin{aligned} &\text{If } h|\ell, \text{ then } k(h) \equiv \ell(\mod 2), \text{ and} \\ &\text{if } h|\ell+1, \text{ then } k(h) \equiv \ell+1(\mod 2). \end{aligned} \tag{19}$$

A further calculation using the details in Proposition 8 now proves that if ℓ is even, then

$$\sum_{i\equiv\ell(\mod 2)} H^{i,\mu_m}(F/G,\mathbb{C}) = \sum_{|\gamma||\ell} \gamma,$$

while if ℓ is odd, then

$$\sum_{i\equiv\ell(\mod 2)} H^{i,\mu_m}(F/G,\mathbb{C}) = \sum_{|\gamma|=2b \text{ where } b|\ell} \gamma.$$

Putting these facts together with the analogous ones for $\ell+1$, one obtains the Corollary. □

Remark 6 Corollary 9 both confirms the formula (18), and shows further that there is no cancellation in arriving at that formula. It is useful to reformulate it as follows.

Corollary 10 *Let $G = \mathrm{Sym}_{\ell+1}$ as above. For any integer i such that $0 \le i \le \ell-1$, define a subset $\mathcal{M}_i \subseteq \widehat{\mu_m}$ as follows. Let*

$$h(i) := \begin{cases} \frac{2\ell}{\ell-i} \text{ if } i \equiv \ell(\mod 2) \\ \frac{2(\ell+1)}{\ell+1-i} \text{ if } i \equiv \ell+1(\mod 2). \end{cases}$$

Define

$$\mathcal{M}_i = \begin{cases} \{\gamma \in \widehat{\mu_m} \mid |\gamma| = h(i)\} \text{ if } h(i) \in \mathbb{Z} \text{ and } h(i) \equiv 0(\mod 4) \\ \{\gamma \in \widehat{\mu_m} \mid |\gamma| = \frac{h(i)}{2}\} \text{ if } h(i) \in \mathbb{Z} \text{ and } h(i) \equiv 2(\mod 4) \\ \{\gamma \in \widehat{\mu_m} \mid |\gamma| = 2h(i)\} \text{ if } h(i) \in \mathbb{Z} \text{ and } h(i) \equiv 1 \text{ or } 3(\mod 4) \\ \emptyset \text{ otherwise.} \end{cases}$$

The 1_G-isotypic part of $P^\Gamma(F,t)$ (see (1)) is given by

$$P^{\mu_m}(F/G,t) = P^\Gamma(F,t)^{1_G} = \sum_{i=0}^{\ell-1}\sum_{\gamma\in\mathcal{M}_i} (1_G \otimes \gamma)t^i.$$

This is an easy consequence of Proposition 8 and Corollary 9.

Remark 7 An example of the way in which Corollary 10 may be used is as follows. For elements $A, B \in R(\Gamma)[t]$ say that $A \ge B$ if $A - B \in R_+(\Gamma)[t]$. Then $P^\Gamma(F,t) \ge 1_G \otimes P^{\mu_m}(F/G,t)$, and $\left(P^\Gamma(F,t) - 1_G \otimes P^{\mu_m}(F/G,t)\right)^{1_G} = 0$.

Example 6 Let $G = \mathrm{Sym}_3$. Then $m = 6$ and we denote by 1, ε and ρ respectively the trivial one dimensional representation, the sign representation and the (irreducible, two dimensional) reflection representation of G. The cohomology of F in this case is completely determined by the formula (16) since there is no cancellation when the Euler characteristic is taken. Using the notation above, the cohomology of Milnor fibre F in this case has the following description as Γ-representation.

(i) $H^0(F) = 1\otimes\gamma_0 + \varepsilon\otimes\gamma_3$.
(ii) $H^1(F) = 1\otimes\gamma_1 + 1\otimes\gamma_5 + \varepsilon\otimes\gamma_2 + \varepsilon\otimes\gamma_4 + \rho\otimes\gamma_0 + \rho\otimes\gamma_3$
$= (1\otimes\gamma_0 + \varepsilon\otimes\gamma_3)(1\otimes\gamma_1 + 1\otimes\gamma_5 + \rho\otimes\gamma_0)$.

Thus in particular, in the notation of Corollary 1

$$P^\Gamma(F,t) = (1\otimes\gamma_0 + \varepsilon\otimes\gamma_3)\,(1\otimes\gamma_0 + (1\otimes(\gamma_1+\gamma_5) + \rho\otimes\gamma_0)t)\,.$$

Note that this shows that $P^{\Gamma_0}(F_0,t) = 1\otimes\gamma_0 + (1\otimes(\gamma_1+\gamma_5)+\rho\otimes\gamma_0)t$, which is consistent with [34, Table 2]. Note that we also have $P^{\Gamma_0}(F_0,t) = 1\otimes\gamma_0 + (1\otimes\gamma_1 + \varepsilon\otimes\gamma_2 + \rho\otimes\gamma_0)t$.

We note further that $H^1(F/G) = H^1(F)^G = \gamma_1 + \gamma_5$, which agrees with the known monodromy of the cusp $F_{\Delta_0} : \Delta_0(y_2,y_3) = 0$ (see Remark 4).

4.5 Application to the Symmetric Groups

We again take $G = \mathrm{Sym}_{\ell+1}$, with Q, Q_0 etc. as in Sect. 2. In this section we shall use our results hitherto to give complete solutions to Problem A for $G = \mathrm{Sym}_4$ and $G = \mathrm{Sym}_5$.

Example 7 We illustrate the use of the above results by computing $P^\Gamma(F,t)$ in the case where $G = \mathrm{Sym}_4$.

Denote the irreducible representations of Sym_4 by $1, \varepsilon, \rho, \varepsilon\rho$ and σ, where $1, \varepsilon$ are the trivial and alternating representations respectively, $\rho, \varepsilon\rho$ are the reflection representation and its tensor with ε respectively, and σ is the 2-dimensional irreducible representation corresponding to the partition $(2,2)$.

Our notation for the characters of the cyclic group $\mu_m \subset \mathbb{C}^\times$ is as above: γ_1 denotes the generating character, which is just the inclusion : $\mu_m \hookrightarrow \mathbb{C}^\times$ and for any $i \in \mathbb{Z}$, $\gamma_i = \gamma_1^i$. Notice that this notation is independent of m, so that restriction of characters from μ_m to μ_d (for $d|m$) is trivial. The distinct irreducible characters of μ_m are $\gamma_0 = 1_{\mu_m}, \gamma_1, \ldots, \gamma_{m-1}$.

Every irreducible representation of $\Gamma = G \times \mu_m$ is of the form $\alpha\otimes\gamma_i$, where $\alpha \in I(G)$.

First we apply (16) with $\ell = 3$ to obtain the following expression for $\chi^\Gamma(F) \in R(\Gamma)$. The factorisation is in accord with Lemma 1.

$$\begin{aligned}\chi^\Gamma(F) = {} & 1\otimes(\gamma_0+\gamma_3+\gamma_9)+\varepsilon\otimes(\gamma_3+\gamma_6+\gamma_9)+\rho\otimes(\gamma_2+\gamma_6+\gamma_{10})+\varepsilon\rho\otimes(\gamma_0+\gamma_4+\gamma_8)\\ & +\sigma\otimes(\gamma_1+\gamma_5+\gamma_7+\gamma_{11})-[1\otimes(\gamma_2+\gamma_{10})+\varepsilon\otimes(\gamma_4+\gamma_8)+\sigma\otimes(\gamma_0+\gamma_6)]\\ = {} & (1\otimes\gamma_0+\varepsilon\otimes\gamma_6)\{1\otimes(\gamma_0+\gamma_3+\gamma_9)+\rho\otimes(\gamma_2+\gamma_6+\gamma_{10})+\sigma\otimes(\gamma_1+\gamma_{11})\\ & -[1\otimes(\gamma_2+\gamma_{10})+\sigma\otimes\gamma_0]\}.\end{aligned} \tag{20}$$

We know that as Γ-module, $H^0(F,\mathbb{C}) = \mathrm{Ind}_{\Gamma_0}^{\Gamma}(1) = 1\otimes\gamma_0+\varepsilon\otimes\gamma_6$. Further, since $\chi^\Gamma(F) = H^0(F,\mathbb{C}) - H^1(F,\mathbb{C}) + H^2(F,\mathbb{C})$ it follows from (20) that as Γ-module, $H^1(F,\mathbb{C})$ contains the module

$$M_1 := 1\otimes(\gamma_2+\gamma_{10})+\varepsilon\otimes(\gamma_4+\gamma_8)+\sigma\otimes(\gamma_0+\gamma_6),$$

and that $H^2(F,\mathbb{C})$ contains the module

$$\begin{aligned}M_2 := {} & 1\otimes(\gamma_3+\gamma_9)+\varepsilon\otimes(\gamma_3+\gamma_9)+\rho\otimes(\gamma_2+\gamma_6+\gamma_{10})+\varepsilon\rho\otimes(\gamma_0+\gamma_4+\gamma_8)\\ & +\sigma\otimes(\gamma_1+\gamma_5+\gamma_7+\gamma_{11}).\end{aligned}$$

Next we note that the G-module structure of $H^*(U,\mathbb{C}) = H^*(F/G,\mathbb{C})$ is explicitly given for all the symmetric groups in [18]. In our case we have

$$P^G(U,t) = 1+(\rho+\sigma)t+(\rho+\varepsilon\rho)t^2. \tag{21}$$

Hence by Lemma 2(1), $H^1(F,\mathbb{C})$ also contains the module $(1\otimes\gamma_0+\varepsilon\otimes\gamma_6)(\rho\otimes\gamma_0)$, which has no common irreducible constituent with M_1, whence $H^1(F,\mathbb{C})$ contains $M_1+(1\otimes\gamma_0+\varepsilon\otimes\gamma_6)(\rho\otimes\gamma_0)$.

But inspection of [34, Table 2] shows that $\dim H^1(F_0,\mathbb{C}) = 7$, whence $\dim H^1(F,\mathbb{C}) = 14$, and by counting dimensions it follows that

$$H^1(F,\mathbb{C}) = M_1+(1\otimes\gamma_0+\varepsilon\otimes\gamma_6)(\rho\otimes\gamma_0).$$

Similarly, applying Lemma 2 to $H^2(F,\mathbb{C})$, it follows that $(1\otimes\gamma_0+\varepsilon\otimes\gamma_6)(\rho\otimes\gamma_0)$ is also a submodule of $H^2(F,\mathbb{C})$, and has no common constituent with M_2. Using an argument similar to that above, we see that

$$H^2(F,\mathbb{C}) = M_2+(1\otimes\gamma_0+\varepsilon\otimes\gamma_6)(\rho\otimes\gamma_0).$$

We have therefore proved the following

Proposition 9 *The Γ-module structure of $H^*(F,\mathbb{C})$ is given by the following equations in $R(\Gamma)$.*

$$H^0(F,\mathbb{C}) \simeq 1\otimes\gamma_0 + \varepsilon\otimes\gamma_6; \tag{22}$$

$$\begin{aligned} H^1(F,\mathbb{C}) \simeq 1\otimes(\gamma_2+\gamma_{10}) &+ \varepsilon\otimes(\gamma_4+\gamma_8) + \sigma\otimes(\gamma_0+\gamma_6) \\ &+\rho\otimes\gamma_0 + \varepsilon\rho\otimes\gamma_6; \end{aligned} \tag{23}$$

$$\begin{aligned} H^2(F,\mathbb{C}) \simeq 1\otimes(\gamma_3+\gamma_9) &+ \varepsilon\otimes(\gamma_3+\gamma_9) + \rho\otimes(\gamma_0+\gamma_2+\gamma_6+\gamma_{10}) \\ &+\varepsilon\rho\otimes(\gamma_0+\gamma_4+\gamma_6+\gamma_8) + \sigma\otimes(\gamma_1+\gamma_5+\gamma_7+\gamma_{11}). \end{aligned} \tag{24}$$

In terms of Poincaré polynomials, this result may be restated as follows. We have $P^\Gamma(F,t) = (1\otimes\gamma_0 + \varepsilon\otimes\gamma_6)P_0^\Gamma(F,t)$, where

$$\begin{aligned} P_0^\Gamma = 1\otimes\gamma_0 &+ [1\otimes(\gamma_2+\gamma_{10}) + (\sigma+\rho)\otimes\gamma_0]t \\ &+\{1\otimes(\gamma_3+\gamma_9) + \rho\otimes(\gamma_0+\gamma_2+\gamma_6+\gamma_{10}) + \sigma\otimes(\gamma_1+\gamma_{11})\}t^2. \end{aligned} \tag{25}$$

Note that by Corollary 1 $P^{\Gamma_0}(F_0,\mathbb{C})$ is the restriction to Γ_0 of P_0^Γ of the above proposition, and that by the proof of Lemma 1, each irreducible representation of Γ restricts to an irreducible representation of Γ_0. The μ_6-structure of $H^*(F_0,\mathbb{C})$, given in [34], is an easy consequence. The formulae above may be seen to be consistent with both [34, Table 2] and Proposition 8.

Example 8 A similar, but slightly longer, calculation yields the following result for the symmetric group Sym_5, which we set out below without details, for comparison with some of the general results we describe later.

As usual, we label the irreducible representations of Sym_5 with the partitions λ of 5. Thus χ^λ corresponds to λ, and in accord with the usual conventions, $\chi^{(5)} = 1$, the trivial representation, $\chi^{1^5} = \varepsilon$, the alternating representation, $\chi^{(4,1)} = \rho$, the reflection representation, and $\varepsilon\otimes\chi^\lambda = \chi^{\lambda'}$, where λ' is the partition dual to λ. The irreducible representations $\chi^{(5)} = 1, \chi^{(1^5)} = \varepsilon, \chi^{(4,1)} = \rho, \chi^{(2,1^3)} = \varepsilon\rho, \chi^{(3,2)}, \chi^{(2^2,1)}$ and $\chi^{(3,1^2)}$ thus have respective dimensions $1,1,4,4,5,5$ and 6. Notation for the irreducible characters of μ_{20} is as above. The character γ_i takes $\zeta\in\mu_{20}$ to ζ^i.

Proposition 10 *Let $\Gamma = \mathrm{Sym}_5\times\mu_{20}$, and let F, F_0, Γ_0 be as above for the hyperplane arrangement of type A_4. The cohomology $H^i = H^i(F,\mathbb{C})$ is given as a Γ-module by the formulae below.*

1. $H^0 = 1\otimes\gamma_0 + \varepsilon\otimes\gamma_{10}$.
2. $H^1 = \rho\otimes\gamma_0 + \varepsilon\rho\otimes\gamma_{10} + \chi^{(3,2)}\otimes\gamma_0 + \chi^{(2^2,1)}\otimes\gamma_{10}$.
3. $H^2 = 1\otimes(\gamma_5+\gamma_{15}) + \varepsilon\otimes(\gamma_5+\gamma_{15}) + \rho\otimes\gamma_0 + \varepsilon\rho\otimes\gamma_{10} + \chi^{(3,2)}\otimes(\gamma_0+\gamma_{10}) + \chi^{(2^2,1)}\otimes(\gamma_0+\gamma_{10}) + 2\chi^{(3,1^2)}\otimes(\gamma_0+\gamma_{10})$.

4. $H^3 = 1\otimes(\gamma_2+\gamma_6+\gamma_{14}+\gamma_{18}) + \varepsilon\otimes(\gamma_4+\gamma_8+\gamma_{12}+\gamma_{16}) + \rho\otimes(\gamma_0+\gamma_5+\gamma_{10}+\gamma_{15}) + \varepsilon\rho\otimes(\gamma_0+\gamma_5+\gamma_{10}+\gamma_{15}) + \chi^{(3,2)}\otimes(\gamma_0+\gamma_4+\gamma_8+\gamma_{10}+\gamma_{12}+\gamma_{16}) + \chi^{(2^2,1)}\otimes(\gamma_0+\gamma_2+\gamma_6+\gamma_{10}+\gamma_{14}+\gamma_{18}) + \chi^{(3,1^2)}\otimes(\gamma_0+\gamma_1+\gamma_3+\gamma_7+\gamma_9+\gamma_{10}+\gamma_{11}+\gamma_{13}+\gamma_{17}+\gamma_{19})$.
5. *In terms of Poincaré polynomials, this result may be stated as follows. We have* $P^\Gamma(F,t) = (1\otimes\gamma_0+\varepsilon\otimes\gamma_{10})P_0^\Gamma(F,t)$, *where* $P_0^\Gamma(F,t) = 1 + [\rho\otimes\gamma_0+\chi^{(3,2)}\otimes\gamma_0]t + [1\otimes(\gamma_5+\gamma_{15}) + \rho\otimes\gamma_0 + \chi^{(3,2)}\otimes(\gamma_0+\gamma_{10}) + \chi^{(3,1^2)}\otimes(\gamma_0+\gamma_{10})]t^2 + [1\otimes(\gamma_2+\gamma_6+\gamma_{14}+\gamma_{18}) + \rho\otimes(\gamma_0+\gamma_5+\gamma_{10}+\gamma_{15}) + \chi^{(3,2)}\otimes(\gamma_0+\gamma_4+\gamma_8+\gamma_{10}+\gamma_{12}+\gamma_{16}) + \chi^{(3,1^2)}\otimes(\gamma_0+\gamma_1+\gamma_3+\gamma_7+\gamma_9)]t^3$.

5 Groups of Rank Two

In this section we shall determine the polynomials $P^\Gamma(F,t)$ and $W_c^\Gamma(F,t)$ for all 'uniform' unitary reflection groups G of rank two, i.e. those which satisfy the condition (26) below, using the fact that in this case, the polynomial $Q_0 := \prod_{H\in\mathcal{A}_G}\ell_H$ has an isolated singularity at 0, and so the theory of [27, 28] applies. For the other groups of rank 2, one may apply an Euler characteristic argument directly.

5.1 Reduction to the Reduced Case

Say that the unitary reflection group G is *uniform* if it satisfies the following condition (for notation see the first paragraph of Sect. 1).

$$\text{For all } H \in \mathcal{A}_G, e_H = |G_H| := e \text{ is independent of } H. \tag{26}$$

We write $d = \deg(Q_0) = |\mathcal{A}_G|$ and $m = \deg(Q)$, where $Q = Q_0^e = \prod_{H\in\mathcal{A}_G}\ell_H^{e_H}$, so that $m = de$. A list of the irreducible uniform groups in dimension two is given in Table 1.

The next result restates Lemma 1 and its consequences. Recall that for any finite group μ_N of roots of unity in $\mathbb{C}$, γ_j is the character of μ_N whose value at $\xi\in\mu_N$ is ξ^j.

Lemma 5 *Suppose that G is a uniform unitary reflection group in V, and that Q, Q_0, d, e and m are as in the previous paragraph. Let $\Gamma = G\times\mu_m$, $F = \{v\in V \mid Q(V) = 1\}$ and for $\zeta\in\mu_e$ define F_ζ by $F_\zeta := \{v\in V \mid Q_0(v) = \zeta\}$. Then*

1. $F = \bigsqcup_{\zeta\in\mu_e} F_\zeta$.
2. *The group Γ permutes the F_ζ ($\zeta\in\mu_e$) transitively, and writing F_0 for F_1, the stabiliser of F_0 in Γ is $\Gamma_0 := \ker(\varepsilon\otimes\gamma_d)$, where ε is the character of G defined by $\varepsilon(g) = \det_V(g)^{-1}\in\mu_e$.*

Table 1 The irreducible uniform reflection groups of rank 2

Group	e	d	Degrees of G
$G(2p, p, 2)$ $(p \geq 2)$	2	$2p+2$	$2p, 4$ (imprimitive)
$G(p, p, 2)$	2	p	$2, p$ (dihedral)
G_4	3	4	4, 6
G_5	3	8	6, 12
G_8	4	6	8, 12
G_{12}	2	12	6, 8
G_{13}	2	18	8, 12
G_{16}	5	12	20, 30
G_{20}	3	20	12, 30
G_{22}	2	30	12, 20

3. *Let* $\Phi = \sum_{i=0}^{e-1} \varepsilon^i \otimes \gamma_{di} \in R_+(\Gamma)$. *Then for each* j, *as elements of* $R_+(\Gamma)$, *we have*

$$H^j(F, \mathbb{C}) = \mathrm{Ind}_{\Gamma_0}^{\Gamma}(H^j(F_0, \mathbb{C})) = \Phi.H_0^j(F_0, \mathbb{C}),$$

where $H_0^j(F_0, \mathbb{C})$ *is any element of* $R_+(\Gamma)$ *which restricts to* $H^j(F_0, \mathbb{C}) \in R_+(\Gamma_0)$.

4. *We have*

$$P^{\Gamma}(F, t) = \mathrm{Ind}_{\Gamma_0}^{\Gamma}(P^{\Gamma_0}(F_0, t) = \Phi.P_0^{\Gamma}(F_0, t),$$

where $P_0^{\Gamma}(F_0, t)$ *is any element of* $R_+(\Gamma)$ *which restricts to* $P^{\Gamma_0}(F_0, t) \in R_+(\Gamma_0)$.

In view of the above lemma, in order to determine $P^{\Gamma}(F, t)$, it suffices to compute $H^j(F_0, \mathbb{C}) \in R(\Gamma_0)$, and since $\dim(F_0) = 1$, $H^j(F_0, \mathbb{C}) = 0$ for $j \geq 2$ and $H^0(F_0, \mathbb{C}) = 1 \otimes \gamma_0$, we need only compute $H^1(F_0, \mathbb{C})$. For this we shall apply the results in [27, 28].

5.2 Determination of $H^1(F_0, \mathbb{C})$

We begin with the following consequences of the results of Orlik and Solomon [27, 28].

Proposition 11 *Let* X, Y *be coordinates in* V^*; *then* $F_0 = F_0(X, Y)$ *is a homogeneous polynomial of degree* d *with an isolated singularity at* $(0, 0)$.

1. *Let* I *be the ideal of* $\mathbb{C}[X, Y]$ *generated by* $\frac{\partial F_0}{\partial X}$ *and* $\frac{\partial F_0}{\partial Y}$. *Then there is an isomorphism* $\theta : \mathbb{C}[X, Y]/I \longrightarrow H^1(F_0, \mathbb{C})$ *which satisfies the following condition. For* $x \in \mathrm{GL}(V)$ *such that* $x.Q_0 = Q_0$, *we have for any element* $f \in \mathbb{C}[X, Y]/I$, $\theta(x.f) = \det_V(x)x.\theta(f)$.

2. *With* $x \in \mathrm{GL}(V)$ *as in (1), we have*

$$\mathrm{Trace}(x, \mathbb{C}[X, Y]/I) = (-1)^{k(x)}(d-1)^{k(x)}\det_V(x), \tag{27}$$

where $k(x) = \dim(\mathrm{Fix}_V(x))$.

Proof The statement (1) follows from [27, Theorem, p. 257], while (2) follows from [28, Theorem (5.4)]. □

The formula below is immediate from (2).

Corollary 11 *The character* χ *of* Γ_0 *on* $H^1(F_0, \mathbb{C})$ *is given by*

$$\chi((g, \zeta)) = \mathrm{Trace}\left((g, \zeta), H^1(F_0, \mathbb{C})\right) = (1-d)^{k(g,\zeta)},$$

where $k(g, \zeta) = \dim(\mathrm{Fix}(g, \zeta))$.

This permits the immediate determination of the monodromy action.

Proposition 12 *The structure of* $H^1(F_0, \mathbb{C})$ *as* μ_d*-module, where* $\mu_d (= 1 \times \mu_d)$ *is the monodromy subgroup of* Γ_0*, is given by*

$$H^1(F_0, \mathbb{C}) = (d-1)\gamma_0 + (d-2)\sum_{i=1}^{d-1} \gamma_i.$$

Proof Note that for $\xi \in \mu_d$, $k(1, \xi) = \begin{cases} 0 \text{ if } \xi \neq 1 \\ 2 \text{ if } \xi = 1. \end{cases}$

It follows from Corollary 11 that

$$\begin{aligned}
(\chi, \gamma_i)_{\mu_d} &= \frac{1}{d}\sum_{\xi \in \mu_d} \chi((1, \xi))\xi^{-i} \\
&= \frac{1}{d}\left((d-1)^2 + \sum_{\xi \in \mu_d,\, \xi \neq 1} \chi((1, \xi))\xi^{-i}\right) \\
&= \begin{cases} d-2 \text{ if } i \neq 0 \\ d-1 \text{ if } i = 0. \end{cases}
\end{aligned}$$

The result is now immediate. □

In Example 10 below we shall give the Poincaré-Deligne polynomial for the monodromy action in general.

5.3 Determination of $H^1(F_0, \mathbb{C})$ for the Dihedral Groups

In this section we give an explicit formula for the first cohomology of F_0 as an element of $R(\Gamma_0)$. For convenience, denote $G(p,p,2)$ by $G(p)$. When $p = 3, 4$ and 6, $G(p)$ is the Weyl group of type A_2, B_2 and G_2 respectively. For other values $p > 3$ $G(p)$ is a unitary (complex) reflection group; in each case, the reflection representation of $G(p)$ may be realised as the set of matrices $\left\{ \begin{pmatrix} \zeta & 0 \\ 0 & \zeta^{-1} \end{pmatrix}, \begin{pmatrix} 0 & \zeta \\ \zeta^{-1} & 0 \end{pmatrix}, \zeta \in \mu_p \right\}$. The description of the irreducible characters of $G(p)$ depends on the parity of p. If p is odd, the character values are given in Table 2 below.

If p is even, the values of the irreducible characters of $G(p)$ are shown in Table 3.

Now the irreducible characters of Γ are all of the form $\rho\otimes\gamma_i$, where ρ is an irreducible character of $G(p)$ and γ_i is the character $\zeta \mapsto \zeta^i$ of μ_{2p}. Now Γ_0 is the kernel of the character $\varepsilon\otimes\gamma_p$. It follows (see Lemma 1) that the irreducible characters of Γ_0 are the (distinct) restrictions to Γ_0 of $\{\rho\otimes\gamma_i \mid 0 \le i \le p-1\}$, where ρ runs over the irreducible characters of $G(p)$. Using this parametrisation of the irreducible characters of Γ_0, we have the following result.

Proposition 13 *Let $G(p)$ $(p \ge 3)$ be the reflection group $G(p,p,2)$ as described above, and let Γ, Γ_0, F, F_0 be as in Lemma 1. Then $H^1(F_0, \mathbb{C})$ is given as an element of $R(\Gamma_0$ as follows.*

1. If p is odd, then

$$H^1(F_0, \mathbb{C}) = \sum_{\substack{0\le i\le p-1 \\ i \text{ odd}}} 1\otimes\gamma_i + \sum_{\substack{0<i\le p-1 \\ i \text{ even}}} \varepsilon\otimes\gamma_i + \sum_{k=1}^{\frac{p-1}{2}} \sum_{\substack{0\le i\le p-1 \\ i\ne k,p-k}} \chi_k\otimes\gamma_i. \tag{28}$$

Table 2 The characters of $G(p)$, p odd

Character	$\begin{pmatrix} \zeta & 0 \\ 0 & \zeta^{-1} \end{pmatrix}$	$\begin{pmatrix} 0 & \zeta \\ \zeta^{-1} & 0 \end{pmatrix}$
1	1	1
ε	1	-1
χ_k $(1 \le k \le \frac{p-1}{2})$	$\zeta^k + \zeta^{-k}$	0

Table 3 The characters of $G(p)$, p even

Character	$\begin{pmatrix} \zeta & 0 \\ 0 & \zeta^{-1} \end{pmatrix}$	$\begin{pmatrix} 0 & \zeta \\ \zeta^{-1} & 0 \end{pmatrix}$
1	1	1
ε	1	-1
δ	$\zeta^{\frac{p}{2}}$	$\zeta^{\frac{p}{2}}$
$\varepsilon\delta$	$\zeta^{\frac{p}{2}}$	$-\zeta^{\frac{p}{2}}$
χ_k $(1 \le k \le \frac{p-2}{2})$	$\zeta^k + \zeta^{-k}$	0

2. *If p is even, then*

$$H^1(F_0,\mathbb{C}) = \varepsilon\otimes\gamma_0 + 2\sum_{\substack{0<i\leq p-1\\ i\ even}} \varepsilon\otimes\gamma_i + \sum_{\substack{0\leq i\leq p-1\\ i+\frac{p}{2}\in 2\mathbb{Z},\ i\neq\frac{p}{2}}} (\varepsilon\delta+\delta)\otimes\gamma_i$$
$$+2\sum_{k=1}^{\frac{p-2}{2}} \sum_{\substack{0\leq i\leq p-1\\ i+k\in 2\mathbb{Z},\ i\neq k,p-k}} \chi_k\otimes\gamma_i + \sum_{k=1}^{\frac{p-2}{2}} \chi_k\otimes(\gamma_k+\gamma_{p-k}). \qquad (29)$$

Proof This is proved by simply carrying out the tedious task of computing the inner product $(\chi,\rho\otimes\gamma_i)_{\Gamma_0}$ of each irreducible character $\rho\otimes\gamma_i$ of Γ_0 with the character χ of the representation of Γ_0 on $H^1(F_0,\mathbb{C})$, as described in Corollary 11 above. □

This last result may be used to compute the weight polynomial $W^{\Gamma_0}(F_0,t)$ (see Proposition 16).

5.4 Non-uniform Groups of Rank Two

The values of $e_H = |G_H|$ for $H \in \mathcal{A}_G$ for the non-uniform groups G of rank two may be found in [23, Table D1]. There may be 1, 2 or 3 distinct values for a given group G; for example the 2-dimensional group G_{19} has e_H values of 2, 3 and 5.

Now we know the Euler characteristic $H^0(F,\mathbb{C}) - H^1(F,\mathbb{C})$ by Corollary 7, and $H^0(F,\mathbb{C})$ is easily determined formally, since $H^0(F_0,\mathbb{C}) = \mathbb{C}$. This suffices to deduce $H^1(F,\mathbb{C})$.

6 Relations with Mixed Hodge Theory

In this section we investigate the relationship between the Γ-action and the mixed Hodge structure (henceforth denoted MHS) on the Milnor fibre cohomology. Refer to [31] for general notions and results concerning the MHS.

6.1 Equivariant Hodge-Deligne Polynomials and Spectra

To do this, we first recall the definition of the equivariant Hodge-Deligne polynomials $HD^{\Gamma}(X)$ associated to a finite group Γ acting (algebraically) on a complex algebraic variety X, see [13] .

More precisely, let X be a quasi-projective variety over $\mathbb{C}$ and consider the Deligne MHS on the rational cohomology groups $H^*(X,\mathbb{Q})$ of X. Since this MHS is functorial with respect to algebraic mappings, if Γ acts algebraically on X, each of the graded pieces

$$H^{p,q}(H^j(X,\mathbb{C})) := Gr_F^p Gr_{p+q}^W H^j(X,\mathbb{C}) \tag{30}$$

becomes a Γ-module in the usual functorial way, and these modules are the building blocks of the polynomial $HD^\Gamma(X; u, v) \in R(\Gamma)[u, v]$, defined by

$$HD^\Gamma(X; u, v) = \sum_{p,q} E^{\Gamma;p,q}(X) u^p v^q, \tag{31}$$

where $E^{\Gamma;p,q}(X) = \sum_j (-1)^j H^{p,q}(H^j(X,\mathbb{C})) \in R(\Gamma)$. Note that

$$HD^\Gamma(X; 1, 1) = \chi^\Gamma(X).$$

One may consider an even finer (and hence harder to determine) invariant, namely the equivariant Poincaré-Deligne polynomial

$$PD^\Gamma(X; u, v, t) = \sum_{p,q,j} H^{p,q}(H^j(X,\mathbb{C})) u^p v^q t^j \in R_+(\Gamma)[u, v, t]. \tag{32}$$

Clearly one has $PD^\Gamma(X; u, v, -1) = HD^\Gamma(X; u, v)$ and $PD^\Gamma(X; 1, 1, t) = P^\Gamma(X; t)$.
Coming back to our setting, Lemma 1 yields the following result.

Proposition 14 *Assume we are in the situation of Sect. 1. Then in the notation of Lemma 1, we have*

$$PD^\Gamma(F; u, v, t) = (1\otimes\gamma_0 + \lambda\otimes\gamma_{\frac{m}{e}} + (\lambda\otimes\gamma_{\frac{m}{e}})^2 + \cdots + (\lambda\otimes\gamma_{\frac{m}{e}})^{e-1}) P_0^\Gamma(F; u, v, t),$$

where $P_0^\Gamma(F; u, v, t)$ is a polynomial in the ring $R_+(\Gamma)[u, v, t]$ whose restriction to Γ_0 is $PD^{\Gamma_0}(F_0; u, v, t)$, and λ is the character of G on Q_0.
In particular,

$$HD^\Gamma(F; u, v)) = (1\otimes\gamma_0 + \lambda\otimes\gamma_{\frac{m}{e}} + (\lambda\otimes\gamma_{\frac{m}{e}})^2 + \cdots + (\lambda\otimes\gamma_{\frac{m}{e}})^{e-1}) HD_0^\Gamma(F; u, v)$$

where $HD_0^\Gamma(F; u, v)$ is any polynomial in the ring $R(\Gamma)[u, v]$ whose restriction to Γ_0 is $HD^{\Gamma_0}(F_0; u, v)$.

It follows that the crucial question is how to compute (or at least get information about) the polynomials $PD^{\Gamma_0}(F_0; u, v, t)$ and $HD^{\Gamma_0}(F_0; u, v)$. Consider the natural inclusion $\mu_d \to \Gamma_0$ given by $\zeta \to (1, \zeta) \in \Gamma_0 \subset G \times \mu_m$, where as in Sect.1, $d = m/e = deq(Q_0)$. It induces a ring homomorphism

$$\theta : R(\Gamma_0) \to R(\mu_d), \tag{33}$$

by restriction of representations, i.e. such that $\theta(W \otimes \gamma_k) = (\dim W)\gamma_k$, for any G-module W and character $\gamma_k \in \hat{\mu}_m$; note that on the right side of this relation, $\gamma_k = (\gamma_1)^k \in \hat{\mu}_d$. This morphism θ can be used to relate known facts on the usual monodromy action (i.e. μ_d-action) on the cohomology of the usual (connected) Milnor fibre F_0 and the Γ_0-action on the same cohomology.

A simple example is that Proposition 2 may be reformulated as follows.

$$\theta(\chi^{\Gamma_0}(F_0)) = \chi(U)\,\mathrm{Reg}_{\mu_d}\,. \tag{34}$$

It is interesting to apply this restriction morphism to the equivariant Hodge-Deligne polynomial $HD^{\Gamma_0}(F_0; u, v)$. We get

$$\theta(HD^{\Gamma_0}(F_0; u, v)) = \sum_{p,q}\Big(\sum_{j,s}(-1)^j \dim H^{p,q}(H^j(F_0, \mathbb{C}))^{h^j}_{\lambda(s,d)}\gamma_s\Big)u^p v^q, \tag{35}$$

with $\lambda(s,d) = \gamma_s(\zeta_d)$ and $H^{p,q}(H^j(F_0, \mathbb{C}))^{h^j}_{\lambda(s,d)}$ denoting the $\lambda(s,d)$-eigenspace of the monodromy operator h^j (which corresponds to multiplication by ζ_d in the μ_d-action on $H^j(F_0, \mathbb{C})$ arising from the obvious inclusion $\mu_d \to \mu_m \to \Gamma$).

Taking $v = 1$ in this expression, we get an element in $R(\mu_d)[u]$, given by

$$\begin{aligned}\theta(HD^{\Gamma_0}(F_0; u, 1)) &= \sum_{p}\Big(\sum_{j,s}(-1)^j \dim Gr^p_F(H^j(F_0, \mathbb{C}))^{h^j}_{\lambda(s,d)}\gamma_s\Big)u^p \\ &= \sum_{p,s} c_{p,s}\gamma_s u^p,\end{aligned} \tag{36}$$

for some coefficients $c_{p,s} \in \mathbb{Z}$.

Recall that $T^j = (h^j)^{-1}$, the inverse of the monodromy operator on the jth cohomology of F_0, is the monodromy of the local system on $\mathbb{C}^*$ coming from the constructible sheaf $R^j Q_{0*}\mathbb{C}_V$, see Sect. 2 in [15]. For any linear operator A on a module M, we denote by $M^A_\xi = M(A, \xi)$ the ξ-eigenspace of A. Then

$$Gr^p_F(H^j(F_0, \mathbb{C}))^{h^j}_{\lambda(s,d)} = Gr^p_F(H^j(F_0, \mathbb{C}))^{T^j}_{\beta(s,d)}, \tag{37}$$

where $\lambda(s,d) = \exp(2\pi i\alpha_0)$ and $\beta(s,d) = \exp(-2\pi i\alpha_0)$, with $\alpha_0 = \frac{s}{d} \in [0,1)$.

Now recall the definition of the spectrum $Sp(\mathcal{A})$ of an essential hyperplane arrangement $\mathcal{A} : Q_0(x) = 0$ in the ℓ-dimensional vector space V:

$$Sp(\mathcal{A}) = \sum_{\alpha \in \mathbb{Q}} m_\alpha t^\alpha, \tag{38}$$

with $m_\alpha = \sum_j (-1)^{j-\ell+1} \dim Gr^p_F \tilde{H}^j(F, \mathbb{C})^{T^j}_\beta$ where $p = [\ell - \alpha]$ and $\beta = \exp(-2\pi i\alpha)$. Here $[y]$ denotes the integral part of a real number y, i.e. the largest integer z such that $z \leq y$. A recent result of Budur and Saito in [1] asserts that $Sp(\mathcal{A})$ is determined by the combinatorics, i.e. by the lattice $L(\mathcal{A})$.

For any vector space $W = \mathbb{C}^N$, linear transformation ϕ of W and $\xi \in \mathbb{C}$ write $W[\phi, \xi] := \{w \in W \mid \phi(w) = \xi w\}$ for the corresponding eigenspace. Writing $\mu_d = \langle h \rangle$, where h is, as above, the monodromy induced by multiplication on F by $\zeta_d = \exp \frac{2\pi i}{d}$, we evidently have, for any μ_d-module M,

$$\dim M(h, \zeta_d^i) = (M, \gamma_i)_{\mu_d},$$

where the right side denotes the multiplicity pairing in $R(\mu_d)$. The same formula applies when M is taken to be any virtual μ_d-module, with dim suitably interpreted.

Proposition 15 *Let $M^{(p)}$ be the virtual Γ_0 module defined by*

$$M^{(p)} = (-1)^{\ell-1} \sum_j (-1)^j Gr_F^p \widetilde{H}^j(F, \mathbb{C}).$$

In particular $M^{(p)}$ is (by restriction) a μ_d-module, and we have the following equation in $\mathbb{Z}[t^{\frac{1}{d}}]$.

$$Sp(\mathcal{A}) = \left(\sum_{p=0}^{\ell-1} M^{(p)} t^{\ell-1-p}, \sum_{j=1}^{d} \gamma_j t^{\frac{j}{d}} \right)_{\mu_d} = \sum_{p=0}^{\ell-1} \sum_{j=1}^{d} \left(M^{(p)}, \gamma_j\right)_{\mu_d} t^{\ell-1-p+\frac{j}{d}},$$

where $(-,-)_{\mu_d}$ denotes the multiplicity pairing, extended to $R(\mu_d)[t^{\frac{1}{d}}]$ in the obvious way.

Proof Note that $M^{(p)} = 0$ unless $0 \leq p \leq \ell-1$ and that $\dim M^{(p)}(h, \exp(2\pi i\alpha) = 0$ unless $\alpha = \frac{k}{d}$ for some $k \in \mathbb{Z}$. From these constraints, it is clear that $1 \leq k \leq d\ell$, and that $Sp(\mathcal{A})$ is a sum of terms of the form $f(t)(M^{(p)}, \gamma_j)_{\mu_d}$, where $f(t)$ is a (fractional) power of t. Moreover a simple calculation shows that for $1 \leq i \leq \ell$ and $1 \leq j \leq d$, the coefficient of $(M^{(\ell-i)}, \gamma_j)_{\mu_d}$ is $t^{\frac{(i-1)d+j}{d}}$. □

Corollary 12 *For $p \in \{0, 1, \dots, \ell-1\}$ and $s \in \{0, 1, \dots, d-1\}$, we have, in the notation of* (36) *and Proposition 15, $c_{p,s} = \left(M^{(p)}, \gamma_s\right)_{\mu_d}$*

In particular, the restriction $\theta(HD^{\Gamma_0}(F_0; u, 1))$ contains exactly the same information as the spectrum $Sp(\mathcal{A})$ and hence it is determined by the combinatorics.

Proof From the remarks preceding Proposition 15, (36) may be written

$$\theta(HD^{\Gamma_0}(F_0; u, 1)) = \sum_p \left(M^{(p)}, \gamma_s\right)_{\mu_d} \gamma_s u^p = \sum_{p,s} c_{p,s} \gamma_s u^p,$$

where $c_{p,s} = \left(M^{(p)}, \gamma_s\right)_{\mu_d}$; this is the first statement. Hence Proposition 15 may be written $Sp(\mathcal{A}) = \sum_{p=0}^{\ell-1}\sum_{j=1}^{d} c'_{\ell-1-p,j} t^{p+\frac{j}{d}}$, where $c'(p,j) = \begin{cases} c_{p,j} \text{ if } j \neq d \\ c_{p,0} \text{ if } j = d. \end{cases}$

Thus $\theta(HD^{\Gamma_0}(F_0; u, 1))$ contains precisely the same information as $Sp(\mathcal{A})$. □

Using the explicit formulas for the spectrum given in [1] , this proposition gives valuable information on the Poincaré-Deligne polynomial $PD^{\Gamma_0}(F_0; u, v))$. We illustrate the use of the above results by computing the polynomial $PD^{\Gamma_0}(F_0; u, v, t)$ in the case where $G = \mathrm{Sym}_3$ or $G = \mathrm{Sym}_4$.

Example 9 Consider first the case $G = \mathrm{Sym}_3$, when $\ell = 2$. Example 6 implies that

$$P^{\Gamma_0}(F_0, t) = 1\otimes\gamma_0 + (1\otimes(\gamma_1 + \gamma_5) + \rho\otimes\gamma_0)t.$$

On the other hand, we know that the usual monodromy action h^1 on $H^1(F_0, \mathbb{C})$ yields a decomposition

$$H^1(F_0, \mathbb{C}) = H^1(F_0, \mathbb{C})_1 \oplus H^1(F_0, \mathbb{C})_{\neq 1}.$$

Moreover, the first summand $H^1(F_0, \mathbb{C})_1 = \rho\otimes\gamma_0$ is a pure Hodge structure of type $(1, 1)$, and $H^1(F_0, \mathbb{C})_{\neq 1} = 1\otimes(\gamma_1 + \gamma_5)$ is a pure Hodge structure of weight 1, see [13], Remark 1.4. (i) and Theorem 1.5 or [14], Theorem 4.1. It follows that the only problem is to decide whether the subspace corresponding to $1\otimes\gamma_1$ has Hodge type $(1, 0)$ or $(0, 1)$.

On the other hand, in this case it is easy to compute the spectrum $Sp(\mathcal{A})$ since it coincides with the spectrum of the isolated hypersurface singularity $(x - y)(2x + y)(x+2y) = 0$ and hence can be computed using the usual formulas, see for instance the formula (2.4.7) in [15]. It follows that

$$Sp(\mathcal{A}) = \left(\frac{t - t^{\frac{1}{3}}}{t^{\frac{1}{3}} - 1}\right)^2 = t^{\frac{2}{3}} + 2t + t^{\frac{4}{3}}.$$

Applying Corollary 12 with $\ell = 2$, $s = 1$, $p = 0$ we get $\alpha = \frac{4}{3}$ and hence the coefficient of $g_1 u^0$ is $c_{0,1} = -m_{\frac{4}{3}} = -1$. This yields that $1\otimes\gamma_1$ has Hodge type $(0, 1)$ and hence we get

$$PD^{\Gamma_0}(F_0; u, v, t) = 1\otimes\gamma_0 + (1\otimes(\gamma_5 u + \gamma_1 v) + \rho\otimes\gamma_0 uv)t.$$

The results of Sect. 5 enable us to generalise this result to all rank 2 reflection groups.

Example 10 We use the notation of Proposition 12; thus G is a rank two reflection group and $d = |\mathcal{A}_G|$. We shall show that

$$PD^{\mu_d}(F_0; u, v, t) = \gamma_0 + \left(\sum_{i=2}^{d-1}(i-1)\gamma_i u + \sum_{j=1}^{d-2}(d-1-j)\gamma_j v + (d-1)\gamma_0 uv\right)t. \tag{39}$$

To see this, note that here we again have by [15, (2.4.7)] that

$$Sp(\mathcal{A}) = \left(\frac{t - t^{\frac{1}{d}}}{t^{\frac{1}{d}} - 1}\right)^2 = \sum_{i=1}^{d-2} i t^{\frac{i+1}{d}} + (d-1)t + \sum_{j=1}^{d-2} j t^{1+\frac{d-1-j}{d}}.$$

We now argue exactly as in the previous example to obtain (39). Note that it is evident that the coefficients of u and v are mutually contragredient in $R(\mu_d)$.

The above argument may also be used to compute the weight polynomial of F_0 in the dihedral case. The result is

Proposition 16 *Let $G = G(p, p, 2)$ be the dihedral group, as in Sect. 5. If p is odd, the weight polynomial of F_0 is given, in the notation of Sect. 5.3, by*

$$W^{\Gamma_0}(F_0, t) = 1\otimes\gamma_0 + \left(\sum_{i=1}^{p-1}\left[\varepsilon^{\otimes(i+1)} + \sum_{\substack{k=1\\ k\neq i, p-i}}^{\frac{p-1}{2}} \chi_k\right]\otimes\gamma_i\right)t + \left(\sum_{k=1}^{\frac{p-1}{2}} \chi_k\otimes\gamma_0\right)t^2.$$

Proof This is proved by combining the information in Proposition 13 and Example 10, as well as our knowledge of the weight 0 and weight 2 parts of the cohomology, which come from the associated hyperplane complement. □

Now we consider the more involved case $G = \mathrm{Sym}_4$, when $\ell = 3$.

Proposition 17 *With notation from Proposition 9 one has the following.*

$$\begin{aligned} PD^{\Gamma_0}(F_0; u, v, t) =& 1\otimes\gamma_0 + [1\otimes(\gamma_{10}u + \gamma_2 v) + (\sigma + \rho)\otimes\gamma_0 uv]t + \{(1\otimes\gamma_9 + \sigma\otimes\gamma_{11})u^2 \\ &+ (1\otimes\gamma_3 + \sigma\otimes\gamma_1)v^2) + \rho\otimes(\gamma_{10}u^2 v + \gamma_2 uv^2 + (\gamma_0 + \gamma_6)u^2v^2)\}t^2. \end{aligned}$$

Proof The associated line arrangement in $\mathbb{P}^2$ consists of six lines, has only double and triple points and has exactly $\nu_3 = 4$ triple points. The formula given in Theorem 3 in [1] yields

$$Sp(\mathcal{A}) = t^{\frac{1}{2}} + 3t^{\frac{2}{3}} + 2t^{\frac{5}{6}} + 6t + 3t^{\frac{4}{3}} - t^{\frac{5}{3}} - 5t^2 + 2t^{\frac{13}{6}} - t^{\frac{7}{3}} + t^{\frac{15}{6}}.$$

Using Corollary 12 as above we get

$$\theta(HD^{\Gamma_0}(F_0;u,1)) = \gamma_0-\gamma_2-\gamma_4u-5\gamma_0u+\gamma_3u^2+\gamma_3+6\gamma_0u^2+3\gamma_2u+3\gamma_4u^2+2\gamma_1+2\gamma_5u^2. \tag{40}$$

As above, we know that the usual monodromy action on $H^1(F_0,\mathbb{C})$ yields a decomposition

$$H^1(F_0,\mathbb{C}) = H^1(F_0,\mathbb{C})_1 \oplus H^1(F_0,\mathbb{C})_{\neq 1}.$$

Using the formula obtained in (25), we see that $H^1(F_0,\mathbb{C})_1 = (\sigma+\rho)\otimes\gamma_0$ is a pure Hodge structure of type $(1,1)$, and $H^1(F_0,\mathbb{C})_{\neq 1} = 1\otimes(\gamma_2+\gamma_{10})$ is a pure Hodge structure of weight 1. Moreover, one has a similar decomposition

$$H^2(F_0,\mathbb{C}) = H^2(F_0,\mathbb{C})_1 \oplus H^2(F_0,\mathbb{C})_{\neq 1}.$$

Here $H^2(F_0,\mathbb{C})_1 = \rho\otimes(\gamma_0+\gamma_6)$ is a pure Hodge structure of type $(2,2)$. And $H^2(F_0,\mathbb{C})_{\neq 1} = 1\otimes(\gamma_3+\gamma_9)+\rho\otimes(\gamma_2+\gamma_{10})+\sigma\otimes(\gamma_1+\gamma_{11})$ has Hodge weights 2 and 3 and the characters $\mathfrak{g}_i \in \hat{\mu}_{12}$ having weight 3 must have order 3 in $\hat{\mu}_6$, see [13]. The above discussion combined with (40) yield the following two possibilities. (The Γ-representations occurring in these formulas should be considered as being Γ_0-representations, exactly as in the discussion after the formula (25).)

$$\begin{aligned} PD^{\Gamma_0}(F_0;u,v,t) =& 1\otimes\gamma_0+[1\otimes(\gamma_{10}u+\gamma_2v)+(\sigma+\rho)\otimes\gamma_0uv]t+\{1\otimes(\gamma_9u^2+\gamma_3v^2)\\ &+\rho\otimes(\gamma_{10}u^2v+\gamma_2uv^2+(\gamma_0+\gamma_6)u^2v^2)+\sigma\otimes(\gamma_{11}u^2+\gamma_1v^2)\}t^2 \end{aligned} \tag{41}$$

or

$$\begin{aligned} PD^{\Gamma_0}(F_0;u,v,t) =& 1\otimes\gamma_0+[1\otimes(\gamma_{10}u+\gamma_2v)+(\sigma+\rho)\otimes\gamma_0uv]t+\{1\otimes(\gamma_9v^2+\gamma_3u^2)\\ &+\rho\otimes(\gamma_{10}u^2v+\gamma_2uv^2+(\gamma_0+\gamma_6)u^2v^2)+\sigma\otimes(\gamma_{11}u^2+\gamma_1v^2)\}t^2. \end{aligned} \tag{42}$$

This indeterminacy is due to the fact that both characters γ_3 and $\mathfrak{g}_9$ induce the same character $\mathfrak{g}_3$ by restriction from μ_{12} to μ_6.

To decide which of the two formulas above is the correct one, we construct a new μ_{12}-action on F_0 as follows. Consider the transposition $\tau = (1,2)$ and the group monomorphism $\mu_{12} \to \Gamma_0$ given by $\zeta_{12} \mapsto (\tau,\zeta_{12})$. This morphism induces as above a ring morphism

$$\theta' : R(\Gamma_0) \to R(\mu_{12}), \tag{43}$$

by restriction of representations, namely

$$\theta'(W\otimes\gamma_h) = (\dim W_0)\gamma_h + (\dim W_-)\gamma_{h+6}$$

where $W_{\pm}$ are the ± 1-eigenspaces of τ in the representation W, and $h + 6$ has to be computed modulo 12. Using this we see that the coefficient of u^2 in the corresponding Hodge-Deligne polynomial $\theta'(HP^{\Gamma_0}(F_0; u, v))$ is either

$$\gamma_5 + \gamma_9 + \gamma_{11} \tag{44}$$

if formula (41) holds, or

$$\gamma_1 + \gamma_3 + \gamma_7, \tag{45}$$

if formula (42) holds. We study now this new action of μ_{12} on the cohomology group $H^2(F_0, \mathbb{C})$ using a similar approach to Sect. 5 in [13]. First note that μ_{12}-module $H^{2,0}(H^2(F_0, \mathbb{C}))$, which is the coefficient of u^2 in $\theta'(HP^{\Gamma_0}(F_0; u, v))$, is isomorphic to the μ_{12}-module $H^{2,0}(H^2_c(F_0, \mathbb{C}))$ (compare with Corollary 1.2 in [13]).

Next, let V be the union of the six lines corresponding to the A_3-arrangement in $\mathbb{P}^2$, and choose $Q_V(x_1, x_2, x_3) = 0$ be a reduced equation for V such that $\tau Q_V = -Q_V$. One may take

$$Q_V(x_1, x_2, x_3) = (x_1 - x_2)(x_1 - x_3)(x_2 - x_3)(2x_1 + x_2 + x_3)(x_1 + 2x_2 + x_3)(x_1 + x_2 + 2x_3).$$

Consider next the surface in $\mathbb{P}^3$ given by

$$X_V : Q_V(x_1, x_2, x_3) - t^6 = 0.$$

The μ_{12}-action on F_0 extends to a μ_{12}-action on X_V given by

$$\zeta_{12} \cdot (x_1 : x_2 : x_3 : t) = (x_2 : x_1 : x_3 : \zeta_{12} t).$$

This implies that for a cohomology class α, one has

$$\zeta_{12} \cdot \alpha = (h')^*(\alpha)$$

with $h'((x_1 : x_2 : x_3 : t)) = (x_2 : x_1 : x_3 : \zeta_{12}^{-1} t)$.

As $F_0 = X_V \setminus V$, the long exact sequence of cohomology with compact supports yields

$$0 \to H^1(V) \to H^2_c(F_0) \to H^2(X_V) \to H^2(V) \to \dots$$

Since $H^1(V)$ has weights at most 1, and $H^2(V)$ has type $(1, 1)$ it follows that we have an isomorphism of μ_{12}-modules

$$H^{2,0}(H^2_c(F_0, \mathbb{C})) \simeq H^{2,0}(H^2(X_V, \mathbb{C})).$$

On the other hand, since $H^2(X_V, \mathbb{C})$ has weights at most 2, it follows that we have $H^{2,0}(H^2(X_V, \mathbb{C})) = Gr_F^2 H^2(X_V, \mathbb{C}))$, and hence we get an inclusion

$$Gr_F^2 H^2(X_V, \mathbb{C})) \to Gr_{F_{SS}}^2 H^2(X_\infty) \tag{46}$$

as in the exact sequence (5.3) in [13]. Here X_∞ is a smooth surface in $\mathbb{P}^3$, nearby X_V, regarded as a generic fibre in a 1-parameter smoothing X_w of X_V. Moreover, $H^2(X_\infty)$ is endowed with the Schmid-Steenbrink limit MHS, whose Hodge filtration will be denoted by F_{SS}. The Hodge filtration F_{SS} on $H^2(X_\infty)$, being the limit of the Deligne Hodge filtration F on $H^2(X_w)$, yields an isomorphism

$$Gr_{F_{SS}}^2 H^2(X_\infty) = Gr_F^2 H^2(X_w) \tag{47}$$

of $\mathbb{C}$-vector spaces (i.e. equality of dimensions). Note that our smoothing X_w can be constructed in a μ_{12}-equivariant way, e.g. just take X_w to be the zeroset in $\mathbb{P}^3$ of a polynomial of the form $Q_V(x) - t^6 + w(x_1^6 - x_2^6 + x_3^6 + t^6)$. With such a choice, the isomorphism (47) becomes an equation in the representation ring $R(\mu_{12})$. Moreover, these representations can be explicitly determined, as explained in a similar context in Example 5.1 in [13]. A direct computation using rational differential forms yields

$$Gr_F^2 H^2(X_w) = \gamma_4 + 2\gamma_5 + 2\gamma_9 + \gamma_{10} + 4\gamma_{11}. \tag{48}$$

Comparing this with the inclusion in (46), it follows that we are in the situation given by (44) and hence the formula (41) holds.

Finally we consider the case of the braid arrangement $\mathcal{A}_4$, and forget (at least for the moment) the Sym_5-action on the Milnor fiber.

The $\mathcal{A}_4$ arrangement is associated with the natural Sym_5-action on $\mathbb{C}^5$, namely it consists of the ten hyperplanes $H_{ij} : x_i - x_j = 0$ for $1 \leq i < j \leq 5$. To have an essential arrangement, we take the intersection with the hyperplane $H : x_1 + \ldots + x_5 = 0$ and get in this way our model for the arrangement $\mathcal{A}_4$ as a central essential arrangement in $\mathbb{C}^4$, consisting of the ten hyperplanes $H'_{ij} = H_{ij} \cap H$. To compute the spectrum we use the approach outlined in [1] and the explicit formulas given in [35].

Note that the edges contained in the non normal crossing part of the central arrangement $\mathcal{A}_4$ in $\mathbb{C}^4$ are the following:

1. Ten codimension two edges X where three hyperplanes come together, e.g. as in the intersection $X = H'_{12} \cap H'_{23} \cap H'_{13}$; hence they have the multiplicity $m_X = 3$.
2. Ten codimension three edges Y which arise when an edge X as above is cut by a transversal hyperplane, e.g. $Y = X \cap H'_{45} = H'_{12} \cap H'_{23} \cap H'_{13} \cap H'_{45}$; hence they have the multiplicity $m_Y = 4$.
3. Five codimension three edges Z where six hyperplanes come together, e.g. as in the intersection $X = H'_{12} \cap H'_{23} \cap H'_{13} \cap H'_{14} \cap H'_{24} \cap H'_{34}$; hence they have the multiplicity $m_Z = 6$.

It is important to note that each edge X of the first type (*i*) contains exactly one edge Y of the second type (*ii*) and two edges Z of the third type (*iii*). In the associated projective space $\mathbb{P}^3$, the edges X give rise to lines, while the edges Y and Z to points. Note that the edges of type X and Z are dense, while the edges of type Y are not.

Using the above description and Theorem 1.1 in [35], we get the following by a direct careful computation.

Proposition 18

$$\begin{aligned} Sp(\mathcal{A}_4) &= t^{\frac{4}{10}} + 4t^{\frac{5}{10}} + 5t^{\frac{6}{10}} + 5t^{\frac{8}{10}} + 6t^{\frac{9}{10}} + 24t \\ &+6t^{\frac{13}{10}} - t^{\frac{15}{10}} + t^{\frac{18}{10}} - 26t^2 \\ &+t^{\frac{22}{10}} + 4t^{\frac{25}{10}} + 6t^{\frac{27}{10}} + 9t^3 \\ &+6t^{\frac{31}{10}} + 5t^{\frac{32}{10}} + 5t^{\frac{34}{10}} - t^{\frac{35}{10}} + t^{\frac{36}{10}}. \end{aligned}$$

Let U_4 be the complement of the corresponding hyperplane arrangement in $\mathbb{P}^3$. Then it is known that $b_0(U_4) = 1$, $b_1(U_4) = 9$, $b_2(U_4) = 26$ and $b_3(U_4) = 24$, and this explains the integral powers of t in the above formula (recall that the spectrum uses the *reduced* cohomology). In particular, $\chi(U_4) = -6$, so the alternating sum of the multiplicities of any eigenvalue $\lambda \neq 1$ should be -6 by our Proposition 2.

In fact, using the computations in Settepanella [34] or in our Proposition 10, we already know these multiplicities. They are as follows.

1. The eigenvalue -1 occurs with multiplicity 2 on $H^2(F_0)$ and with multiplicity 8 on $H^3(F_0)$.
2. All the other eigenvalues $\exp(2\pi ik/10)$ for $k = 1, 2, 3, 4, 6, 7, 8, 9$ occur only on the top cohomology group with multiplicity 6.

Using the spectrum computation above, we can describe the Hodge-Deligne μ_{10}-equivariant polynomial of F_0 as follows.

Corollary 13

$$\begin{aligned} DP^{\mu_{10}}(F_0; u, v, t) &= \gamma_0 + 9\gamma_0 uvt + [\gamma_5 u^2 + \gamma_5 v^2 + 26\gamma_0 u^2 v^2]t^2 + \\ &+[(\gamma_4 + 5\gamma_6 + 5\gamma_8 + 6\gamma_9)u^3 + (6\gamma_3 + \gamma_8)u^2 v + (\gamma_2 + 6\gamma_7)uv^2 \\ &+(6\gamma_1 + 5\gamma_2 + 5\gamma_4 + \gamma_6)v^3 + 4\gamma_5 u^3 v + 4\gamma_5 uv^3 + 24\gamma_0 u^3 v^3]t^3. \end{aligned}$$

Proof We will prove only the claims concerning the character $\mathfrak{g}_5$, since the proof of the remaining claims is quite similar. Using the definition of the spectrum, the coefficient of $t^{\frac{15}{10}}$ gives

$$\dim Gr_F^2 H^3(F_0, \mathbb{C})_{\gamma_5} - \dim Gr_F^2 H^2(F_0, \mathbb{C})_{\gamma_5} = -1. \tag{49}$$

Since $H^2(F_0, \mathbb{C})_{\gamma_5}$ is 2-dimensional (and contains no elements of type $(2,2)$, see Theorem 1.3 in [13]), the only possibility is $\dim Gr_F^2 H^3(F_0, \mathbb{C})_{\gamma_5} = 0$ and on the other hand $\dim Gr_F^2 H^2(F_0, \mathbb{C})_{\gamma_5} = 1$, which yields the claim concerning $H^2(F_0)_{\gamma_5}$. In a similar way, the coefficient of $t^{\frac{25}{10}}$ yields the equality $Gr_F^0 H^3(F_0, \mathbb{C})_{\gamma_5} = 0$.

Next, $H^3(F_0, \mathbb{C})_{\gamma_5}$ is 8-dimensional and the coefficient of $t^{\frac{5}{10}}$ gives the equality $\dim Gr_F^3 H^3(F_0, \mathbb{C})_{\gamma_5} = 4$. This fact combined with the above vanishings shows that the eigenspace $H^3(F_0, \mathbb{C})_{\gamma_5}$ has only Hodge classes of type $(3,1)$ and $(1,3)$. This proves the claim concerning $H^3(F_0)_{\gamma_5}$.

Remark 8 The spectrum of $\mathcal{A}_4$ has a non-cancellation property, since the only eigenvalue $\neq 1$ occurring in two distinct cohomology groups is -1, but the Hodge types (p,q) have $p = 0, 2$ on H^2 and $p = 1, 3$ on H^3, so no cancellation is possible.

Note that this non cancellation property is quite subtle, since it involves either the knowledge of the multiplicities of (-1), or the knowledge of the Hodge types of the eigenspaces $H^2(F_0, \mathbb{C})_{\gamma_5}$ and $H^3(F_0, \mathbb{C})_{\gamma_5}$ and the fact that they correspond to distinct p's.

The corresponding formula for the $\mathcal{A}_3$-arrangement is given below; the characters $\mathfrak{g}_2$ and $\mathfrak{g}_4$ occur in both H^1 and H^2. The corresponding Hodge types each have $p = 1$ (once for $\mathfrak{g}_4$ in H^1 and once for $\mathfrak{g}_2$ in H^2), and similarly for $p = 2$. This again yields a non-cancellation property for the corresponding spectrum.

It seems a difficult task to prove such a non cancellation result in general. The non-cancellation property of the spectrum is not a consequence of Remark 6 (since it refers to larger eigenspaces) and does not imply Remark 6 (since it contains additional information coming from Hodge theory, which prevents possible cancellations at the level of $\chi^{\mu_m}(F)$.

For the sake of completeness, we give below the corresponding μ_d-equivariant Hodge-Deligne polynomials for the arrangements $\mathcal{A}_2$ and $\mathcal{A}_3$, to be derived in an obvious way from the results in Example 9 and Proposition 17.

Corollary 14 *Let $\mathcal{A}_n$ be the essential braid arrangement in $\mathbb{C}^n$. Then one has the following.*

(i) The μ_3-equivariant Hodge-Deligne polynomial for the braid arrangement $\mathcal{A}_2$ is given by

$$DP^{\mu_3}(F_0; u, v, t) = \gamma_0 + [\gamma_2 u + \gamma_1 v + 2\gamma_0 uv]t.$$

(ii) The μ_6-equivariant Hodge-Deligne polynomial for the braid arrangement $\mathcal{A}_3$ is given by

$$DP^{\mu_6}(F_0; u, v, t) = \gamma_0 + [\gamma_4 u + \gamma_2 v + 5\gamma_0 uv]t +$$
$$+[(\gamma_3 + 2\gamma_5)u^2 + (\gamma_3 + 2\gamma_1)v^2 + 3(\gamma_4 u^2 v + \gamma_2 uv^2 + 2\gamma_0 u^2 v^2]t^2.$$

6.2 A Purity Result for Eigenspaces in Milnor Fiber Cohomology

Let $\mathcal{A}$ be an essential central arrangement of d hyperplanes in $\mathbb{C}^\ell$, with $\ell \geq 2$, given by a reduced equation $Q_0(x) = 0$. Then clearly $H^m(F_0, \mathbb{Q})_1$ and $H^m(F_0, \mathbb{C})_{-1}$ are mixed Hodge substructures in $H^m(F_0, \mathbb{Q})$ for any positive integer m. Moreover, for $\beta \in \mu_d$, $\beta \neq \pm 1$, the same is true for the subspace

$$H^m(F, \mathbb{C})_{\beta,\overline{\beta}} = H^m(F, \mathbb{C})_\beta \oplus H^m(F, \mathbb{C})_{\overline{\beta}} = \ker[(h^m)^2 - 2Re(\beta)h^m + Id] \qquad (50)$$

which is in fact defined over $\mathbb{R}$ (as the last equality shows). For $\beta = -1$, we set $H^m(F, \mathbb{C})_{\beta,\overline{\beta}} = H^m(F, \mathbb{C})_{-1}$ for uniformity of notation.

Let $D = Q_0^{-1}(0) = \cup_{H \in \mathcal{A}} H$ and for any dense edge X of the arrangement $\mathcal{A}$ with $\dim X > 0$ let m_X be the multiplicity of X, i.e. the number of hyperplanes in $\mathcal{A}$ containing X.

With this notation we have the following result, which complements our Proposition 4 with Hodge theoretic information.

Proposition 19 *Let $\beta \in \mu_d$, $\beta \neq 1$ be a d-root of unity such that $\beta^{m_X} \neq 1$ for any dense edge X with* $\dim X > 0$*. Then the corresponding eigenspace $H^{\ell-1}(F_0, \mathbb{C})_{\beta,\overline{\beta}}$ is a pure Hodge structure of weight $\ell - 1$ and $H^m(F_0, \mathbb{C})_{\beta,\overline{\beta}} = 0$ for $m < \ell - 1$.*

In particular, if $\beta = \exp(-2\pi i\alpha)$ for some $\alpha \in \mathbb{Q}$, then the coefficients in the corresponding spectrum $Sp(\mathcal{A})$ have the following symmetry property:

$$m_\alpha = m_{\ell-\alpha}. \qquad (51)$$

Proof For a point $x \in D$, $x \neq 0$, let $L_x = \cap_{H \in \mathcal{A}, x \in H} H$ and denote by $\mathcal{A}_x$ the central hyperplane arrangement induced by $\mathcal{A}$ on a linear subspace T_x, passing through x and transversal to L_x. We may choose $\dim T_x = \operatorname{codim} L_x$ and identify x with the origin in the linear space T_x. Let $h_x^* : H^*(F_x, \mathbb{C}) \to H^*(F_x, \mathbb{C})$ be the corresponding monodromy operator at x. If $x \in X$, with X a dense edge, it follows from the formula of the zeta-function $Z(t)$ of the monodromy given in Proposition 2 that the eigenvalues of h_x^* are exactly the $(m_X)^{th}$-roots of unity. Indeed, it is known that an edge is dense if and only if the Euler characteristic of the projective complement associated to $\mathcal{A}_x$ is nonzero.

Then we apply Proposition 4.1 in [11] and Proposition 4.

Example 11 Consider the essential arrangement $\mathcal{A}_4$ in $\mathbb{C}^4$. If we consider the list of the dense edges given at the end of the section above, we see that $m_X = 3$ and $m_Z = 6$

It follows that for any $\beta \in \mu_{10}$ such that $\beta^6 \neq 1$, i.e. $\beta \neq \pm 1$, the corresponding eigenspace $H^3(F_0, \mathbb{C})_{\beta,\overline{\beta}}$ is a pure Hodge structure of weight 3. Moreover, we have the symmetry $m_\alpha = m_{4-\alpha}$ for all $\alpha \in \mathbb{Q}$ such that $2\alpha \notin \mathbb{Z}$.

Note also that for $\beta = -1$, $H^2(F_0, \mathbb{C})_{\beta,\overline{\beta}}$ is a pure Hodge structure of weight 2. The following result generalises this property.

Proposition 20 *Let $\beta \in \mu_d$, $\beta \neq 1$ be a d-root of unity such that $\beta^{m_X} \neq 1$ for any dense edge X with* $\operatorname{codim} X < c(\beta)$ *and there is at least one dense edge $Y(\beta)$ with* $\operatorname{codim} Y(\beta) = c(\beta)$ *such that $\beta^{m_Y} = 1$. Then the corresponding eigenspace $H^{d(\beta)}(F_0, \mathbb{C})_{\beta,\overline{\beta}}$ is a pure Hodge structure of weight $d(\beta)$ and $H^m(F_0, \mathbb{C})_{\beta,\overline{\beta}} = 0$ for $m < d(\beta)$, with $d(\beta) = c(\beta) - 1$.*

Proof Let E be a generic linear subspace of dimension $c(\beta)$ in $\mathbb{C}^{n+1}$. Then the dense edges of the hyperplane arrangement $\mathcal{A}|E$ in E obtained by taking all the traces $H \cap E$ for $H \in \mathcal{A}$ are exactly the intersections $X \cap E$ for X a dense edge in $\mathcal{A}$ of codimension at most $c(\beta)$. The inclusion of Milnor fibers

$$\iota : F_0(\mathcal{A}|E) \to F_0(\mathcal{A})$$

induces an isomorphism $\iota^* : H^m(F_0(\mathcal{A})) \to H^m(F_0(\mathcal{A}|E))$ for $m < \dim F_0(\mathcal{A}|E) - 1 = d(\beta) - 1$ and a monomorphism for $m = d(\beta)$ preserving the mixed Hodge structures and compatible with the monodromy actions.

The result now follows from Proposition 19. □

As an application, we obtain a new proof of the following known result, see [13].

Corollary 15 *Let $\mathcal{A}$ be central essential arrangement in $\mathbb{C}^3$. Then $H^1(F_0)_{\neq 1}$ is a pure Hodge structure of weight* 1.

Indeed, $H^1(F_0)_{\neq 1}$ is a direct sum of pure Hodge structures of the type $H^1(F_0, \mathbb{C})_{\beta,\overline{\beta}}$, each of weight 1, since associated to dense edges of codimension $c(\beta) = 2$.

Example 12 Consider the essential arrangement $\mathcal{A}_5$ in $\mathbb{C}^5$. The dense edges X of codimension k are such that the corresponding arrangement $\mathcal{A}_X$ is the braid arrangement of type $\mathcal{A}_k$, for $k = 1, 2, 3, 4$, see Example 2.

Take β a primitive root of unity of order 5. Settepanella's computations in [34] imply that $H^3(F_0, \mathbb{C})_{\beta,\overline{\beta}}$ is 12-dimensional, and our Proposition 20 implies that it is a pure Hodge structure of weight 3.

However, it also follows from [34] that in $H^2(F_0)$ there is a 2-dimensional eigenspace $H^2(F_0, \mathbb{C})_{\beta,\overline{\beta}}$, where $\beta \neq 1$ is a cubic root of unity. Our Proposition 20 can say nothing about the corresponding mixed Hodge structure. Indeed, such a β is related to a codimension 2 dense edge $Y(\beta)$.

Acknowledgements The author "Alexandru Dimca" was partially supported by Institut Universitaire de France. The authors "Alexandru Dimca and Gus Lehrer" were partially supported by Australian Research Council Grants DP0559325 and DP110103451.

References

1. N. Budur, M. Saito, Jumping coefficients and spectrum of a hyperplane arrangement. Math. Ann. **347**, 545–579 (2010)
2. F. Callegaro, On the cohomology of Artin groups in local systems and the associated Milnor fibre. J. Pure Appl. Algebra **197**(1–3), 323–332 (2005)
3. D. Cohen, P. Orlik, A. Dimca, Nonresonance conditions for arrangements. Ann. Institut Fourier (Grenoble), **53**, 1883–1896 (2003)
4. P. Deligne, G. Lusztig, Representations of reductive groups over finite fields. Ann. Math. (2) **103**(1), 103–161 (1976)
5. J. Denef, F. Loeser, Regular elements and monodromy of discriminants of finite reflection groups. Indag. Math. (N.S.) **6**(2), 129–143 (1995)
6. G. Denham, N. Lemire, Equivariant Euler characteristics of discriminants of reflection groups. Indag. Math. (N.S.) **13**(4), 441–458 (2002)
7. C. De Concini and C. Procesi, Wonderful models of subspace arrangements. Sel. Math. (N.S.) **1**(3), 459–494 (1995)
8. C. De Concini, C. Procesi, M. Salvetti, Arithmetic properties of the cohomology of braid groups. Topology **40**(4), 739–751 (2001)
9. A. Dimca, *Singularities and Topology of Hypersurfaces*, Universitext (Springer-Verlag, New York, 1992)
10. A. Dimca, *Sheaves in Topology*, Universitext (Springer-Verlag, New York, 2004)
11. A. Dimca, Tate properties, polynomial-count varieties, and monodromy of hyperplane arrangements. Nagoya Math. J. **206**, 75–97 (2012)
12. A. Dimca, G.I. Lehrer, Purity and equivariant weight polynomials, in *Algebraic Groups and Lie Groups*, ed. by G.I. Lehrer (Cambridge University Press, Cambridge, 1997)
13. A. Dimca, G. Lehrer, Hodge-Deligne equivariant polynomials and monodromy of hyperplane arrangements, in *Configuration Spaces, Geometry, Combinatorics and Topology*, ed. by A. Björner, F. Cohen, C. De Concini, C. Procesi, M. Salvetti. Publications of Scuola Normale Superiore, vol. 14 (Edizioni della Normale, Pisa, 2012), pp. 231–253
14. A. Dimca, S. Papadima, Finite Galois covers, cohomology jump loci, formality properties, and multinets. Ann. Sc. Norm. Super. Pisa Cl. Sci (5), **X**, 253–268 (2011)
15. A. Dimca, M. Saito, Some remarks on limit mixed Hodge structure and spectrum. An. Şt. Univ. Ovidius Constanţa (2012, to appear), arXiv:1210.3971
16. M.T. Karkar, J.A. Green, A theorem on the restriction of group characters, and its application to the character theory of $SL(n, q)$. Math. Ann. **215**, 131–134 (1975)
17. G.I. Lehrer, The characters of the finite special linear groups. J. Algebra **26**, 564–583 (1973)
18. G.I. Lehrer, On the Poincaré series associated with Coxeter group actions on complements of hyperplanes. J. London Math. Soc. (2) **36**(2), 275–294 (1987)
19. G.I. Lehrer, The ℓ-adic cohomology of hyperplane complements. Bull. Lond. Math. Soc. **24**, 76–82 (1992)
20. G.I. Lehrer, Poincaré polynomials for unitary reflection groups. Invent. Math. **120**(3), 411–425 (1995)
21. G.I. Lehrer, Remarks concerning linear characters of reflection groups. Proc. Am. Math. Soc. **133**(11), 3163–3169 (2005)
22. G.I. Lehrer, T.A. Springer, Reflection subquotients of unitary reflection groups. Dedicated to H. S. M. Coxeter on the occasion of his 90th birthday. Can. J. Math. **51**(6), 1175–1193 (1999)
23. G. I. Lehrer, D.E. Taylor, *Unitary Reflection Groups*. Australian Mathematical Society Lecture Series, vol. 20 (Cambridge University Press, Cambridge, 2009)
24. A. Libgober, Eigenvalues for the monodromy of the Milnor fibers of arrangements, in *Trends in Mathematics: Trends in Singularities*, ed. by A. Libgober, M. Tibăr (Birkhäuser, Basel, 2002)
25. J. Milnor, *Singular Points of Complex Hypersurfaces*. Annals of Mathematics Studies, 61, (Princeton Univ. Press, Princeton, 1968)

26. M. Oka, On the cohomology structure of projective varieties, in: *Manifolds, Tokyo 1973, Proceedings* (University of Tokyo Press, Tokyo, 1975), pp. 137–143
27. P. Orlik, L. Solomon, Singularities. I. Hypersurfaces with an isolated singularity. Adv. Math. **27**(3), 256–272 (1978)
28. P. Orlik, L. Solomon, Singularities. II. Automorphisms of forms. Math. Ann. **231**(3), 229–240 (1977/78)
29. P. Orlik, H. Terao, Arrangements of Hyperplanes (Springer, Berlin, 1992)
30. P. Orlik, H. Terao, *Arrangements and Hypergeometric Integrals*. MSJ Memoirs, 9 (Mathematical Society of Japan, Tokyo, 2001)
31. C. Peters, J. Steenbrink, *Mixed Hodge Structures*. Ergeb. der Math. und ihrer Grenz. 3. Folge 52 (Springer, New York, 2008)
32. V. Schechtman, H. Terao, A. Varchenko, Local systems over complements of hyperplanes and the Kac-Kazhdan condition for singular vectors. J. Pure Appl. Algebra **100**, 93–102 (1995)
33. S. Settepanella, A stability like theorem for cohomology of pure braid groups of the series A, B and D. Topology Appl. **139**(1), 37–47 (2004)
34. S. Settepanella, Cohomology of pure braid groups of exceptional cases. Topology Appl. **156**(5), 1008–1012 (2009)
35. Y. Yoon, Spectrum of hyperplane arrangements in four variables (2012), arXiv:1211.1689
36. A. Zarelua, On finite groups of transformations, in *1969 Proc. Internat. Sympos. on Topology and its Applications (Herceg-Novi, 1968)* Savez Drustava Mat. Fiz. i Astronom., Belgrade, pp. 334–339

On the Twisted Cohomology of Affine Line Arrangements

Mario Salvetti and Matteo Serventi

Abstract Let $\mathcal{A}$ be an affine line arrangement in $\mathbb{C}^2$, with complement $\mathcal{M}(\mathcal{A})$. The twisted (co)homology of $\mathcal{M}(\mathcal{A})$ is an interesting object which has been considered by many authors. In this paper we give a vanishing conjecture of a different nature with respect to the known results: namely, we conjecture that if the graph of double points of the arrangement is connected then there is no nontrivial monodromy. This conjecture is obviously combinatorial (meaning that it depends only on the lattice of the intersections). We prove it in some cases with stronger hypotheses. We also consider the integral case, relating the property of having trivial monodromy over $\mathbb{Z}$ with a certain property of "commutativity" of the fundamental group up to some subgroup. At the end, we give several examples and computations.

Keywords Line arrangements • Milnor fibre • Twisted cohomology

1 Introduction

Let $\mathcal{A} := \{\ell_1, \ldots, \ell_n\}$ be an affine line arrangement in $\mathbb{C}^2$, with complement $\mathcal{M}(\mathcal{A})$. Let $\mathcal{L}$ be a rank-1 local system on $\mathcal{M}(\mathcal{A})$, which is defined by a unitary commutative ring R and an assignment of an invertible element $t_i \in R^*$ for each line $\ell_i \in \mathcal{A}$. Equivalently, $\mathcal{L}$ is defined by a module structure on R over the fundamental group of $\mathcal{M}(\mathcal{A})$ (such structure factorizes through the first homology of $\mathcal{M}(\mathcal{A})$). A big amount of work has been done when $R = \mathbb{C}$, in that case the t_i's being non-zero complex numbers, trying to understand the jump-loci (in $(\mathbb{C}^*)^n$) of the cohomology (see for example [5, 6, 11, 12, 18, 25]). By "coning" $\mathcal{A}$ one obtains a three-dimensional central arrangement, with complement fibering over $\mathbb{C}^*$. The Milnor fiber F of such fibration is a surface of degree $n+1$, endowed with a natural monodromy automorphism of order $n+1$. It is well known that the trivial (co)homology of F with coefficients in a commutative ring A, as a module over the monodromy action, is obtained by the (co)homology of $\mathcal{M}(\mathcal{A})$ with coefficients in $R := A[t^{\pm 1}]$, where here the structure of R as a $\pi_1(\mathcal{M}(\mathcal{A}))$-module is given by taking

M. Salvetti (✉) • M. Serventi
Department of Mathematics, University of Pisa, Pisa, Italy
e-mail: salvetti@dm.unipi.it; serventi@mail.dm.unipi.it

F. Callegaro et al. (eds.), *Configuration Spaces*, Springer INdAM Series 14,
DOI 10.1007/978-3-319-31580-5_11

all the t_i's equal to t and the monodromy action corresponds to $t-$multiplication. For reflection arrangements, relative to a Coxeter group $\mathbf{W}$, many computations were done, especially for the *orbit space* $\mathcal{M}_{\mathbf{W}}(\mathcal{A}) := \mathcal{M}(\mathcal{A})/\mathbf{W}$, which has an associated Milnor fiber $F_{\mathbf{W}} := F/\mathbf{W}$: in this case we know a complete answer for $R = \mathbb{Q}[t^{\pm 1}]$, for all groups of finite type (see [8, 9, 15]), and for some groups of affine type [4, 19] (see also [7, 22]). For $R = \mathbb{Z}[t^{\pm 1}]$ a complete answer is known in case A_n (see [3]). Some results are known for (non quotiented) reflection arrangements (see [24]).

Some algebraic complexes computing the twisted cohomology of $\mathcal{M}(\mathcal{A})$ are known (see for example the above cited papers). In [16] the minimal cell structure of the complement constructed in [23] (see [10, 20]) was used to find an algebraic complex which computes the twisted cohomology, in the case of real defined arrangements (see also [17]). The form of the boundary maps depends not only on the lattice of the intersections which is associated to $\mathcal{A}$ but also on its *oriented matroid:* for each singular point P of multiplicity m there are $m-1$ generators in dimension 2 whose boundary has non vanishing components along the lines contained in the "cone" of P and passing above P.

Many of the specific examples of arrangements with non-trivial cohomology (i.e., having non-trivial monodromy) which are known are based on the theory of *nets* and *multinets* (see [13]): it seems that there are few arrangements with non trivial monodromy in cohomology and some conjecture claim very strict restrictions for line arrangements (see [26]).

In this paper we state a vanishing conjecture of a very different nature, which is very easily stated and which involves only the lattice associated to the arrangement. Let Γ be the graph with vertex set $\mathcal{A}$ and edge set which is given by taking an edge (ℓ_i, ℓ_j) iff $\ell_i \cap \ell_j$ is a double point. Then our conjecture is as follows:

Conjecture 1 Assume that Γ is connected; then $\mathcal{A}$ has trivial monodromy.

This conjecture is supported by "experiments", since all computations we made confirm it. Also, all non-trivial monodromy examples which we know have disconnected graph Γ.

Unfortunately, we are not able to prove the conjecture in general: we give here a proof holding with further restrictions. Our proof uses the algebraic complex given in [16] so our arrangements are real.

An arrangement with trivial monodromy will be called *a-monodromic.* We also introduce a notion of monodromic triviality over $\mathbb{Z}$. By using free differential calculus, we show that $\mathcal{A}$ is a-monodromic over $\mathbb{Z}$ iff the fundamental group of the complement $\mathcal{M}(A)$ of the arrangement is commutative modulo the commutator subgroup of the length-zero subgroup of the free group F_n. As a consequence, we deduce that if $\pi_1(\mathcal{M}(\mathcal{A}))$ modulo its second derived group is commutative, then $\mathcal{A}$ has trivial monodromy over $\mathbb{Z}$.

In the final part we give several examples which describe the notions introduced in Sect. 2. Some explicit computations were done by using suitable software which was developed by using Axiom computer algebra system.

2 Algebraic Complexes

Here we recall an algebraic complex which was obtained in [16], computing local system cohomology for affine arrangements. Such complex was obtained as a simplification of another one in [23], where the authors used the explicit construction of a *minimal cell complex* which models the complement.

Of course, there are other algebraic complexes computing local system cohomology (see the references listed in the introduction). The one in [16] seemed to us particularly suitable to attack the present problem (even if we were not able to solve it in general).

First, the complex depends on a fixed and generic system of "polar coordinates". In the present situation, this just means to take an oriented affine real line ℓ which is transverse to the arrangement. We also assume (even if it is not strictly necessary) that ℓ is "far away" from $\mathcal{A}$, meaning that it does not intersect the closure of the bounded facets of the arrangement. This is clearly possible because the union of bounded chambers is a compact set (the arrangement is finite). The choice of ℓ induces a labelling on the lines $\{\ell_1, \ldots, \ell_n\}$ in $\mathcal{A}$, where the indices of the lines agree with the ordering of the intersection points with ℓ, induced by the orientation of ℓ.

Let us choose a basepoint $O \in \ell$, coming before all the intersection points of ℓ with $\mathcal{A}$ (with respect to the just introduced ordering). We recall the construction in [16] in the case of abelian local systems: namely, we take R a commutative unitary ring, and the representation

$$\pi_1(\mathcal{M}(\mathcal{A}), O) \to H_1(\mathcal{M}(\mathcal{A})) \to Aut(R)$$

takes an elementary loop in $\mathcal{M}(\mathcal{A})$, based in O and turning counterclockwise around a line ℓ_i, into the multiplication by an invertible element $t_i \in R^*$. As stated in the introduction, of particular interest is the case when $R = A[t^{\pm 1}]$ and $t_i = t$, $i = 1, \ldots, n$, since this local system is strictly related to the homology of the Milnor fibre, as a module over the monodromy operator.

Let $Sing(\mathcal{A})$ be the set of singular points of the arrangement. For any point $P \in Sing(\mathcal{A})$, let $S(P) := \{\ell \in \mathcal{A} : P \in \ell\}$; so $m(P) = |S(P)|$ is the multiplicity of P.

Let i_P, i^P be the minimum and maximum index of the lines in $S(P)$ (so $i_P < i^P$). We denote by $C(P)$ the subset of lines in $\mathcal{A}$ whose indices belong to the closed interval $[i_P, i^P]$. We also denote by

$$U(P) := \{\ell \in \mathcal{A} : \ell \text{ does not separates } P \text{ from the basepoint } O\}.$$

Let (C_*, ∂_*) be the 2–dimensional algebraic complex of free R–modules having one 0–dimensional basis element e^0, n 1–dimensional basis elements e^1_j, $j = 1, \ldots, n$, (e^1_j corresponding to the line ℓ_j) and $\nu_2 = \sum_{P \in Sing(\mathcal{A})} m(P) - 1$ 2–dimensional basis elements: to the singular point P of multiplicity $m(P)$ we associate generators $e^2_{P,h}$, $h = 1, \ldots, m(P)-1$. The lines through P will be indicized as $\ell_{j_{P,1}}, \ldots, \ell_{j_{P,m(P)}}$ (with growing indices).

As a dual statement to [16], Theorem 2, we obtain:

Theorem 1 *The local system homology $H_*(\mathcal{M}(\mathcal{A}); R)$ is computed by the complex (C_*, ∂_*) above, where*

$$\partial_1(e^1_j) = (t_j - 1)\, e^0$$

and

$$\begin{aligned}\partial_2(e^2_{P,h}) = & \sum_{\ell_j \in S(P)} \left(\prod_{\substack{i<j \text{ s.t.} \\ \ell_i \in U(P)}} t_i \right) \left(\prod_{i \in [j_{P,h+1} \to j)} t_i - \prod_{\substack{i<j \text{ s.t.} \\ \ell_i \in S(P)}} t_i \right) e^1_j + \\ & + \sum_{\ell_j \in C(P) \cap U(P)} \left(\prod_{\substack{i<j \text{ s.t.} \\ \ell_i \in U(P)}} t_i \right) \left(1 - \prod_{\substack{i \le j_{P,h},\, i<j \\ \ell_i \in S(P)}} t_i \right) \left(\prod_{\substack{i \ge j_{P,h+1},\, i<j \\ \ell_i \in S(P)}} t_i - \prod_{\substack{i \ge j_{P,h+1} \\ \ell_i \in S(P)}} t_i \right) e^1_j \end{aligned} \tag{1}$$

where $[j_{P,h+1} \to j)$ is the set of indices of the lines in $S(P)$ which run from $j_{P,h+1}$ (included) to j (excluded) in the cyclic ordering of $1, \ldots, n$.

By convention, a product over an empty set of indices equals 1.

When $R = A[t^{\pm 1}]$ and $t_i = t$, $i = 1, \ldots, n$, we obtain the local homology $H_*(\mathcal{M}(\mathcal{A}); R)$ by using an analog algebraic complex, where all t_i's equal t in the formulas. In particular (1) becomes

$$\begin{aligned}\partial_2(e^2_{P,h}) = & \sum_{\ell_j \in S(P)} t^{\#\{\ell_i \in U(P):\ i<j\}} \left(t^{\#[j_{P,h+1} \to j)} - t^{\#\{\ell_i \in S(P):\ i<j\}} \right) e^1_j + \\ & + \sum_{\ell_j \in C(P) \cap U(P)} t^{\#\{\ell_i \in U(P):\ i<j\} + \#\{\ell_i \in S(P):\ i \ge j_{P,h+1},\ i<j\}} \left(1 - t^{\#\{\ell_i \in S(P):\ i \le j_{P,h},\ i<j\}} \right) \cdot \\ & \cdot \left(1 - t^{\#\{\ell_i \in S(P):\ i \ge j_{P,h+1},\ i \ge j\}} \right) e^1_j\ . \end{aligned} \tag{2}$$

By separating in the first sum the case $j \geq j_{P,h+1}$ from the case $j \leq j_{P,h}$ we have:

$$\partial_2(e^2_{P,h}) = \sum_{\substack{\ell_j \in S(P) \\ j \geq j_{P,h+1}}} t^{\#\{\ell_i \in U(P):\ i<j\}+\#\{\ell_i \in S(P):\ j_{P,h+1} \leq i<j\}} \left(1 - t^{\#\{\ell_i \in S(P):\ i \leq j_{P,h}\}}\right) e^1_j +$$

$$+ \sum_{\substack{\ell_j \in S(P) \\ j \leq j_{P,h}}} t^{\#\{\ell_i \in U(P):\ i<j\}+\#\{\ell_i \in S(P):\ i<j\}} \left(t^{\#\{\ell_i \in S(P):\ j_{P,h+1} \leq i\}} - 1\right) e^1_j +$$

$$+ \sum_{\ell_j \in C(P) \cap U(P)} t^{\#\{\ell_i \in U(P):\ i<j\}+\#\{\ell_i \in S(P):\ i \geq j_{P,h+1},\ i<j\}} \left(1 - t^{\#\{\ell_i \in S(P):\ i \leq j_{P,h},\ i<j\}}\right) \cdot$$

$$\cdot \left(1 - t^{\#\{\ell_i \in S(P):\ i \geq j_{P,h+1},\ i \geq j\}}\right) e^1_j \ . \tag{3}$$

In particular, let P be a double point. Then h takes only the value 1, and $j_{P,1}$, $j_{P,2}$ are the indices of the two lines passing through P. So formula (3) becomes

$$\begin{aligned} \partial_2(e^2_{P,1}) &= t^{\#\{\ell_i \in U(P):\ i<j_{P,2}\}} (1-t)\ e^1_{j_{P,2}} + t^{\#\{\ell_i \in U(P):\ i<j_{P,1}\}} (t-1)\ e^1_{j_{P,1}} + \\ &+ \sum_{\ell_j \in C(P) \cap U(P)} t^{\#\{\ell_i \in U(P):\ i<j\}} (t-1)^2\ e^1_j \ . \end{aligned} \tag{4}$$

Since ∂_2 is divisible by $t-1$ we can rewrite (4) as

$$\partial_2(e^2_{P,1}) = (t-1)\ \tilde{\partial}_2(e^2_{P,1}) \tag{5}$$

where

$$\begin{aligned} \tilde{\partial}_2(e^2_{P,1}) &= t^{\#\{\ell_i \in U(P):\ i<j_{P,2}\}}\ e^1_{j_{P,2}} - t^{\#\{\ell_i \in U(P):\ i<j_{P,1}\}}\ e^1_{j_{P,1}} + \\ &+ \sum_{\ell_j \in C(P) \cap U(P)} t^{\#\{\ell_i \in U(P):\ i<j\}} (1-t)\ e^1_j \ . \end{aligned} \tag{6}$$

3 A Proof in Particular Cases

We give a proof of Conjecture 1 with further hypotheses on $\mathscr{A}$.

Notice that the rank of ∂_2 is $n-1$ (the sum of all rows vanishes). Then the arrangement has no monodromy iff the only elementary divisor of ∂_2 is $\varphi_1 := t-1$, so ∂_2 diagonalizes to $\oplus_{i=1}^{n-1} \varphi_1$. This is equivalent to the reduced boundary $\tilde{\partial}_2$ having an invertible minor of order $n-1$.

Let T be a spanning tree of the connected graph Γ. A choice of an admissible coordinate system, including a choice of a generic oriented line ℓ, induces a total ordering on the lines: we just consider the natural ordering of the intersection points of the lines with ℓ. Therefore we have an induced labelling, varying between 1 and n, on the set of vertices $V\Gamma$ of Γ (so, on $VT = V\Gamma$).

Definition 1 We say that the induced labelling on $VT = V\Gamma$ is *very good* (with respect to the given coordinate system) if the sequence $n, \ldots, 1$ is a *collapsing ordering* on T. In other words, the graph obtained by T by removing all vertices with label $\geq i$ and all edges having both vertices with label $\geq i$, is a tree, for all $i = n, \ldots, 1$.

We say that the spanning tree T is *very good* if there exists an admissible coordinate system such that the induced labelling on VT is very good.

Remark 1

1. A labelling over a spanning tree T gives a collapsing ordering iff for each vertex v, the number of adjacent vertices with lower label is ≤ 1. In this case, only the vertex labelled with 1 has no lower labelled adjacent vertices (by the connectness of T).
2. Given a collapsing ordering over T, for each vertex v with label $i_v > 1$, let $\ell(v)$ be the edge which connects v with the unique adjacent vertex with lower label; by giving to $\ell(v)$ the label $i_v + \frac{1}{2}$, we obtain a discrete Morse function on the graph T (see [14]) with unique critical cell given by the vertex with label 1. The set of all pairs $(v, \ell(v))$ is the acyclic matching which is associated to this Morse function.

Let us indicate by Γ_0 the linear tree with n vertices: we think as Γ_0 as a *CW*-decomposition of the real segment $[1, n]$, with vertices $\{j\}$, $j = 1, \ldots, n$, and edges the segments $[j, j+1]$, $j = 1, \ldots, n-1$.

Definition 2 We say that a labelling induced by some coordinate system on the tree T is *good* if there exists a permutation $i_1, \ldots, i_n$ of $1, \ldots, n$ which gives a collapsing sequence both for T and for Γ_0. In other words, at each step we always remove either the maximum labelled vertex or the minimum, and this is a collapsing sequence for T.

We say that T is *good* if there exists an admissible coordinate system such that the induced labelling on VT is good.

Notice that a very good labelling is a good labelling where at each step one removes the maximum vertex.

Consider some arrangement $\mathcal{A}$ with graph Γ and labels on the vertices which are induced by some coordinate system. Notice that changes of coordinates act on the labels by giving all possible *cyclic permutations*, which are generated by the transformation $i \to i + 1 \bmod n$. So, given a labelled tree T, checking if T is very good (resp. good) consists in verifying if some cyclic permutation of the labels is very good (resp. good). This property depends not only on the "shape" of the tree, but also on how the lines are disposed in $\mathbb{R}^2$ (the associated oriented matroid). In fact, one can easily find arrangements where some "linear" tree is very good, and others where some linear tree is not good.

Definition 3 We say that an arrangement $\mathcal{A}$ is *very good* (resp. *good*) if Γ is connected and has a very good (resp. good) spanning tree.

It is not clear if this property is combinatorial, i.e. if it depends only on the lattice. Of course, $\mathcal{A}$ very good implies $\mathcal{A}$ good.

Theorem 2 *Let $\mathcal{A}$ be a good arrangement. Then $\mathcal{A}$ is a-monodromic.*

Proof We use induction on the number n of lines, the claim being trivial for $n = 1$.

Take a suitable coordinate system as in definition (3), such that the graph Γ has a spanning tree T with good labelling. Assume for example that at the first step we remove the last line, so the graph Γ' of the arrangement $\mathcal{A}' := \mathcal{A}\setminus\{\ell_n\}$ is connected and the spanning tree T' obtained by removing the vertex $\{\ell_n\}$ and the "leaf-edge" (ℓ_n, ℓ_j) (for some $j < n$) has a good labelling. Then by induction $\mathcal{A}'$ is good, so it is a-monodromic.

There are $n-1$ double points which correspond to the edges of T : only one of these is contained in ℓ_n, namely $\ell_n \cap \ell_j$ (see remark 1). Let $\mathcal{D} := \{d_1, \ldots, d_{n-1}\}$ be the set of such double points, with $d_{n-1} = \ell_n \cap \ell_j$. Let also $\mathcal{D}' := \{d_1, \ldots, d_{n-2}\}$, which corresponds to the edges of T'. Let $(C(\mathcal{D})_*, \partial_*)$ (resp. $(C(\mathcal{D}')_*, \partial'_*)$) be the subcomplex of $C(\mathcal{A})_*$ generated by the 2-cells which correspond to $\mathcal{D}$ (resp. $\mathcal{D}'$): then $C(\mathcal{D})_2 = \oplus_{1\le i\le n-1} Re_i$, and $C(\mathcal{D}')_2 = \oplus_{1\le i\le n-2} Re'_j$. Notice that, by the explicit formulas given in Sect. 2, the boundary $\partial_2(e_j)$, $j = 1, \ldots, n-2$, has no non-vanishing components along the 1-dimensional generator corresponding to ℓ_n. Actually, the natural map taking e'_j into e_j, $j = 1, \ldots, n-2$, identifies $C(\mathcal{D}')_*$ with the sub complex of $C(\mathcal{D})_*$ generated by the e_j's, $j = 1, \ldots, n-2$

$$\partial_2 = \left[\begin{array}{c|c} \partial'_2 & * \\ \hline 0 & -\varphi_1 \end{array}\right] \tag{7}$$

Since by induction $\mathcal{A}'$ is a-monodromic then ∂'_2 diagonalizes to $\oplus_{j=1}^{n-2} \varphi_1$. Therefore ∂_2 diagonalizes to $\oplus_{j=1}^{n-1} \varphi_1$, which gives the thesis.

If at the first step we remove the first line, the argument is similar, because $\partial_2(e_j)$ has no non-vanishing components along the generator corresponding to ℓ_1. □

Let us consider a different situation.

Definition 4 We say that a subset Σ of the set of singular points $Sing(\mathcal{A})$ of the arrangement $\mathcal{A}$ is *conjugate-free* (with respect to a given admissible coordinate system) if $\forall P \in \Sigma$ the set $U(P) \cap C(P)$ is empty.

An arrangement $\mathcal{A}$ will be called *conjugate-free* if Γ is connected and contains a spanning tree T such that the set of points in $Sing(\mathcal{A})$ that correspond to the edges ET of T is conjugate-free.

Let Σ be conjugate-free: it follows from formula (3) that the boundary of all generators $e^2_{P,h}$, $P \in \Sigma$, can have non-vanishing components only along the lines which contain P.

Theorem 3 *Assume that $\mathcal{A}$ is conjugate-free. Then $\mathcal{A}$ is a-monodromic.*

Proof The sub matrix of ∂_2 which corresponds to the double points ET is φ_1-times the incidence matrix of the tree T. Such matrix is the boundary matrix of the complex which computes the $\mathbb{Z}$-homology of T : it is a unimodular rank-$(n-1)$ integral matrix (see for example [1]). From this the result follows straightforward. □

We can have a mixed situation between Definitions 3 and 4.

Theorem 4 *Assume that Γ is connected and contains a spanning tree T which reduces, after a sequence of moves where we remove either the maximum or the minimum labelled vertex, to a subtree T' which is conjugate-free. Then $\mathcal{A}$ is a-monodromic.*

Proof The thesis follows easily by induction on the number n of lines. In fact, either T is conjugate-free, and we use Theorem 3, or one of the subtrees $T \setminus \{\ell_n\}, T \setminus \{\ell_1\}$ satisfies again the hypotheses of the theorem. Assume that it is $T'' = T \setminus \{\ell_n\}$. Then the boundary map ∂_2 restricted to the 2-cells corresponding to ET'' has a shape similar to (7). Therefore by induction we conclude. □

Some examples are given in Sect. 5.

4 Free Calculus

In this section we reformulate our conjecture in terms of the Fox calculus.

Let $\mathcal{A} = \{l_1, \dots, l_n\}$ be as above; if we denote by β_i an elementary loop around l_i we have that the fundamental group $\pi_1(\mathcal{M}(\mathcal{A}))$ is generated by $\beta_1, \dots, \beta_n$ and a presentation of this group is given in [21]. Let $R = \mathbb{Q}[t^{\pm 1}]$ be as above with the given structure of $\pi_1(\mathcal{M}(\mathcal{A}))$–module.

We denote by F_n the free group generated by $\beta_1, \dots, \beta_n$ and let $\varphi : F_n \longrightarrow <t>$ the group homomorphism defined by $\varphi(\beta_i) = t$ for every $1 \le i \le n$ where $<t>$ is the multiplicative subgroup of R generated by t. As in [2], if w is a word in the β_j's, we use the notation w^{φ} for $\varphi(w)$.

Consider the algebraic complex which computes the local homology of $\mathcal{M}(\mathcal{A})$ introduced in Sect. 2. The following remark is crucial for the rest of this section: if $e^2_{P,j}$ is a two-dimensional generator corresponding to a two-cell which is attached along the word w in the β_j's, then $\left(\frac{\partial w}{\partial \beta_i}\right)^{\varphi}$ is the coefficient of e^1_i of the border of $e^2_{P,j}$. This follows from the combinatorial calculation of local system homology.

Let $l : F_n \longrightarrow \mathbb{Z}$ be the length function, given by

$$l(\beta_{i_1}^{\epsilon_1} \cdots \beta_{i_r}^{\epsilon_r}) = \sum_{k=1}^{r} \epsilon_k.$$

Then kerφ is the normal subgroup of F_n given by the words of length 0.

Each relation in the fundamental group $\pi_1(\mathcal{M}(\mathcal{A}))$ is a commutator (cfr. [21]), so it lies in kerφ. So, in the sequel, we consider only words in kerφ.

Remark 2 The arrangement $\mathcal{A}$ is a-monodromic iff (by definition) the $\mathbb{Q}[t^{\pm 1}]$-module generated by $\partial_2(e_j^2)$, $j = 1, \ldots, \nu_2$, equals $(t-1)\ker\partial_1$. One has: $\ker\partial_1 = \{\sum_{j=1}^{n} x_j\, e_j^1 \;:\; \sum_{j=1}^{n} x_j = 0\}$.

Let R_j, $j = 1 \ldots, \nu_2$, be all the relations in $\pi_1(\mathcal{M}(\mathcal{A}))$. We use now e_j^2 to indicate the two-dimensional generator corresponding to a two-cell which is attached along the word R_j. Then the boundary of e_j^2 is given by

$$\partial_2(e_j^2) = \sum_{i=1}^{n} \left(\frac{\partial R_j}{\partial \beta_i}\right)^{\varphi} e_i^1, \quad j = 1, \ldots, \nu_2. \tag{8}$$

Then by Remark 2 $\mathcal{A}$ is a-monodromic iff each element of the shape

$$P(t) := (1-t)\sum_{i=1}^{n} P_i(t)\, e_i^1, \quad \sum_{i=1}^{n} P_i(t) = 0, \quad (P_i(t) \in \mathbb{Q}[t^{\pm 1}], i = 1 \ldots, n) \tag{9}$$

is linear combination with coefficients in $\mathbb{Q}[t^{\pm 1}]$ of the elements in (8), i.e.:

$$P(t) = \sum_{j=1}^{\nu_2} Q_j(t)\partial_2(e_j^2), \quad Q_j(t) \in \mathbb{Q}[t^{\pm 1}]. \tag{10}$$

It is natural to ask about solutions with coefficients in $\mathbb{Z}[t^{\pm 1}]$ instead of $\mathbb{Q}[t^{\pm 1}]$. We say that $\mathcal{A}$ is a-monodromic over $\mathbb{Z}$ if there is solution of (10) over $\mathbb{Z}[t^{\pm 1}]$ (when all the $P_i(t)$'s in (9) are in $\mathbb{Z}[t^{\pm 1}]$).

Theorem 5 *The arrangement $\mathcal{A}$ is a-monodromic over $\mathbb{Z}$ iff $\pi_1(\mathcal{M}(\mathcal{A}))$ is commutative modulo $[ker\varphi, ker\varphi]$. More precisely, $\mathcal{A}$ is a-monodromic over $\mathbb{Z}$ iff*

$$[F_n, F_n] = N[ker\varphi, ker\varphi],$$

where N is the normal subgroup generated by the relations R_j's .

Proof A set of generators for $(t-1)\ker\partial_1$ as $\mathbb{Z}[t^{\pm 1}]$-modulo is given by all elements of the type

$$P_{rs} := (1-t)(e_r^1 - e_s^1)\ , \quad r \neq s.$$

Such element can be re-written in the form (8) as

$$P_{rs} = \sum_{i=1}^{n} \left(\frac{\partial[\beta_r, \beta_s]}{\partial \beta_i} \right)^{\varphi} e_i^1$$

where $[\beta_r, \beta_s] = \beta_r\beta_s\beta_r^{-1}\beta_s^{-1}$. Now there exists an expression as in (10) for P_{rs}, with all $Q_j(t) \in \mathbb{Z}[t^{\pm 1}]$ iff

$$\left(\frac{\partial[\beta_r, \beta_s]}{\partial \beta_i} \right)^{\varphi} = \left(\frac{\partial \prod_{j=1}^{\nu_2} R_j^{Q_j(\beta_1)}}{\partial \beta_i} \right)^{\varphi}, \quad i = 1, \ldots, n. \tag{11}$$

Here $Q_j(\beta_1) \in \mathbb{Z}[F_n]$ is obtained by substituting t with β_1 (any word of length one would give the same here). Moreover, for R, w any words in $\ker\varphi$ we set $R^w := wRw^{-1}$, and for $a \in \mathbb{Z}$ we set $R^{aw} := R^w \ldots R^w$ (a factors) if $a > 0$ and $(R^{-1})^w \ldots (R^{-1})^w$ ($|a|$ factors) for $a < 0$. Also, we set $R^{aw+bu} := R^{aw}R^{bu}$. Then equalities (11) come from standard Fox calculus.

Then from Blanchfield theorem (see [2], Chap. 3) it follows that

$$[F_n, F_n] \subset N[\ker\varphi, \ker\varphi].$$

The opposite inclusion follows because, as we said before Remark 2, for any arrangement one has $N \subset [F_n, F_n]$. □

Remark 3 The condition in Theorem 5 is equivalent to saying that there is an equality

$$\frac{F_n}{N[\ker\varphi, \ker\varphi]} = \frac{F_n}{[F_n, F_n]} = H_1(\mathcal{M}(\mathcal{A}); \mathbb{Z}).$$

Since $\ker\varphi \supset [F_n, F_n]$, so $[\ker\varphi, \ker\varphi] \supset [[F_n, F_n], [F_n, F_n]]$, it follows immediately from Theorem 5

Corollary 1 *Assume that $\pi_1(\mathcal{M}(\mathcal{A}))/\pi_1(\mathcal{M}(\mathcal{A}))^{(2)}$ is abelian (iff $\pi_1(\mathcal{M}(\mathcal{A}))^{(1)} = \pi_1(\mathcal{M}(\mathcal{A}))^{(2)}$) where $\pi_1(\mathcal{M}(\mathcal{A}))^{(i)}$ is the i-th element of the derived series of $\pi_1(\mathcal{M}(\mathcal{A}))$, $i \geq 0$. Then $\mathcal{A}$ is a-monodromic over $\mathbb{Z}$.*

□

We know that $\ker\varphi$ is a free group being a subgroup of the free group F_n. We use the Reidemeister-Schreier method to write an explicit list of generators. Notice that for any fixed $1 \leq j \leq n$, the set $\{\beta_j^k : k \in \mathbb{Z}\}$ is a Schreier right coset representative system for $F_n/\ker\varphi$. Denote briefly by $s_{k,i}$ the element $s_{\beta_j^k,\beta_i} = \beta_j^k\beta_i\overline{(\beta_j^k\beta_i)}^{-1} = \beta_j^k\beta_i\beta_j^{-(k+1)}$. Then

$$\ker\varphi =< \{s_{k,i} : 1 \leq i \leq n,\ k \in \mathbb{Z}\}; s_{k,i} >$$

where $s_{k,i}$ is a relation if and only if $\beta_j^k\beta_i$ is freely equal to β_j^{k+1}; this happens if and only if $i = j$. So $\ker\varphi$ is the free group generated by $\{s_{k,i} : k \in \mathbb{Z}, 1 \leq i \leq n, i \neq j\}$. Its abelianization

$$\text{ab}\,(\ker\varphi) = \ker\varphi/[\ker\varphi, \ker\varphi]$$

is the free abelian group on the classes $\bar{s}_{k,i}$ of the generators $s_{k,i}$'s, $i \neq j$. Let

$$\text{ab} : \ker\varphi \longrightarrow \text{ab}\,(\ker\varphi) = \ker\varphi/[\ker\varphi, \ker\varphi]$$

be the abelianization homomorphism.

Now we define the automorphism σ of $\ker\varphi$ by

$$\sigma(s_{k,i}) = s_{k+1,i}$$

which passes to the quotient, so it defines an automorphism (call it again σ) of ab $(\ker\varphi)$. Therefore we may view ab $(\ker\varphi)$ as a finitely genereted free $\mathbb{Z}[\sigma^{\pm 1}]$-module, with basis $\bar{s}_{0,i}$ with $1 \leq i \leq n$ and $i \neq j$.

In this language theorem (5) translates as

Theorem 6 *The arrangement $\mathcal{A}$ is a-monodromic over $\mathbb{Z}$ iff the submodule* $(1 - \sigma)ab(\ker\varphi)$ *of* $ab(\ker\varphi)$ *is generated by* $ab(R_j)$, $j = 1, \ldots, \nu_2$, *as* $\mathbb{Z}[\sigma^{\pm 1}]$*-module.* □

Of course, one can give a conjecture holding over $\mathbb{Z}$.

Conjecture 2 Assume that Γ is connected; then $\mathcal{A}$ is a-monodromic over $\mathbb{Z}$.

Conjectureclearly implies Conjecture 1.

Our experiments agree with this stronger conjecture.

5 Examples

We give here a list of significant examples, chosen according to the definitions given in Sect. 3. For each example, we show the sequence of contractions which correspond to the definition which the given arrangement satisfies. For sake of completeness, we also give the twisted homology of the complement over $\mathbb{Q}[t^{\pm 1}]$.

In this first example (Fig. 1) we see that Γ is connected and the tree built with the selected double points is good. We also indicate the sequence of contractions.

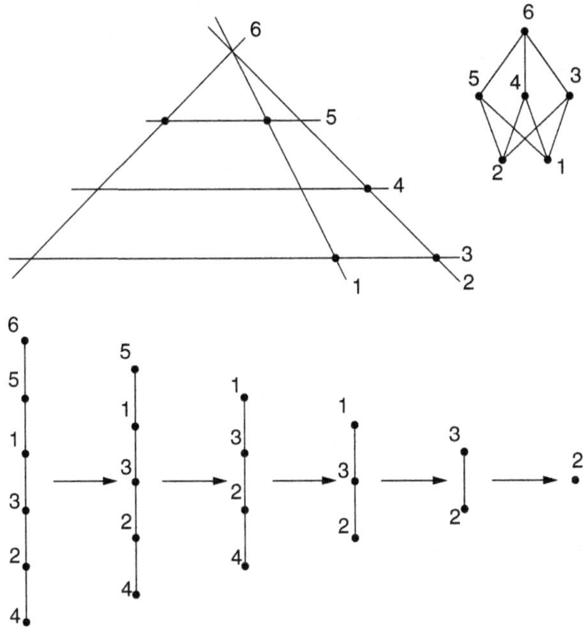

Fig. 1 Connected graph and sequence of contractions

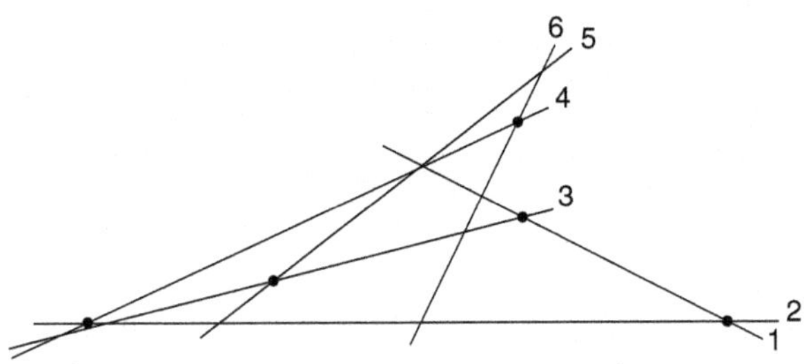

Fig. 2 A very good tree

The twisted homology for this arrangement is given by

$$H_2(\mathcal{M}(\mathcal{A}), \mathbb{Q}[t^{\pm 1}]) \simeq \left(\mathbb{Q}[t^{\pm 1}]\right)^6;$$

$$H_1(\mathcal{M}(\mathcal{A}), \mathbb{Q}[t^{\pm 1}]) \simeq \left(\frac{\mathbb{Q}[t^{\pm 1}]}{(t-1)}\right)^5 \simeq \mathbb{Q}^5,$$

$$H_0(\mathcal{M}(\mathcal{A}), \mathbb{Q}[t^{\pm 1}]) \simeq \frac{\mathbb{Q}[t^{\pm 1}]}{(t-1)} \simeq \mathbb{Q}.$$

In Fig. 2 we give an example of an arrangement having a very good tree (Fig. 3).

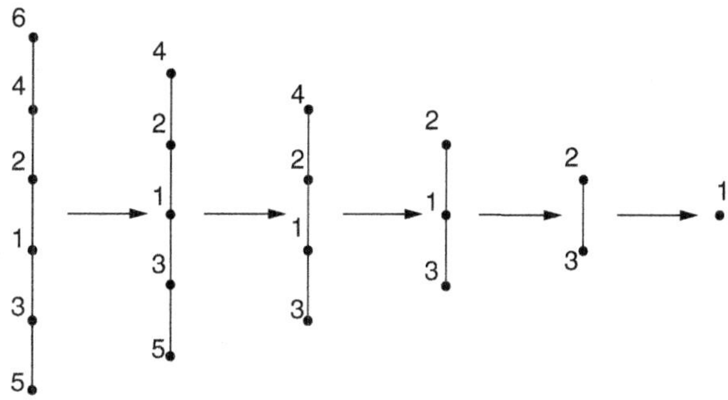

Fig. 3 Contractions of the selected very good tree

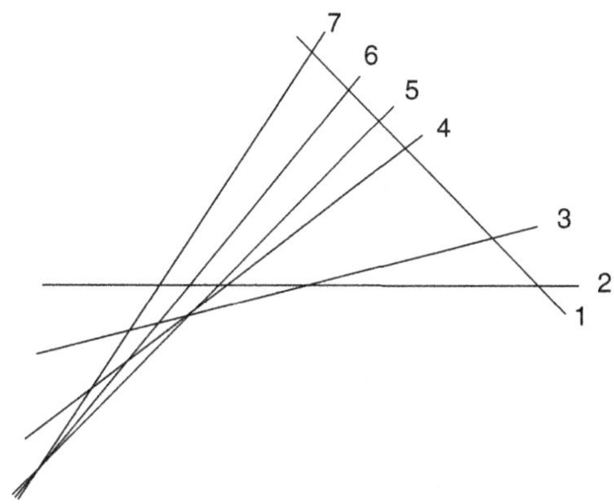

Fig. 4 A conjugate-free arrangement

For this arrangement we have

$$H_2(\mathcal{M}(\mathcal{A}), \mathbb{Q}[t^{\pm 1}]) \simeq \left(\mathbb{Q}[t^{\pm 1}]\right)^9;$$

$$H_1(\mathcal{M}(\mathcal{A}), \mathbb{Q}[t^{\pm 1}]) \simeq \left(\frac{\mathbb{Q}[t^{\pm 1}]}{(t-1)}\right)^5 \simeq \mathbb{Q}^5;$$

$$H_0(\mathcal{M}(\mathcal{A}), \mathbb{Q}[t^{\pm 1}]) \simeq \frac{\mathbb{Q}[t^{\pm 1}]}{(t-1)} \simeq \mathbb{Q}.$$

The next example (Fig. 4) is a conjugate-free arrangement. In this case every spanning tree is conjugate-free. The homology here is given by

$$H_2(\mathcal{M}(\mathcal{A}), \mathbb{Q}[t^{\pm 1}]) \simeq \left(\mathbb{Q}[t^{\pm 1}]\right)^{13};$$

$$H_1(\mathcal{M}(\mathcal{A}), \mathbb{Q}[t^{\pm 1}]) \simeq \left(\frac{\mathbb{Q}[t^{\pm 1}]}{(t-1)}\right)^6 \simeq \mathbb{Q}^6;$$

$$H_0(\mathcal{M}(\mathcal{A}), \mathbb{Q}[t^{\pm 1}]) \simeq \frac{\mathbb{Q}[t^{\pm 1}]}{(t-1)} \simeq \mathbb{Q}.$$

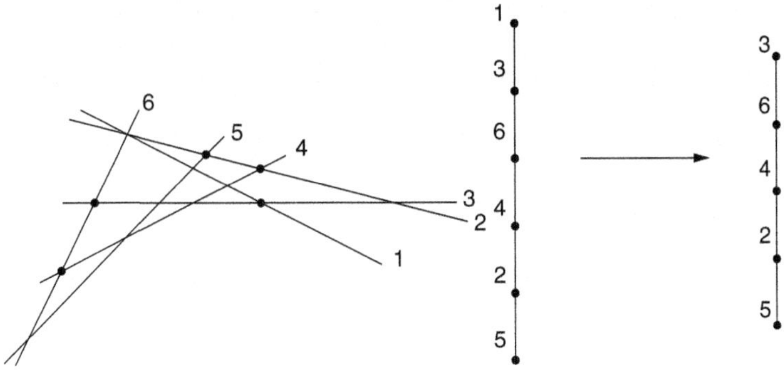

Fig. 5 A not good tree which reduces to a conjugate free tree

In the next example (Fig. 5) we show an arrangement having a tree which is not good but that can be reduced to a conjugate-free tree. We can see that the selected tree admits one contraction and the resulting tree admits no contractions but it is conjugate free.

We have here the following homology

$$H_2(\mathcal{M}(\mathcal{A}), \mathbb{Q}[t^{\pm 1}]) \simeq \left(\mathbb{Q}[t^{\pm 1}]\right)^8;$$

$$H_1(\mathcal{M}(\mathcal{A}), \mathbb{Q}[t^{\pm 1}]) \simeq \left(\frac{\mathbb{Q}[t^{\pm 1}]}{(t-1)}\right)^5 \simeq \mathbb{Q}^5;$$

$$H_0(\mathcal{M}(\mathcal{A}), \mathbb{Q}[t^{\pm 1}]) \simeq \frac{\mathbb{Q}[t^{\pm 1}]}{(t-1)} \simeq \mathbb{Q}.$$

To conclude, we give a couple of arrangements with non-trivial monodromy. They have (as all example that we know) disconnected graph Γ.

The first one (Fig. 6) is the usual deconing of A_3.

For this arrangement we have

$$H_2(\mathcal{M}(\mathcal{A}), \mathbb{Q}[t^{\pm 1}]) \simeq \left(\mathbb{Q}[t^{\pm 1}]\right)^2;$$

$$H_1(\mathcal{M}(\mathcal{A}), \mathbb{Q}[t^{\pm 1}]) \simeq \left(\frac{\mathbb{Q}[t^{\pm 1}]}{(t-1)}\right)^3 \oplus \frac{\mathbb{Q}[t^{\pm 1}]}{(t^3-1)} \simeq \mathbb{Q}^3 \oplus \frac{\mathbb{Q}[t^{\pm 1}]}{(t^3-1)};$$

$$H_0(\mathcal{M}(\mathcal{A}), \mathbb{Q}[t^{\pm 1}]) \simeq \frac{\mathbb{Q}[t^{\pm 1}]}{(t \quad 1)} \simeq \mathbb{Q}.$$

The last arrangement is the "complete triangle" (Fig. 7), which also has non-trivial monodromy. The associated graph Γ is not connected.

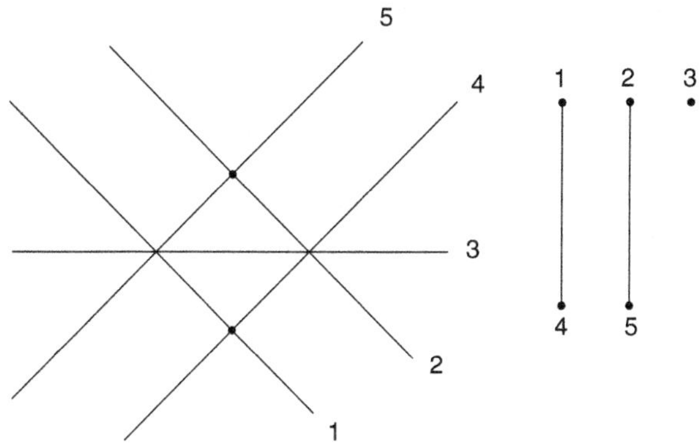

Fig. 6 Deconing A_3 and the graph Γ

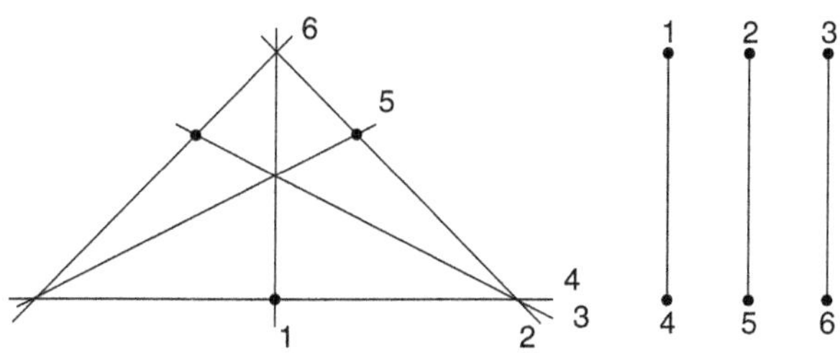

Fig. 7 The affine "complete triangle" has non-trivial monodromy and disconnected graph

Here the twisted homology is

$$H_2(\mathcal{M}(\mathcal{A}), \mathbb{Q}[t^{\pm 1}]) \simeq \left(\mathbb{Q}[t^{\pm 1}]\right)^6;$$

$$H_1(\mathcal{M}(\mathcal{A}), \mathbb{Q}[t^{\pm 1}]) \simeq \left(\frac{\mathbb{Q}[t^{\pm 1}]}{(t-1)}\right)^4 \oplus \frac{\mathbb{Q}[t^{\pm 1}]}{(t^3-1)} \simeq \mathbb{Q}^4 \oplus \frac{\mathbb{Q}[t^{\pm 1}]}{(t^3-1)};$$

$$H_0(\mathcal{M}(\mathcal{A}), \mathbb{Q}[t^{\pm 1}]) \simeq \frac{\mathbb{Q}[t^{\pm 1}]}{(t-1)} \simeq \mathbb{Q}.$$

References

1. N. Biggs, *Algebraic Graph Theory* (Cambridge University Press, Cambridge, 1974)
2. J. Birman, *Braids, Links and Mapping Class Groups* (Princeton University Press, Princeton, 1975)
3. F. Callegaro, The homology of the Milnor fibre for classical braid groups. Algebra Geom. Topol. **6**, 1903–1923 (2006)
4. F. Callegaro, D. Moroni, M. Salvetti, The $K(\pi, 1)$ problem for the affine Artin group of type $\widetilde{B}_n$ and its cohomology. J. Eur. Math. Soc. **12**, 1–22 (2010)

5. D. Cohen, P. Orlik, Arrangements and local systems. Math. Res. Lett. **7**(2–2), 299–316 (2000)
6. D. Cohen, A. Suciu, Characteristic varieties of arrangements. Math. Proc. Camb. Philos. Soc. **127**, 33–53 (1999)
7. C. De Concini, M. Salvetti, Cohomology of Artin groups. Math. Res. Lett. **3**(2), 293–297 (1996)
8. C. De Concini, C. Procesi, M. Salvetti, F. Stumbo, Arithmetic properties of the cohomology of Artin groups. Ann. Scuola Norm. Sup. Pisa Cl. Sci. (4) **28**(4), 695–717 (1999)
9. C. De Concini, C. Procesi, M. Salvetti, Arithmetic properties of the cohomology of braid groups. Topology **40**(4), 739–751 (2001)
10. A. Dimca, S. Papadima, Hypersurface complements, Milnor fibers and higher homotopy groups of arrangements. Ann. Math. (2) **158**(2), 473–507 (2003)
11. A. Dimca, S. Papadima, A. Suciu Topology and geometry of cohomology jump loci. Duke Math. J. **148**, 405–457 (2009)
12. M.J. Falk, Arrangements and cohomology. Ann. Comb. **1**(2), 135–157 (1997)
13. M.J. Falk, S. Yuzvinsky Multinets, resonance varieties and pencils of plane curves. Compos. Math. **143**, 10069–1088 (2007)
14. R. Forman, Morse theory for cell complexes. Adv. Math. **134**(1), 90–145 (1998)
15. E.V. Frenkel, Cohomology of the commutator subgroup of the braid group. Func. Anal. Appl. **22**(13), 248–250 (1988)
16. G. Gaiffi, M. Salvetti The Morse complex of a line arrangement. J. Algebra **321**, 316–337 (2009)
17. G. Gaiffi, F. Mori, M. Salvetti Minimal CW-complexes for complement to line arrangements via discrete Morse theory, in *Topology of Algebraic Varieties and Singularities*. Contemporary Mathematics, vol. 538 (American Mathematical Society, Providence, 2011), pp. 293–308
18. A. Libgober, S. Yuzvinsky, Cohomology of the Orlik-Solomon algebras and local systems. Compos. Math. **21**, 337–361 (2000)
19. C. Okonek, Das $K(\pi, 1)$-Problem für die affinen Wurzelsysteme vom Typ A_n, C_n. Math. Z. **168**, 143–148 (1979)
20. R. Randell, Morse theory, Milnor fibers and minimality of a complex hyperplane arrangement. Proc. Am. Math. Soc. **130**(9), 2737–2743 (2002)
21. M. Salvetti, Topology of the complement of real hyperplanes in $\mathbb{C}^n$. Invent. Math. **88**(3), 603–618 (1987)
22. M. Salvetti, The homotopy type of Artin groups. Math. Res. Lett. **1**, 567–577 (1994)
23. M. Salvetti, S. Settepanella, Combinatorial Morse theory and minimality of hyperplane arrangements. Geom. Topol. **11**, 1733–1766 (2007)
24. S. Settepanella, Cohomology of pure Braid groups of exceptional cases. Topol. Appl. **156**, 1008–1012 (2009)
25. A. Suciu, Translated tori in the characteristic varieties of complex hyperplane arrangements. Topol. Appl. **118**, 209–223 (2002)
26. M. Yoshinaga, Resonant bands and local system cohomology groups for real line arrangements. Vietnam J. Math. **42**(3), 377–392 (2014)

Braid Groups and Euclidean Simplices

Elizabeth Leyton Chisholm and Jon McCammond

Abstract When Daan Krammer and Stephen Bigelow independently proved that braid groups are linear, they used the Lawrence–Krammer–Bigelow representation for generic values of its variables q and t. The t variable is closely connected to the traditional Garside structure of the braid group and plays a major role in Krammer's algebraic proof. The q variable, associated with the dual Garside structure of the braid group, has received less attention.

In this article we give a geometric interpretation of the q portion of the LKB representation in terms of an action of the braid group on the space of non-degenerate euclidean simplices. In our interpretation, braid group elements act by systematically reshaping (and relabeling) euclidean simplices. The reshapings associated to the simple elements in the dual Garside structure of the braid group are of an especially elementary type that we call relabeling and rescaling.

Keywords Braid groups • Dual Garside structures • Euclidean simplices • Lawrence-Krammer-Bigelow representation • Non-crossing partitions

At the turn of the millenium three papers on the linearity of braid groups appeared in rapid succession and all three used what is now known as the Lawrence–Krammer–Bigelow or LKB representation [3, 10, 11]. Its two variables, q and t, are connected to two different Garside structures on the braid group. The t variable is closely connected to the traditional Garside structure of the braid group and plays a major role in Krammer's algebraic proof [11]. The q variable is associated with the dual Garside structure and has received less attention. In this article, we introduce an elegant geometric interpretation of the q variable in the special case where $t = 1$, q is real, and the matrices of the representation are written with respect to the original basis used by Krammer in [10]. We call this special case the simplicial representation because of our first main result.

E.L. Chisholm • J. McCammond (✉)
Department of Mathematics, University of California, Santa Barbara, CA 93106, USA
e-mail: eeleyton@math.ucsb.edu; jon.mccammond@math.ucsb.edu

F. Callegaro et al. (eds.), *Configuration Spaces*, Springer INdAM Series 14,
DOI 10.1007/978-3-319-31580-5_12

Theorem 1 (Braids Reshape Simplices) *The simplicial representation of the n-string braid group preserves the set of $\binom{n}{2}$-tuples of positive reals that represent the squared edge lengths of a nondegenerate euclidean simplex with n labeled vertices.*

Thus braid group elements can be viewed as acting on and systematically reshaping the space of all nondegenerate euclidean simplices. Moreover, the dual simple braids in the braid group reshape simplices in an extremely elementary way that we call relabeling and rescaling. In this language we prove the following result; for a more precise statement, see Sect. 3.

Theorem 2 (Dual Simple Braids Relabel and Rescale) *Under the simplicial representation of the braid group, each dual simple braid acts by relabeling the vertices and rescaling specific edges.*

The article is structured around the three distinct contexts that play a role in these results: discs, simplices and matrices. In Sect. 1 convexly punctured discs are used to define noncrossing partitions and dual simple braids. In Sect. 2 we describe various ways to systematically reshape euclidean simplices, including the type of reshaping that we call edge rescaling. In Sect. 3 we connect these two relatively elementary discussions with the explicit matrices of the simplicial representation of the braid group to establish our main results. Finally, Sect. 4 explains our motivation for pursuing this line of investigation and some ideas for future work.

1 Discs

In this section metric discs with a finite number of labeled points are used to define the lattice of noncrossing partitions and the finite set of dual simple braids. We begin by recalling the notion of a convexly punctured disc.

Definition 1 (Convexly Punctured Disc) Let $\mathbf{D}_n$ be a topological disc in the euclidean plane with a distinguished n-element subset that we call its *punctures* or *vertices*. When the disc $\mathbf{D}_n$ is a convex subset of $\mathbb{R}^2$ and the convex hull of its n punctures is an n-gon (i.e. every puncture occurs as a vertex of the convex hull) then we say that $\mathbf{D}_n$ is a *convexly punctured disc*. There is a natural cyclic ordering of the vertices corresponding to the clockwise orientation of the boundary cycle of the n-gon. A labeling of the vertices is said to be *standard* if it uses the set $[n] := \{1,2,\ldots,n\}$ (or better yet $\mathbb{Z}/n\mathbb{Z}$) and the vertices are labeled in the natural cyclic order. More generally, when the vertices p_i are bijectively labeled by elements i in a finite set A, we refer to the convexly punctured disc as $\mathbf{D}_A$.

The 2-elements subsets are of particular interest.

Definition 2 (Edges) Let $\mathbf{D}_n$ be a convexly punctured disc. For each two element subset $\{i,j\} \subset [n]$, the convex hull of the corresponding points p_i and p_j in $\mathbf{D}_n$ is called an *edge* and denoted $e_{i,j} = e_{j,i}$, or even e_{ij} when the comma is not needed for clarity. When a *standard name* is needed we insist $i < j$. The number of edges is $\binom{n}{2}$ and we consistently use N for this number throughout the article. For later use, it is also convenient to impose a *standard order* on the set of all $N = \binom{n}{2}$ edges. We do so by lexicographically ordering them by their standard names. In $\mathbf{D}_4$, for example, the standard names of its $6 = \binom{4}{2}$ edges in their standard order are e_{12}, e_{13}, e_{14}, e_{23}, e_{24} and e_{34}.

We also need words to describe the position of one edge relative to another.

Definition 3 (Pairs of Edges) Let (e_{ij}, e_{kl}) be an ordered pair of edges in a convexly punctured disc $\mathbf{D}_n$ and let $B = \{i, j, k, l\}$. If we restrict our attention to the convex subdisc $\mathbf{D}_B$ (where $\mathbf{D}_B$ is an ϵ-neighborhood of the convex hull of the vertices indexed by the elements in B) then there are exactly five distinct possible configurations. We call the possibilities crossing, noncrossing, identical, clockwise and counterclockwise. When all four endpoints are distinct (i.e. when $|B| = 4$) these edges are either *crossing* or *noncrossing* depending on whether or not they intersect. At the other extreme, when e_{ij} and e_{kl} have both endpoints in common, they are *identical*. Finally, when these edges have exactly one endpoint in common, the convex hull of the three endpoints is a triangle and the edges occur as consecutive edges in its boundary cycle. We call this arrangement *clockwise* or *counterclockwise* depending on the orientation of the boundary which ensures that e_{kl} is the edge that occurs immediately after e_{ij}. More colloquially we say that e_{kl} is *to the right (left)* of e_{ij} and e_{ij} is *to the left (right)* of e_{kl} when the ordered pair (e_{ij}, e_{kl}) is clockwise (counterclockwise). Examples are shown in Fig. 1.

Noncrossing partitions are defined in a convexly punctured disc.

Definition 4 (Noncrossing Partitions) Let $\mathbf{D}_n$ be a convexly punctured disc. We say that two subsets $B, B' \subset [n]$ are *noncrossing* when the convex hulls of the corresponding sets of vertices in $\mathbf{D}_n$ are completely disjoint. See Fig. 2. More generally, a partition σ of the set $[n]$ is *noncrossing* when its blocks are pairwise noncrossing. Noncrossing partitions are usually ordered by refinement, so that $\sigma < \tau$ if and only if each block of σ is contained in some block of τ. Under this ordering, the set of all noncrossing partitions form a bounded graded lattice denoted NC_n. The poset NC_4 is shown in Fig. 3. The number of noncrossing partitions in NC_n is given by the n-th Catalan number $C_n = \frac{1}{n+1}\binom{2n}{n}$. For further information about noncrossing partitions see [1, 13, 15].

To each noncrossing partition there is a corresponding permutation.

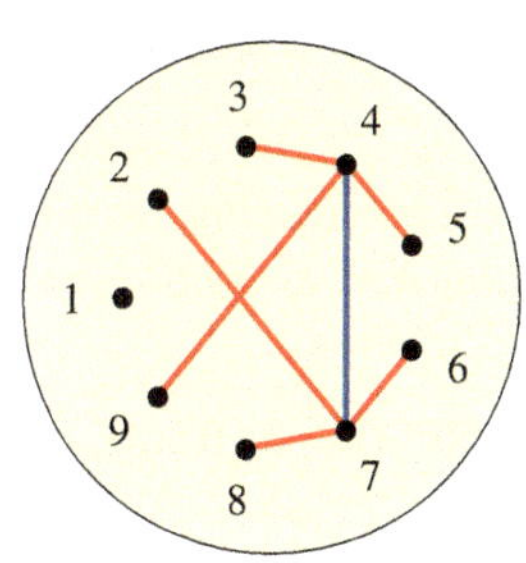

Fig. 1 The edges e_{34}, e_{49} and e_{67} are to the *left* of the edge e_{47} and the edges e_{27}, e_{78} and e_{45} are to the *right*. This is because ordered pairs such as (e_{34}, e_{47}) and (e_{67}, e_{47}) are clockwise while the ordered pair (e_{27}, e_{47}) is counterclockwise

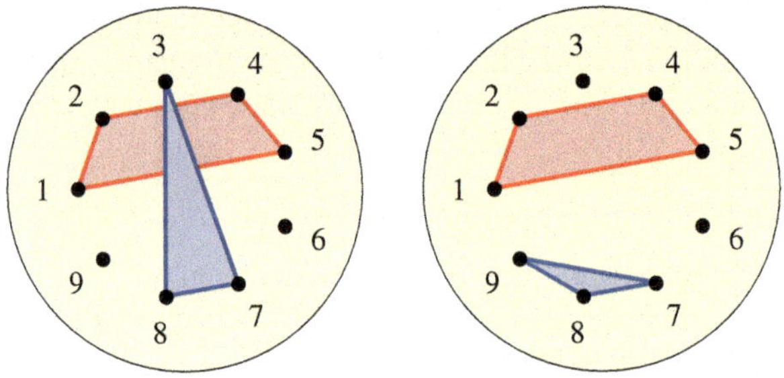

Fig. 2 The subsets $\{1, 2, 4, 5\}$ and $\{3, 7, 8\}$ are crossing. The subsets $\{1, 2, 4, 5\}$ and $\{7, 8, 9\}$ are noncrossing

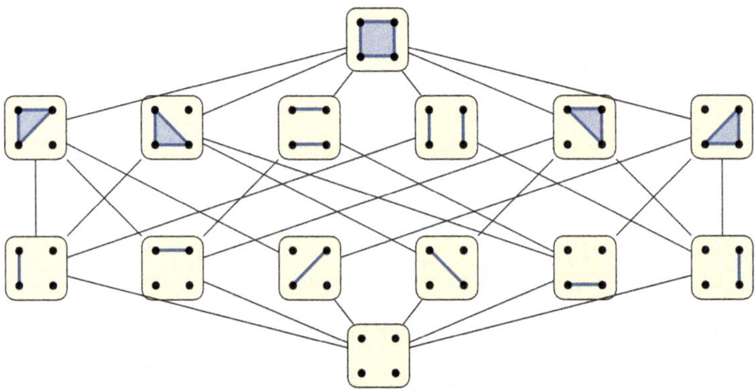

Fig. 3 Noncrossing partition lattice NC_4

Definition 5 (Noncrossing Permutations) To every subset $B \subset [n]$ with at least two elements we associate a permutation in SYM_n obtained by linearly ordering the elements in B. More generally, we associate a permutation to each noncrossing partition by multiplying the permutations associated to each block and we call the result a *noncrossing permutation*. Since distinct blocks are disjoint, their permutations commute and the product is well-defined. Thus $B = \{1, 3, 4\}$ becomes the permutation $(1, 3, 4)$ and the partition $\sigma = \{\{1, 3, 4\}, \{2\}, \{5, 6, 7, 8, 9\}\}$ becomes the permutation $(1, 3, 4)(5, 6, 7, 8, 9)$. We identify each noncrossing partition with its corresponding noncrossing permutation, using the same symbol for both. The permutation associated to the full set $[n]$ is an important n-cycle that we call δ.

As is well-known, the elements of the braid group can be identified with (equivalence classes of) motions of n distinct labeled points in a disc such as $\mathbf{D}_n$. The dual simple braids are a finite set of braids indexed by the noncrossing permutations as follows.

Definition 6 (Rotations) The *dual Garside element* s_δ of the n-string braid group is the motion where each labeled point in $\mathbf{D}_n$ moves clockwise along the boundary of the convex hull of all n points to the next vertex. More generally, for each set $B \subset \{1, \ldots, n\}$, let P_B be the convex hull of the vertices indexed by B and let $\mathbf{D}_B$ be an ϵ-neighborhood of P_B. The braid group element s_B is a similar motion restricted to the subdisc $\mathbf{D}_B$, i.e. the vertices in the subdisc move clockwise along one side of the polygon P_B to the next vertex, leaving all other vertices fixed. See Fig. 4. When B has at most one element, the motion is trivial. When B has two elements, the points avoid collisions by passing on the left.

Rotations can be used to assign a braid to each noncrossing partition.

Definition 7 (Dual Simple Braids) The *dual simple braids* are elements of the braid group in one-to-one correspondence with the set of noncrossing partitions NC_n. More precisely, for each noncrossing partition σ, we associate the product of the rotations corresponding to each of its blocks and call the result s_σ. Because rotations of noncrossing blocks take place in disjoint subdiscs they commute and the resulting element in the braid group is well-defined. Note that the noncrossing permutation σ is the permutation of the vertices induced by s_σ. The dual simple braids in BRAID_4 written as products of rotations are shown in Fig. 5.

It is useful to have specific names for four sets of dual simple braids.

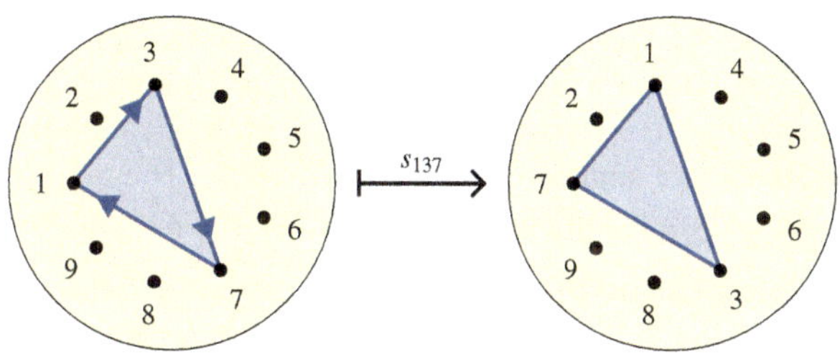

Fig. 4 The rotation s_{137}

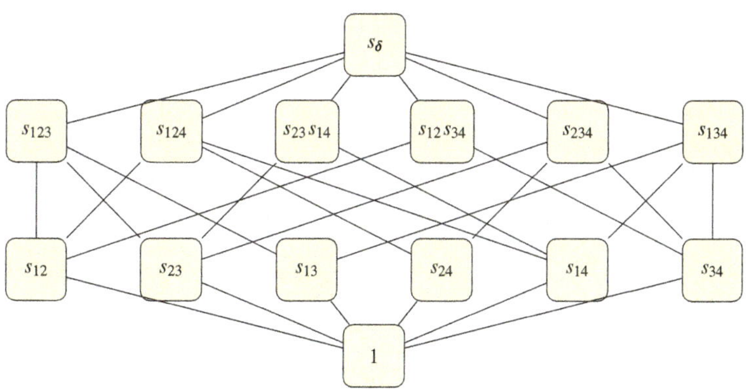

Fig. 5 The dual simple elements in BRAID_4

Definition 8 (Four Sets of Simple Braids) The *standard generating set* for the braid group BRAID_n consists of the $n-1$ rotations of the form s_{ij} with $1 \leq i < n$ and $j = i+1$. The *dual generators* of BRAID_n are the set of all $N = \binom{n}{2}$ rotations s_B where B has exactly two elements. The *rotations* s_B with $|B| \neq 1$ form a third set, and the full set of all dual simple braids form a fourth set. We write $\text{STD}_n \subset \text{GEN}_n \subset \text{ROT}_n \subset \text{SIMP}_n$ for these four nested sets, whose sizes are $n-1$, N, $2^n - n$ and C_n (the n-th Catalan number). When $n = 4$, these sets have 3, 6, 11 and 14 elements.

The multiplication in the symmetric group can be used to extract other information about noncrossing partitions.

Definition 9 (Multiplication) For consistency with our latter conventions we view permutations as functions and thus we multiply them from *right to left*. For example, the product $(1,2,3) \cdot (3,4,5)$ is $(1,2,3,4,5)$. More generally, if B, $\{i\}$ and C are pairwise disjoint subsets of $[n]$ such that there is a place to start reading the boundary cycle of the convex hull of the points corresponding to the elements in $B \cup \{i\} \cup C$ so that, reading clockwise, one encounters all of the vertices indexed by elements in B, followed by p_i, followed by all of the vertices indexed by the elements in C, then $s_{Bi}s_{iC} = s_{BiC}$. Here we follow the conventions of [12], removing parentheses from singletons, and using juxtaposition to indicate union.

The reader should be careful to note that our multiplication convention differs from much of the literature on the braid groups where multiplication is from left to right. Thus extra vigilance is required. The general multiplication rule given above, for example, is stated in a slightly different form in [12]. The multiplication can be used to define left and right complements of noncrossing permutations.

Definition 10 (Complements) Let σ be a noncrossing permutation in SYM_n and recall that δ is the n-cycle $(1,2,\ldots,n)$. The *left complement* of σ is the unique element σ' such that $\sigma'\sigma = \delta$ and its *right complement* is the unique element σ'' such that $\sigma\sigma'' = \delta$. We denote these permutations by $\sigma' = \text{lc}(\sigma)$ and $\sigma'' = \text{rc}(\sigma)$. The permutation $\text{lc}(\sigma)$ is always also a noncrossing permutation and, in fact, the edges in its blocks are precisely those that are to the left or noncrossing with respect to each of the edges in the blocks of σ. Similarly $\text{rc}(\sigma)$ is a noncrossing permutation whose blocks are formed by the edges that are to the right or noncrossing with respect to each edge in a block of σ. See Fig. 6.

The following observation is not crucial to our results, but we sometimes use this language.

Remark 1 (Hypertrees) A *hypergraph* is a generalization of a graph where its *hyperedges* are allowed to span more than two vertices, and a *hypertree* is the natural generalization of a tree. As can be seen in Fig. 6, the blocks of the noncrossing partition associated to a dual simple element and the blocks of one of its complements together form the hyperedges of a planar hypertree.

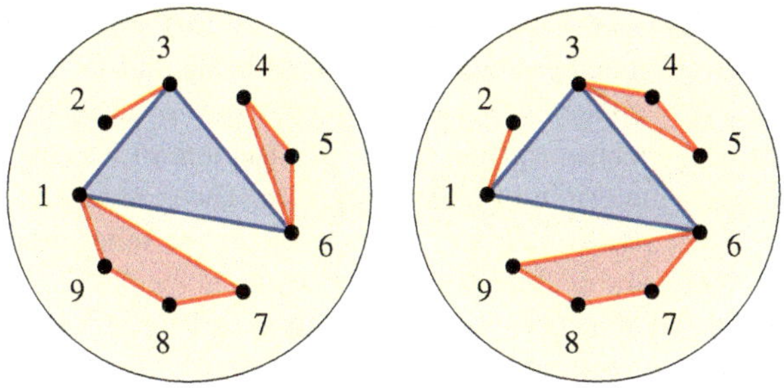

Fig. 6 (23)(456)(1789) is the *left* complement of (136) in SYM_9 and (12)(345)(6789) is its *right* complement

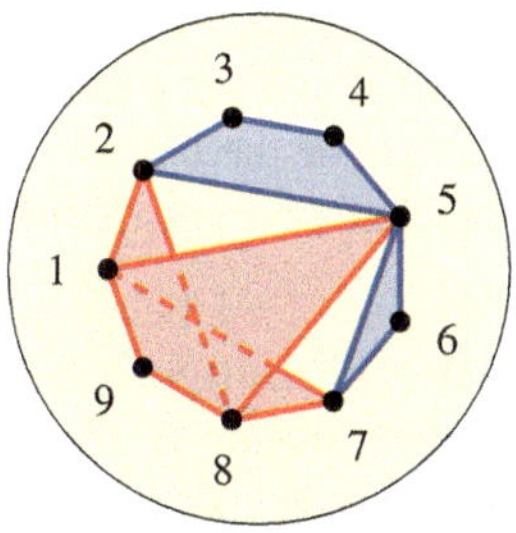

Fig. 7 If $\sigma_1 = (2,3,4,5)$ and $\sigma_2 = (5,6,7)$, then the permutations σ_3, σ_4 and σ_5 defined in Definition 11 are $\sigma_3 = (1,7,8,9)$, $\sigma_4 = (1,5,8,9)$ and $\sigma_5 = (1,2,8,9)$

The Hasse diagram of the noncrossing partition lattice can actually be viewed as a portion of the Cayley graph of SYM_n with respect the generating set of all transpositions, specifically the portion between the identity and the element δ. This way of looking at the noncrossing partition lattice combined with the fact that the set of transpositions in SYM_n is closed under conjugation, helps to explain why factorizations of δ as a product of noncrossing permutations are so flexible. The only aspect of this flexibility that we need here is the following.

Definition 11 (Five Permutations) If σ_1 and σ_2 are permutations in SYM_n such that σ_1, σ_2 and their product $\sigma_1\sigma_2$ are all three noncrossing, then there exist noncrossing permutations σ_3, σ_4 and σ_5 such that $\delta = \sigma_1\sigma_2\sigma_3 = \sigma_1\sigma_4\sigma_2 = \sigma_5\sigma_1\sigma_2$. The permutations σ_5 and σ_3 are simply the left and right complements of the product $\sigma_1\sigma_2$, while σ_4 is obtained by conjugation. An example is shown in Fig. 7.

2 Simplices

In this section we discuss the geometry of euclidean simplices with n labeled vertices and, in particular, how this geometry changes under certain carefully controlled deformations. As in [6], we start by distinguishing between points and vectors.

Definition 12 (Points) Let V be an $(n-1)$-dimensional *real vector space* with a fixed positive definite inner product but no fixed basis and let E be an $(n-1)$-dimensional *euclidean space*, which may be defined as a set with a fixed simply-transitive action of the additive group of V. The structure of E is essentially that of V but the location of the origin has been forgotten. The elements of V are *vectors*, the elements of E are *points*, and we write $\langle u, v\rangle$ for the inner product of vectors u and v. Two points p and p' determine a line segment e called an *edge* and p and p' are its *endpoints*. By the simply-transitive action of V on E, they also determine two vectors: the unique vector v that sends p to p' and the vector $-v$ which sends p' to p. The pair $\pm v$ is a *lax vector*. When the points involved are labeled, say p_i and p_j, we write $e_{ij} = e_{ji}$ for the edge they span, and $v = v_{ij}$ and $-v = v_{ji}$ for the two vectors they determine. The *norm* of a vector v is $\langle v, v\rangle$, which is also the square of the length of the corresponding edge e. We write $\textsc{Norm} : V \to \mathbb{R}$ for the norm map. Note that the norm of a lax vector is well-defined since $\textsc{Norm}(v) = \textsc{Norm}(-v)$ and the norm of an edge is the norm of the lax vector determined by its endpoints. When the points involved are labeled we write a_{ij} for $\textsc{Norm}(v_{ij})$.

Definition 13 (Simplices) A set $\{p_i\}$ of n labeled points in E is in *general position* if this set is not contained in any proper affine subspace of E, and the convex hull of such a set is a *labeled euclidean simplex* Δ of dimension $(n-1)$. For any such labeled euclidean simplex Δ we use subsets of punctures in the convexly punctured disc $\mathbf{D}_n$ to describe various simplicial faces of Δ via their vertex labelings. For example, the three blocks of the left complement of s_{136} shown in Fig. 6 correspond to an edge, a triangle and a tetrahedron in any 8-dimensional simplex Δ with nine labeled vertices.

We are primarily interested in the isometry class of a labeled euclidean simplex Δ and this is completely determined by the ordered list of the norms of its edges.

Definition 14 (Edge Norm Vectors) Let Δ be a labeled euclidean simplex with n vertices. The *edge norm vector* of Δ is a column vector $\mathbf{v}$ of the $N = \binom{n}{2}$ positive real numbers a_{ij} which are the norms of its edges e_{ij}, listed in the standard lexicographic order of the edges as discussed in Definition 2.

Edge norm vectors characterize isometry classes of labeled euclidean simplices and as a result when we reshape a labeled euclidean simplex, these changes to its geometry are captured by the modifications that occur in its edge norm vector. The well-known formula $2\langle u, v\rangle = \textsc{Norm}(u+v) - \textsc{Norm}(u) - \textsc{Norm}(v)$ shows that inner products of vectors can be calculated in terms of their norms, but we need a slightly more general formula that computes the inner product of two vectors determined by four possibly distinct points in a euclidean space E.

Proposition 1 (Inner Products and Norms) *If p_i, p_j, p_k and p_l are four not necessarily distinct points in a euclidean space E, then the inner product* $2\langle v_{ij}, v_{kl}\rangle = a_{il} + a_{jk} - a_{ik} - a_{jl}$.

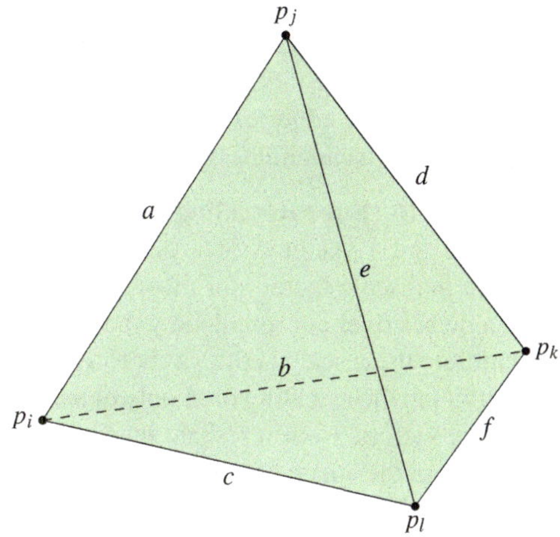

Fig. 8 Tetrahedron determined by four points, edges labeled by norm

Proof To improve readability, we write a, b, c, d, e and f for the norms a_{ij}, a_{ik}, a_{il}, a_{jk}, a_{jl} and a_{kl}, respectively. The situation under discussion is shown in Fig. 8. Expanding the norms of $v_{ik} = v_{ij} + v_{jk}$ and $v_{jl} = v_{jk} + v_{kl}$ produces the identities $b = a + d + 2\langle v_{ij}, v_{jk}\rangle$ and $e = d + f + 2\langle v_{jk}, v_{kl}\rangle$. Expanding the norm of $v_{il} = v_{ij} + v_{jk} + v_{kl}$ produces $c = a + d + f + 2\langle v_{ij}, v_{jk}\rangle + 2\langle v_{jk}, v_{kl}\rangle + 2\langle v_{ij}, v_{kl}\rangle$. Thus $2\langle v_{ij}, v_{jk}\rangle = b - a - d$, $2\langle v_{jk}, v_{kl}\rangle = e - d - f$, and $2\langle v_{ij}, v_{kl}\rangle = c - a - d - f - (b - a - d) - (e - d - f) = c + d - b - e$. □

The following definition identifies a class of geometric reshapings of labeled euclidean simplices that are particularly elegant and easy to describe. We call them edge rescalings and, as far as we are aware, they have not been previously discussed in the literature.

Definition 15 (Edge Rescaling) Let Δ and Δ' be two labeled euclidean simplices with n vertices situated in a common euclidean space. We say that an edge e_{ij} in Δ is merely *rescaled* if it and the corresponding edge e'_{ij} in Δ' point in the same direction. More generally, we say that Δ' is an *edge rescaling* of Δ if there exist enough pairs of corresponding edges pointing in the same direction (but with possibly different lengths) to form a basis for the vector space out of these common direction vectors.

Remark 2 (Spanning Trees) Let Δ' be a labeled euclidean simplex which is an edge rescaling of Δ. By definition there are sufficiently many edges that are merely rescaled to form a basis out of the corresponding vectors and a minimal set of rescaled edges in the 1-skeleton of Δ would form a spanning tree in this complete graph. There might, however, be more than one such spanning tree of merely rescaled edges. When two edges in Δ share a common endpoint and are both rescaled by the same scale factor, the triangle they span, and in particular the third edge in that triangle, is also rescaled by the same scale factor. Thus any two of these three edges could be included in the spanning tree. In fact, it would be more canonical to identify the *maximal* simplicial faces that are merely rescaled. For each scale factor, there would be a partition of the vertices into maximal subsimplices rescaled by that factor and the blocks in all of these partitions together would form a spanning hypertree in the sense of Remark 1. The full variety of spanning trees which satisfy the edge rescaling definition are selected from the edges inside the blocks of such a canonical spanning hypertree.

Definition 16 (Edge Rescaling Maps) One thing to note is that for every spanning tree T in the 1-skeleton of a labeled euclidean simplex Δ and for every set of positive real scale factors for these edges, there does exist a rescaled simplex Δ' in which these edges are rescaled by these factors since the rescaled tree T' formed by assembling the rescaled edges as before tells us how the vertices should be arranged. In particular, the rescaling of Δ only depends on the set of merely rescaled edges and the scale factors used to rescale them. Thus there is a well-defined *edge rescaling map* R from the space of labeled euclidean simplices to itself which, based only on such data, rescales each Δ in the set to a new simplex Δ'. Such a map R is clearly invertible since rescaling the same edges by the multiplicative inverse of each scale factor returns each Δ' to Δ. The edge rescaling maps we are primarily interested in are those where every scale factor is 1 or q. We call these *q-rescalings* and we introduce a special notation for them. Let $R = R^{\sigma}_{\tau}$ denote the q-rescaling where the blocks of the partition σ index the subsimplices rescaled by q and the blocks of the partition τ index the subsimplices which are fixed, i.e. rescaled by a factor of 1.

The next proposition shows that the effect of R is encoded in a matrix.

Proposition 2 (Edge Rescaling Matrices) *The effect of an edge rescaling map R on an edge norm vector $\mathbf{v}$ for a labeled euclidean simplex Δ is captured by an N by N matrix M whose entries only depend on the map R and not on the vector $\mathbf{v}$ or the simplex Δ. In particular, $R(\mathbf{v}) = M \cdot \mathbf{v}$ for all $\mathbf{v}$ and Δ.*

Proof Let T be a spanning set of edges that are rescaled by R. For each edge e_{ij} we can use paths in the spanning tree T to find a linear combination of vectors associated with rescaled edges whose sum is v_{ij}. The new vector v'_{ij}, by definition, is a similar sum where the vectors in the sum are rescaled according to the scale factors of R. Thus the norm of v'_{ij} can be expanded as a linear combination of inner products of vectors whose edges belong to the tree T and the coefficients of this linear combination are independent of the original edge norms. Next, by Proposition 1, the inner products can be rewritten as linear combinations of the original edge norms. Substituting these in produces a formula for each new edge norm a'_{ij} as a linear combination of the old edge norms in $\mathbf{v}$ with coefficients that are independent of $\mathbf{v}$. The matrix M is formed by collecting these coefficients. □

For simplicity we use the same symbol R to denote both the edge rescaling map and the edge rescaling matrix that was called M in the proposition. As an explicit example of this process, consider the q-rescaling of a triangle shown in Fig. 9.

Proposition 3 (Edge Rescaling a Triangle) *If Δ is a labeled euclidean triangle and Δ' is the labeled euclidean triangle obtained by the edge rescaling $R = R^{12}_{23}$, then the edge norms of Δ' can be computed from the edge norms of Δ as follows:*

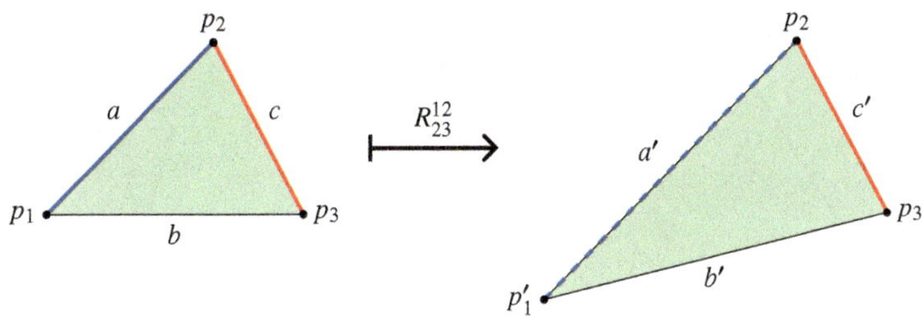

Fig. 9 An edge reshaping $R = R^{12}_{23}$ that rescales e_{12} by a factor of q and fixes e_{23} (i.e. rescales e_{23} by a factor of 1)

$a'_{12} = q^2 a_{12}$, $a'_{13} = (q^2 - q)a_{12} + qa_{13} + (1-q)a_{23}$, *and* $a'_{23} = a_{23}$. *In particular, R has the effect of multiplying the edge norm vector of* Δ *by a matrix with entries in* $\mathbb{Z}[q]$.

$$\mathbf{v}' = \begin{bmatrix} a'_{12} \\ a'_{13} \\ a'_{23} \end{bmatrix} = \begin{bmatrix} q^2 & 0 & 0 \\ q^2 - q & q & 1-q \\ 0 & 0 & 1 \end{bmatrix} \cdot \begin{bmatrix} a_{12} \\ a_{13} \\ a_{23} \end{bmatrix} = R \cdot \begin{bmatrix} a_{12} \\ a_{13} \\ a_{23} \end{bmatrix} = R \cdot \mathbf{v}. \qquad (1)$$

Proof For clarity let a, b and c be the edge norms a_{12}, a_{13} and a_{23} in Δ and add primes for the corresponding edge norms in Δ'. In the original triangle we have $a = \langle v_{12}, v_{12} \rangle$, $c = \langle v_{23}, v_{23} \rangle$ and $b = \langle v_{13}, v_{13} \rangle = \langle v_{12} + v_{23}, v_{12} + v_{23} \rangle = a + 2\langle v_{12}, v_{23} \rangle + c$. Thus $2\langle v_{12}, v_{23} \rangle = b - a - c$. In the new triangle we have $c' = \langle v_{23}, v_{23} \rangle = c$, $a' = \langle q \cdot v_{12}, q \cdot v_{12} \rangle = q^2 a$, and $b' = \langle q \cdot v_{12} + v_{23}, q \cdot v_{12} + v_{23} \rangle = q^2 a + 2q\langle v_{12}, v_{23} \rangle + c = q^2 a + q(b - a - c) + c = (q^2 - q)a + qb + (1-q)c$ as required. □

Many of the properties of the matrix $R = R^{12}_{23}$ extend to all q-rescalings.

Proposition 4 (Quadratic Matrices) *Let* $R = R^{\sigma}_{\tau}$ *be a q-rescaling of a labeled euclidean simplex* Δ *with n vertices. The effect of R on the edge norm vector* $\mathbf{v}$ *of* Δ *is to multiply from the left by an N by N matrix with entries in* $\mathbb{Z}[q]$ *of degree at most* 2.

Proof The proof is the same as that of Proposition 2 but with the additional observation that the coefficients are at most quadratic polynomials in q since there is at most one q coming from each side of the inner product. □

Our second example is very similar to the first.

Example 1 (Rescaling a Tetrahedron) Consider the edge rescaling R^{12}_{234} as shown in Fig. 10 and note that we can compute all of the new edge norms using Proposition 3. For clarity we write a through f for a_{12} through a_{34} in lexicographic order. The effect

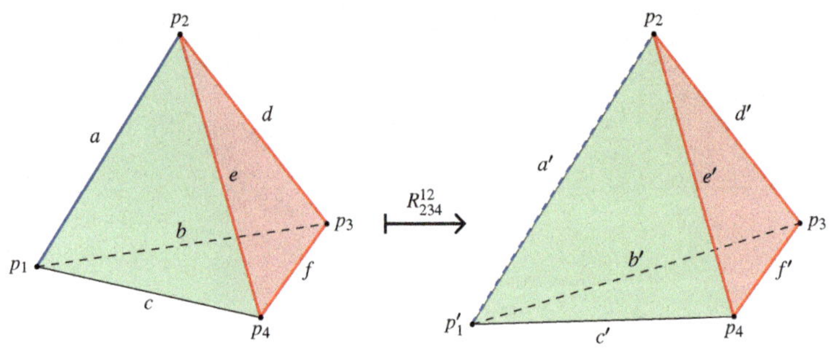

Fig. 10 The edge rescaling R^{12}_{234} which fixes the triangle Δ_{234} and rescales edge e_{12} by a factor of q

of the map $R = R^{12}_{234}$ is as follows:

$$R \cdot \begin{bmatrix} a \\ b \\ c \\ d \\ e \\ f \end{bmatrix} = \begin{bmatrix} q^2 a \\ (q^2 - q)a + qb + (1-q)d \\ (q^2 - q)a + qc + (1-q)e \\ d \\ e \\ f \end{bmatrix} = \begin{bmatrix} a' \\ b' \\ c' \\ d' \\ e' \\ f' \end{bmatrix}. \tag{2}$$

Thus the matrix that encodes the rescaling R is

$$R = \begin{bmatrix} q^2 & 0 & 0 & 0 & 0 & 0 \\ q^2 - q & q & 0 & 1-q & 0 & 0 \\ q^2 - q & 0 & q & 0 & 1-q & 0 \\ 0 & 0 & 0 & 1 & 0 & 0 \\ 0 & 0 & 0 & 0 & 1 & 0 \\ 0 & 0 & 0 & 0 & 0 & 1 \end{bmatrix}. \tag{3}$$

When the simplices are higher-dimensional, additional notation is needed.

Definition 17 (Row Descriptions) Since the new edge norms are determined by the rows of the matrix $R = R^{\sigma}_{\tau}$, we introduce a way to describe these rows. Recall that e_{ij} denotes an edge in the punctured disc $\mathbf{D}_n$ and an edge in a labeled euclidean simplex Δ. We now add a third interpretation: as an element of the canonical basis of the vector space $\mathbb{R}^N$ containing the edge norm vectors. Using this interpretation of e_{ij} as basis vectors, the second row of the matrix for $R = R^{12}_{23}$ as given in Proposition 3 is the row vector $(q^2 - q, q, 1 - q)$ or, equivalently, it is the linear combination $(q^2 - q)e_{12} + qe_{13} + (1 - q)e_{23}$. To select the second row one would multiply the matrix R on the right by the row vector $(0, 1, 0)$ which is just the vector e_{13}. In other words, $(e_{13})R = (q^2 - q)e_{12} + qe_{13} + (1 - q)e_{23}$.

Remark 3 (Left and Right) The linear combination that describes the image of a basis vector acted on by a rescaling matrix R from the right looks very similar to the corresponding entry in the edge norm vector when acted on by R from the left precisely because both are essentially encoding the entries in one row of R. Nevertheless, this switch between left and right has the potential to be slightly confusing.

It turns out that the linear combination that describes the e_{kl} row of the matrix $R^{ij}_{\mathrm{rc}(ij)}$ (or $R^{ij}_{\mathrm{lc}(ij)}$) only depends on the geometric relationship between the edges e_{ij} and e_{kl} in the punctured disc $\mathbf{D}_n$. Thus, in our row descriptions of these matrices the final column uses the language of Definition 3 to describe how e_{kl} is situated relative to e_{ij}.

Example 2 (Rescaling a Boundary Edge) In this example we give explicit row descriptions for the q-rescalings which stretch a single boundary edge (i.e. an edge that corresponds to an edge in the boundary of the convex polygon in the disk) while fixing either its left or its right complement. We start with the row description of the matrix $R = R^{12}_{\mathrm{rc}(12)}$.

$$(e_{kl})R = \begin{cases} q^2 e_{kl} & \text{identical } (k = 1, l = 2) \\ e_{kl} & \text{noncrossing } (k, l > 2) \\ e_{kl} & \text{to the right } (k = 2) \\ (q^2 - q)e_{12} + qe_{kl} + (1 - q)e_{2l} & \text{to the left } (k = 1). \end{cases} \tag{4}$$

This is a natural generalization of the triangular and tetrahedral examples in the new notation. The row description of $R = R^{ij}_{\mathrm{rc}(ij)}$ with $j = i + 1 \mod n$ is only slightly more complicated. We write e_{new} for the third edge of the triangle when e_{ij} and e_{kl} have exactly one endpoint in common.

$$(e_{kl})R = \begin{cases} q^2 e_{kl} & \text{identical} \\ e_{kl} & \text{noncrossing} \\ e_{kl} & \text{to the right} \\ (q^2 - q)e_{ij} + qe_{kl} + (1 - q)e_{\mathrm{new}} & \text{to the left.} \end{cases} \tag{5}$$

Switching from the right complement to the left complement causes only very minor changes. The row description of the matrix $R = R^{12}_{\mathrm{lc}(12)}$ is as follows

$$(e_{kl})R = \begin{cases} q^2 e_{kl} & \text{identical } (k = 1, l = 2) \\ e_{kl} & \text{noncrossing } (k, l > 2) \\ e_{kl} & \text{to the left } (k = 1) \\ (q^2 - q)e_{12} + qe_{kl} + (1 - q)e_{1l} & \text{to the right } (k = 2). \end{cases} \tag{6}$$

And finally, we list the row description of $R^{ij}_{\mathrm{lc}(ij)}$ with $j = i + 1 \mod n$, with the same convention that e_{new} denotes the third edge of the triangle when e_{ij} and e_{kl}

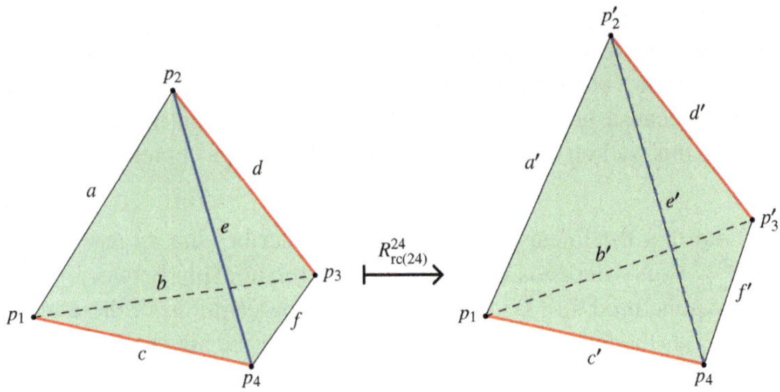

Fig. 11 The edge rescaling $R^{24}_{\mathrm{rc}(24)}$ which rescales the edge e_{24} while fixing the edges e_{23} and e_{14}

have exactly one endpoint in common.

$$(e_{kl})R = \begin{cases} q^2 e_{kl} & \text{identical} \\ e_{kl} & \text{noncrossing} \\ e_{kl} & \text{to the left} \\ (q^2-q)e_{ij} + qe_{kl} + (1-q)e_{\mathrm{new}} & \text{to the right.} \end{cases} \tag{7}$$

In order to extend Example 2 to more general situations, we need one more computation that is illustrated in (Fig. 11).

Proposition 5 (Diagonal Edges) *Let Δ be a labeled euclidean tetrahedron. If Δ' is the labeled euclidean tetrahedron obtained by the edge rescaling $R = R^{24}_{\mathrm{rc}(24)}$, then the new edge norm a'_{13} can be computed from the edge norms of Δ as follows:*

$$a'_{13} = a_{13} + (q-1)^2 a_{24} + (q-1)(a_{14} + a_{23}) + (1-q)(a_{12} + a_{34})$$

Proof Since $v_{13} = v_{12} + v_{24} + v_{43}$, we have $v'_{13} = v_{12} + qv_{24} + v_{43}$. Expanding $a'_{13} = \langle v'_{13}, v'_{13} \rangle$ we find that a'_{13} is equal to $a_{12} + q^2 a_{24} + a_{34} + 2q\langle v_{12}, v_{24}\rangle + 2q\langle v_{24}, v_{43}\rangle + 2\langle v_{12}, v_{34}\rangle$. Using Proposition 1 we find that $\langle v_{12}, v_{24}\rangle = a_{14} - a_{12} - a_{24}$, $\langle v_{24}, v_{43}\rangle = a_{23} - a_{24} - a_{34}$ and $\langle v_{12}, v_{43}\rangle = a_{13} + a_{24} - a_{14} - a_{23}$. Substituting and simplifying yields the result. □

Example 3 (Rescaling a Diagonal Edge) The row description for the rescaling matrix $R^{ij}_{\mathrm{rc}(ij)}$ is essentially identical to the one listed in Example 2. In particular, the formulas for the cases where e_{kl} is identical to, noncrossing, to the left or to the right of e_{ij} are the same as before. The final geometric configuration that is possible when e_{ij} is not a boundary edge is that e_{ij} and e_{kl} might be crossing. When this happens $(e_{kl})R^{ij}_{\mathrm{rc}(ij)}$ can be computed using Proposition 5. If the clockwise ordering of i, j, k and l is (k, i, l, j) then the answer is $e_{kl} + (q-1)^2 e_{ij} + (q-1)e_{kj} + (q-1)e_{il} + (1-q)e_{ki} + (1-q)e_{lj}$. A more intrinsic geometric description would use the convex hull of $\{p_i, p_j, p_k, p_l\}$ in $\mathbf{D}_n$. The answer obtained is e_{kl} plus $(q-1)^2 e_{ij}$ plus $(q-1)$ times the two boundary edges of the convex hull which are simultaneously to the right of e_{ij} and to the left of e_{kl} plus $(1-q)$ times the two boundary edges which are simultaneously to the right of e_{ij} and to the left of e_{kl}. A row description for $R^{ij}_{\mathrm{lc}(ij)}$ for arbitrary i and j can be computed in a similar fashion.

3 Matrices

In this section, we define three explicit representations of the braid group and then we prove our main results. The first representation, and the most complicated, is the Lawrence–Krammer–Bigelow or LKB representation.

Definition 18 (LKB Representation) Let q and t be nonzero positive real numbers, let $\mathcal{E}$ be the set $\{e_{ij}\}$ with $1 \leq i < j \leq n$ of size $N = \binom{n}{2}$ and let $\mathbb{R}^N$ be the $N = \binom{n}{2}$-dimensional real vector space with $\mathcal{E}$ as its ordered basis. The *LKB representation* of the braid group is the map $\rho : \mathrm{BRAID}_n \to GL_N(\mathbb{R})$ defined by the following action (from the right) of the standard braid group generators s_{ij} (with $j = i+1$ and $1 \leq i < n$) on elements of $\mathcal{E}$.

$$(e_{kl})\rho(s_{ij}) = \begin{cases} tq^2 e_{kl} & i = k, j = l \\ e_{kl} & i,j \notin \{k,l\} \\ e_{jl} & i = k, j < l \\ e_{kj} & i = l \\ t(q^2-q)e_{ij} + qe_{ki} + (1-q)e_{kl} & k < i, j = l \\ (q^2-q)e_{ij} + qe_{il} + (1-q)e_{kl} & j = k. \end{cases} \tag{8}$$

We make two remarks about this definition.

Remark 4 (Left/Right Actions) In the literature, this action is written as an action from the left but our alteration only has the effect of transposing the relevant matrices. Our choice is dictated by our desire to match up (a simplified version of) this representation with the obviously very similar edge rescaling matrices discussed in the previous section.

Remark 5 (The t Variable and Its Sign) The t variable depends on the linear ordering of the vertices, it is associated the standard presentation of the braid group, and its presence obscures the fundamentally cyclically symmetric nature of the dependence on q. To highlight this cyclic symmetry, we shall consider the specialization with $t = 1$ below. We should also note, however, that there are inconsistencies in the literature regarding the sign of t. The variable t in [10] corresponds to $-t$ in [3] (with an additional sign correction in [4]) and [9]. We have written the LKB representation using Krammer's sign convention. If we had followed Bigelow's we would be setting t equal to -1.

In light of our first main theorem, we call the simplified LKB representation the *simplicial representation* of the braid group.

Definition 19 (Simplicial Representation) The *simplicial representation* of the braid group is the specialization of the LKB representation with t set equal to 1. Concretely, let q be a nonzero positive real number, let $\mathcal{E}$ be the set $\{e_{ij}\}$ with $1 \leq i < j \leq n$ in lexicographic order and let $\mathbb{R}^N$ be the $N = \binom{n}{2}$-dimensional real vector space with $\mathcal{E}$ as its ordered basis. The *simplicial representation* of the braid group is defined by the following action (from the right) of the standard braid group generators s_{ij} (with $j = i + 1$ and $1 \leq i < n$) on elements of $\mathcal{E}$. We write S_σ for the matrix that represents s_σ with respect to the ordered basis $\mathcal{E}$ and we have introduced the notation e_{new} to denote the third side of the triangle when e_{ij} and e_{kl} have exactly one endpoint in common as in the previous section.

$$(e_{kl})S_{ij} = \begin{cases} q^2 e_{kl} & i = k, j = l \\ e_{kl} & i,j \notin \{k,l\} \\ e_{\text{new}} & i = k, j < l \\ e_{\text{new}} & i = l \\ (q^2 - q)e_{ij} + qe_{\text{new}} + (1-q)e_{kl} & k < i, j = l \\ (q^2 - q)e_{ij} + qe_{\text{new}} + (1-q)e_{kl} & j = k. \end{cases} \tag{9}$$

Now that the t variable has been eliminated, some of the rows are identical and they can be rewritten more elegantly using the language of Definition 3.

$$(e_{kl})S_{ij} = \begin{cases} q^2 e_{kl} & \text{identical} \\ e_{kl} & \text{noncrossing} \\ e_{\text{new}} & \text{to the left} \\ (q^2 - q)e_{ij} + qe_{\text{new}} + (1-q)e_{kl} & \text{to the right.} \end{cases} \tag{10}$$

Our third representation is obtained by also eliminating the q variable.

Definition 20 (Permutation Representation) The *permutation representation* of the braid group that we are interested in is the one obtained from the simplicial representation by setting $q = 1$ (or both $t = q = 1$ in the LKB representation). This encodes the permutation of the edges induced by the corresponding permutation of the vertices. We write P_σ for the matrix corresponding to s_σ. Its row description is as follows.

$$(e_{kl})P_{ij} = \begin{cases} e_{kl} & \text{identical, crossing or noncrossing} \\ e_{\text{new}} & \text{to thc lcft or right.} \end{cases} \tag{11}$$

It should be clear that the simplicial representation matrix S_{ij} is very closely connected, but not quite identical to the rescaling matrix $R^{ij}_{\text{rc}(ij)}$ given in Example 2. The difference is the permutation matrix P_{ij}.

Example 4 (Geometry of S_{12}) The matrices corresponding to the first standard generator s_{12} of the four string braid group in the simplicial representation and the permutation representation are as follows:

$$S_{12} = \begin{bmatrix} q^2 & 0 & 0 & 0 & 0 & 0 \\ 0 & 0 & 0 & 1 & 0 & 0 \\ 0 & 0 & 0 & 0 & 1 & 0 \\ q^2-q & q & 0 & 1-q & 0 & 0 \\ q^2-q & 0 & q & 0 & 1-q & 0 \\ 0 & 0 & 0 & 0 & 0 & 1 \end{bmatrix} \quad P_{12} = \begin{bmatrix} 1 & 0 & 0 & 0 & 0 & 0 \\ 0 & 0 & 0 & 1 & 0 & 0 \\ 0 & 0 & 0 & 0 & 1 & 0 \\ 0 & 1 & 0 & 0 & 0 & 0 \\ 0 & 0 & 1 & 0 & 0 & 0 \\ 0 & 0 & 0 & 0 & 0 & 0 \end{bmatrix}. \tag{12}$$

It is now straightforward to check that $S_{12} = P_{12}R^{12}_{234} = R^{12}_{134}P_{12}$. The matrix $R^{12}_{234} = R^{12}_{\mathrm{rc}(12)}$ is listed explicitly in Example 1 and the row description of $R^{12}_{134} = R^{12}_{\mathrm{lc}(12)}$ is given in Example 2.

Before proceeding to the proofs of our main results, we introduce one final definition which makes our precise results easier to state.

Definition 21 (Relabeling and Rescaling) Let σ be a noncrossing permutation in SYM_n and let S_σ be the explicit matrix representing s_σ under the simplicial representation. We say that S_σ *relabels and rescales* if $S_\sigma = P_\sigma R^{\sigma}_{\mathrm{rc}(\sigma)} = R^{\sigma}_{\mathrm{lc}(\sigma)}P_\sigma$, where $\mathrm{lc}(\sigma)$ and $\mathrm{rc}(\sigma)$ are the left/right complements of σ.

For clarity we also say what this definition means geometrically. To say that the matrix S_σ relabels and rescales means that the effect it has on labeled euclidean simplices (keeping in mind that we multiply from right to left) is to first rescale the edges (fixing the length and direction of the edges in the right complement of σ while multiplying the lengths of the edges in σ by a factor of q) followed by the edge relabeling induced by the way σ permutes the vertices. Alternatively, the edge relabeling P_σ can be performed first (i.e. on the right), in which case it is the edges of the left complement of σ whose length and direction are fixed while the lengths of the edges in σ are multiplied by a factor of q. In this language, Example 4 shows that the matrix S_{12} in the simplicial representation of BRAID_4 relabels and rescales. The next proposition shows that the standard generators in all of the braid groups share this property.

Proposition 6 (Standard Generators) *For every standard generator s_{ij} of the braid group BRAID_n, the corresponding matrix S_{ij} in the simplicial representation relabels and rescales.*

Proof This is essentially immediate at this point once we compare the row description of S_{ij} in Definition 19 with the row descriptions of $R^{ij}_{\mathrm{rc}(ij)}$ and $R^{ij}_{\mathrm{lc}(ij)}$ in Example 2 and note that multiplying by P_{ij} on the left switches the rows to the left of e_{ij} with the rows to the right of e_{ij} (which has the effect of switching which edge is denoted e_{kl} and which is e_{new}), while multiplying by P_{ij} on the right permutes columns and thus the subscripts on the e's that occur in the various terms of the row descriptions. □

From Proposition 6 we deduce our first result.

Theorem 3 (Braids Reshape Simplices) *The simplicial representation of the n-string braid group preserves the set of* $\binom{n}{2}$*-tuples of positive reals that represent the squared edge lengths of a nondegenerate euclidean simplex with n labeled vertices.*

Proof First note that it suffices to prove that this holds for some generating set of BRAID_n. Next, both vertex relabelings and edge rescalings clearly preserve the set of $\binom{n}{2}$-tuples that describe the squared edge lengths of a nondegenerate euclidean simplex with n labeled vertices, so Proposition 6 completes the proof. □

Once we know that some dual simple braids relabel and rescale, it is straightforward to show that their products (at least those which remain dual simple braids) do so as well.

Proposition 7 (Products) *Let* σ_1 *and* σ_2 *be noncrossing permutations in* SYM_n *such that* s_{σ_1}, s_{σ_2} *and* $s_{\sigma_1\sigma_2}$ *are dual simple braids. If both* S_{σ_1} *and* S_{σ_2} *relabel and rescale, then* $S_{\sigma_1\sigma_2}$ *relabels and rescales.*

Proof One consequence of the fact that s_{σ_1}, s_{σ_2} and $s_{\sigma_1\sigma_2}$ are dual simple braids is that there are noncrossing permutations σ_3, σ_4 and σ_5 such that $\delta = \sigma_1\sigma_2\sigma_3 = \sigma_1\sigma_4\sigma_2 = \sigma_5\sigma_1\sigma_2$ (Definition 11). For $S_{\sigma_1\sigma_2} = S_{\sigma_1}S_{\sigma_2}$ we have the following equalities.

$$\begin{aligned} S_{\sigma_1}S_{\sigma_2} &= (P_{\sigma_1}R^{\sigma_1}_{\sigma_4\sigma_2})(R^{\sigma_2}_{\sigma_1\sigma_4}P_{\sigma_2}) \\ &= P_{\sigma_1}R^{\sigma_1\sigma_2}_{\sigma_4}P_{\sigma_2} \\ &= P_{\sigma_1}P_{\sigma_2}R^{\sigma_1\sigma_2}_{\sigma_3} \\ &= R^{\sigma_1\sigma_2}_{\sigma_5}P_{\sigma_1}P_{\sigma_2}. \end{aligned}$$

The first line uses the hypotheses on S_{σ_1} and S_{σ_2}. The second combines the two edge scalings into a single rescaling. In particular, the rescaling $R^{\sigma_1}_{\sigma_4\sigma_2}$ rescales the edges in σ_1 by q and fixes the edges in σ_4 and σ_2, as well as the rest of the edges in the product $\sigma_4\sigma_2$. Similarly, the rescaling $R^{\sigma_2}_{\sigma_1\sigma_4}$ fixes the edges in σ_1 and σ_4 and rescales those in σ_2 by a factor of q. Thus, in the product of these two edge rescalings, the edges in σ_1 and σ_2 are rescaled by q and those in σ_4 are fixed. The third and fourth lines simply conjugate the points involved. Since $P_{\sigma_1}P_{\sigma_2} = P_{\sigma_1\sigma_2}$ and σ_5 and σ_3 are the left/right complements of $\sigma_1\sigma_2$, this completes the proof. □

Note that the same equalities used in the proof, slightly rearranged, would show that if any two of S_{σ_1}, S_{σ_2} and $S_{\sigma_1\sigma_2}$ relabel and rescale then so does the third. Propositions 6 and 7 are not quite enough to prove our second main result because not all dual simple braids are products of standard generators. We need to extend Proposition 6 to the full set of dual generators.

Proposition 8 (Dual Generators) *For every dual generator* $s_{ij} \in \text{GEN}_n$ *of the braid group* BRAID_n*, the corresponding matrix* S_{ij} *in the simplicial representation relabels and rescales.*

Proof An explicit description of the matrix for s_{ij} under the LKB representation is given in Krammer's earlier paper [10]. If we set $t = 1$ in that description, we find that the matrix S_{ij} has the exact same description as it does when $j = i + 1$ as given in Definition 19 except that a new case must be added that gives the result of $(e_{kl})S_{ij}$ when e_{ij} and e_{kl} cross. This simplification of the answer listed in [10] agrees with the corresponding row of $P_{ij}R^{ij}_{\mathrm{rc}(ij)}$, which is the same as the corresponding row of $R^{ij}_{\mathrm{lc}(ij)}P_{ij}$ obtained by combining Example 3 and Definition 20. □

Our second main result is an immediate corollary.

Theorem 4 (Dual Simple Braids Relabel and Rescale) *Under the simplicial representation of the braid group, each dual simple braid relabels and rescales. Concretely, for each* $\sigma \in NC_n$, *we have* $S_\sigma = P_\sigma R^\sigma_{\mathrm{rc}(\sigma)} = R^\sigma_{\mathrm{lc}(\sigma)}P_\sigma$.

Proof Proposition 8 shows that the assertion is true for each dual generator. Then Proposition 7 and a simple induction extends this fact to all of the dual simple braids. □

4 Final Remarks

In this final section we make a few remarks about the origins of these results and some promising directions for further investigation.

To explain how we stumbled upon this point of view, we first need to review the differences between the three papers by Krammer and Bigelow establishing the linearity of the braid groups. In his earlier article Daan Krammer used a ping-pong type argument on the dual Garside structure of the 4 string braid group to establish that the LKB representation is faithful for suitably generic values of q and t [10]. Next Stephen Bigelow replaced Krammer's algebraic approach with a more topological one and succeeded in showing that these representations are faithful for all n [3]. In his later article Krammer used an alternative version of his original approach that also succeeded in establishing linearity for every n [11]. The main difference between the two papers by Krammer is that the first uses the dual Garside presentation of the braid groups (indexed by the q variable and closely related to the Birman–Ko–Lee presentation introduced in [5]) while the second reverts to the standard presentation of the braid group (indexed by the t variable).

Shortly after these articles appeared, their results were extended to prove linearity results for various other Artin groups, all based more or less on the approach used in Krammer's second article [11]. François Digne extended linearity to the Artin groups of crystallographic type [8] and Arjeh Cohen and David Wales proved linearity for all spherical Artin groups [7]. Finally Luis Paris generalized these results further by proving that the Artin *monoid* is linear for all Artin groups [14]. In fact, all of these proofs establish the linearity of the positive monoid. When the Artin group is spherical, the Artin group is the group of fractions of the positive monoid and thus linearity of the positive monoid implies linearity for the group.

For general Artin groups it is known that the positive monoid generated by the standard minimal generating set is not large enough for this implication to hold. On the other hand, the positive monoid generated by the larger dual generating set might have this property. This led us to attempt to generalize Krammer's initial argument using the dual generating set. If one could find a proof of linearity for all the braid groups focused on the q variable as in [10], then there is a chance that it might generalize as in the papers by Digne, Cohen and Wales and Paris to eventually produce linearity results for non-spherical Artin groups. This was our initial motivation.

We first wrote code in `sage` to investigate the properties of the LKB matrices and found, experimentally, nice ways to decompose them and we began to isolate the changes that each factor was making. The interpretation of the simplified version as modifications of edge norms of simplices was one of the final steps in our evolving understanding of these representations.

We conclude this article with three directions for additional research.

Problem 1 (Linearity) The simplified simplicial representation is known not to be faithful for large n because it is also known as the symmetric tensor square of the Burau representation (which is known to not be faithful for $n \geq 5$ [2]). Thus, establishing a new q based proof of braid group linearity requires the variable t to remain generic. One project is to find a way to extend the geometric interpretation given here so that the t variable can remain generic and then to use this extended interpretation to give an alternative proof of braid group linearity focused on the q variable. In particular, one should try to show directly that the dual positive monoid generated by full dual generating set acts faithfully using a ping-pong type argument similar to the one in the original Krammer article on BRAID_4 [10].

Problem 2 (Dual Garside Length) The q variable in the LKB representation has received very little attention since Krammer's earlier paper primarily because it was the t variable which was the focus for the more general proof. Recently, however, Tetsuya Ito and Bert Wiest posted an article that proves one of the facts originally conjectured by Krammer in [10], namely, that the highest power of q in the LKB representation of a dual positive braid is twice its dual Garside length [9]. It seems quite likely that one could give an alternative and elementary proof of their results using the geometric understanding of the q variable introduced in this article.

Problem 3 (Spherical Artin Groups) The set of labeled euclidean simplices, with dilated simplices identified, is one of the standard parameterizations of the higher rank symmetric space $SL(V)/SO(V)$ and the simplicial representation appears to act on this space by isometries. Once this action is made explicit, it should be possible to define a similar construction and to give a similar interpretation for all of the spherical Artin groups once the focus on labeled euclidean simplices is replaced by linear transformations of root systems.

References

1. D. Armstrong, Generalized noncrossing partitions and combinatorics of Coxeter groups. Mem. Am. Math. Soc. **202**(949), x+159 (2009)
2. S. Bigelow, The Burau representation is not faithful for $n = 5$. Geom. Topol. **3**, 397–404 (1999)
3. S.J. Bigelow, Braid groups are linear. J. Am. Math. Soc. **14**(2), 471–486 (2001)
4. S. Bigelow, The Lawrence–Krammer representation, in *Topology and Geometry of Manifolds (Athens, GA, 2001)*. Proceedings of Symposia in Pure Mathematics, vol. 71 (American Mathematical Society, Providence, RI, 2003), pp. 51–68.
5. J. Birman, K.H. Ko, S.J. Lee, A new approach to the word and conjugacy problems in the braid groups. Adv. Math. **139**(2), 322–353 (1998)
6. N. Brady, J. McCammond, Factoring Euclidean isometries. Int. J. Algebra Comput. **25**(1–2), 325–347 (2015). doi:10.1142/S0218196715400135
7. A.M. Cohen, D.B. Wales, Linearity of Artin groups of finite type. Isr. J. Math. **131**, 101–123 (2002)
8. F. Digne, On the linearity of Artin braid groups. J. Algebra **268**(1), 39–57 (2003)
9. T. Ito, B. Wiest, Lawrence–Krammer–Bigelow representations and dual Garside length of braids. Geom. Topol. **19**(3), 1361–1381 (2015). doi:10.2140/gt.2015.19.1361
10. D. Krammer, The braid group B_4 is linear. Invent. Math. **142**(3), 451–486 (2000)
11. D. Krammer, Braid groups are linear. Ann. Math. (2) **155**(1), 131–156 (2002)
12. D. Margalit, J. McCammond, Geometric presentations for the pure braid group. J. Knot Theor. Ramif. **18**(1), 1–20 (2009)
13. J. McCammond, Noncrossing partitions in surprising locations. Am. Math. Mon. **113**(7), 598–610 (2006)
14. L. Paris, Artin monoids inject in their groups. Comment. Math. Helv. **77**(3), 609–637 (2002)
15. R. Simion, D. Ullman, On the structure of the lattice of noncrossing partitions. Discrete Math. **98**(3), 193–206 (1991)

The Cohomology of $\mathcal{M}_{0,n}$ as an FI-Module

Rita Jiménez Rolland

Abstract In this paper we revisit the cohomology groups of the moduli space of n-pointed curves of genus zero using the FI-module perspective introduced by Church–Ellenberg–Farb. We recover known results about the corresponding representations of the symmetric group.

Keywords Cohomology • Configuration spaces • FI-modules • Moduli space

1 Introduction

Our space of interest is $\mathcal{M}_{0,n}$, the *moduli space of n-pointed curves of genus zero*. It is defined as the quotient

$$\mathcal{M}_{0,n} := \mathcal{F}\left(\mathbb{P}^1(\mathbb{C}), n\right) / \operatorname{Aut}\left(\mathbb{P}^1(\mathbb{C})\right),$$

where $\mathcal{F}(\mathbb{P}^1(\mathbb{C}), n)$ is the configuration space of n-ordered points in the projective line $\mathbb{P}^1(\mathbb{C})$ and the automorphism group of the projective line $\operatorname{Aut}\left(\mathbb{P}^1(\mathbb{C})\right) = \mathrm{PGL}_2(\mathbb{C})$ acts componentwise on $\mathcal{F}(\mathbb{P}^1(\mathbb{C}), n)$. For $n \geq 3$, $\mathcal{M}_{0,n}$ is a *fine moduli space* for the problem of classifying smooth n-pointed rational curves up to isomorphism [16, Proposition 1.1.2].

The space $\mathcal{M}_{0,n}$ carries a natural action of the symmetric group S_n. The cohomology ring of $\mathcal{M}_{0,n}$ is known and the S_n-representations $H^i(\mathcal{M}_{0,n}; \mathbb{C})$ are well-understood (see for example [9, 11, 15]).

In this paper, we will consider the sequence of S_n-representations $H^i(\mathcal{M}_{0,n}; \mathbb{C})$ as a single object, an *FI-module over* $\mathbb{C}$. Via this example, we introduce the basics of the FI-module theory developed by Church, Ellenberg and Farb in [4]. We then use a well-known description of the cohomology ring of $\mathcal{M}_{0,n}$ to show in Theorem 5 that a finite generation property is satisfied which allows us to recover information about the S_n-representations in Theorem 6. Specifically, we obtain a stability result concerning the decomposition of $H^i(\mathcal{M}_{0,n}; \mathbb{C})$ into irreducible S_n-

R. Jiménez Rolland (✉)
Centro de Ciencias Matemáticas, Universidad Nacional Autónoma de México, Morelia, Michoacán 58089, Mexico
e-mail: rita@matmor.unam.mx

F. Callegaro et al. (eds.), *Configuration Spaces*, Springer INdAM Series 14,
DOI 10.1007/978-3-319-31580-5_13

representations, we exhibit a bound on the lengths of the representations and show that their characters have a highly constrained "polynomial" form.

2 The Co-FI-Spaces $\mathcal{M}_{0,\bullet}$ and $\mathcal{M}_{0,\bullet+1}$

Let **FI** be the category whose objects are natural numbers **n** and whose morphisms $\mathbf{m} \to \mathbf{n}$ are injections from $[m] := \{1, \ldots, m\}$ to $[n] := \{1, \ldots, n\}$.

We are interested in the *co-FI-space* $\mathcal{M}_{0,\bullet}$: the functor from $\mathbf{FI}^{op}$ to the category **Top** of topological spaces given by given by $\mathbf{n} \mapsto \mathcal{M}_{0,n}$ that assigns to $f : [m] \hookrightarrow [n]$ in $\mathrm{Hom}_{\mathrm{FI}}(\mathbf{m}, \mathbf{n})$ the morphism $f^* : \mathcal{M}_{0,n} \to \mathcal{M}_{0,m}$ defined by $f^*\big([(p_1, p_2, \ldots, p_n)]\big) = [(p_{f(1)}, p_{f(2)}, \ldots, p_{f(m)})]$. This is a particular case of the co-FI-space $\mathcal{M}_{g,\bullet}$ considered in [12] which is the functor given by $\mathbf{n} \mapsto \mathcal{M}_{g,n}$, the moduli space of Riemann surfaces of genus g with n marked points.

An *FI-module over* $\mathbb{C}$ is a functor V from **FI** to the category of $\mathbb{C}$-vector spaces $\mathbf{Vec}_{\mathbb{C}}$. Below, we denote $V(\mathbf{n})$ by V_n. Church, Ellenberg and Farb used FI-modules in [4] to encode sequences of S_n-representations in single algebraic objects and with this added structure significantly strengthened the representation stability theory introduced in [3]. FI-modules translate the representation stability property into a finite generation condition.

By composing the co-FI-space $\mathcal{M}_{0,\bullet}$ with the cohomology functor $H^i(-;\mathbb{C})$, we obtain the FI-module $H^i(\mathcal{M}_{0,\bullet}) := H^i(\mathcal{M}_{0,\bullet};\mathbb{C})$. We can also consider the graded version $H^*(\mathcal{M}_{0,\bullet}) := H^*(\mathcal{M}_{0,\bullet};\mathbb{C})$, we call this a *graded FI-module* over $\mathbb{C}$.

The co-FI-space $\mathcal{F}(\mathbb{C}, \bullet)$ given by $\mathbf{n} \mapsto \mathcal{F}(\mathbb{C}, n)$, the configuration space of n ordered points in $\mathbb{C}$, and the corresponding FI-modules $H^i\big(\mathcal{F}(\mathbb{C}, \bullet)\big)$ are key in our discussion below. In the expository paper [7], representation stability and FI-modules are motivated mainly through this example. A formal discussion of FI-modules and their properties is given in [4]. In [5] the theory of FI-modules is extended to modules over arbitrary Noetherian rings.

The "Shifted" Co-FI-Space $\mathcal{M}_{0,\bullet+1}$ Consider the functor Ξ_1 from **FI** to **FI** given by $[n] \mapsto [n] \sqcup \{0\}$. Notice that this functor induces the inclusion of groups

$$J_n : S_n = \mathrm{End}_{\mathrm{FI}}[\mathbf{n}] \hookrightarrow \mathrm{End}_{\mathrm{FI}}[\mathbf{n}+\mathbf{1}] = S_{n+1}$$

that sends the generator $(i\, i+1)$ of S_n to the transposition $(i+1\; i+2)$ of S_{n+1}. In our discussion below we are interested in the "shifted" co-FI-space $\mathcal{M}_{0,\bullet+1}$ obtained by $\mathcal{M}_{0,\bullet} \circ \Xi_1$. Notice that this co-FI-space is given by $\mathbf{n} \mapsto \mathcal{M}_{0,n+1}$. In the notation from [5, Sect. 2], this means that the FI-module

$$H^i(\mathcal{M}_{0,\bullet+1}) = S_{+1}\big(H^i(\mathcal{M}_{0,\bullet})\big),$$

where $S_{+1} : \mathbf{FI\text{-}Mod} \to \mathbf{FI\text{-}Mod}$ is the shift functor given by $S_{+1} := - \circ \Xi_1$. The functor S_{+1} performs the restriction, from an S_{n+1}-representation to an S_n-

representation, consistently for all n so that the resulting sequence of representations still has the structure of an FI-module. Comparing the S_n-representation $H^i(\mathcal{M}_{0,\bullet+1})_n$ with the S_{n+1}-representation $H^i(\mathcal{M}_{0,\bullet})_{n+1} = H^i(\mathcal{M}_{0,n+1})$ we have an isomorphism of S_n-representations

$$H^i(\mathcal{M}_{0,\bullet+1})_n \cong \mathrm{Res}_{S_n}^{S_{n+1}} H^i(\mathcal{M}_{0,n+1}).$$

3 Relation with the Configuration Space

We will understand the cohomology ring of $\mathcal{M}_{0,n}$ through its relation with the configuration space $\mathcal{F}(\mathbb{C}, n)$ of n ordered points in $\mathbb{C}$.

In our descriptions below, we consider $\mathbb{P}^1(\mathbb{C})$ with coordinates $[t : z]$ and the embedding $\mathbb{C} \hookrightarrow \mathbb{P}^1(\mathbb{C})$, given by $z \mapsto [1 : z]$ and let $[0 : 1] = \infty$. We use the brackets to indicate "equivalence class of". Since there is a unique element in $\mathrm{PGL}_2(\mathbb{C})$ that takes any three distinct points in $\mathbb{P}^1(\mathbb{C})$ to $\big([0 : 1], [1 : 0], [1 : 1]\big) = (\infty, 0, 1)$, every element in $\mathcal{M}_{0,n+1}$ can be written canonically as $\big[\big([0 : 1], [1 : 0], [1 : 1], [t_1 : z_1], \cdots, [t_{n-2} : z_{n-2}]\big)\big]$. Hence, $\mathcal{M}_{0,4} \cong \mathbb{P}^1(\mathbb{C})\backslash\{\infty, 0, 1\}$ and $\mathcal{M}_{0,n+1} \cong \mathcal{F}\big(\mathcal{M}_{0,4}, n-2\big)$.

Define the map $\psi : \mathcal{F}(\mathbb{C}, n) \longrightarrow \mathcal{M}_{0,n+1}$ by

$$\psi(z_1, z_2, \ldots, z_n) = \left[\left(\infty, 0, 1, \frac{z_3 - z_1}{z_2 - z_1}, \frac{z_4 - z_1}{z_2 - z_1}, \cdots, \frac{z_n - z_1}{z_2 - z_1}\right)\right].$$

The symmetric group S_n acts on $\mathcal{F}(\mathbb{C}, n)$ by permuting the coordinates. Let $(1\,2), (2\,3), \ldots (n-1\ n)$ be transpositions generating S_n and notice that

$$\begin{aligned}\psi\big((1\,2)\cdot(z_1, z_2, \ldots, z_n)\big) &= \left[\left(\infty, 0, 1, \tfrac{z_3-z_2}{z_1-z_2}, \tfrac{z_4-z_2}{z_1-z_2}, \cdots, \tfrac{z_n-z_2}{z_1-z_2}\right)\right] \\ = \left[\begin{bmatrix}1 & 0\\ 1 & -1\end{bmatrix}\cdot\left([0:1], [1:0], [1:1], [1:\tfrac{z_3-z_2}{z_1-z_2}], [1:\tfrac{z_4-z_2}{z_1-z_2}], \cdots, [1:\tfrac{z_n-z_2}{z_1-z_2}]\right)\right] \\ = \left[\left([0:1], [1:1], [1:0], [1:\tfrac{z_3-z_1}{z_2-z_1}], [1:\tfrac{z_4-z_1}{z_2-z_1}], \cdots, [1:\tfrac{z_n-z_1}{z_2-z_1}]\right)\right] \\ = (2\,3)\cdot\psi(z_1, z_2, \ldots, z_n)\end{aligned}$$

and in general

$$\psi\big((i\ i+1)\cdot(z_1, z_2, \ldots, z_n)\big) = (i+1\ i+2)\cdot\psi(z_1, z_2, \ldots, z_n) \ \text{ for } i \geq 2.$$

Therefore the map $\psi : \mathcal{F}(\mathbb{C}, n) \longrightarrow \mathcal{M}_{0,n+1}$ is equivariant with respect to the inclusion $J_n : S_n \hookrightarrow S_{n+1}$. In other words, $\psi : \mathcal{F}(\mathbb{C}, \bullet) \longrightarrow \mathcal{M}_{0,\bullet+1}$ is a map of co-FI-spaces.

Relation with the Coxeter Arrangement of Type A_{n-1} The *complement of the complexified Coxeter arrangement of hyperplanes type* A_{n-1} is $M(\mathcal{A}_{n-1})$, the image of $\mathcal{F}(\mathbb{C}, n)$ under the quotient map $\mathbb{C}^n \to \mathbb{C}^n/N$, where $N = \{(z_1, \ldots, z_n) \in \mathbb{C}^n : z_i = z_j \text{ for } 1 \leq i, j \leq n\}$.

As explained in [9], it turns out that the moduli space $\mathcal{M}_{0,n+1}$ is also in bijective correspondence with the projective arrangement

$$M(d\mathcal{A}_{n-1}) := \pi\big(M(\mathcal{A}_{n-1})\big) \cong M(\mathcal{A}_{n-1})/\mathbb{C}^*,$$

where $\pi : \mathbb{C}^{n-1}\backslash\{0\} \to \mathbb{P}^{n-2}(\mathbb{C})$ is the Hopf bundle projection, which takes $z \in \mathbb{C}^{n-1}\backslash\{0\}$ to λz for $\lambda \in \mathbb{C}^*$. Moreover, the map ψ factors through $M(\mathcal{A}_{n-1})$ and $M(d\mathcal{A}_{n-1})$.

$$\begin{array}{ccc}
\mathcal{F}(\mathbb{C}, n) & \longrightarrow & M(\mathcal{A}_{n-1}) \\
 & \searrow & \downarrow \pi \\
 & & M(d\mathcal{A}_{n-1}) \\
 & \psi & \downarrow \cong \\
 & & \mathcal{M}_{0,n+1}
\end{array}$$

In [9], Gaiffi extends the natural S_n-action on $H^*\big(M(\mathcal{A}_{n-1}); \mathbb{C}\big)$ to an S_{n+1}-action using the vertical map in the diagram above and the natural S_{n+1}-action on $H^*\big(\mathcal{M}_{0,n+1}; \mathbb{C}\big)$.

The Cohomology Rings As proved in [9, Proposition 2.2 and Theorem 3.2] , the map ψ allows us to relate the cohomology rings of $\mathcal{M}_{0,n+1}$ and $\mathcal{F}(\mathbb{C}; n)$. See also [11, Corollary 3.1].

Proposition 1 *The maps ψ induces an isomorphism of cohomology rings*

$$H^*\big(\mathcal{F}(\mathbb{C}, n); \mathbb{C}\big) \cong H^*\big(\mathcal{M}_{0,n+1}; \mathbb{C}\big) \otimes H^*(\mathbb{C}^*; \mathbb{C})$$

as S_n-modules. The symmetric group S_n acts trivially on $H^*(\mathbb{C}^*; \mathbb{C})$ and acts on $H^*\big(\mathcal{M}_{0,n+1}; \mathbb{C}\big)$ through the inclusion $J_n : S_n \hookrightarrow S_{n+1}$ that sends the generator $(i\; i+1)$ of S_n to the transposition $(i+1\; i+2)$ of S_{n+1}.

This means that the map of co-FI-spaces $\psi : \mathcal{F}(\mathbb{C}, \bullet) \longrightarrow \mathcal{M}_{0,\bullet+1}$ induces an isomorphism of graded FI-modules

$$H^*\big(\mathcal{F}(\mathbb{C}, \bullet)\big) \cong H^*\big(\mathcal{M}_{0,\bullet+1}\big) \otimes H^*(\mathbb{C}^*),$$

where $H^*(\mathbb{C}^*)$ is the trivial graded FI-module given by $\mathbf{n} \mapsto H^*(\mathbb{C}^*; \mathbb{C})$.

Furthermore, Arnol'd obtained a presentation of the cohomology ring of $\mathcal{F}(\mathbb{C}, n)$ in [1].

Theorem 1 *The cohomology ring $H^*(\mathcal{F}(\mathbb{C},n);\mathbb{C})$ is isomorphic to the $\mathbb{C}$-algebra $\mathcal{R}_n$ generated by 1 and forms $\omega_{i,j} := \frac{d\log(z_j-z_i)}{2\pi i}$, $1 \le i \ne j \le n$, with relations $\omega_{i,j} = \omega_{j,i}$, $\omega_{i,j}\omega_{k,l} = -\omega_{k,l}\omega_{i,j}$ and $\omega_{i,j}\omega_{j,k} + \omega_{j,k}\omega_{k,i} + \omega_{k,i}\omega_{i,j} = 0$. The action S_n is given by $\sigma \cdot \omega_{i,j} = \omega_{\sigma(i)\sigma(j)}$ for $\sigma \in S_n$.*

As a consequence of the isomorphism in Proposition 1, we also have a concrete description of the cohomology ring $H^*(\mathcal{M}_{0,n+1};\mathbb{C})$. We refer the reader to [11, Corollary 3.1], [9, Theorem 3.4] and references therein.

Theorem 2 *The cohomology ring $H^*(\mathcal{M}_{0,n+1};\mathbb{C})$ is isomorphic to the subalgebra of $\mathcal{R}_n$ generated by 1 and elements $\theta_{i,j} := \omega_{i,j} - \omega_{12}$ for $\{i,j\} \ne \{1,2\}$. The S_n-action given by $\sigma \cdot \theta_{i,j} = \theta_{\sigma(i),\sigma(j)} - \theta_{\sigma(1),\sigma(2)}$, for $\sigma \in S_n$.*

4 Finite Generation

We can use the explicit presentations in Theorems 1 and 2 to understand the FI-modules $H^i(\mathcal{F}(\mathbb{C},\bullet))$ and $H^i(\mathcal{M}_{0,\bullet+1})$.

An FI-module V over $\mathbb{C}$ is said to be *finitely generated in degree $\le m$* if there exist $v_1,\dots,v_s$, with each $v_i \in V_{n_i}$ and $n_i \le m$, such that V is the minimal sub-FI-module of V containing $v_1,\dots,v_s$. Finitely generated FI-modules have strong closure properties: extensions and quotients of finitely generated FI-modules are still finitely generated and finite generation is preserved when taking sub-FI-modules.

Notice that from Theorem 1 it follows that $H^1(\mathcal{F}(\mathbb{C},n);\mathbb{C})$ is generated as an S_n-module by the class $\omega_{1,2}$. Therefore, the FI-module $H^1(\mathcal{F}(\mathbb{C},\bullet))$ is finitely generated in degree 2 by the class $\omega_{1,2}$ in $H^1(\mathcal{F}(\mathbb{C},2))$.

Similarly, from Theorem 2 we know that $H^1(\mathcal{M}_{0,n+1})$ is generated by the $\theta_{i,j}$ classes and notice that $\theta_{1,j} = (j\ 3)\cdot\theta_{1,3}$ for $j \ne \{1,2\}$; $\theta_{2,j} = (j\ 3)\cdot\theta_{2,3}$ for $j \ne \{1,2\}$ and $\theta_{i,j} = (i\ 3)(j\ 4)\cdot\theta_{3,4}$ for $\{i,j\} \ne \{1,2\}$. Therefore, $H^1(\mathcal{M}_{0,n+1})$ is generated by $\theta_{1,3}$, $\theta_{2,3}$ and $\theta_{3,4}$ as an S_n-module. This means that the FI-module $H^1(\mathcal{M}_{0,\bullet+1})$ is finitely generated by the classes $\theta_{1,3}$, $\theta_{2,3}$ and $\theta_{3,4}$ in $H^1(\mathcal{M}_{0,\bullet+1})_4$, hence in degree 4.

An FI-module V encodes the information of the sequence V_n of S_n-representations. Finite generation of V puts strong constraints on the decomposition of each V_n into irreducible representations and its character.

Notation for Representations of S_n The irreducible representations of S_n over $\mathbb{C}$ are classified by partitions of n. A partition λ of n is a set of positive integers $\lambda_1 \ge \dots \ge \lambda_l > 0$ where $l \in \mathbb{Z}$ and $\lambda_1 + \dots + \lambda_l = n$. We write $|\lambda| = n$. The corresponding irreducible S_n-representation will be denoted by V_λ. Every V_λ is defined over $\mathbb{C}$ and any S_n-representation decomposes over $\mathbb{C}$ into a direct sum of irreducibles.

If λ is any partition of m, i.e. $|\lambda| = m$, then for any $n \ge |\lambda| + \lambda_1$ the *padded partition* $\lambda[n]$ of n is given by $n - |\lambda| \ge \lambda_1 \ge \dots \ge \lambda_l > 0$. Keeping the notation from [3], we set $V(\lambda)_n = V_{\lambda[n]}$ for any $n \ge |\lambda| + \lambda_1$.

Every irreducible S_n-representation is of the form $V(\lambda)_n$ for a unique partition λ. We define the *length* of an irreducible representation of S_n to be the number of parts in the corresponding partition of n. The trivial representation has length 1, and the alternating representation has length n. We define the *length* $\ell(V)$ of a finite dimensional representation V of S_n to be the maximum of the lengths of the irreducible constituents.

We say that an FI-module V over $\mathbb{C}$ has *weight* $\leq d$ if for every $n \geq 0$ and every irreducible constituent $V(\lambda)_n$ we have $|\lambda| \leq d$. The degree of generation of an FI-module V gives an upper bound for the weight [4, Proposition 3.2.5]. The weight of an FI-module is closed under subquotients and extensions. Moreover, if a finitely generated FI-module V has weight $\leq d$, by definition, $\ell(V_n) \leq d+1$ for all n and the alternating representation cannot not appear in the decomposition into irreducibles of V_n once $n > d+1$.

Notice that the FI-module $H^1\big(\mathcal{F}(\mathbb{C}, \bullet)\big)$ has weight at most 2 and so does $H^1(\mathcal{M}_{0,\bullet+1})$, since it is a sub-FI-module of $H^1\big(\mathcal{F}(\mathbb{C}, \bullet)\big)$.

An FI-module V has *stability degree* $\leq N$, if for every $a \geq 0$ and $n \geq N + a$ the map of coinvariants

$$(I_n)_* : (V_n)_{S_{n-a}} \to (V_{n+1})_{S_{(n+1)-a}} \tag{1}$$

induced by the standard inclusion $I_n : \{1, \ldots n\} \hookrightarrow \{1, \ldots, n, n+1\}$, is an isomorphism of S_a-modules (see [4, Definition 3.1.3] for a more general definition). Here, S_{n-a} is the subgroup of S_n that permutes $\{a+1, \ldots, n\}$ and acts trivially on $\{1, 2, \ldots, a\}$. The coinvariant quotient $(V_n)_{S_{n-a}}$ is the S_a-module $V_n \otimes_{\mathbb{C}[S_{n-a}]} \mathbb{C}$, the largest quotient of V_n on which S_{n-a} acts trivially.

The finite generation properties of the FI-modules $H^i\big(\mathcal{F}(\mathbb{C}, \bullet)\big)$ have already been discussed in [4, Example 5.1.A].

Proposition 2 *The FI-module $H^i\big(\mathcal{F}(\bullet; \mathbb{C})\big)$ is finitely generated with weight $\leq 2i$ and has stability degree $\leq 2i$*

Proof From Theorem 1 the graded FI-module $H^*\big(\mathcal{F}(\mathbb{C}, \bullet)\big)$ is generated by the FI-module $H^1\big(\mathcal{F}(\mathbb{C}, \bullet)\big)$ that has weight ≤ 2. It follows by [4, Theorem 4.2.3] that $H^i\big(\mathcal{F}(\mathbb{C}, \bullet)\big)$ is finitely generated with weight $\leq 2i$. Moreover, in [4] it is shown that $H^i\big(\mathcal{F}(\mathbb{C}, \bullet)\big)$ has the additional structure of what [4] calls an FI#-module, which implies that it has stability degree bounded above by the weight (see proof of [4, Corollary 4.1.8]). □

Finite generation for the FI-modules $H^i(\mathcal{M}_{0,\bullet+1})$ follows from Theorem 2 and Proposition 2.

Theorem 3 *The FI-module $H^i(\mathcal{M}_{0,\bullet+1})$ is finitely generated in degree $\leq 4i$, with weight $\leq 2i$ and has stability degree $\leq 2i$.*

Proof By Theorem 2 the graded FI-module $H^*(\mathcal{M}_{0,\bullet+1})$ is generated by the FI-module $H^1(\mathcal{M}_{0,\bullet+1})$, which is finitely generated in degree ≤ 4 and has weight $\leq$

2. It follows from [4, Proposition 2.3.6] that the FI-module $H^i(\mathcal{M}_{0,\bullet+1})$ is finitely generated in degree $\leq 4i$. By [4, Corollary 4.2.A] it has weight $\leq 2i$.

Moreover, from [4, Lemma 3.1.6] we have that the stability degree of $H^i(\mathcal{M}_{0,\bullet+1})$ is bounded above by the stability degree of $H^i(\mathcal{F}(\mathbb{C},\bullet))$. □

From $H^i(\mathcal{M}_{0,\bullet+1})$ to the FI-Module $H^i(\mathcal{M}_{0,\bullet})$ The relation between the degree of generation of an FI-module V and its "shift" $S_{+1}V$ was established in [5, Corollary 2.13]. We can also relate the weights and stability degrees using the classical branching rule (see e.g. [8]).

Proposition 3 *Let λ be a partition of $n+1$ and V_λ the corresponding irreducible S_{n+1}-representation, then as S_n-representations we have the decomposition*

$$Res_{S_n}^{S_n+1} V_\lambda \cong \bigoplus_\nu V_\nu$$

over those partitions ν of n obtained from λ by removing one box from one of the columns of the corresponding Young diagram.

Theorem 4 (Finite Generation and "Shifted" FI-Modules) *Let V be a finitely generated FI-module generated in degree $\leq d$, then $S_{+1}V$ is finitely generated in degree $\leq d$. Conversely, if the FI-module $S_{+1}V$ is finitely generated in degree $\leq d$, then V is finitely generated in degree $\leq d+1$.*

Furthermore, if $S_{+1}V$ has weight $\leq M$ and stability degree $\leq N$, then V has weight $\leq M+1$ and stability degree $\leq N+1$. Conversely, if V has weight $\leq M$ and stability degree $\leq N$, then $S_{+1}V$ has weight $\leq M$ and stability degree $\leq N$.

Proof If V has weight $\leq M$, then for all $n \geq 0$, the irreducible components $V(\mu)_{n+1}$ of V_{n+1} have $|\mu| \leq M$. From Proposition 3 it follows that $\mathrm{Res}_{S_n}^{S_n+1} V(\mu)_{n+1}$ will be a direct sum of irreducibles $V(\lambda)_n$, with $|\lambda| \leq |\mu| \leq M$. Conversely, if $S_{+1}V$ has weight $\leq M$, then each irreducible component $V(\lambda)_n$ of $S_{+1}V_n$ has $|\lambda| \leq M$. By Proposition 3, it comes from the restriction of some $V(\mu)_{n+1}$ with $|\mu| \leq |\lambda|+1 \leq M+1$.

On the other hand, the functor Ξ_1 sends $\{1,\ldots,a\}$ into $\{2,\ldots,a+1\}$ and $\{a+1,\ldots,n\}$ into $\{a+2,\ldots,n+1\}$. Therefore, the inclusion $J_n : S_n \hookrightarrow S_{n+1}$ maps the subgroup S_{n-a} of S_n onto the subgroup $S_{(n+1)-(a+1)}$ of S_{n+1} and we have that

$$(V_{n+1})_{S_{(n+1)-(a+1)}} = V_{n+1} \otimes_{\mathbb{C}[S_{(n+1)-(a+1)}]} \mathbb{C} = S_{+1}(V)_n \otimes_{\mathbb{C}[S_{n-a}]} \mathbb{C} = \left(S_{+1}(V)_n\right)_{S_{n-a}},$$

which implies the statement about stability degrees. □

In [12] we proved finite generation for the FI-modules $H^i(\mathcal{M}_{g,\bullet})$ when $g \geq 2$. The case when $g=0$ follows from Theorems 4 and 3.

Theorem 5 *The FI-module $H^i(\mathcal{M}_{0,\bullet})$ is finitely generated with weight $\leq 2i+1$ and has stability degree $\leq 4i$.*

The First Cohomology Group Recall that $H^1(\mathcal{M}_{0,\bullet+1})_n$ is generated by the classes $\theta_{i,j} = \omega_{i,j} - \omega_{1,2}$ and it is a subrepresentation of $H^1(\mathcal{F}(\mathbb{C}, n))$ which has a basis given by the classes $\omega_{i,j}$. In particular, notice that $\dim H^1(\mathcal{M}_{0,\bullet+1})_n = \dim H^1(\mathcal{F}(\mathbb{C}, n)) - 1$. Moreover for $n \geq 4$, we have the decomposition

$$H^1(\mathcal{F}(\mathbb{C}, n)) = V(0)_n \oplus V(1)_n \oplus V(2)_n.$$

Then, for $n \geq 4$ the S_n-representation

$$H^1(\mathcal{M}_{0,\bullet+1})_n = V(1)_n \oplus V(2)_n \cong \mathrm{Res}_{S_n}^{S_{n+1}} H^1(\mathcal{M}_{0,n+1}).$$

Proposition 3 implies that for $n \geq 4$, we have that $H^1(\mathcal{M}_{0,n+1}) = V(2)_{n+1}$ as a representation of S_{n+1}. Moreover, notice that $H^1(\mathcal{M}_{0,n+1})$ is finitely generated by the classes $\theta_{1,3}$, $\theta_{2,3}$ and $\theta_{3,4}$ in $H^1(\mathcal{M}_{0,5})$ not only an S_n-module, but also as an S_{n+1}-module. Therefore, the FI-module $H^1(\mathcal{M}_{0,\bullet})$ is finitely generated in degree≤ 5 and has weight ≤ 2.

5 The S_n-Representations $H^i(\mathcal{M}_{0,n}; \mathbb{C})$

At this point we can apply the theory of FI-modules to the finitely generated FI-modules $H^i(\mathcal{M}_{0,\bullet})$ and $H^i(\mathcal{M}_{0,\bullet+1})$ to obtain information about the corresponding sequences of S_n-representations and their characters. The following result is a direct consequence from [4, Proposition 3.3.3 and Theorem 3.3.4] and Theorems 3 and 5.

Theorem 6 *Let $i \geq 0$. For $n \geq 4i + 2$, the sequence $\{H^i(\mathcal{M}_{0,n})\}$ of representations of S_n and the sequence $\{H^i(\mathcal{M}_{0,\bullet+1})_{n-1}\}$ of S_{n-1}-representations satisfy the following:*

(a) *The decomposition into irreducibles of $H^i(\mathcal{M}_{0,n}; \mathbb{C})$ and of $H^i(\mathcal{M}_{0,\bullet+1}; \mathbb{C})_{n-1}$ stabilize in the sense of uniform representation stability [3] with stable range $n \geq 4i + 2$.*
(b) *The length of $H^i(\mathcal{M}_{0,\bullet+1}; \mathbb{C})_{n-1}$ is bounded above by $2i$ and the length of $H^i(\mathcal{M}_{0,n}; \mathbb{C})$ is bounded above by $2i + 1$.*
(c) *The sequence of characters of the representations $H^i(\mathcal{M}_{0,\bullet+1}; \mathbb{C})_{n-1}$ and $H^i(\mathcal{M}_{0,n}; \mathbb{C})$ are eventually polynomial, in the sense that there exist character polynomials $P_i(X_1, X_2, \ldots, X_r)$ and $Q_i(X_1, X_2, \ldots, X_s)$ in the cycle-counting functions $X_k(\sigma)$:=(number of k-cycles in σ) such that for all $n \geq 4i + 2$:*

$$\chi_{H^i(\mathcal{M}_{0,\bullet+1};\mathbb{C})_{n-1}}(\sigma) = P_i(X_1, X_2, \ldots, X_r)(\sigma) \quad \textit{for all} \quad \sigma \in S_{n-1}, \quad \textit{and}$$

$$\chi_{H^i(\mathcal{M}_{0,n};\mathbb{C})}(\sigma) = Q_i(X_1, X_2, \ldots, X_s)(\sigma) \quad \textit{for all} \quad \sigma \in S_n.$$

Moreover, the degree of P_i is $\leq 2i$ and the degree of Q_i is $\leq 2i + 1$, where we take $\deg X_k = k$. In particular, $r \leq 2i$ and $s \leq 2i + 1$.

If $e \in S_{n-1}$ is the identity element, from Theorem 6(c), we obtain that the dimensions

$$\dim_{\mathbb{C}}\big(H^i(\mathcal{M}_{0,n};\mathbb{C})\big)=\chi_{H^i(\mathcal{M}_{0,\bullet+1};\mathbb{C})_{n-1}}(e) = P_i(X_1(e),\dots,X_r(e)) = P_i(n-1,\dots,0)$$

are polynomials in n of degree $\leq 2i$. This agrees with the known Poincaré polynomial of $\mathcal{M}_{0,n}$ (see[15, Corollary 2.10] and also [11, 5.5(8)]).

From Theorem 5 and the definition of weight, we recover the fact that the alternating representation does not appear in the cohomology of $\mathcal{M}_{0,n}$ [15, Proposition 2.16].

Theorem 6(a) implies that the dimensions of the vector spaces $H^i(\mathcal{M}_{0,n}/S_n;\mathbb{C})$ and $H^i(\mathcal{M}_{0,n+1}/S_n;\mathbb{C})$ are constant. For the sequence $\{\mathcal{M}_{0,n}/S_n\}$, this is actually a trivial consequence from the fact that $\mathcal{M}_{0,n}/S_n$ has the cohomology of a point as shown in [15, Theorem 2.3].

Recursive Relation for Characters In [9, Theorem 4.1], Gaiffi obtained a recursive formula that connects the characters of the S_n-representations $H^*(\mathcal{M}_{0,\bullet+1})_n$ and $H^*(\mathcal{M}_{0,n})$ as follows

$$\chi_{H^i(\mathcal{M}_{0,\bullet+1})_n} = \chi_{H^i(\mathcal{M}_{0,n})} + \big(X_1 - 1\big)\cdot \chi_{H^{i-1}(\mathcal{M}_{0,n})} \quad \text{for } n \geq 3. \tag{2}$$

In particular, we know that $\chi_{H^1(\mathcal{M}_{0,\bullet+1})_n} = \chi_{H^1(\mathcal{F}(n,\mathbb{C}))} - 1 = \binom{X_1}{2} + X_2 - 1$ when $n \geq 4$. Therefore, for $i = 1$, the recursive formula (2) gives us the character polynomial of degree 2

$$\chi_{H^1(\mathcal{M}_{0,n})} = \chi_{H^1(\mathcal{M}_{0,\bullet+1})_n} - \big(X_1 - 1\big)\cdot\chi_{H^0(\mathcal{M}_{0,n})} = \binom{X_1}{2} + X_2 - X_1 = \chi_{V(2)}$$

as expected since $H^1(\mathcal{M}_{0,n}) = V(2)_n$.

Furthermore, if P_i and Q_i are the character polynomials of $H^i(\mathcal{M}_{0,\bullet+1})_n$ and $H^i(\mathcal{M}_{0,n})$ from Theorem 6 (c) for $n \geq 4i+2$, then formula (2) can be written as $Q_i = P_i - (X_1 - 1)\cdot Q_{i-1}$ and $\deg Q_i \leq max\big(\deg P_i, 1+\deg Q_{i-1}\big) \leq 2i$. As a consequence of this and Theorem 6 (c) we have that, for $n \geq 4i+2$, the values of $\chi_{H^i(\mathcal{M}_{0,\bullet+1};\mathbb{C})_n}(\sigma)$ and $\chi_{H^i(\mathcal{M}_{0,n};\mathbb{C})}(\sigma)$ depend only on "short cycles", i.e. cycles on σ of length $\leq 2i$.

More is Known About the S_n-Representations In this paper we were mainly interested in highlighting the methods, since more precise information about the characters of the S_n-representations is known. The moduli space $\mathcal{M}_{0,n}$ can be represented by a finite type $\mathbb{Z}$-scheme and the manifold $\mathcal{M}_{0,n}(\mathbb{C})$ of $\mathbb{C}$-points of this scheme corresponds to the definition in Sect. 1. In [15] Kisin and Lehrer used an equivariant comparison theorem in ℓ-adic cohomology and the Grothendieck-Lefschetz's fixed point formula to obtain explicit descriptions of the graded character of the S_n-action on the cohomology of $\mathcal{M}_{0,n}(\mathbb{C})$ via counts of number of points of varieties over finite fields. With their techniques they obtain the Poincaré polynomial of a permutation in S_n of a specific cycle type acting on $H^*(\mathcal{M}_{0,n};\mathbb{C})$ [15, Theorem 2.9] and a description of the top cohomology $H^{n-3}(\mathcal{M}_{0,n};\mathbb{C})$ [15, Proposition 2.18]. Furthermore, Getzler uses the language of operads in [11] to obtain formulas for the characters of the S_n-modules $H^i(\mathcal{M}_{0,n};\mathbb{C})$.

The Cohomology of $\overline{\mathcal{M}}_{0,n}$ A related space of interest is $\overline{\mathcal{M}}_{0,n}$, the *Deligne-Mumford compactification* of $\mathcal{M}_{0,n}$. It is a *fine moduli space* for stable n-pointed rational curves for $n \geq 3$ (see [16, Chap. 1] and reference therein). It can also be constructed from $M(d\mathcal{A}_{n-1})$ using the theory of wonderful models of hyperplanes arrangements developed by De Concini and Procesi (see for example [10, Chap. 2]). The space $\overline{\mathcal{M}}_{0,n}$ also carries a natural action of the symmetric group S_n. Hence, a natural question to ask is whether the FI-module theory could tell us something about its cohomology groups as S_n-representations.

Explicit presentations of the cohomology ring of the manifold of complex points $\overline{\mathcal{M}}_{0,n}(\mathbb{C})$ have been obtained by Keel [14] and Yuzvinsky [19]. Moreover, several recursive and generating formulas for the Poincaré polynomials have been computed (for instance see [2, 11, 17, 19]). The sequence $H^i(\overline{\mathcal{M}}_{0,n}(\mathbb{C});\mathbb{C})$ has the structure of an FI-module, however, the Betti numbers of $\overline{\mathcal{M}}_{0,n}(\mathbb{C})$ grow exponentially in n, which precludes finite generation. Therefore an analogue of Theorem 6 cannot be obtained for this space.

On the other hand, as observed in [6], the manifold $\overline{\mathcal{M}}_{0,n}(\mathbb{R})$ of real points of $\overline{\mathcal{M}}_{0,n}$ is topologically similar to $\mathcal{F}(\mathbb{C}, n-1)$, the configuration space of $n-1$ ordered points in $\mathbb{C}$, in the sense that both are $K(\pi, 1)$-spaces, have Poincaré polynomials with a simple factorization and Betti numbers that grow polynomially in n. The cohomology ring of the real locus $\overline{\mathcal{M}}_{0,n}(\mathbb{R})$ was completely determined in [6] and an explicit formula for the graded character of the S_n-action was obtained in [18]. The presentation of the cohomology ring given in [6] can be used to prove finite generation for the FI-modules $H^i(\overline{\mathcal{M}}_{0,n}(\mathbb{R});\mathbb{C})$ and to obtain an analogue of Theorem 6 for this space (see [13]).

Acknowledgements I would like to thank to Alex Suciu for pointing out a relevant reference and Jennifer Wilson for useful comments. I am grateful to the Department of Mathematics at Northeastern University for providing such appropriate working conditions that allowed the completion of this paper.

References

1. V.I. Arnol'd, On some topological invariants of algebraic functions. Trans. Moscow Math. Soc. **21**, 30–51 (1970)
2. F. Callegaro, G. Gaiffi, On models of the braid arrangement and their hidden symmetries. Int. Math. Res. Not. IMRN (21), 11117–11149 (2015)
3. T. Church, B. Farb, Representation theory and homological stability. Adv. Math. **245**, 250–314 (2013)
4. T. Church, J.S. Ellenberg, B. Farb, FI-modules and stability for representations of symmetric groups. Duke Math. J. **164**(9), 1833–1910 (2015)

5. T. Church, J.S. Ellenberg, B. Farb, R. Nagpal, FI-modules over Noetherian rings. Geom. Topol. **18**(5), 2951–2984 (2014)
6. P. Etingof, A. Henriques, J. Kamnitzer, E.M. Rains, The cohomology ring of the real locus of the moduli space of stable curves of genus 0 with marked points. Ann. Math. (2) **171**(2), 731–777 (2010)
7. B. Farb, Representation stability, in *Proceedings of the International Congress of Mathematicians*, Seoul 2014, vol. II (2014), pp. 1173–1196
8. W. Fulton, J. Harris, *Representation Theory: A First Course*. Graduate Texts in Mathematics, vol. 129 (Springer, New York, 1991). Readings in Mathematics
9. G. Gaiffi, The actions of S_{n+1} and S_n on the cohomology ring of a Coxeter arrangement of type A_{n-1}. Manuscripta Math. **91**(1), 83–94 (1996)
10. G. Gaiffi, De Concini Procesi models of arrangements and symmetric group actions. Tesi di Perfezionamento della Scuola Normale Superiore di Pisa (1999)
11. E. Getzler, Operads and moduli spaces of genus 0 Riemann surfaces, in *The Moduli Space of Curves (Texel Island, 1994)*. Progress in Mathematics, vol. 129 (Birkhäuser, Boston, MA, 1995), pp. 199–230
12. R. Jiménez Rolland, On the cohomology of pure mapping class groups as FI-modules. J. Homotopy Relat. Struct. **10**(3), 401–424 (2015)
13. R. Jiménez Rolland, J. Maya Duque, Representation stability for the pure cactus group. Preprint (2015). arXiv:1501.02835
14. S. Keel, Intersection theory of moduli space of stable n-pointed curves of genus zero. Trans. Am. Math. Soc. **330**(2), 545–574 (1992)
15. M. Kisin, G.I. Lehrer, Equivariant Poincaré polynomials and counting points over finite fields. J. Algebra **247**(2), 435–451 (2002)
16. J. Kock, I. Vainsencher, *An Invitation to Quantum Cohomology*. Progress in Mathematics, vol. 249 (Birkhäuser, Basel, 2007)
17. Y.I. Manin, Generating functions in algebraic geometry and sums over trees, in *The Moduli Space of Curves (Texel Island, 1994)*. Progress in Mathematics, vol. 129 (Birkhäuser, Boston, MA, 1995), pp. 401–417
18. E.M. Rains, The action of S_n on the cohomology of $\overline{M}_{0,n}(\mathbb{R})$. Selecta Math. (N.S.) **15**(1), 171–188 (2009)
19. S. Yuzvinsky, Cohomology bases for the De Concini-Procesi models of hyperplane arrangements and sums over trees. Invent. Math. **127**(2), 319–335 (1997)

Syzygies in Equivariant Cohomology for Non-abelian Lie Groups

Matthias Franz

Abstract We extend the work of Allday–Franz–Puppe on syzygies in equivariant cohomology from tori to arbitrary compact connected Lie groups G. In particular, we show that for a compact orientable G-manifold X the analogue of the Chang–Skjelbred sequence is exact if and only if the equivariant cohomology of X is reflexive, if and only if the equivariant Poincaré pairing for X is perfect. Along the way we establish that the equivariant cohomology modules arising from the orbit filtration of X are Cohen–Macaulay. We allow singular spaces and introduce a Cartan model for their equivariant cohomology. We also develop a criterion for the finiteness of the number of infinitesimal orbit types of a G-manifold.

Keywords Cohen–Macaulay module • Equivariant cohomology • Infinitesimal orbit type • Non-abelian Lie group • Syzygy

1 Introduction

Let R be a polynomial ring in r indeterminates, and let M be a finitely generated module over R. Then M is called a *jth syzygy* if there is an exact sequence

$$0 \to M \to F_1 \to \cdots \to F_j \tag{1}$$

with finitely generated free R-modules $F_1, \ldots, F_j$. The first syzygies are exactly the torsion-free modules, the second syzygies the reflexive ones, and the rth syzygies are free. In this sense, syzygies interpolate between torsion-freeness and freeness. Allday, Puppe and the author initiated the study of syzygies in the context of torus-equivariant cohomology [3, 4]. To illustrate their results, let us focus on the second syzygies. We use Alexander–Spanier cohomology with real coefficients.

The author was supported by an NSERC Discovery Grant.

M. Franz (✉)
Department of Mathematics, University of Western Ontario, London, ON, N6A 5B7 Canada
e-mail: mfranz@uwo.ca

F. Callegaro et al. (eds.), *Configuration Spaces*, Springer INdAM Series 14,
DOI 10.1007/978-3-319-31580-5_14

Let $T \cong (S^1)^r$ be a torus, and let X be a T-space with finite Betti sum and satisfying some other mild assumptions. Then $R_T = H^*(BT)$ is a polynomial ring in r indeterminates of degree 2, and the equivariant cohomology $H_T^*(X)$ of X is an R_T-module. Allday–Franz–Puppe showed that the Chang–Skjelbred sequence

$$0 \to H_T^*(X) \to H_T^*(X^T) \to H_T^{*+1}(X_{1,T}, X^T) \tag{2}$$

is exact if and only if $H_T^*(X)$ is a reflexive R_T-module. Here X^T denotes the fixed point set, and $X_{1,T}$ is the union of X^T and all 1-dimensional orbits. The sequence (2) often permits an efficient computation of $H_T^*(X)$. Moreover, if X is a Poincaré duality space, then the reflexivity of $H_T^*(X)$ is also equivalent to the perfection of the equivariant Poincaré pairing

$$H_T^*(X) \times H_T^*(X) \to R_T, \qquad (\alpha, \beta) \mapsto \langle \alpha \cup \beta, o_T \rangle, \tag{3}$$

which is the composition of the cup product and evaluation on an equivariant orientation o_T of X.

The main purpose of the present paper is to extend the results of [3] and [4] to actions of arbitrary compact connected Lie groups. It turns out that essentially all results carry over to this more general setting. We achieve this by combining the techniques of Allday–Franz–Puppe with those of Goertsches and Rollenske [22], whose study of Cohen–Macaulay actions gave a first hint at the possibility of such an extension. Let us describe our results in more detail.

Let G be a compact connected Lie group with maximal torus T and corresponding Weyl group W. Then r is the rank of G, and $R_G = H^*(BG)$ is again a polynomial ring in r indeterminates of even degrees. Let X be a G-space; we always assume that it admits an equivariant closed embedding into some smooth G-manifold or G-orbifold. This allows us to define a Cartan model for the equivariant cohomology $H_G^*(X)$ of X and also for a suitably defined *equivariant homology* $H_*^G(X)$. The latter is *not* the homology of the Borel construction of X; it is related to equivariant cohomology via a universal coefficient spectral sequence (Proposition 6) and for manifolds and locally orientable orbifolds also via equivariant Poincaré duality (Proposition 8). Here "locally orientable" means that X is a rational homology manifold, see Sect. 3.3 for the precise definition. For most results, we also assume that only finitely many infinitesimal orbit types occur in X. (The infinitesimal orbit type of $x \in X$ is the orbit of the Lie algebra $\mathfrak{g}_x$ of the stabilizer G_x under the adjoint action of G.) This condition is often redundant, see below.

The *G-orbit filtration* of X is defined by

$$X_{i,G} = \{ x \in X \mid \operatorname{rank} G_x \geq r - i \} \tag{4}$$

for $-1 \leq i \leq r$. Then $X_{-1,G} = \emptyset$, $X_{0,G}$ is the maximal-rank stratum and $X_{r,G} = X$. All $X_{i,G}$ are G-stable and closed in X. Note that if $G = T$, then $X_{0,T} = X^T$. The maximal-rank stratum plays the role of the fixed point set in the non-abelian context.

Based on the work of Goertsches–Rollenske, we show:

Proposition 1 *Let X be a G-space. For $0 \leq i \leq r$, there is an isomorphism of R_G-modules*

$$H_G^*(X_{i,G}, X_{i-1,G}) = H_T^*(X_{i,T}, X_{i,T} \cap X_{i-1,G})^W.$$

An immediate consequence is the following:

Corollary 1 *Let X be a G-space with only finitely many infinitesimal orbit types and such that $H^*(X)$ is finite-dimensional. The R_G-module $H_G^*(X_{i,G}, X_{i-1,G})$ is zero or Cohen–Macaulay of dimension $r - i$ for $0 \leq i \leq r$.*

For $i = 0$, both results were obtained independently by Baird [5, Sect. 3] and Goertsches and Rollenske [22, Sect. 3]. The latter authors proved Corollary 1 also for compact G-manifolds X such that $X_{i-1} = \emptyset$ [22, Corollary 4.3].

As for torus actions, we consider the *Atiyah–Bredon sequence*

$$\begin{aligned} 0 \longrightarrow H_G^*(X) \xrightarrow{\iota^*} H_G^*(X_{0,G}) \xrightarrow{\delta_0} H_G^{*+1}(X_{1,G}, X_{0,G}) \xrightarrow{\delta_1} \cdots \\ \xrightarrow{\delta_{r-2}} H_G^{*+r-1}(X_{r-1,G}, X_{r-2,G}) \xrightarrow{\delta_{r-1}} H_G^{*+r}(X_{r,G}, X_{r-1,G}) \longrightarrow 0. \end{aligned} \tag{5}$$

Here ι^* is induced by the inclusion $\iota: X_{0,G} \hookrightarrow X$, and δ_i for $i \geq 0$ is the connecting homomorphism in the long exact sequence for the triple $(X_{i+1,G}, X_{i,G}, X_{i-1,G})$. The *Atiyah–Bredon complex* $AB_G^*(X)$ with $AB_G^i(X) = H_G^{*+i}(X_{i,G}, X_{i-1,G})$ is obtained from (5) by dropping the leading term.

Theorem 1 *Let X be a G-space with only finitely many infinitesimal orbit types and such that $H^*(X)$ is finite-dimensional. For any $j \geq 0$, the jth cohomology of the Atiyah–Bredon complex is*

$$H^j(AB_G^*(X)) = \operatorname{Ext}_{R_G}^j(H_*^G(X), R_G).$$

This allows us to characterize syzygies in equivariant cohomology:

Theorem 2 *Let X be a G-space with only finitely many infinitesimal orbit types and such that $H^*(X)$ is finite-dimensional, and let $1 \leq j \leq r$. Then $H_G^*(X)$ is a jth syzygy over R_G if and only if the part*

$$0 \to H_G^*(X) \to H_G^*(X_{0,G}) \to \cdots \to H_G^{*+j-1}(X_{j-1,G}, X_{j-2,G})$$

of the Atiyah–Bredon sequence is exact.

In the torus case this result is due to Allday et al. [3, Theorem 5.7], [4, Theorem 4.8]. For general G and compact X, the case $j = 1$ has been established by Allday (unpublished, based on the case $G = SU(2)$ solved by Chen) and Goertsches and

Rollenske [22, Theorem 3.9]. Whether it extends to higher syzygies in the non-abelian setting has been an open question, see [22, Question 4.7].

In the important special case $j = 2$ we get:

Corollary 2 *Let X be a compact orientable G-manifold or G-orbifold. The following are equivalent:*

1. *The R_G-module $H_G^*(X)$ is reflexive.*
2. *The sequence*

$$0 \to H_G^*(X) \to H_G^*(X_{0,G}) \to H_G^{*+1}(X_{1,G}, X_{0,G})$$

   is exact.
3. *The equivariant Poincaré pairing $H_G^*(X) \times H_G^*(X) \to R_G$ is perfect.*

That the freeness of $H_G^*(X)$ over R_G implies the exactness of the non-abelian Chang–Skjelbred sequence is due to Brion [9, Theorem 9] for compact multiplicity-free G-spaces and to Goertsches and Mare [21, Theorem 2.2] for compact G-manifolds. The perfection of the equivariant Poincaré pairing was shown by Ginzburg [20, Corollary 3.9] and Brion [10, Proposition 1] again under the assumption that $H_G^*(X)$ is free over R_G.

Let X be a G-space. In [3] and [4], T-spaces were assumed to have finite Betti sum and also finitely many infinitesimal orbit types, and as mentioned above, we often require the same for G-spaces. As another application of equivariant homology, we show that the latter condition is redundant for manifolds and locally orientable orbifolds.

Theorem 3 *Let X be a G-manifold or locally orientable G-orbifold. If $H^*(X)$ is finite-dimensional, then only finitely many infinitesimal orbit types occur in X.*

If X is compact, it follows easily from the differentiable slice theorem that there are actually only finitely many orbits types in X, cf. [7, Proposition VIII.3.13]. (For the orbifold version of the differentiable slice theorem, see [27, Proposition 2.3].) By a result of Mann [28, Theorem 3.5], the same conclusion holds if X is an orientable manifold (or cohomology manifold) with finitely generated *integral* cohomology. While Theorem 3 is in the same spirit as Mann's result, the proof is very different.

The paper is organized as follows: After discussing Cohen–Macaulay modules, syzygies and the algebraic aspects of Cartan models in Sect. 2, we define equivariant de Rham homology and cohomology for possibly singular spaces in Sect. 3. In Sect. 4 we discuss the behaviour of syzygies under restriction to and induction from a maximal torus. To prepare for the proof of the main results, we study spaces with isotropy groups of constant rank in Sect. 5. The results of [3] and [4] are generalized in Sect. 6 where we also introduce the notion of a Cohen–Macaulay filtration of a G-space. In the final Sect. 7 we prove Theorem 3.

2 Algebraic Preliminaries

2.1 Notation and Standing Assumptions

Throughout this paper, the letter G denotes a compact connected Lie group of rank r with maximal torus $T \cong (S^1)^r$ and corresponding Weyl group $W = N_G(T)/T$. The Lie algebra of G is written as $\mathfrak{g}$ and its dual as $\mathfrak{g}^*$. We adopt this naming scheme for all Lie groups.

All manifolds and orbifolds are assumed to be paracompact. (See [1, Sect. 1.1] for the definition of an orbifold.) All (co)homology is taken with real coefficients.

We adopt a cohomological grading for all complexes, so that differentials always have degree $+1$. The degree of an element c of a complex is denoted by $|c|$. Homological complexes are turned into cohomological ones by grading them negatively. For example, elements in the nth homology group $H_n(X)$ of a space X have degree $-n$. A quasi-isomorphism is a map of complexes inducing an isomorphism in cohomology.

Let A and B be complexes. The nth degree of the complex $\mathrm{Hom}_{\mathbb{R}}(A, B)$ consists of all linear maps $f: A \to B$ that raise degrees by $n \in \mathbb{Z}$. The differential is defined by

$$d(f) = d_B \circ f - (-1)^n f \circ d_A. \tag{6}$$

This generalizes to differential graded (dg) modules over some dg algebra. The complex $A^\vee = \mathrm{Hom}_{\mathbb{R}}(A, \mathbb{R})$ is the *dual complex* of A. Moreover, for $k \in \mathbb{Z}$, the nth degree of the *shifted complex* $A[k]$ is equal to A^{n+k}; the differential of $A[k]$ is $(-1)^k d_A$. We also write $H^{*+k}(X)$ instead of $H^*(X)[k]$.

Unless specified otherwise, all tensor products are over $\mathbb{R}$.

2.2 Cohen–Macaulay Modules

Let R be a polynomial ring over a field in r indeterminates. We assume that R is graded by assigning a positive degree to each indeterminate. Recall that a graded R-module $M \neq 0$ is called *Cohen–Macaulay* if it is finitely generated and

$$\operatorname{depth} M = \dim M. \tag{7}$$

Proposition 2 *Let M be a finitely generated graded R-module, and let $d \in \mathbb{N}$.*

1. *M is Cohen–Macaulay of dimension d if and only if*

$$\mathrm{Ext}^i_R(M, R) \neq 0 \quad \Leftrightarrow \quad i = r - d.$$

2. *In this case, $\mathrm{Ext}^{r-d}_R(M, R)$ is again Cohen–Macaulay of dimension d.*

Proof See [17, Proposition A1.16] for the first part. A version of the second part for local rings can be found in [11, Proposition 3.3.3 (b)]; our claim follows from this by localizing $\operatorname{Ext}^*_R(\operatorname{Ext}^i_R(M,R))$ at all primes $\mathfrak{p} \lhd R$, noting that $R_\mathfrak{p}$ is a Cohen–Macaulay local ring. □

2.3 Syzygies

Let R be a polynomial ring in r indeterminates over a field. The definition of a syzygy has been given in the introduction. We observe that any finitely generated R-module is a zeroth syzygy. From [12, Appendix E] we note:

Proposition 3 *Let M be a finitely generated R-module. The following are equivalent for any $0 \le j \le r$:*

1. *M is a jth syzygy.*
2. *Any regular sequence $f_1, \ldots, f_j \in R$ is M-regular.*
3. $\operatorname{depth}_{R_\mathfrak{p}} M_\mathfrak{p} \ge \min(j, \dim R_\mathfrak{p})$ *for any prime ideal* $\mathfrak{p} \lhd R$.

2.4 Cartan Models

Recall that a $G^\star$-module is a complex A of G-modules with operations L_ξ (of degree 0) and ι_ξ (of degree -1) for $\xi \in \mathfrak{g}$ satisfying the same relations as the Lie derivative and the contraction operator for differential forms; see [23, Definition 2.3.1] for details.

The Cartan model of a $G^\star$-module A is denoted by

$$C^*_G(A) = (\mathbb{R}[\mathfrak{g}^*] \otimes A)^G \tag{8}$$

where $\mathbb{R}[\mathfrak{g}^*]$ denotes the real-valued polynomials on $\mathfrak{g}$, graded by twice the polynomial degree. The differential is

$$d(f \otimes a) = f \otimes da - \sum_{k=1}^{r} f x_k \otimes \iota_{\xi_k} a\,; \tag{9}$$

here (ξ_k) is a basis for $\mathfrak{g}$ with dual basis (x_k). The cohomology of $C^*_G(A)$ is the *equivariant cohomology* of A and denoted by $H^*_G(A)$. The coefficient field $\mathbb{R}$ is a $G^\star$-module with trivial operations; we set $R_G = H^*_G(\mathbb{R}) = C^*_G(\mathbb{R}) = \mathbb{R}[\mathfrak{g}^*]^G$. The Cartan model (8) is a dg R_G-module, hence $H^*_G(A)$ is an R_G-module.

Remark 1 If A is a $G^\star$-algebra in the sense of [23, Definition 2.3.1], then $C^*_G(A)$ is a dg R_G-algebra by componentwise multiplication, and $H^*_G(A)$ is an R_G-algebra. However, most of the time we will not be concerned with the multiplicative structure in equivariant cohomology. The only exceptions are Sect. 3.3, where we use the cup product in the equivariant de Rham cohomology of an orientable manifold or orbifold to establish equivariant Poincaré duality, and Sect. 6.4, where we define the equivariant Poincaré pairing.

Lemma 1 *Let A be a bounded below $G^\star$-module. There is a first-quadrant spectral sequence, natural in A, with*

$$E_1 = R_G \otimes H^*(A) \;\Rightarrow\; H^*_G(A).$$

Proof See [23, Theorem 6.5.2]. □

Let A be a $G^\star$-module. The inclusion $K \hookrightarrow G$ of a closed subgroup induces a canonical restriction map $C^*_G(A) \to C^*_K(A)$, which is a chain map compatible with the restriction map $R_G \to R_K$. For $K = 1$ we get a map $C^*_G(A) \to A$, which induces the restriction map

$$H^*_G(A) \to H^*(A). \tag{10}$$

Let $\mathcal{W} = \mathbb{R}[\mathfrak{g}^*] \otimes \bigwedge \mathfrak{g}^*$ be the Weil algebra of $\mathfrak{g}^*$, cf. [23, Sect. 3.2], and let A be a $\mathcal{W}^\star$-module, see [23, Definition 3.4.1] for the definition. For example, if G acts locally freely on a manifold X, then the de Rham complex $\Omega^*(X)$ is a $\mathcal{W}^\star$-module, cf. Lemma 9 below. There is a canonical isomorphism (with some twisted differential on the right-hand side)

$$A = \bigwedge \mathfrak{g}^* \otimes A_{\mathrm{hor}} \tag{11}$$

where $A_{\mathrm{hor}} \subset A$ denotes the subcomplex of horizontal elements, cf. [23, Theorem 3.4.1]. The basic subcomplex $A_{\mathrm{bas}} = (A_{\mathrm{hor}})^G$ is canonically a dg module over $\mathcal{W}_{\mathrm{bas}} = R_G$.

Lemma 2 *Let A be a $\mathcal{W}^\star$-module. As a complex, A_{bas} is a deformation retract of $C^*_G(A)$. There is a natural deformation retraction $C^*_G(A) \to A_{\mathrm{bas}}$ which is a morphism of dg R_G-modules. In particular, there is a natural isomorphism of R_G-modules*

$$H^*_G(A) = H^*(A_{\mathrm{bas}}).$$

Proof See in particular [31], or [23, Chap. 5]. □

The dual complex $A^\vee$ of a $G^\star$-module A is again a $G^\star$-module via the assignments

$$\langle g\varphi, a\rangle = \langle \varphi, g^{-1}a\rangle, \quad \langle L_\xi\varphi, a\rangle = -\langle \varphi, L_\xi a\rangle, \quad \langle \iota_\xi\varphi, a\rangle = -(-1)^{|\varphi|}\langle \varphi, \iota_\xi a\rangle \tag{12}$$

for $\varphi \in A^\vee$, $a \in A$, $g \in G$ and $\xi \in \mathfrak{g}$. Here $\langle \varphi, a\rangle$ denotes the pairing between $A^\vee$ and A. Moreover, if A is a $\mathcal{W}^\star$-module, then so is $A^\vee$, where the $\mathcal{W}$-module structure is defined by

$$\langle w\varphi, a\rangle = (-1)^{|w|\,|\varphi|}\langle \varphi, wa\rangle \tag{13}$$

for $\varphi \in A^\vee$, $a \in A$ and $w \in \mathcal{W}$.

Lemma 3 *For any $\mathcal{W}^\star$-module A there is an isomorphism of dg R_G-modules*

$$(A^\vee)_{\mathrm{bas}} \cong (A_{\mathrm{bas}})^\vee\,[-\dim G].$$

Proof Dualizing (11), we get an isomorphism of $\mathcal{W}^\star$-modules

$$A^\vee = \bigwedge \mathfrak{g} \otimes (A_{\mathrm{hor}})^\vee \tag{14}$$

where $\bigwedge \mathfrak{g}^* \subset \mathcal{W}^\star$ acts on the right-hand side through the pairing with $\bigwedge \mathfrak{g}$, and the operators ι_ξ act on $\bigwedge \mathfrak{g}$ by exterior multiplication. Hence

$$(A^\vee)_{\mathrm{hor}} = \bigwedge^{\dim G} \mathfrak{g} \otimes (A_{\mathrm{hor}})^\vee \cong (A_{\mathrm{hor}})^\vee[-\dim G], \tag{15}$$

and the claim follows by taking G-invariants, given that $\bigwedge^{\dim G} \mathfrak{g}$ is G-invariant. □

2.5 Universal Coefficient Theorem

Let A be a bounded $G^\star$-module. We want to relate the equivariant cohomology of A to that of $A^\vee$. The following result will be crucial for us.

Lemma 4 (Kostant) *There is a G-submodule $\mathcal{H} \subset \mathbb{R}[\mathfrak{g}^*]$ such that the restricted multiplication map*

$$R_G \otimes \mathcal{H} = \mathbb{R}[\mathfrak{g}^*]^G \otimes \mathcal{H} \to \mathbb{R}[\mathfrak{g}^*]$$

is bijective, hence an isomorphism of G-modules and R_G-modules.

The R_G-action on $R_G \otimes \mathcal{H}$ is on the first factor only. Note that we necessarily have $\mathcal{H}^G = \mathbb{R}$, the constant polynomials.

Proof See [26, Theorem 0.2]; the submodule $\mathcal{H}$ is given by the harmonic polynomials on $\mathfrak{g}$. The case of complex coefficients that Kostant considers is obtained from the real case by extension of scalars, so the result holds already with real coefficients.

□

From Lemma 4 we get isomorphisms of R_G-modules

$$C^*_G(A) = \big(\mathbb{R}[\mathfrak{g}^*] \otimes A\big)^G = R_G \otimes (\mathcal{H} \otimes A)^G, \tag{16}$$

$$\mathrm{Hom}_{R_G}(C^*_G(A), R_G) = \mathrm{Hom}_{\mathbb{R}}((\mathcal{H} \otimes A)^G, R_G). \tag{17}$$

Lemma 5 *As a dg R_G-module, $C^*_G(A)$ is homotopy equivalent to $R_G \otimes H^*(A)$ with some twisted differential.*

Proof Consider the complex

$$B = \mathbb{R} \otimes_{R_G} C^*_G(A). \tag{18}$$

Because $C^*_G(A)$ is free over R_G, it is R_G-homotopy equivalent to $R_G \otimes H^*(B)$ with some twisted differential by [2, Corollary B.2.4], cf. also [3, Remark 3.3]. We claim that $H^*(B) \cong H^*(A)$.

By (16) we have $B = (\mathcal{H} \otimes A)^G$ with some differential. We filter this complex by degree in A. The E_1 page is

$$E_1 = (\mathcal{H} \otimes H^*(A))^G = \mathcal{H}^G \otimes H^*(A) = H^*(A) \tag{19}$$

because taking G-invariants and cohomology commute and G acts trivially in cohomology. The spectral sequence therefore degenerates, and $H^*(B) = H^*(A)$. □

The canonical pairing $A^\vee \times A \to \mathbb{R}$, $(\varphi, a) \mapsto \langle \varphi, a\rangle$, has the equivariant extension

$$C^*_G(A^\vee) \times C^*_G(A) \to C^*_G(\mathbb{R}) = R_G, \tag{20}$$

which is the restriction of the canonical pairing

$$\mathbb{R}[\mathfrak{g}^*] \otimes A^\vee \times \mathbb{R}[\mathfrak{g}^*] \otimes A \to \mathbb{R}[\mathfrak{g}^*], \qquad (f_1 \otimes \varphi, f_2 \otimes a) \mapsto \langle \varphi, a\rangle f_1 f_2. \tag{21}$$

Hence we get a morphism of dg R_G-modules

$$\Phi\colon C^*_G(A^\vee) \to \mathrm{Hom}_{R_G}(C^*_G(A), R_G). \tag{22}$$

For $G = T$ a torus, it is an isomorphism as one can see by comparing (16) with (17). In general we have the following:

Lemma 6 *The map Φ is a quasi-isomorphism of dg R_G-modules.*

Proof We filter (16) (with $A^\vee$ instead of A) and (17) by degree in R_G; this is compatible with the map (22). The differentials d_0 on both E_0 pages come from the differential in A, hence the E_1 pages are

$$E_1 = R_G \otimes H^*\big((\mathcal{H} \otimes A^\vee)^G\big) = R_G \otimes H^*(A)^\vee \tag{23}$$

in the first case, and

$$E_1 = \mathrm{Hom}_{\mathbb{R}}(H^*(A), R_G) = R_G \otimes H^*(A)^\vee \tag{24}$$

in the second. We claim that the map $E_1(\Phi)$ between them is an isomorphism: Represent an element of (23) by $f \otimes \varphi$ where φ is a G-invariant cycle in $A^\vee$ and $f \in R_G$. It maps again to $f \otimes [\varphi]$ in (24), proving the claim. We therefore get an isomorphism between the E_∞ pages.

By assumption, $C^*_G(A^\vee)$ is bounded below, so that the spectral sequence (23) converges naively. The R_G-homotopy equivalence give by Lemma 5 induces one between (17) and

$$\mathrm{Hom}_{R_G}(R_G \otimes H^*(A), R_G) = \mathrm{Hom}_{\mathbb{R}}(H^*(A), R_G), \tag{25}$$

which is compatible with the filtration by R_G-degree. Moreover, the map between the E_0 pages of the associated spectral sequences comes from the induced homotopy equivalence between $(\mathcal{H} \otimes A^\vee)^G$ and $H^*(A)$, hence is itself a homotopy equivalence. From the E_1 pages on, the spectral sequence (24) therefore coincides with the one for (25), for which convergence is again naive. Thus no convergence issues arise, and we conclude that $H^*(\Phi)$ is an isomorphism. □

Proposition 4 (Universal Coefficient Theorem) *Let A be a bounded $G^\star$-module. There is a spectral sequence, natural in A, with*

$$E_2 = \mathrm{Ext}_{R_G}(H^*_G(A), R_G) \;\Rightarrow\; H^*_G(A^\vee). \tag{26}$$

Proof By Lemma 5, $C^*_G(A)$ is R_G-homotopy equivalent to a dg R_G-module which is free over R_G on generators of bounded degrees. For such dg R_G-modules the proof of [3, Proposition 3.5] carries over and establishes a spectral sequence with the given E_2 page and converging to the cohomology of $\mathrm{Hom}_{R_G}(C^*_G(A), R_G)$, which is equal to $H^*_G(A^\vee)$ by Lemma 6. □

3 Equivariant Homology and Cohomology

3.1 Equivariant de Rham Cohomology

Although we are primarily interested in manifolds, singular spaces will inevitably come up in our discussion of the orbit filtration. Because the Cartan model for equivariant cohomology has technical advantages, we discuss how to extend it to singular spaces embedded in manifolds and, more generally, in orbifolds.

3.1.1 Closed Supports

Let $Y \subset X$ be closed subsets of an orbifold Z. We write $H^*(X, Y)$ for the Alexander–Spanier cohomology of the pair (X, Y) with closed supports. Our starting point for a de Rham model is the tautness property of Alexander–Spanier cohomology,

$$H^*(X, Y) = \varinjlim H^*(U, V), \tag{27}$$

where the direct limit is taken over all open neighbourhood pairs (U, V) of (X, Y), see [32, Corollary 6.6.3]. Since U and V are orbifolds, $H^*(U, V)$ may be interpreted as singular cohomology or, if $V = \emptyset$, as de Rham cohomology.

Because direct limits are exact functors, we have $H^*(X) = H^*(\Omega_Z^*(X))$, where the dg algebra $\Omega_Z^*(X)$ of germs at X of differential forms on Z is defined as

$$\Omega_Z^*(X) = \varinjlim \Omega^*(U)\,; \tag{28}$$

the direct limit ranges over all open neighbourhoods U of X in Z; $\Omega_Z^*(Y)$ is defined analogously. Let $\Omega_Z^*(X, Y)$ be the kernel of the restriction map $\Omega_Z^*(X) \to \Omega_Z^*(Y)$; it is a dg module over $\Omega_Z^*(X)$.

Lemma 7 *As a graded vector space, $H^*(\Omega_Z^*(X, Y))$ is naturally isomorphic to $H^*(X, Y)$, the Alexander–Spanier cohomology of the pair (X, Y). This isomorphism is compatible with the module structure over $H^*(\Omega_Z^*(X)) = H^*(X)$.*

Proof The Alexander–Spanier cohomology of an open pair (U, V) in Z is naturally isomorphic to the cohomology $H_\infty^*(U, V)$ of the complex $S_\infty^*(U, V)$ of smooth singular cochains. (For the manifold case see [33, Sect. 5.32], for instance. It follows for orbifolds by looking at uniformizing charts.) We therefore have

$$H^*(X, Y) = \varinjlim H_\infty^*(U, V) = H^*\big(\varinjlim S_\infty^*(U, V)\big). \tag{29}$$

From the de Rham theorem, we obtain a natural quasi-isomorphism

$$\Omega_Z(X) = \varinjlim \Omega^*(U) \to \varinjlim S_\infty^*(U). \tag{30}$$

Naturality yields the left vertical arrow in the commutative diagram

$$\begin{array}{ccccccccc} 0 & \longrightarrow & \Omega_Z^*(X, Y) & \longrightarrow & \Omega_Z^*(X) & \longrightarrow & \Omega_Z^*(Y) & \longrightarrow & 0 \\ & & \downarrow & & \downarrow & & \downarrow & & \\ 0 & \longrightarrow & \varinjlim S_\infty^*(U, V) & \longrightarrow & \varinjlim S_\infty^*(U) & \longrightarrow & \varinjlim S_\infty^*(V) & \longrightarrow & 0. \end{array} \tag{31}$$

The top row is exact: Given a differential form α on $V \supset Y$, choose a smooth function f on Z with support in V and identically equal to 1 in a smaller neighbourhood V' of Y. Then $f\alpha$ is defined on all of Z and restricts to the same germ at Y as α.

Since the bottom row of (31) is also exact, we conclude from the five lemma that the left vertical arrow is a quasi-isomorphism. Together with (29) this establishes the claimed isomorphism of graded vector spaces.

That the module structures coincide follows from the fact that the isomorphisms induced by the vertical arrows in (31) are compatible with (sheaf-theoretic) cup products, cf. [33, Theorem 5.45] and [8, Sect. II.7, Theorem III.3.1]. □

Now assume that Z is a G-orbifold and the pair (X,Y) G-stable. We can always assume germs of differential forms at X to be defined on G-stable open neighbourhoods, which are cofinal among all open neighbourhoods of X. Hence $\Omega_Z^*(X)$ is a $G^\star$-module, and the same applies to $\Omega_Z^*(Y)$ and $\Omega_Z^*(X,Y)$. So we can consider their Cartan models. (See [29, p. 245] for the Cartan model for orbifolds.) We compare the cohomology of $\Omega_Z^*(X,Y)$ to $H_G^*(X,Y)$, the Borel G-equivariant Alexander–Spanier cohomology of the pair (X,Y).

Lemma 8 *As R_G-module, $H_G^*(\Omega_Z^*(X,Y))$ is naturally isomorphic to $H_G^*(X,Y)$.*

In particular, R_G is isomorphic to $H^*(BG)$, the cohomology algebra of the classifying space of G.

Proof Recall the natural isomorphism

$$H_G^n(X,Y) = H^n(X \times_G EG_k, Y \times_G EG_k) \tag{32}$$

where (EG_k) is a family of compact free G-manifolds approximating EG and k is chosen sufficiently large compared to n. By Lemma 7, this implies

$$H_G^n(X,Y) = H^n(\Omega_{Z\times_G EG_k}(X \times_G EG_k, Y \times_G EG_k)) \tag{33}$$

for $k \gg n$. Also by Lemma 7 and naturality, this isomorphism is compatible with the module structures over $H_G^*(\mathrm{pt}) = H^*(BG)$.

It therefore suffices to find a family of natural morphisms of dg algebras

$$C_G(\Omega_Z^*(X,Y)) \to \Omega^*_{Z\times_G EG_k}(X \times_G EG_k, Y \times_G EG_k) \tag{34}$$

that induce isomorphisms in any given cohomological degree $n \ll k$.

For the case $(X,Y) = (Z,\emptyset)$, this can be found in [23, Sect. 2.5], modulo the natural isomorphism between the Cartan model and the Weil model for equivariant cohomology, cf. [23, Theorem 4.2.1]. This implies in particular $R_G = H^*(BG)$. Noting that due to the compactness of EG_k neighbourhoods of the form $U \times_G EG_k$ are cofinal among all neighbourhoods of $X \times_G EG_k$, we get the absolute case for arbitrary X by taking direct limits as in the proof of Lemma 7. The relative case again follows by the five lemma. □

Remark 2 Lemma 1 implies that $H_G^*(X,Y)$ is finitely generated over R_G if $H^*(X,Y)$ is finite-dimensional. The converse is a consequence of the Eilenberg–Moore theorem.

Lemma 9 *If G acts locally freely on $X \setminus Y$, then $\Omega_Z^*(X, Y)$ is a $\mathcal{W}^\star$-module, and*

$$\Omega_Z^*(X, Y)_{\mathrm{bas}} = \Omega_{Z/G}^*(X/G, Y/G).$$

In particular, $H_G^*(X, Y) = H^*(X/G, Y/G)$.

If G acts locally freely on all of Z, then Z/G is an orbifold and $(X/G, Y/G)$ is a closed pair in Z/G, so that $\Omega_{Z/G}^*(X/G, Y/G)$ has been defined above. The definition of $\Omega_{Z/G}^*(X/G, Y/G)$ in the general case will be given in the proof.

Proof Assume first that G acts locally freely on Z, and choose a connection form for Z. Then the de Rham complex $\Omega^*(U)$ of any G-stable open subset $U \subset Z$ is naturally a $\mathcal{W}^\star$-module, hence so are $\Omega_Z^*(X)$ and $\Omega_Z^*(Y)$, and the restriction map $\Omega_Z^*(X) \to \Omega_Z^*(Y)$ is a morphism of $\mathcal{W}^\star$-module. Thus, its kernel $\Omega_Z^*(X, Y)$ is also a $\mathcal{W}^\star$-module.

We have $\Omega^*(U)_{\mathrm{bas}} = \Omega^*(U/G)$, cf. [23, Sect. 2.3.5], hence $\Omega_Z^*(X)_{\mathrm{bas}} = \Omega_{Z/G}^*(X/G)$. The same holds again for Y, and by naturality we find

$$\Omega_Z^*(X, Y)_{\mathrm{bas}} = \Omega_{Z/G}^*(X/G, Y/G). \tag{35}$$

This together with Lemma 2 gives

$$H_G^*(X, Y) = H^*(X/G, Y/G). \tag{36}$$

For the general case we observe that X is metrizable, hence normal. We can therefore write

$$\Omega_Z^*(X, Y) = \varinjlim \Omega_Z^*(X, B) \tag{37}$$

where the direct limit is taken over all G-stable closed neighbourhoods B of Y in X,

$$= \varinjlim \Omega_{Z \setminus Y}^*(X \setminus Y, B \setminus Y) \tag{38}$$

because any germ of a differential form on $X \setminus Y$ that vanishes on $B \setminus Y$ can be extended to a germ on X vanishing on Y: Assume that the differential form α is defined on the neighbourhood U of $X \setminus Y$ in $Z \setminus Y$ and vanishes on the neighbourhood $V \subset U$ of $B \setminus Y$. Choose disjoint neighbourhoods $U' \subset U$ of $X \setminus V$ and V' of Y in Z, and let β be the form on $U' \cup V \cup V'$ that agrees with α on $U' \cup V$ and vanishes on V'. Then $\alpha = \beta$ as germs on $X \setminus Y$.

Because G acts locally freely on $X \setminus Y$, it does so on some G-stable open neighbourhood $W \subset Z \setminus Y$, and we can write

$$\Omega_Z^*(X, Y) = \varinjlim \Omega_W^*(X \setminus Y, B \setminus Y). \tag{39}$$

By what we have said above, this is a direct limit of $\mathcal{W}^\star$-modules, hence a $\mathcal{W}^\star$-module itself, and

$$\Omega_Z^*(X,Y)_{\mathrm{bas}} = \varinjlim \Omega_{W/G}^*(X/G \setminus Y/G, B' \setminus Y/G), \tag{40}$$

where B' runs through the closed neighbourhoods of Y/G in X/G. The right-hand side is our definition of $\Omega_{Z/G}^*(X/G, Y/G)$ (which is independent of the choice of W).

Moreover, using Lemma 2 and arguments similar to the ones above, we find

$$H_G^*(\Omega_Z^*(X,Y)) = H^*(\Omega_Z^*(X,Y)_{\mathrm{bas}}) = \varinjlim H^*(X/G \setminus Y/G, B' \setminus Y/G) \tag{41}$$

$$= \varinjlim H^*(X/G, B') = H^*(X/G, Y/G). \tag{42}$$

□

3.1.2 Compact Supports

We write $H_c^*(X,Y)$ for the Alexander–Spanier cohomology of the pair (X,Y) with compact supports. There is a natural isomorphism

$$H_c^*(X,Y) = \varinjlim H^*(X,B) \tag{43}$$

where the direct limit ranges over all neighbourhoods B of Y such that $X \setminus B$ has compact closure in X, see [32, Theorem 6.6.16]. Since X is metrizable, it is enough to consider closed neighbourhoods of B. We therefore set

$$\Omega_{Z,c}^*(X,Y) = \varinjlim \Omega_Z^*(X,B) \tag{44}$$

where B ranges over all closed neighbourhoods of Y such that $X \setminus B$ has compact closure. Then everything said about closed supports carries over to compact supports. In particular:

Lemma 10 *$H^*(\Omega_{Z,c}^*(X,Y)) = H_c^*(X,Y)$, the Alexander–Spanier cohomology of the pair (X,Y) with compact supports.*

Lemma 11 *$H_G^*(\Omega_{Z,c}^*(X,Y)) = H_{G,c}^*(X,Y)$, the G-equivariant Alexander–Spanier cohomology of the pair (X,Y) with compact supports.*

Lemma 12 *If G acts locally freely on $X \setminus Y$, then $\Omega_{Z,c}^*(X,Y)$ is a $\mathcal{W}^\star$-module, and*

$$\Omega_{Z,c}^*(X,Y)_{\mathrm{bas}} = \Omega_{Z/G,c}^*(X/G, Y/G).$$

In particular, $H_{G,c}^*(X,Y) = H_c^*(X/G, Y/G)$.

Restriction of forms and extension by 0 give an isomorphism of $G^\star$-modules

$$\Omega^*_{Z,c}(X,Y) = \Omega^*_{Z\setminus Y,c}(X\setminus Y), \tag{45}$$

which confirms that cohomology with compact supports is a 'single space theory'. From it we recover the natural isomorphism (see [32, Theorem 6.6.11])

$$H^*_c(X,Y) = H^*_c(X\setminus Y) \tag{46}$$

and, using Lemma 1, its equivariant analogue,

$$H^*_{G,c}(X,Y) = H^*_{G,c}(X\setminus Y). \tag{47}$$

The special case $X = Z$ of (45) together with the exact top row of diagram (31) gives a natural isomorphism

$$\Omega^*_{Z,c}(Y) = \Omega^*_c(Z) \mathbin{/} \Omega^*_{Z,c}(Z,Y) = \Omega^*_c(Z) \mathbin{/} \Omega^*_c(Z\setminus Y), \tag{48}$$

which could be used as the definition of $\Omega^*_{Z,c}(Y)$; again via (45) it can be extended to pairs. This is the approach taken for example by Guillemin and Sternberg [23, Sect. 11.1] and also by De Concini et al. [13, Sect. 1, Appendix A], who prove Lemmas 10 and 11 under the assumption that $X\setminus Y$ is locally contractible [13, Propositions A.4 & A.8].

3.2 Equivariant Homology

We want to define equivariant homology by appropriately dualizing the Cartan model of differential forms. The most conceptual approach would be to use currents, as done by Metzler [23, p. 169] and Meinrenken [30, Sect. 4]. Because we are ultimately only interested in spaces with finite-dimensional cohomology, we will work with the algebraic instead of the topological dual of the de Rham complex.

For ease of notation, we only write out results for cohomology with closed supports and homology with compact supports in this section. All results remain valid for (co)homology with the other pair of supports.

Let $Y \subset X$ be closed G-stable subsets of an orbifold Z. We say in this case that (X,Y) is a *G-pair* in Z. Often we do not mention the ambient orbifold Z explicitly; we say that X and Y are *G-spaces*, and we write $\Omega^*(X,Y)$ instead of $\Omega^*_Z(X,Y)$.

Assumption 4 *From now on, whenever we consider the homology or equivariant homology (with compact supports) of a G-pair (X,Y), we assume that $H_*(X,Y)$ or, equivalently, $H^*(X,Y)$ is finite-dimensional. In this case there is a natural* *isomorphism*

$$H_*(X, Y) = H^*(X, Y)^\vee. \tag{49}$$

We make no such assumption for $H^c_*(X, Y)$, the homology of (X, Y) with closed supports, and in fact

$$H^c_*(X, Y) = H^*_c(X, Y)^\vee \tag{50}$$

is always true.

We define the *equivariant homology* $H^G_*(X, Y)$ of the G-pair (X, Y) as the cohomology of the Cartan model

$$C^G_*(X, Y) = C^*_G(\Omega^*(X, Y)^\vee), \tag{51}$$

where $\Omega^*(X, Y)^\vee$ denotes the dual of the $G^\star$-module $\Omega^*(X, Y)$ as defined in Sect. 2.4.[1]

Remark 3 Assume that $G = T$ is a torus. As remarked after (22), the canonical map

$$C^T_*(X, Y) \to \operatorname{Hom}_{R_T}(C^*_T(X, Y), R_T) \tag{52}$$

is an isomorphism of dg R_T-modules in this case. We thus recover the definition of torus-equivariant homology given in [3] and [4], up to the substitution of the Cartan model for the "singular Cartan model" used there. For general G, the analogue of (52) still is a quasi-isomorphism by Lemma 6.

The cohomology of $\Omega^*(X, Y)^\vee$ is $H^*(X, Y)^\vee = H_*(X, Y)$ by Assumption 4. As a special case of Lemma 1 we therefore get:

Proposition 5 *There is a first-quadrant spectral sequence, natural in (X, Y), with*

$$E_1 = R_G \otimes H_*(X, Y) \Rightarrow H^G_*(X, Y).$$

Proposition 6 (Universal Coefficient Theorem) *There is a spectral sequence, natural in (X, Y), with*

$$E_2 = \operatorname{Hom}_{R_G}(H^*_G(X, Y), R_G) \Rightarrow H^G_*(X, Y).$$

If $H_*(X, Y)$ is finite-dimensional, then there is another natural spectral sequence with

$$E_2 = \operatorname{Hom}_{R_G}(H^G_*(X, Y), R_G) \Rightarrow H^*_G(X, Y).$$

[1]Using Propositions 5 and 7, one can show that $H^G_*(X)$ is isomorphic to the equivariant Borel–Moore homology $H^G_{\mathrm{BM},*}(X)$ as introduced by Edidin and Graham [16, Sect. 2.8].

Proof The first spectral sequence is a special case of the algebraic universal coefficient theorem (Proposition 4). The same result also gives a spectral sequence of the second form converging to $H_G^*(\Omega^*(X,Y)^{\vee\vee})$. The latter is equal to $H_G^*(X,Y)$ by Lemma 1 because the canonical map $\Omega^*(X,Y) \to \Omega^*(X,Y)^{\vee\vee}$ is a quasi-isomorphism by assumption. □

Proposition 7 *If G acts locally freely on $X \setminus Y$, then there is an isomorphism of R_G-modules*

$$H_*^G(X,Y) \cong H_{*+\dim G}(X/G, Y/G),$$

where the R_G-module structure of $H_*(X/G, Y/G)$ is dual to the one in cohomology.

Proof By Lemma 9 (or Lemma 12, depending on supports) we have

$$\Omega^*(X,Y)_{\mathrm{bas}} = \Omega^*(X/G, Y/G), \tag{53}$$

hence

$$(\Omega^*(X,Y)^{\vee})_{\mathrm{bas}} \cong \Omega^*(X/G, Y/G)^{\vee}[-\dim G] \tag{54}$$

by Lemma 3. We conclude with Lemma 2. □

3.3 *Equivariant Poincaré Duality*

In this section the supports of the (co)homology groups do matter, and we carefully distinguish between them.

Let Z be an orbifold of dimension n. We say that Z is *locally orientable* if it is locally the quotient of some $\mathbb{R}^n$ by a finite subgroup of $SO(n)$. In this case, it is a rational homology manifold, and orientable if and only if $H_n^c(Z) \cong \mathbb{R}$. Any manifold is a locally orientable orbifold.

Now assume that Z is a locally orientable G-orbifold, and let $\pi: \tilde{Z} \to Z$ be the orientable two-fold cover of Z with a fixed orientation. If Z is orientable, then $\tilde{Z}$ consists of two copies of Z with opposite orientations. The G-action on Z lifts to $\tilde{Z}$, cf. [4, Lemma 2.13]. The non-trivial deck transformation τ commutes with G and reverses the orientation.

For any G-pair (X,Y) in Z, the $(+1)$-eigenspace of the induced map τ^* on $H_G^*(\pi^{-1}(X), \pi^{-1}(Y))$ is isomorphic to $H_G^*(X,Y)$ as an R_G-module. We define $H_G^*(X,Y;\tilde{\mathbb{R}})$, the equivariant cohomology of (X,Y) with twisted coefficients, to be the (-1)-eigenspace. Equivalently, it is the equivariant cohomology of $\Omega_Z^*(X,Y;\tilde{\mathbb{R}})$, the (-1)-eigenspace of τ^* on $\Omega_{\tilde{Z}}^*(\pi^{-1}(X), \pi^{-1}(Y))$. There are analogous definitions for non-equivariant cohomology, for homology and for other supports.

The integration map

$$I\colon \Omega_c^*(\tilde{Z}) \to \mathbb{R}, \quad \alpha \mapsto \int_{\tilde{Z}} \alpha \tag{55}$$

(which gives 0 if $|\alpha| < n$) defines an orientation $o = [I] \in H_c^n(Z;\tilde{\mathbb{R}})^\vee = H_n^c(Z;\tilde{\mathbb{R}})$. Moreover, it is a morphism of $G^\star$-modules of degree $-n$, where $\mathbb{R}$ is considered as a trivial $G^\star$-module. Hence $I_G = 1 \otimes I \in C_G^*(\Omega_c^*(\tilde{Z})^\vee)$ is a cycle and defines an equivariant homology class $o_G \in H_n^{G,c}(Z;\tilde{\mathbb{R}})$, which restricts to o under the map (10). In fact, it is the unique lift:

Lemma 13 *Any orientation $o \in H_n^c(Z)$ lifts uniquely to an equivariant orientation $o_G \in H_n^{G,c}(Z)$.*

Proof The proof of [3, Proposition 3.7] carries over. □

The integration map I also induces the morphism of $G^\star$-modules

$$PD\colon \Omega^*(\tilde{Z}) \to \Omega_c^*(\tilde{Z})^\vee, \quad \alpha \mapsto \Big(\beta \mapsto I(\alpha \wedge \beta)\Big) \tag{56}$$

which interchanges the (± 1)-eigenspaces. Based on it, we obtain equivariant Poincaré duality in the same way as Metzler [23, Theorem 10.10.1]:

Proposition 8 *The map PD induces an isomorphism $H_G^*(Z) \to H_{n-*}^{G,c}(Z;\tilde{\mathbb{R}})$ of R_G-modules of degree $-n$.*

Proof Between the E_1 pages of the spectral sequences from Lemma 1, the map induced by PD is the R_G-linear extension of the non-equivariant map $H^*(PD)$, hence an isomorphism. Thus so is PD itself. □

4 Restriction and Induction

For ease of notation, we again stick to cohomology with closed supports and homology with compact supports in this section. All results remain valid for (co)homology with the other pair of supports and/or with twisted coefficients.

Recall that $T \cong (S^1)^r$ is a maximal torus of G and W the corresponding Weyl group.

4.1 Restriction

The restriction $R_G \to R_T$ is an isomorphism onto the subalgebra $(R_T)^W$ of Weyl group invariants. Moreover, there is a canonical isomorphism of R_G-modules

$$R_T = R_G \otimes H^*(G/T) \tag{57}$$

where R_G acts only on the first factor of the tensor product. In particular, R_T is finitely generated and free over R_G. As a consequence, any R_T-module is finitely generated over R_T if and only if it is finitely generated over R_G.

Let (X, Y) be a G-pair. It is canonically a T-pair by restriction of the action. The following proposition is a special case of general results about restrictions of $G^\star$-modules, see [23, Sect. 6.8].

Proposition 9 *There are the following isomorphisms, natural in* (X, Y)*:*

1. *As* R_G*-modules,*

$$H_G^*(X,Y) = H_T^*(X,Y)^W \quad \text{and} \quad H_*^G(X,Y) = H_*^T(X,Y)^W.$$

2. *As* R_T*-modules,*

$$H_T^*(X,Y) = R_T \otimes_{R_G} H_G^*(X,Y) \quad \text{and} \quad H_*^T(X,Y) = R_T \otimes_{R_G} H_*^G(X,Y).$$

The cohomological parts are well-known and imply in particular that $H_G^*(X, Y)$ is finitely generated over R_G if and only if $H_T^*(X, Y)$ is finitely generated over R_T. This also follows from the fact that both conditions are equivalent to $H^*(X, Y)$ being finite-dimensional.

Proposition 10 *Let* (X, Y) *be a G-pair such that* $H_G^*(X, Y)$ *is finitely generated over* R_G*, and let* $0 \leq j \leq r$*. Then* $H_G^*(X, Y)$ *is a jth syzygy over* R_G *if and only if* $H_T^*(X, Y)$ *is a jth syzygy over* R_T*.*

Proof Assume that $H_G^*(X, Y)$ is a jth syzygy over R_G. By definition, there is an exact sequence

$$0 \to H_G^*(X,Y) \to F_1 \to \cdots \to F_j \tag{58}$$

with finitely generated free R_G-modules F_i. Because R_T is free over R_G, we obtain the exact sequence

$$0 \to R_T \otimes_{R_G} H_G^*(X,Y) \to R_T \otimes_{R_G} F_1 \to \cdots \to R_T \otimes_{R_G} F_j \tag{59}$$

with finitely generated free R_T-modules $R_T \otimes_{R_G} F_i$. This shows that $H_T^*(X, Y) = R_T \otimes_{R_G} H_G^*(X, Y)$ is a jth syzygy over R_T.

Now assume that $H_T^*(X, Y)$ is a jth syzygy over R_T. This time there is an exact sequence

$$0 \to H_T^*(X,Y) \to F_1 \to \cdots \to F_j \tag{60}$$

with finitely generated free R_T-modules F_i. Since the F_i are also finitely generated and free over R_G, this shows that

$$H_T^*(X,Y) = H_G^*(X,Y) \otimes H^*(G/T) \tag{61}$$

is a jth syzygy over R_G. It now follows from criterion (2) or (3) of Proposition 3 that the same holds for $H_G^*(X, Y)$ itself. □

4.2 Induction

Let (X, Y) be a T-pair, say contained in the T-orbifold Z. Then

$$\hat{X} = G \times_T X \quad \text{and} \quad \hat{Y} = G \times_T Y \tag{62}$$

are closed G-stable subsets of the G-orbifold $\hat{Z} = G \times_T Z$, hence $(\hat{X}, \hat{Y})$ is a G-pair. There is a canonical inclusion $Z \hookrightarrow \hat{Z}$ sending z to $[1, z]$, equivariant with respect to the inclusion $T \hookrightarrow G$. Also note that any R_T-module is canonically an R_G-module via the restriction map $R_G \to R_T$.

Lemma 14 *The inclusion of pairs $(X, Y) \hookrightarrow (\hat{X}, \hat{Y})$ induces an isomorphism of R_G-modules*

$$H_G^*(\hat{X}, \hat{Y}) = H_T^*(X, Y)$$

Proof See [14, Théorème 24] for a proof using equivariant de Rham theory. □

In particular, $H_G^*(\hat{X}, \hat{Y})$ is finitely generated over R_G if and only if $H_T^*(X, Y)$ is finitely generated over R_T.

To study the behaviour of syzygies under induction from T to G, we need the following simple algebraic fact.

Lemma 15 *Let $A \subset B$ be an extension of commutative rings, $\mathfrak{q} \lhd B$ be a prime ideal and $\mathfrak{p} = \mathfrak{q} \cap A$. Then for any B-module M,*

$$\operatorname{depth}_{B_\mathfrak{q}} M_\mathfrak{q} \geq \operatorname{depth}_{A_\mathfrak{p}} M_\mathfrak{p}.$$

Proof Let $a_1, \ldots, a_k \in \mathfrak{p}_\mathfrak{p} \subset \mathfrak{q}_\mathfrak{q}$ be an $M_\mathfrak{p}$-regular sequence. By induction on k we show that this sequence is also $M_\mathfrak{q}$-regular. The case $k = 0$ is void.

Assume

$$a_k \frac{m_k}{s_k} = a_1 \frac{m_1}{s_1} + \cdots + a_{k-1} \frac{m_{k-1}}{s_{k-1}} \tag{63}$$

for some $m_1, \ldots, m_k \in M$ and some $s_1, \ldots, s_k \in B \setminus \mathfrak{q}$. Then, for some $s \in B \setminus \mathfrak{q}$,

$$a_k(ss_1, \ldots, s_{k-1}m_k) \in (a_1, \ldots, a_{k-1})M_\mathfrak{p}. \tag{64}$$

Since $a_1, \ldots, a_k$ is $M_\mathfrak{p}$-regular, this implies $ss_1 \cdots s_{k-1} m_k \in (a_1, \ldots, a_{k-1})M_\mathfrak{p}$ and therefore

$$\frac{m_k}{s_k} \in (a_1, \ldots, a_{k-1})M_\mathfrak{q}. \tag{65}$$

Hence multiplication by a_k is injective on $M_\mathfrak{q}/(a_1, \ldots, a_{k-1})M_\mathfrak{q}$, which means that the sequence is $M_\mathfrak{q}$-regular. □

Proposition 11 *Let (X, Y) and $(\hat{X}, \hat{Y})$ be as before, and assume that $H_T^*(X, Y)$ is finitely generated over R_T. Then $H_G^*(\hat{X}, \hat{Y})$ is a jth syzygy over R_G if and only if $H_T^*(X, Y)$ is a jth syzygy over R_T.*

Proof Assume that $H_T^*(X, Y)$ is a jth syzygy over R_T. Then there is an exact sequence

$$0 \to H_T^*(X, Y) \to F_1 \to \cdots \to F_j \tag{66}$$

with finitely generated free R_T-modules F_i. Since the F_i are also finitely generated and free over R_G, Lemma 14 implies that $H_G^*(\hat{X}, \hat{Y})$ is a jth syzygy over R_G.

Now assume that $H_G^*(\hat{X}, \hat{Y})$ is a jth syzygy over R_G. Let $\mathfrak{q} \lhd R_T$ be a prime ideal and set $\mathfrak{p} = \mathfrak{q} \cap R_G$. Because $R_T \supset R_G$ is an integral extension of commutative rings, Cohen–Seidenberg's going-up theorem implies

$$\dim(R_G)_\mathfrak{q} = \operatorname{ht}_{R_G} \mathfrak{q} = \operatorname{ht}_{R_T} \mathfrak{p} = \dim(R_T)_\mathfrak{p}\,; \tag{67}$$

we also have

$$\operatorname{depth} H_T^*(X, Y)_\mathfrak{q} \geq \operatorname{depth} H_G^*(\hat{X}, \hat{Y})_\mathfrak{p} \tag{68}$$

by combining Lemmas 14 and 15. Using Proposition 3 and the assumption that $H_G^*(\hat{X}, \hat{Y})$ is a jth syzygy, we conclude

$$\operatorname{depth} H_T^*(X, Y)_\mathfrak{q} \geq \min(j, \dim(R_G)_\mathfrak{p}) = \min(j, \dim(R_T)_\mathfrak{q}). \tag{69}$$

Thus, $H_T^*(X, Y)$ is a jth syzygy over R_T. □

Remark 4 The case $j = 1$ of Proposition 11 can be shown more easily, see the proof of [24, Theorem C.70] or [22, Theorem 3.9]: If $f \in R_T$ is non-regular for $H_T^*(X, Y) = H_G^*(\hat{X}, \hat{Y})$, then the product $\prod_{w \in W} w \cdot f$ gives another such element in $(R_T)^W = R_G$.

Example 1 Let X be a projective toric manifold of dimension $2r$ with moment polytope $P = X/T$. It is well-known that $H_T^*(X)$ is free over R_T in this case.

Choose two distinct fixed points $x, y \in X$ and set $Y = X \setminus \{x, y\}$. Let $F \subset P$ be the smallest face of P containing (the images of) x and y. As shown in [18,

Sect. 8.1], the equivariant cohomology $H_T^*(Y)$ with closed supports is a syzygy of order exactly $\dim F - 1$ over R_T; the case $X = (S^2)^r$ with two diametrically opposite vertices x and y of the r-cube P appeared already in [3, Sect. 6.1].

From Proposition 11 we see that the equivariant cohomology with closed supports of the induced G-manifold $G \times_T Y$ is a syzygy of order exactly $\dim F - 1$ over R_G. In particular, syzygies of any order can appear as the equivariant cohomology of G-manifolds.

5 The Constant Rank Case

Let X be a G-space such that all G-isotropy groups in X have the same rank, say equal to $b \in \mathbb{N}$. Let

$$Y = \{\, x \in X \mid \operatorname{rank} T_x = b \,\} \tag{70}$$

be the highest-rank stratum for the T-action on X; it is $N_G(T)$-stable and, by the slice theorem, closed in X. For a subtorus $L \subset T$ of rank b, we define

$$X(L) = \{\, x \in X \mid L \text{ is conjugate in } G \text{ to a maximal torus of } G_x \,\}, \tag{71}$$

$$Y(L) = \{\, x \in X \mid L \text{ is a maximal torus of } G_x \,\} = X^L. \tag{72}$$

Note that X is the union of all such $X(L)$, and Y is the disjoint union of all such $Y(L)$. Each $Y(L)$ is closed in X.

Lemma 16

1. *The sets $X(L)$ partition X.*
2. *Each $X(L)$ is open in X.*
3. *Each $X(L)$ is a union of connected components of X.*

Proof Let $x \in X(L)$. If in addition $x \in X(L')$, then $X(L) = X(L')$ because all maximal tori of G_x are conjugate. By the slice theorem, any y in a sufficiently small neighbourhood of x has an isotropy group that is conjugate to a subgroup of G_x. By our assumption on X, this means that G_y contains a maximal torus conjugate to a maximal torus of G_x, whence $y \in X(L)$. The last claim follows from the first two. □

The following result is due to Goertsches–Rollenske; it appears in the proof of [22, Proposition 4.2].

Lemma 17 *$G \cdot y \cap Y(L) = N_G(L) \cdot y$ for any $y \in Y(L)$.*

Proof $\supset$: because $Y(L)$ is stable under $N_G(L)$.
$\subset$: Assume $gy \in Y(L)$ for some $g \in G$. Then L is a maximal torus of both G_y and G_{gy}. In other words, L and gLg^{-1} are both maximal tori of G_{gy}, hence there

is an $h \in G_{gy}$ such that $h^{-1}Lh = gLg^{-1}$ or $(hg)L(hg)^{-1} = L$. This shows that $hg \in N_G(L)$, and therefore $gy = hgy \in N_G(L) \cdot y$. □

Let us write $K = N_G(L) \supset T$.

Lemma 18 *The map $q: N_G(T) \times_{N_K(T)} Y(L) \to Y \cap X(L)$ induced by the G-action is an $N_G(T)$-equivariant homeomorphism.*

Proof Clearly, q is $N_G(T)$-equivariant. Note that $Y \cap X(L)$ is the disjoint union of the $Y(L')$ where L' runs through all subtori of T which are conjugate in G to L. We claim that

$$\{ L' \subset T \mid L' \text{ is conjugate in } G \text{ to } L \} = N_G(T) \cdot L = W \cdot L. \tag{73}$$

$\supset$: is trivial.

$\subset$: Assume $L' = gLg^{-1}$ or, equivalently, $\mathfrak{l}' = \operatorname{Ad}_g \mathfrak{l}$ for some $g \in G$. This means that $\mathfrak{l}'$ lies in the G-orbit of $\mathfrak{l}$ for the induced action on the Grassmannian $\operatorname{Gr}_b(\mathfrak{g})$. The set B of all b-dimensional abelian Lie subalgebras of $\mathfrak{g}$ is closed in $\operatorname{Gr}_b(\mathfrak{g})$ and G-stable, and all isotropy groups in B have rank r. Because $\mathfrak{l}$ and $\mathfrak{l}'$ are contained in B^T, Lemma 17 (with B instead of X and $L = T$) implies that $\mathfrak{l}' \in N_G(T) \cdot \mathfrak{l}$.

Coming back to the original claim, we observe that $N_K(T) = N_G(T) \cap K$ is the isotropy group of L for the conjugation action of $N_G(T)$ on the subtori of T. Hence $Y \cap X(L)$ consists, like $N_G(T) \times_{N_K(T)} Y(L)$, of copies of $Y(L)$ indexed by the cosets $N_G(T)/N_K(T)$. Thus, q is a homeomorphism. □

Lemma 19 *The map $f_1: G \times_K Y(L) \to X(L)$ induced by the G-action is a quasi-isomorphism (for cohomology with closed or compact supports and possibly twisted coefficients).*

For compact X this is again contained in the proof of [22, Proposition 4.2], and also in the proof of [6, Theorem 1.2].

Proof We are going to use the Vietoris–Begle theorem, see [32, Theorem 6.9.15] for a precise statement for Alexander–Spanier cohomology with closed supports. We extend it to relative cohomology by the five lemma and then to cohomology with compact supports by (43). The case of twisted coefficients follows by looking at the eigenspaces of the deck transformation τ.

Because G is compact, the action map $G \times X(L) \to X(L)$ is closed, hence so is f_1. Surjectivity is clear by construction. To apply the Vietoris–Begle theorem, it is therefore enough to show that the fibres of f_1 are acyclic. By G-equivariance it suffices to study the fibre over some $x \in Y(L)$.

Assume $f_1([g, y]) = x$, so that $gy = x = hy$ for some $h \in K$ by Lemma 17. Then $[g, y] = [gh^{-1}, hy] = [gh^{-1}, x]$ and $gh^{-1} \in G_x$. Hence

$$f_1^{-1}(x) \cong G_x/(K \cap G_x) = G_x/N_{G_x}(L), \tag{74}$$

which is acyclic by [22, Lemma 3.2] given that L is a maximal torus of G_x. □

Proposition 12 *The inclusion $Y \hookrightarrow X$ induces an isomorphism of R_G-modules*

$$H_G^*(X) = H_T^*(Y)^W$$

(for equivariant cohomology with closed or compact supports and possibly twisted coefficients).

Proof By Lemma 16 we may assume $X = X(L)$ for some subtorus $L \subset T$ of rank b.
Consider the commutative diagram

$$\begin{array}{ccc} G \times_{N_K(T)} Y(L) & \xrightarrow{f_2} & G \times_K Y(L) \\ {\scriptstyle f_3}\big\downarrow & & \big\downarrow{\scriptstyle f_1} \\ G \times_{N_G(T)} Y & \xrightarrow{f} & X, \end{array} \tag{75}$$

where f_2 comes from the inclusion $N_K(T) \hookrightarrow K$, and f_3 is induced from the $N_G(T)$-equivariant map q defined in Lemma 18,

$$f_3\colon G \times_{N_K(T)} Y(L) = G \times_{N_G(T)} \Big(N_G(T) \times_{N_K(T)} Y(L)\Big) \xrightarrow{(\mathrm{id},q)} G \times_{N_G(T)} Y. \tag{76}$$

Like q, the map f_3 is a homeomorphism.

Again by [22, Lemma 3.2], the fibre $K/N_K(T)$ of the bundle map f_2 is acyclic, so that f_2 is a quasi-isomorphism. We finally know from Lemma 19 that f_1 is a quasi-isomorphism, too. Hence f is a G-equivariant quasi-isomorphism. It follows that

$$H_G^*(X) = H_G^*(G \times_{N_G(T)} Y) = H_{N_G(T)}^*(Y) = H_T^*(Y)^W. \tag{77}$$

□

6 Cohen–Macaulay Filtrations

The results in this section hold for (co)homology with either pair of supports and/or with twisted coefficients. For simplicity we only state them for cohomology with closed supports/homology with compact supports and constant coefficients.

6.1 Definition and First Properties

Let X be a G-space, and let $\mathcal{F}$ be a filtration

$$\emptyset = X_{-1} \subset X_0 \subset \cdots \subset X_r = X \tag{78}$$

of X by G-stable closed subsets. We call $\mathcal{F}$ *Cohen–Macaulay* if $H_*^G(X_i, X_{i-1})$ is zero or Cohen–Macaulay of dimension $r - i$ for $0 \leq i \leq r$. Because of the following observation, we could substitute equivariant cohomology for equivariant homology in the definition.

Lemma 20 *Let (X, Y) be a G-pair and $d \in \mathbb{N}$. Then $H_*^G(X, Y)$ is Cohen–Macaulay of dimension d if and only if $H_G^*(X, Y)$ is so.*

Proof It follows from the universal coefficient theorem (Proposition 6) that $H_*^G(X, Y)$ is zero if and only if $H_G^*(X, Y)$ is so. Moreover, if $H_G^*(X, Y)$ is Cohen–Macaulay of dimension d, then the spectral sequence converging to $H_*^G(X, Y)$ collapses at the E_2 page by Proposition 2, and

$$H_*^G(X, Y) = \operatorname{Ext}_{R_G}^{r-d}(H_G^*(X, Y), R_G)[r - d] \tag{79}$$

is again Cohen–Macaulay of dimension d. The other direction is analogous. □

We will see in Sect. 6.3 that Cohen–Macaulay filtrations exist. For the moment, we record several properties.

Proposition 13 *Let (X_i) be a Cohen–Macaulay filtration of X. For any $0 \leq i \leq r$ there is a short exact sequence*

$$0 \to H_*^G(X_i) \to H_*^G(X) \to H_*^G(X, X_i) \to 0.$$

Proof As in [4, Proposition 4.3]. □

Proposition 14 *Let (X_i) be a Cohen–Macaulay filtration of X. The associated spectral sequence converging to $H_*^G(X)$ degenerates at $E_i^1 = H_{*+i}^G(X_i, X_{i-1})$.*

Proof As in [3, Corollary 4.4]. □

Proposition 15 *Let $\mathcal{F}$ be a G-stable filtration of X. Then $\mathcal{F}$ is a Cohen–Macaulay filtration for X, considered as a G-space, if and only if it is Cohen–Macaulay for X, considered as a T-space.*

Proof Write $\mathcal{F} = (X_i)$. Proposition 9 (2) gives, for $0 \leq i \leq r$,

$$H_*^T(X_i, X_{i-1}) = R_T \otimes_{R_G} H_*^G(X_i, X_{i-1}), \tag{80}$$

which as an R_G-module consists of finitely many copies of $H_*^G(X_i, X_{i-1})$. Hence $H_*^G(X_i, X_{i-1})$ is zero or Cohen–Macaulay of dimension $r - i$ over R_G if and only if

$H_*^T(X_i, X_{i-1})$ is so. But $H_*^T(X_i, X_{i-1})$ is Cohen–Macaulay of dimension $r-i$ over R_G if and only if it is so over R_T, cf. [22, Lemma 2.6]. □

6.2 The Atiyah–Bredon Sequence

Let $\mathcal{F} = (X_i)$ be a Cohen–Macaulay filtration of X. It gives rise to an *Atiyah–Bredon sequence*

$$0 \longrightarrow H_G^*(X) \xrightarrow{\iota^*} H_G^*(X_0) \xrightarrow{\delta_0} H_G^{*+1}(X_1, X_0) \xrightarrow{\delta_1} \cdots \\ \xrightarrow{\delta_{r-2}} H_G^{*+r-1}(X_{r-1}, X_{r-2}) \xrightarrow{\delta_{r-1}} H_G^{*+r}(X_r, X_{r-1}) \longrightarrow 0. \quad (81)$$

Here ι^* is induced by the inclusion $\iota\colon X_0 \hookrightarrow X$, and δ_i for $i \geq 0$ is the connecting homomorphism in the long exact sequences for the triples (X_{i+1}, X_i, X_{i-1}). We consider this sequence as a dg R_G-module $\overline{AB}_G^*(\mathcal{F})$ with

$$\overline{AB}_G^i(\mathcal{F}) = \begin{cases} H_G^*(X) & \text{if } i = -1, \\ H_G^{*+i}(X_i, X_{i-1}) & \text{if } 0 \leq i \leq r \end{cases} \quad (82)$$

and differentials δ_i as above plus $\delta_{-1} = \iota^*$. We call $\overline{AB}_G^*(\mathcal{F})$ the *augmented Atiyah–Bredon complex* of $\mathcal{F}$. The (non-augmented) *Atiyah–Bredon complex* $AB_G^*(\mathcal{F})$ is obtained by dropping the leading term $\overline{AB}_G^{-1}(\mathcal{F}) = H_G^*(X)$. It is the E_1 page of the spectral sequence arising from the filtration $\mathcal{F}$ and converging to $H_G^*(X)$.

Theorem 5 *The cohomology of the Atiyah–Bredon complex is*

$$H^j(AB_G^*(\mathcal{F})) = \operatorname{Ext}_{R_G}^j(H_*^G(X), R_G)$$

for $j \geq 0$. Under this isomorphism, the map $H_G^*(X) \to H^0(AB_G^*(\mathcal{F}))$ corresponds to the canonical map $H_G^*(X) \to \operatorname{Hom}_{R_G}(H_*^G(X), R_G)$. In particular, the cohomology of $AB_G^*(\mathcal{F})$ and $\overline{AB}_G^*(\mathcal{F})$ is independent of the Cohen–Macaulay filtration $\mathcal{F}$.

Proof This was proven in [3, Theorem 5.1] for the orbit filtration of a T-space; see [4, Sect. 5] for an alternative argument. Both proofs generalize to the present context. The independence of the Cohen–Macaulay filtration $\mathcal{F}$ was already pointed out in [3, Remark 4.9]. □

6.3 The Orbit Filtration

Let X be a G-space such that $H^*(X)$ is finite-dimensional and such that only finitely many infinitesimal orbit types occur in X. We will see in Sect. 7 that the second assumption is redundant if X is a manifold or locally orientable orbifold.

Remark 5 We note that X has finitely many infinitesimal G-orbit types if and only if it has finitely many infinitesimal T-orbit types. That each infinitesimal G-orbit type restricts to only finitely many infinitesimal T-orbit types can be seen as follows, cf. [7, Proposition VIII.3.14]: Let $K \subset G$ be an isotropy group occurring in X. Then G/K is a compact differentiable T-manifold, hence has only finitely many T-orbit types. Moreover, the infinitesimal T-orbit types in G/K depend only on the infinitesimal G-orbit type determined by K.

The converse is due to Mostow, cf. [7, Theorem VII.3.1]. Note that the proof given there only uses that (in the notation of [7, Theorem VII.2.1]) there are finitely many groups $(S \cap T)^0$ for $S \in \mathcal{S}$, see [7, p. 97].

The *G-orbit filtration* $(X_{i,G})$ of X is defined by

$$X_{i,G} = \{ x \in X \mid \operatorname{rank} G_x \geq r - i \} \tag{83}$$

for $-1 \leq i \leq r$. Each $X_{i,G}$ is G-stable and, by the slice theorem, closed in X. If $G = T$ is a torus, then the strata $X_{i,T} \setminus X_{i-1,T}$ are disjoint unions of fixed point sets of subtori, hence smooth if X is so. This may fail for non-commutative G, see [22, Remark 3.1] for an example.

We need a relative version of Proposition 12.

Proposition 16 *Let $0 \leq i \leq r$. The inclusion of pairs $(X_{i,T}, X_{i,T} \cap X_{i-1,G}) \hookrightarrow (X_{i,G}, X_{i-1,G})$ induces an isomorphism of R_G-modules*

$$H_G^*(X_{i,G}, X_{i-1,G}) = H_T^*(X_{i,T}, X_{i,T} \cap X_{i-1,G})^W.$$

Proof For cohomology with compact supports, the claim follows immediately from Proposition 12 and the natural isomorphism (46).

To see that it holds for cohomology with closed supports, we observe first that Proposition 12 generalizes by the five lemma to pairs (X, U) where $U \subset X$ is open (and all isotropy groups in X have rank b). Back to the more general case we are considering, we have by the tautness of Alexander–Spanier cohomology

$$H_G^*(X_{i,G}, X_{i-1,G}) = \varinjlim H_G^*(X_{i,G}, U) \tag{84}$$

where the direct limit is taken over all G-stable open neighbourhoods of $X_{i-1,G}$ in $X_{i,G}$. By excision and Proposition 12 for the pair $(X_{i,G}, U)$,

$$= \varinjlim H_G^*(X_{i,G} \setminus X_{i-1,G}, U \setminus X_{i-1,G}) \tag{85}$$

$$= \varinjlim \big(H_T^*(X_{i,T} \setminus X_{i-1,G}, V \setminus X_{i-1,G})^W\big), \tag{86}$$

this time the direct limit being over all $N_G(T)$-stable open neighbourhoods V of $X_{i,T} \cap X_{i-1,G}$ in $X_{i,T}$. Interchanging the direct limit and taking W-invariants and reversing the previous arguments, we finally arrive at

$$= H_T^*(X_{i,T}, X_{i,T} \cap X_{i-1,G})^W. \tag{87}$$

□

Lemma 21 *The vector spaces* $H^*(X_{i,T}, X_{i,T} \cap X_{i-1,G})$ *and* $H^*(X_{i,G}, X_{i-1,G})$ *are finite-dimensional for* $0 \le i \le r$.

Proof By induction we can assume that $H^*(X_{j,G}, X_{j-1,G})$ is finite-dimensional for $j < i$ and hence so is $H^*(X_{i-1,G})$. Thus, $X_{i-1,G}$ satisfies again our assumptions for X. By [2, Proposition 4.1.14], both $H^*(X_{i,T})$ and

$$H^*(X_{i,T} \cap X_{i-1,G}) = H^*((X_{i-1,G})_{i,T}) \tag{88}$$

are finite-dimensional and therefore also the relative cohomology $H^*(X_{i,T}, X_{i,T} \cap X_{i-1,G})$. Hence $H_T^*(X_{i,T}, X_{i,T} \cap X_{i-1,G})$ is finitely generated over R_T and R_G, and so is the submodule of W-invariants over R_G. The claim now follows from Proposition 16 and Remark 2. □

Proposition 17 *The orbit filtration of X is Cohen–Macaulay.*

Proof The torus case was established in [4, Proposition 4.1].[2] In fact, the proof in [4] shows the following: Let $Y \subset X_{i,T}$ be a T-stable closed subset containing $X_{i-1,T}$ and such that $H^*(X_{i,T}, Y)$ is finite-dimensional. Then $H_T^*(X_{i,T}, Y)$ and $H_*^T(X_{i,T}, Y)$ are zero or Cohen–Macaulay of dimension $r - i$.

By Remark 5, only finitely many infinitesimal T-orbit types occur in $X_{i,T}$, and $H^*(X_{i,T}, X_{i,T} \cap X_{i-1,G})$ is finite-dimensional by Lemma 21. By what we have said above, this implies that the R_T-module $H_T^*(X_{i,T}, X_{i,T} \cap X_{i-1,G})$ is zero or Cohen–Macaulay of dimension $r - i$. By [22, Lemma 2.7] the same holds for the R_G-submodule of W-invariants. Proposition 16 now shows that also $H_G^*(X_i, X_{i-1})$ is Cohen–Macaulay of dimension $r - i$. Hence the filtration is Cohen–Macaulay by Lemma 20. □

[2]The local contractability assumptions made there were needed to ensure that singular and Alexander–Spanier cohomology coincide. This is not necessary in our setting, cf. [4, Remark 2.17].

6.4 *Partial Exactness*

Let X be a G-space such that $H^*(X)$ is finite-dimensional and such that only finitely many infinitesimal orbit types occur in X. Moreover, let $\mathcal{F} = (X_i)$ be a Cohen–Macaulay filtration of X, for example the orbit filtration.

Theorem 6 *Let* $0 \le j \le r$. *Then* $H_G^*(X)$ *is a jth syzygy over* R_G *if and only if* $H^i(\overline{AB}_G^*(\mathcal{F})) = 0$ *for all* $i \le j-2$, *i.e., if and only if the part*

$$0 \to H_G^*(X) \to H_G^*(X_0) \to \cdots \to H_G^{*+j-1}(X_{j-1}, X_{j-2})$$

of the Atiyah–Bredon sequence for $\mathcal{F}$ *is exact.*

Proof The torus case was established in [3, Theorem 5.7] (or [4, Theorem 4.8]). It is stated there for the orbit filtration of a T-space, but we know from Theorem 5 that $H^*(\overline{AB}_T^*(\mathcal{F}))$ is independent of the Cohen–Macaulay filtration $\mathcal{F}$.

To reduce the general case to the torus case, we recall from Proposition 15 that $\mathcal{F}$ is also a Cohen–Macaulay filtration of X, considered as a T-space. In addition, we observe that

$$H^i(\overline{AB}_T^*(\mathcal{F})) = R_T \otimes_{R_G} H^i(\overline{AB}_G^*(\mathcal{F})) \tag{89}$$

for $0 \le i \le r$, which follows from the isomorphism

$$H_T^*(X_i, X_{i-1}) = R_T \otimes_{R_G} H_G^*(X_i, X_{i-1}) \tag{90}$$

given by Proposition 9 (2) and the fact that R_T is free over R_G.

By Proposition 10, $H_G^*(X)$ is a jth syzygy over R_G if and only if $H_T^*(X)$ is a jth syzygy over R_T. As remarked above, this latter condition is equivalent to the vanishing of $H^i(\overline{AB}_T^*(\mathcal{F}))$ for $i \le j-2$, which in turn is equivalent to the vanishing of $H^i(\overline{AB}_G^*(\mathcal{F}))$ for $i \le j-2$ by (89) and again the freeness of R_T over R_G. □

Corollary 3 *Assume that* X *is a compact oriented* G*-orbifold. The* R_G*-module* $H_G^*(X)$ *is torsion-free (reflexive) if and only if the equivariant Poincaré pairing*

$$H_G^*(X) \times H_G^*(X) \to R_G, \quad (\alpha, \beta) \mapsto \int_X \alpha\beta$$

is non-degenerate (perfect).

This characterization of the non-degeneracy of the equivariant Poincaré pairing has been given by Guillemin et al. [24, Theorem C.70] under the assumption that the maximal-rank stratum $X_{0,G}$ is smooth and orientable.

Proof This follows from Theorem 6 by observing that the map

$$H_G^*(X) \to \operatorname{Hom}_{R_G}(H_G^*(X), R_G) \tag{91}$$

induced by the equivariant Poincaré pairing is the composition of the canonical map $H_G^*(X) \to \operatorname{Hom}_{R_G}(H_*^G(X), R_G) = H^0(AB_G^*(\mathcal{F}))$ with the R_G-transpose of the Poincaré duality isomorphism $H_G^*(X) \to H_*^G(X)$ from Proposition 8.

It can alternatively be deduced from Proposition 10 together with the observation that the G-equivariant Poincaré pairing is non-degenerate or perfect if and only if the T-equivariant one is so. We leave the details to the reader. □

Proposition 18 *Assume that X is a compact orientable G-orbifold. If $H_G^*(X)$ is a syzygy of order $\geq r/2$, then it is free over R_G.*

Proof This follows from Theorems 5 and 6 in the same way as for the torus case [3, Proposition 5.12]. Using Proposition 10, one could also deduce it directly from the torus case. □

The bound "$r/2$" in Proposition 18 is optimal for any G:

Example 2 Let $a, b \geq 1$ and $0 \leq m \leq (r-1)/2$. Let Y be the *big polygon space* defined by

$$\|u_j\|^2 + \|z_j\|^2 = 1 \qquad (1 \leq j \leq r), \tag{92}$$

$$u_1 + \cdots + u_{2m+1} = 0 \tag{93}$$

where $u_1, \ldots, u_r \in \mathbb{C}^a$ and $z_1, \ldots, z_r \in \mathbb{C}^b$. The torus $T \cong (S^1)^r$ acts on Y by rotating the variables z_j. In [19, Sect. 5] it is shown that Y is a compact orientable T-manifold and that $H_T^*(Y)$ is a syzygy of order exactly m over R_T.

The induced G-manifold $X = G \times_T Y$ is again compact orientable. By Proposition 11, $H_G^*(X)$ is a syzygy of order exactly m over R_G. In particular, we see that any syzygy order less than $r/2$ can occur among the equivariant cohomology modules of compact orientable G-manifolds.

7 Infinitesimal Orbit Types

7.1 Preliminaries on Torus Actions

Recall that $T \cong (S^1)^r$ is a torus. Several properties of T-equivariant cohomology with compact supports and homology with closed supports were established in [4] for T-spaces X with finite-dimensional cohomology and/or finitely many infinitesimal orbit types[3]; we now drop these restrictions and indicate how to adapt the proofs.

Let X be a T-space.

Proposition 19 *Let $K \subset T$ be a subtorus and set $L = T/K$. If K acts trivially on X, then there are isomorphisms of R_T-modules*

$$H^*_{T,c}(X) = R_T \otimes_{R_L} H^*_{L,c}(X), \qquad H^{T,c}_*(X) = R_T \otimes_{R_L} H^{L,c}_*(X).$$

Proof Choose a splitting $T \cong K \times L$, which induces an isomorphism $R_T \cong R_K \otimes R_L$. Since K acts trivially on X, we have

$$\Omega^*_{T,c}(X) = R_K \otimes \Omega^*_{L,c}(X) = R_T \otimes_{R_L} \Omega^*_{L,c}(X), \tag{94}$$

which gives the first isomorphism. The homological case is analogous. □

Let X_α, $\alpha \in A$, be the pieces of the partition of X into infinitesimal T-orbit types. For $\alpha \in A$ let $T_\alpha \subset T$ be the common identity component of the T_x with $x \in X_\alpha$. Moreover, let $\mathfrak{t}_\alpha$ be the Lie algebra of T_α, and let $r_\alpha = r - \dim T_\alpha$ be the common dimension of the T-orbits in X_α.

Regarding the orbit filtration $(X_p) = (X_{p,T})$ of X, we have

$$X_p = \bigcup_{r_\alpha \le p} X_\alpha \tag{95}$$

for $-1 \le p \le r$. Hence $X_p \setminus X_{p-1}$ is the disjoint union of the X_α with $r_\alpha = p$.

Lemma 22 *We have*

$$H^*_{T,c}(X_p \setminus X_{p-1}) = \bigoplus_{\substack{\alpha \in A \\ r_\alpha = p}} H^*_{T,c}(X_\alpha),$$

$$H^{T,c}_*(X_p \setminus X_{p-1}) = \prod_{\substack{\alpha \in A \\ r_\alpha = p}} H^{T,c}_*(X_\alpha).$$

Proof The first isomorphism is a consequence of the fact that $\Omega^*_c(X_p \setminus X_{p-1})$ is the direct sum of the $\Omega^*_c(X_\alpha)$. Its dual $\Omega^*_c(X_p \setminus X_{p-1})^\vee$ therefore is the direct product of the $\Omega^*_c(X_\alpha)^\vee$, which implies the second identity. □

For a multiplicative subset $S \subset R_T$ set

$$A(S) = \{\alpha \in A \mid S \cap \ker(R_T \to \mathbb{R}[\mathfrak{t}^*_\alpha]) = \emptyset\}, \tag{96}$$

---

[3]The stronger assumption of finitely many *orbit types* made in [4, Sect. 2.1] only serves to allow the use of singular cohomology for the quotient X/T instead of Alexander–Spanier cohomology.

and let X^S be the union of the X_α with $\alpha \in A(S)$. By the slice theorem, X^S is closed in X.

Proposition 20 (Localization Theorem) *Let $S \subset R_T$ be a multiplicative subset. The inclusion $X^S \hookrightarrow X$ induces isomorphisms of $S^{-1}R_T$-modules*

$$S^{-1}H^*_{T,c}(X) \to S^{-1}H^*_{T,c}(X^S),$$
$$S^{-1}H^{T,c}_*(X^S) \to S^{-1}H^{T,c}_*(X).$$

Note that we put no restriction on the number of infinitesimal orbit types in X.

Proof For the cohomological claim we assume first $X = X_\alpha$ and $A(S) = \emptyset$, say $f \in S \cap \ker(R_T \to \mathbb{R}[\mathfrak{t}^*_\alpha])$. Since $L = T/T_\alpha$ acts locally freely on X_α, we have $H^*_{L,c}(X_\alpha) = H^*_c(X_\alpha/L)$ by Lemma 12. In particular, $H^*_{L,c}(X_\alpha)$ is bounded. Because the kernel of the restriction map $R_T \to \mathbb{R}[\mathfrak{t}^*_\alpha]$ is generated by $\mathfrak{l}^* \subset R_T$, this together with Proposition 19 shows that a power of f annihilates $H^*_{T,c}(X_\alpha)$, so that $S^{-1}H^*_{T,c}(X_\alpha) = 0$.

For general X, the orbit filtration gives rise to the decreasing filtration $\Omega^*_{T,c}(X \setminus X_{p-1})$ of $\Omega^*_{T,c}(X)$, hence to a spectral sequence converging to $H^*_{T,c}(X)$ with

$$E^p_0(X) = C^*_T(\Omega^*_{X,c}(X_p, X_{p-1})) = \Omega^*_{T,c}(X_p, X_{p-1}), \tag{97}$$
$$E^p_1(X) = H^*_{T,c}(X_p, X_{p-1}) = H^*_{T,c}(X_p \setminus X_{p-1}), \tag{98}$$

where we have used the isomorphism (47). Note that $H^*_{T,c}(X_p \setminus X_{p-1})$ is given by Lemma 22. Everything said about X applies equally to X^S with $A(S)$ taking the role of A. Since X^S is closed in X, the inclusion $X^S \hookrightarrow X$ induces a map of spectral sequences $E_*(X) \to E_*(X^S)$. Localizing $E_1(X)$ at S eliminates all terms $H^*_{T,c}(X_\alpha)$ with $\alpha \notin A(S)$. The localized map $S^{-1}E_1(X) \to S^{-1}E_1(X^S)$ therefore is an isomorphism, hence so is the map $S^{-1}H^*_{T,c}(X) \to S^{-1}H^*_{T,c}(X^S)$.

The homological claim follows from this and the universal coefficient theorem (Proposition 6) as in [4, Proposition 2.5]. □

Lemma 23 *For $0 \le p \le r$, there is a multiplicative subset $S \subset R_T$ such that $A(S) = \{\alpha \in A \mid r_\alpha \le p\}$. For such an S, the localization map $H^{T,c}_*(X_q, X_{q-1}) \to S^{-1}H^{T,c}_*(X_q, X_{q-1})$ is injective for $q \le p$, and $S^{-1}H^{T,c}_*(X_q, X_{q-1}) = 0$ for $q > p$.*

Proof Note first that T has only countably many subtori, so that A is countable. For $\alpha \in A$, the dimension of the dual $\mathfrak{l}^*_\alpha$ of the Lie algebra of $L_\alpha = T/T_\alpha$ is r_α. Assume $r_\alpha > p$. Since $\mathbb{R}$ is uncountable,

$$\mathfrak{l}^*_\alpha \setminus \bigcup_{r_\beta \le p} \mathfrak{l}^*_\beta \tag{99}$$

cannot be empty. Pick a t_α from this set and let $S \subset R_T$ be the multiplicative subset generated by all such t_α. Then, by construction, $A(S)$ is of the claimed form.

From Lemma 22 and Proposition 19 we see that $H_*^{T,c}(X_q, X_{q-1})$ is S-torsion-free for $q \leq p$, which means that the localization map is injective. Moreover, Proposition 20 gives $S^{-1}H_*^{T,c}(X_q, X_{q-1}) = 0$ for $q > p$ as $(X_q \setminus X_{q-1})^S = \emptyset$. □

The following result generalizes [4, Corollary 4.4] to our setting. Note that unlike [3] and [4], the proof below does not use the theory of Cohen–Macaulay modules.

Proposition 21 *The spectral sequence induced by the orbit filtration and converging to $H_*^{T,c}(X)$ degenerates at the E_1 page, which is given by Lemma 22.*

We could alternatively state this result by saying that certain short sequences are exact, cf. Proposition 13.

Proof We prove the claim by contradiction. Let d_k, $k \geq 1$, be the first non-vanishing differential, say with non-zero value $a = d_k(b) \in H_*^{T,c}(X_p, X_{p-1})$. Choose an $S \subset R_T$ as in Lemma 23 and localize the E_k page of the spectral sequence at S. In the localized spectral sequence the differential vanishes as $S^{-1}H_*^{T,c}(X_q, X_{q-1}) = 0$ for all $q > p$. But $H_*^{T,c}(X_p, X_{p-1})$ injects into its localization $S^{-1}H_*^{T,c}(X_p, X_{p-1})$, which implies $a = 0$. □

Everything done so far in this section goes through for twisted coefficients with respect to a fixed orientation cover $\tilde{Z} \to Z$ in case the ambient T-orbifold Z is locally orientable. In addition, one has the following:

Lemma 24 *Let X be a locally orientable T-orbifold. Then each component Y of X^T is again locally orientable. Moreover, the restriction of the orientation cover for X to Y is the orientation cover for Y.*

Proof The first part is well-known, cf. [7, Theorem V.3.2]. With our tools, it can be seen as follows: Choose a uniformizing chart $\tilde{U} \to U$ of X at $y \in Y$ such that $\tilde{U}$ is an open ball with linearized torus action. Because X is locally orientable, $H_c^*(U)$ is non-zero and concentrated in top degree. Thus $H_{T,c}^*(U) = R_T \otimes H_c^*(U)$. By the localization theorem, this implies that $H_{T,c}^*(U^T) = R_T \otimes H_c^*(U^T)$ is non-zero. Hence U^T is the quotient of $\tilde{U}^T$ by an orientation-preserving action since otherwise we would have $H_c^*(U^T) = 0$.

For the second claim it suffices to observe that Y is orientable if and only if X is so. This is a consequence of the fact that the normal (orbi)bundle of Y in X is always orientable. See [15, Corollary 2] for the case of manifolds, from which the orbifold case follows. □

7.2 The Number of Infinitesimal Orbit Types

Theorem 7 *Let X be a locally orientable G-orbifold. If $H^*(X)$ is finite-dimensional, then only finitely many infinitesimal orbit types occur in X.*

Proof By Mostow's argument we may assume that $G = T$ is a torus, see Remark 5.

Assume that there are infinitely many infinitesimal orbit types. By Proposition 21 and the previous section, the spectral sequence induced by the orbit filtration and converging to $H_*^{T,c}(X;\tilde{\mathbb{R}})$ degenerates at the E_1 page. Each of the infinitely many terms $H_*^{T,c}(X_\alpha;\tilde{\mathbb{R}})$ appearing in the twisted version of Lemma 22 is non-zero as it contains an equivariant orientation by Lemma 24. Thus, $E_\infty(X) = E_1(X)$ is not finitely generated. This implies that $H_*^{T,c}(X;\tilde{\mathbb{R}})$ cannot be finitely generated either because $E_\infty(X)$ arises from a filtration of it and R_T is Noetherian.

On the other hand, $H^*(X)$ is finite-dimensional by assumption, hence $H_T^*(X)$ is finitely generated over R_T. By equivariant Poincaré duality (Proposition 8), the same holds for $H_*^{T,c}(X;\tilde{\mathbb{R}})$. Contradiction. □

We conclude with several examples. The first one, due to Montgomery [34, Sect. 3], illustrates the difference between our result and Mann's [28, Theorem 3.5]. Recall that Mann showed that any orientable cohomology manifold with finitely generated integral cohomology has only finitely many orbit types.

Example 3 Consider a circle acting on $\mathbb{R}^3$ via rotations about some axis. For each $m \in \mathbb{N}$, choose an open solid torus equivariantly diffeomorphic to $A_m \times S^1$, where $A_m \subset \mathbb{R}^3$ is an open disc with centre b_m. (These tori have to be disjoint.) Remove the central circle $b_m \times S^1$ of each torus and glue the remaining manifold to copies of the $A_m \times S^1$ via the diffeomorphisms f_m of $(A_m \setminus b_m) \times S^1 \cong (0,1) \times S^1 \times S^1$ given by $f_m(r,\alpha,\beta) = (r, \alpha\beta^{1-m}, \beta^m/\alpha)$. Each subgroup $\mathbb{Z}_m \subset S^1$ occurs as an isotropy group in the orientable manifold X thus obtained. One can check that $H_1(X;\mathbb{Z})$ is isomorphic to the direct sum of all $\mathbb{Z}_m$ and that $H_2(X;\mathbb{Z})$ vanishes. In particular, $H_*(X;\mathbb{Z})$ is not finitely generated. On the other hand, $H_*(X;\mathbb{R})$ is finite-dimensional, and Theorem 7 holds (trivially).

Example 4 We now imitate this idea for the standard action of $T = S^1 \times S^1$ on $\mathbb{C}^2$. In the free part of the action, choose disjoint open subsets of the form $A_m \times T$, where the slice $A_m \subset \mathbb{C}^2$ is again an open disc with centre b_m. Remove $b_m \times T$ from the manifold and glue the rest to a copy of $A_m \times T$ via the diffeomorphism f_m of $(A_m \setminus b_m) \times T \cong (0,1) \times (S^1)^3$ given by $f_m(r,\alpha,\beta,\gamma) = (r, \alpha\beta, \beta^m\gamma, \alpha)$. Each such operation increases the Betti sum of the manifold by 4, and it produces the isotropy group $\{(g, g^{-m}) \mid g \in S^1\} \subset T$. If this is done for all $m \in \mathbb{Z}$, we obtain a manifold with infinitely many distinct infinitesimal isotropy groups and infinite Betti sum.

We finally recall an example of Kister and Mann [25, Sect. 1] showing that Theorem 7 cannot be naively extended to actions on more general spaces.

Example 5 Take countably many closed discs, connected by a line through their centres. This space is properly homotopy-equivalent to $\mathbb{R}$, hence contractible. On the other hand, the discs can be rotated independently via characters $T \to S^1$ to produce infinitely many distinct infinitesimal orbit types. Note that an argument as in the proof of Theorem 7 fails here because the non-fixed points do not contribute to $H_*^{T,c}(X)$. In fact, for a closed disc D with centre x one has $H_*^c(D,\{x\}) = 0$, hence also $H_*^{T,c}(D,\{x\}) = 0$ by Proposition 5.

Acknowledgements I would like to thank Chris Allday, Andrey Minchenko, Volker Puppe, Reyer Sjamaar and Andrzej Weber for helpful discussions, and the organizers of the conference "Configuration Spaces" for creating a stimulating environment in an extraordinary place. I am also indebted to the referee for a very careful reading and for several valuable suggestions and corrections.

References

1. A. Adem, J. Leida, Y. Ruan, *Orbifolds and Stringy Topology* (Cambridge University Press, Cambridge 2007). doi:10.1017/CBO9780511543081
2. C. Allday, V. Puppe, *Cohomological Methods in Transformation Groups* (Cambridge University Press, Cambridge 1993). doi:10.1017/CBO9780511526275
3. C. Allday, M. Franz, V. Puppe, Equivariant cohomology, syzygies and orbit structure. Trans. Am. Math. Soc. **366**, 6567–6589 (2014). doi:10.1090/S0002-9947-2014-06165-5
4. C. Allday, M. Franz, V. Puppe, Equivariant Poincaré–Alexander–Lefschetz duality and the Cohen–Macaulay property. Algebr. Geom. Topol. **14**, 1339–1375 (2014). doi:10.2140/agt.2014.14.1339
5. T.J. Baird, Cohomology of the space of commuting n-tuples in a compact Lie group, Algebr. Geom. Topol. **7**, 737–754 (2007). doi:10.2140/agt.2007.7.737
6. T.J. Baird, Classifying spaces of twisted loop groups. Algebr. Geom. Topol. **16**, 211–229 (2016). doi:10.2140/agt.2016.16.211
7. A. Borel, *Seminar on Transformation Groups* (Princeton University Press, Princeton, 1960)
8. G.E. Bredon, *Sheaf Theory*, 2nd edn. GTM, vol. 170 (Springer, New York, 1997). doi:10.1007/978-1-4612-0647-7
9. M. Brion, Equivariant cohomology and equivariant intersection theory, pp. 1–37 in *Representation Theories and Algebraic Geometry* edited by A. Broer, et al. (Montreal, 1997). NATO Adv. Sci. Inst. Ser. C, vol.514, (Kluwer, Dordrecht, 1998). doi:10.1007/978-94-015-9131-7_1
10. M. Brion, Poincaré duality and equivariant (co)homology, Mich. Math. J. **48**, 77–92 (2000). doi:10.1307/mmj/1030132709
11. W. Bruns, J. Herzog, *Cohen–Macaulay Rings*, 2nd edn., (Cambridge University Press, Cambridge 1998). doi:10.1017/CBO9780511608681
12. W. Bruns, U. Vetter, *Determinantal Rings*. LNM, vol. 1327 (Springer, Berlin 1988). doi:10.1007/BFb0080378
13. C. De Concini, C. Procesi, M. Vergne, The infinitesimal index. J. Inst. Math. Jussieu **12**, 297–334 (2013). doi:10.1017/S1474748012000722
14. M. Duflo, M. Vergne, Cohomologie équivariante et descente. Astérisque **215**, 5–108 (1993)
15. J. Duflot, Smooth toral actions. Topology **22**, 253–265 (1983). doi:10.1016/0040-9383(83)90012-5
16. D. Edidin, W. Graham, Equivariant intersection theory. Invent. Math. **131**, 595–634 (1998). doi:10.1007/s002220050214
17. D. Eisenbud, *The Geometry of Syzygies*. GTM, vol. 229 (Springer, New York, 2005). doi:10.1007/b137572
18. M. Franz, A quotient criterion for syzygies in equivariant cohomology. Transform. Groups. arXiv:1205.4462v3 (to appear)
19. M. Franz, Big polygon spaces. Int. Math. Res. Not. **2015**, 13379–13405 (2015)
20. V.A. Ginzburg, Equivariant cohomology and Kähler's geometry. Funct. Anal. Appl. **21**, 271–283 (1987). doi:10.1007/BF01077801
21. O. Goertsches, A.-L. Mare, Non-abelian GKM theory. Math. Z. **277**, 1–27 (2014). doi:10.1007/s00209-013-1242-x
22. O. Goertsches, S. Rollenske, Torsion in equivariant cohomology and Cohen–Macaulay G-actions. Transform. Groups **16**, 1063–1080 (2011). doi:10.1007/s00031-011-9154-5

23. V. Guillemin, S. Sternberg, *Supersymmetry and Equivariant de Rham Theory* (Springer, Berlin, 1999). doi:10.1007/978-3-662-03992-2
24. V. Guillemin, V. Ginzburg, Y. Karshon, *Moment Maps, Cobordisms, and Hamiltonian Group Actions* (AMS, Providence, RI, 2002). doi:10.1090/surv/098
25. J.M. Kister, L.N. Mann, Isotropy structure of compact Lie groups on complexes. *Mich. Math. J.* **9**, 93–96 (1962). doi:10.1307/mmj/1028998627
26. B. Kostant, Lie group representations on polynomial rings. *Am. J. Math.* **85**, 327–404 (1963). doi:10.2307/2373130
27. E. Lerman, S. Tolman, Hamiltonian torus actions on symplectic orbifolds and toric varieties. Trans. Am. Math. Soc. **349**, 4201–4230 (1997). doi:10.1090/S0002-9947-97-01821-7
28. L.N. Mann, Finite orbit structure on locally compact manifolds. Mich. Math. J. **9**, 87–92 (1962). doi:10.1307/mmj/1028998626
29. E. Meinrenken, Symplectic surgery and the Spin^c-Dirac operator. Adv. Math. **134**, 240–277 (1998). doi:10.1006/aima.1997.1701
30. E. Meinrenken, Witten's formulas for intersection pairings on moduli spaces of flat G-bundles. Adv. Math. **197**, 140–197 (2005). doi:10.1016/j.aim.2004.10.002
31. L.I. Nicolaescu, On a theorem of Henri Cartan concerning the equivariant cohomology. An. Ştiinţ. Univ. Al. I. Cuza Iaşi. Mat. (N.S.) **45**, 17–38 (1999), http://www.math.uaic.ro/~annalsmath/F1-1999.htm
32. E.H. Spanier, *Algebraic Topology* (McGraw-Hill, New York, 1966)
33. F.W. Warner, *Foundations of Differentiable Manifolds and Lie Groups* (Springer, New York, 1983). doi:10.1007/978-1-4757-1799-0
34. C.-T. Yang, On a problem of Montgomery. Proc. Am. Math. Soc. **8**, 255–257 (1957). doi:10.1090/S0002-9939-1957-0087040-4

A Survey on Spaces of Homomorphisms to Lie Groups

Frederick R. Cohen and Mentor Stafa

Dedicated to Samuel Gitler Hammer,
who brought us much joy and interest in Mathematics.

Abstract The purpose of this article is to give an exposition of topological properties of spaces of homomorphisms from certain finitely generated discrete groups to Lie groups G, and to describe their connections to classical representation theory, as well as other structures. Various properties are given when G is replaced by a small category, or the discrete group is given by a right-angled Artin group.

Keywords Representation variety • Right-angled Artin group • Small category • Space of homomorphisms

1 Introduction

Let π be a finitely generated discrete group and G a Lie group. In this paper we describe features of spaces of group homomorphisms from π to G denoted by $Hom(\pi, G)$, which is given the subspace topology of the Cartesian product of a finite number of copies of G.

Spaces of group homomorphisms $Hom(\pi, G)$ exhibit interesting properties. They have played an important role in mathematical work related to physics going back to work of Witten [31, 32], who considered spaces of commuting pairs and triples in certain Lie groups. Following the same lines but using different methods Borel et al. developed the space of commuting as well as *almost commuting* pairs and triples [9], i.e. pairs and triples commuting up to an element in the center. The spaces have

F.R. Cohen
University of Rochester, Rochester, NY 14627, USA
e-mail: fred.cohen@rochester.edu

M. Stafa (✉)
Tulane University, New Orleans, LA 70118, USA

Departement Mathematik, ETH Zurich, 8092, Switzerland
e-mail: mstafa@tulane.edu

F. Callegaro et al. (eds.), *Configuration Spaces*, Springer INdAM Series 14,
DOI 10.1007/978-3-319-31580-5_15

also been studied by Kac and Smilga [19] and Goldman [16] as well as Friedmann and Stanley [14].

The group $G = G^{ad}$ acts by conjugation on the space $Hom(\pi, G)$. The quotient space $\mathcal{M}(\pi, g) = Hom(\pi, G)/G^{ad}$ is called the *representation space* (or *moduli space* or *representation variety, character variety*) of π in G. Representation varieties for fundamental groups of surfaces in compact Lie groups have been studied in the context of gauge theories and hyperbolic geometry when G is the complex or real projective special linear group. These spaces also admit natural symplectic structures as discovered by Goldman [15].

Recently the subject experienced mathematical development in a different direction. Here, we restrict to homotopy theoretic properties of $Hom(\pi, G)$. Adem and Cohen [1] considered the problem in the framework of spaces of homomorphisms $Hom(\pi, G)$, where π is the free abelian group of any rank n. These spaces split when suspended once as given in Theorem 1.

Properties of $Hom(\pi, G)$ have been developed in [14] in case G is a finite group. Using a different approach, the authors Stafa [27] and Cohen and Stafa [12] assemble all the spaces $Hom(\mathbb{Z}^n, G)$ into a larger, more tractable space $Comm(G)$ reminiscent of a Stiefel manifold. This is discussed in more detail in the following sections. A few problems at the interface of these subjects are posed at the end of this article.

The spaces $Hom(\mathbb{Z}^n, G)$ also admit the structure of simplicial spaces. Their geometric realization, denoted by $B(2, G)$ was introduced by Adem et al. [3]. One curious connection arises by turning the natural map

$$B(2, G) \to BG$$

into a fibration with homotopy fibre denoted $E(2, G)$. Basic properties of the monodromy representation of G on the first homology group of the fibre are curious. For example, if G is a finite group of odd order, then the classical Feit-Thompson theorem is equivalent to the map $H_1(E(2, G))_G \to H_1(B(2, G))$, where $H_1(E(2, G))_G$ denotes the module of coinvariants under the monodromy action, failing to be surjective. Properties of the monodromy representation may be interesting, but are currently not well understood [3, 28]. This setting suggests that the spaces $B(2, G)$ contain compelling information. Thus a categorical interpretation is given here in Sect. 9.

Much beautiful work has been done by many people, some of whom are not mentioned here because of the brevity of this article. Some of this work is surveyed below.

2 Spaces of Homomorphisms and Their Topology

Let π be a finitely generated discrete group with n generators $x_1, \dots, x_n$. Then a homomorphism $f \in Hom(\pi, G)$ is determined by the images of the generators $x_1, \dots, x_n$ in G. Thus there is the natural identification of a homomorphism f with the n-tuple $(f(x_i),\ldots,f(x_n))$ using the natural inclusion

$$Hom(\pi, G) \hookrightarrow G^n.$$

The space of homomorphisms can be endowed with the subspace topology of G^n. Equivalently, this association of the subspace topology can be seen from a choice of surjection from the free group with n letters to π

$$F_n \to \pi,$$

which induces an injection on the level of spaces

$$Hom(\pi, G) \hookrightarrow Hom(F_n, G) = G^n.$$

In particular, the image of this inclusion is homeomorphic to $Hom(\pi, G)$ in case G is a Lie group. The space $Hom(\pi, G)$ is regarded as a pointed space with base-point given by the constant function f_0 with $f_0(x) = 1_G \in G$, for all $x \in \pi$.

3 Path-Components, and Fundamental Groups of Spaces of Homomorphisms and Their Topology

Kac and Smilga [19] addressed the number of path-components in the case of $Hom(\mathbb{Z}^n, G)$ for $G = Spin(7)$. The number of path components for the case when $G = SO(k)$, or $O(k)$ was calculated by Rojo [25]. He shows that if k is large enough, the number of path components of $Hom(\mathbb{Z}^n, G)$ no longer depends on k.

Applying the *classifying space functor* gives a map

$$B : Hom(\pi, G) \to Map_*(B\pi, BG),$$

for which each group homomorphism f gives a pointed map $Bf : B\pi \to BG$. It should be noted that if G is compactly generated as a topological space, then the classifying space functor is a continuous functor, thus B is continuous [29].

Applying π_0, the path components, there is an induced map B_0

$$B_0 : \pi_0(Hom(\pi, G)) \to [B\pi, BG],$$

where the homotopy classes of maps $[B\pi, BG]$ is in one-to-one correspondence with all principal G-bundles over $B\pi$. Hence, properties of $\pi_0(Hom(\pi, G))$ inform on principal G-bundles over $B\pi$, even though the map B_0 is neither an injection nor a surjection, in general.

Moreover, the space of homomorphisms can exhibit curious properties such as being disconnected even if G is 1-connected. The case for which every abelian subgroup of a compact and connected Lie group G can be conjugated to a subgroup of the maximal torus T of G is special. In this case, the space of ordered commuting n-tuples $Hom(\mathbb{Z}^n, G)$ is path-connected as given in [1, Proposition 2.3].

A related natural conjecture was that if G is a finite, discrete group, then $B(2, G)$ is a $K(\pi, 1)$ based on examples in [3]. Okay in his thesis [23] gave a counter-example in the special case where G is an extra-special 2-group of order 32.

More recent work of Lawton and Ramras [20] addressed the fundamental group of the character variety of $Hom(\pi, G)/G^{ad}$ where π is the fundamental group of a compact orientable surface. Tom Baird studied the individual spaces $Hom(\mathbb{Z}^n, G)$ via the Leray spectral sequence of the inclusion $Hom(\mathbb{Z}^n, G) \hookrightarrow G^n$, where G is a compact Lie group [6]. Baird gave the additive structure for the rational cohomology of $Hom(\mathbb{Z}^n, G)$, which is also an isomorphism of graded rings, in case G is compact, and connected.

4 Spaces of Pairwise Commuting Elements for Not Necessarily Compact *G*

Suppose that $\pi = \mathbb{Z}^n$ and that G is a reductive algebraic group. Pettet and Souto [24] showed that if K is a maximal compact subgroup of a reductive algebraic group G, then the inclusion

$$Hom(\mathbb{Z}^n, K) \subset Hom(\mathbb{Z}^n, G)$$

is a strong deformation retract. This result has been generalized by Bergeron [8] to finitely generated nilpotent groups. One of the corollaries of [24] is the isomorphism

$$\pi_1(Hom(\mathbb{Z}^n, G)) \cong \pi_1(G)^n,$$

which was first proven in [17]. Therefore we will usually restrict attention to G given by a compact and connected Lie group.

The above feature is special for the case of $Hom(\mathbb{Z}^n, G)$. Namely, if K is a maximal compact subgroup of G, the natural map $Hom(\pi, K) \to Hom(\pi, G)$ where K is a maximal subgroup of G may not induce an isomorphism on the set of path-components. Many such examples arise with $G = PSL(2, \mathbb{R})$, and $K = SO(2)$ where π is given by the fundamental group of a closed, orientable surface of genus $g > 1$. Other examples arise by setting $G = PSL(2, \mathbb{C})$ with $K = SO(3)$ and π equal to certain mapping class groups [1].

5 Stable Decompositions of Spaces of Homomorphisms

A general method of stable decompositions was given in [1]. It was shown in [1, Theorem 1.6] that the space $Hom(\mathbb{Z}^n, G)$ admits a stable decomposition as a wedge sum of certain spaces resembling a fat wedge with appropriate definitions given next.

Definition 1 Define $S_k(G) \subset Hom(\mathbb{Z}^k, G)$ to be the subspace of ordered n-tuples having at least one coordinate equal to the identity element. Define

$$\widehat{Hom}(\mathbb{Z}^k, G) = Hom(\mathbb{Z}^k, G)/S_k(G).$$

It was observed in [1] that the inclusion $S_k(G) \subset Hom(\mathbb{Z}^k, G)$ is a cofibration in case G is a Lie group. Thus there is a cofibre sequence

$$S_k(G) \to Hom(\mathbb{Z}^k, G) \to \widehat{Hom}(\mathbb{Z}^k, G).$$

This basic point-set topological property is useful in what follows.

Theorem 1 (Adem and Cohen [1, Theorem 1.6]) *Assume that G is a closed subgroup of $GL_m(\mathbb{C})$. Then (i) the cofibre sequence $S_k(G) \to Hom(\mathbb{Z}^k, G) \to \widehat{Hom}(\mathbb{Z}^k, G)$ after suspending once is split, and (ii) there is a homotopy equivalence*

$$\Sigma\,(Hom(\mathbb{Z}^n, G)) \to \bigvee_{1 \le k \le n} \Sigma\left(\bigvee_{\binom{n}{k}} \widehat{Hom}(\mathbb{Z}^k, G)\right).$$

This decomposition follows from the natural decomposition of $Hom(A \times B, G)$ where A, and B are discrete groups. This observation is addressed in Lemma 2 below.

The spaces $Hom(\mathbb{Z}^n, G)$ can be assembled to form a new space called $Comm(G)$ as a subspace inside the free associative monoid generated by G, the *James reduced product* $J(G)$ of G [18], as follows.

Definition 2 (Stafa [27, Definition 1.1]) Let G be a topological group. Define

$$Comm(G) = \left(\bigsqcup_{k \ge 0} Hom(\mathbb{Z}^k, G)\right) / \sim$$

where (i) $Hom(\mathbb{Z}^0, G) = \{1_G\}$, and (ii) the equivalence relation $\sim$ is generated by the relation

$$(g_1, \ldots, g_k) \sim (g_1, \ldots, \widehat{g_i}, \ldots, g_k) \text{ if } g_i = 1_G.$$

One feature of the space $Comm(G)$ is that it can be regarded as a universal construction which contains all of the spaces $Hom(\mathbb{Z}^k, G)$, for all k, and is also computationally tractable. Namely, homological properties of the individual spaces $Hom(\mathbb{Z}^n, G)$ are obtained directly from $Comm(G)$. That follows as $Comm(G)$ admits a stable decomposition which directly implies information about $Hom(\mathbb{Z}^k, G)$ for all k. In addition, an equivariant function space is given below which gives the homology groups of $Comm(G)$.

Theorem 2 (Stafa [27, Theorem 1.12]) *Let G be a Lie group. Then there is a homotopy equivalence*

$$\Sigma(Comm(G)) \to \bigvee_{n\geq 1} \Sigma(\widehat{Hom}(\mathbb{Z}^n, G)).$$

The statement of Theorem 2 is that the suspension of $Comm(G)$ contains exactly one copy of the suspension of the quotient

$$\widehat{Hom}(\mathbb{Z}^n, G) = Hom(\mathbb{Z}^n, G)/S_n(G)$$

that appears in the decomposition of $\Sigma(Hom(\mathbb{Z}^n, G))$, for each integer $n > 0$.

Corollary 1 *Let G be a Lie group, and let $E_*(-)$ be any reduced homology theory. Then there are isomorphisms*

1. $E_*(Comm(G)) \to \oplus_{1\leq j<\infty} E_*(\widehat{Hom}(\mathbb{Z}^j, G))$, *and*
2. $E_*(Hom(\mathbb{Z}^n, G)) \to \oplus_{1\leq j\leq n} \oplus_{\binom{n}{j}} E_*(\widehat{Hom}(\mathbb{Z}^j, G))$.

Next assume that G is a compact and connected Lie group with maximal torus T, and Weyl group W such that $Comm(G)$ is path-connected. Then the homology groups of $Comm(G)$ are organized into a relatively simple construction. Let $\mathcal{T}[V]$ denote the tensor algebra generated by the module V where all modules are assumed to be free over the ring $R = \mathbb{Z}[1/|W|]$. With these conditions on G, for which the order of the Weyl group is a unit in *ungraded* singular reduced homology $H_*(-; R)$, then the following holds where $UH_*(X; R)$ denotes *ungraded homology*.

Corollary 2 *Let G be a compact and connected Lie group with maximal torus T, and Weyl group W such that Comm(G) is path-connected. Consider ungraded singular reduced homology $UH_*(-; R)$ where R is the ring $R = \mathbb{Z}[1/|W|]$. Then the ungraded homology $UH_*(Comm(G); R)$ is a free module over $R = \mathbb{Z}[1/|W|]$ which is isomorphic to $\mathcal{T}(\widetilde{H}_*(T; R))$.*

This corollary follows from Theorem 8 and is restated more generally in Sect. 8 (see also [12, Theorem 1.17]). Some examples are the exceptional Lie groups G_2, F_4, E_6, E_7, E_8 and $SU(n)$, $U(n)$, $Sp(n)$, $Spin(n)$, where the ungraded rational homology of $Comm(G)$ is isomorphic to a tensor algebra generated by a module of rank equal to the rank of the reduced rational homology of the maximal torus.

A more precise hold on $Comm(G)$ will be discussed in the next few sections. For example, the next theorem exhibits an equivalence from an equivariant function space to $Comm(G)$.

Theorem 3 *If G is compact, simply-connected Lie group, then there is a map*

$$G \times_{NT} \Omega\Sigma(T) \to Comm(G),$$

which induces an equivalence in homology after the order of the Weyl group W has been inverted. Thus the cohomology of Comm(G) with the order of W inverted is the ring of W-invariants in the cohomology of

$$G/T \times \Omega\Sigma(T).$$

Baird gave the first computation of the rational homology of $Hom(\mathbb{Z}^k, G)$ in case G is simply-connected, and compact in [6]. Sjerve and Torres-Giese gave (1) a complete computation for the homology of $Hom(\mathbb{Z}^k, G)$ in case $G = SU(2)$, and (2) they also gave a stable decomposition in terms of Thom spaces of multiples of the classical Hopf line bundle over $\mathbb{RP}^2$ [26]. Baird et al. gave an independent analogous result [7].

6 Related Constructions Associated to *G*

Suppose that $\pi = F_n/\Gamma^q$ where Γ^q denotes the q-th stage of the descending central series for F_n. Then there is a generalization of $Comm(G)$ as follows.

Definition 3 (Stafa [27, Definition 1.1]) Let G be a topological group. Define

$$X(q, G) = \left(\bigsqcup_{k \geq 1} Hom(F_n/\Gamma^q, G) \right) / \sim$$

where the relation $\sim$ is generated by the relation

$$(g_1, \dots, g_k) \sim (g_1, \dots, \widehat{g_i}, \dots, g_k) \text{ if } g_i = 1_G.$$

Note that $X(2, G) = Comm(G)$. A natural filtration of $J(G)$ is obtained

$$X(2, G) \subset X(3, G) \subset \cdots \subset X(q, G) \subset \cdots \subset J(G).$$

There are analogues where (1) the descending central series is replaced by the mod-p descending central series, or (2) F_n is replaced by either the pro-finite completion, or pro-p completion. Similar results are satisfied, but not addressed here.

Little is known about the features of $X(q, G)$ at the moment other than the existence of a similar stable decomposition as in the case of $Comm(G)$. Namely, each such space of homomorphisms $Hom(F_n/\Gamma^q, G)$ again splits after suspending into (1) a "singular summand", and (2) a complementary part. The next Theorem was stated, but without full details of proof in [1]. It was proven by Villarreal (private communication) noting that the "singular set" $S_k(G) \subset Hom(F_n/\Gamma^q, G)$ is a closed subspace of a real algebraic set which is triangulable.

Theorem 4 *Assume G is a closed subgroup of $GL_m(\mathbb{C})$. Then there is a homotopy equivalence*

$$\Sigma\left(Hom(F_n/\Gamma^q, G)\right) \to \bigvee_{1\leq k\leq n} \Sigma\left(\bigvee_{\binom{n}{k}} Hom(F_k/\Gamma^q, G)/S_k(G)\right),$$

where $S_k(G) \subset Hom(F_k/\Gamma^q, G)$ is equal to the subset with elements having at least one coordinate the identity element.

The next decomposition was given in [12, 27]:

Theorem 5 (Stafa [27] and Cohen and Stafa [12]) *Let G be a Lie group. Then there is a homotopy equivalence*

$$\Sigma(X(q,G)) \to \bigvee_{n\geq 1} \Sigma(Hom(F_n/\Gamma^q, G)/S_n(G)).$$

One compelling reason for considering these spaces is that their homology provides interesting representations of $Aut(F_n/\Gamma^q)$ especially interesting in case G is a finite group. Related problems are considered in the problem section here.

7 Approximations of *Comm*(*G*)

For any abelian subgroup $A \subset G$, there is a map

$$G \times A^n \to Hom(\mathbb{Z}^n, G)$$
$$(g, g_1, \ldots, g_n) \mapsto (g_1^g, \ldots, g_n^g)$$

gotten by conjugating the n-tuple $(g_1, \ldots, g_n)$ by g. This map factors through the quotient obtained from the natural action of the normalizer NA of A to give a map

$$G \times_{NA} A^n \to Hom(\mathbb{Z}^n, G).$$

Finally, applying the construction of Definition 2 one obtains a map

$$\alpha : G \times_{NA} J(A) \to Comm(G),$$

which is central to our study of the spaces $Hom(\mathbb{Z}^n, G)$ for the case when A is the maximal torus T of G. Let W be the *Weyl group* of G defined as the quotient $W = NT/T$. Then the map α can be rewritten as

$$\alpha : G/T \times_W J(T) \to Comm(G).$$

Now let $Hom(\mathbb{Z}^n, G)_1$ denote the path component of the trivial homomorphism. Define $Comm(G)_1$ to be the space

$$Comm(G)_1 = \left(\bigsqcup_{n \geq 1} Hom(\mathbb{Z}^n, G)_1 \right) / \sim$$

with the same relation. Then the map α restricted to $Comm(G)_1$ is a surjection. The space $G \times_{NT} J(T)$ also admits stable decomposition of an equivariant function space [27, Theorem 1.9] a stable decomposition as follows.

Theorem 6 (Stafa[27, Theorem 1.13]) *Let G be a Lie group. Then there are homotopy equivalences*

$$\Sigma\big(G \times_{NT} J(T)\big) \to \Sigma\left(G/T \vee \bigvee_{q \geq 1} (G \times_{NT} \widehat{T^q})/(G/NT) \right),$$

where $\widehat{T^q}$ is the q-fold smash product of T.

With theorems 2 and 6 one can show the following.

Theorem 7 (Stafa [27, Theorem 1.13]) *Let G be a compact Lie group. Then the map*

$$\alpha : G/T \times_W J(T) \to Comm(G)_1$$

induces an isomorphism in homology if the order of the Weyl group is inverted.

Thus $Comm(G)$ can be approximated by simultaneously conjugating products of the maximal torus T, equivalently by conjugating elements of $J(T)$. This description of $Comm(G)$ and the approximation allow for a direct investigation of the spaces $Hom(\mathbb{Z}^n, G)$ individually. In particular, the additive homology of these spaces can be obtained this way as shown in the next section.

8 Homological Results

Recall that the map $\alpha : G/T \times_W J(T) \to Comm(G)_1$ induces an equivalence in homology if the order of the Weyl group is inverted. Therefore one obtains the homology of $Comm(G)_1$ with coefficients in $R = \mathbb{Z}[|W|^{-1}]$ from the homology of $G/T \times_W J(T)$, where $\mathbb{Z}[|W|^{-1}]$ is the ring of integers with the order of the Weyl group inverted. Let $R[W]$ denote the group ring, and let V be the reduced homology of T as an $R[W]$-module with $\mathcal{T}[V]$ the tensor algebra of V. Using the Leray-Serre spectral sequence for homology, the homology of $Comm(G)_1$ with coefficients in R is obtained as follows.

Theorem 8 *Let G be a compact and connected Lie group with maximal torus T and Weyl group W. Then there is an isomorphism in homology*

$$H_*(Comm(G)_1; R) \cong H_*(G/T; R) \otimes_{R[W]} \mathcal{T}[V].$$

Note that in Theorem 8, if the grading of homology is not considered, namely, ungraded homology is taken, where the order of the Weyl group is inverted, then the ungraded homology of the flag variety G/T, denoted $UH_*(G/T)$, is isomorphic as an $R[W]$-module to the group ring RW itself [6]. Let $UH_*(X; R)$ denote the *ungraded homology* of the space X, the usual singular homology, but regraded with $UH_*(X; R) = \oplus_{0 \leq j} H_j(X; R)$. Since $Comm(G)_1$ is not necessarily equal to $Comm(G)$, Corollary 2 is a special case of the following immediate corollary of Theorem 8.

Corollary 3 *There is an isomorphism in ungraded homology given by*

$$UH_*(Comm(G)_1; R) \cong \mathcal{T}[V].$$

Thus the homology groups of $Comm(G)$ with real coefficients are, roughly speaking, reassembling the tensor algebra by regrading the tensor algebra.

Definition 4 Consider cohomology of $Comm(G)$ with real coefficients. A tri-graded Poincaré series

$$\sum_{i,j,m} A(i,j,m) q^i s^j t^m$$

is defined for the real cohomology of $Comm(G)_1$ obtained from the module of invariants

$$\left(H^*(G/T; \mathbb{R}) \otimes \mathcal{T}^*[V]\right)^W.$$

Here $\mathcal{T}^*$ denotes the dual of the tensor algebra, and the coefficient $A(i,j,m)$ is the rank of the invariants of the module obtained by tensoring the i-degree cohomology of G/T with the dual of an m-fold tensor in $H^*(J(T); \mathbb{R}) = \mathcal{T}[V]$ in homological degree j.

Equivalently, this module $H^*(G/T; \mathbb{R}) \otimes \mathcal{T}^*[V]$ is given as a direct sum of modules

$$M_{(i,j,m)} = \sum_{\substack{j=k_1+\cdots+k_m \\ k_q>0}} H^i \otimes (\Lambda^{k_1}\mathbb{R}^n \otimes \cdots \otimes \Lambda^{k_m}\mathbb{R}^n).$$

Consider cohomology with real coefficients. Denote the module $H_*(G/T; \mathbb{R})$ by H. An application of Molien's theorem gives the coefficients $A(i,j,m)$ in the

Poincaré series with the following closed form as worked out by Vic Reiner in the appendix to [12].

Theorem 9 (Reiner [12, Theorem 1.20]) *If G is a compact and connected Lie group with maximal torus T and Weyl group W, then the tri-graded Poincaré series of Comm(G) is*

$$P((H \otimes \mathcal{T}^*[V])^W; q, s, t)$$
$$= \frac{\prod_{i=1}^n (1 - q^{2d_i})}{|W|} \sum_{w \in W} \frac{1}{\det(1 - q^2 w)(1 - t(\det(1 + sw) - 1))}.$$

A corollary of this theorem gives the reduced homology of all spaces $Hom(\mathbb{Z}^n, G)$ with coefficients in R.

Corollary 4 *Let G be a compact and connected Lie group. Then the reduced cohomology of* $Hom(\mathbb{Z}^m, G)$ *is given additively by*

$$\widetilde{H}^d(Hom(\mathbb{Z}^m, G); R) \cong \sum_{1 \le s \le m} \sum_{i+j=d} \left(\sum_{\substack{j=k_1+\cdots+k_s \\ i \ge 0}} \bigoplus_{\binom{m}{s}} \left(M_{(i,j,s)}\right)^W \right).$$

9 The Spaces $B(q, G)$

A special case of natural subspaces of the classifying space BG were defined in [3] using $Hom(\mathbb{Z}^m, G)$ as follows.

Milgram's construction of BG is the geometric realization of a simplicial space which has objects in degree n given by G^n [21]. The subspaces $Hom(\mathbb{Z}^n, G) \subset G^n$ are preserved by the face and degeneracy operations. Thus there is an associated geometric realization denoted $B(2, G)$ [3].

Similar constructions apply in the case of $Hom(F_n/\Gamma^q, G)$ to obtain a space $B(q, G)$. There are inclusions

$$B(2, G) \subset B(3, G) \subset \cdots \subset BG.$$

Consider the homotopy theoretic fibre $B(q, G) \to BG$ to obtain a fibration

$$E(q, G) \to B(q, G) \to BG.$$

The Serre exact sequence for the fibration

$$E(q, G) \to B(q, G) \to BG$$

has several natural properties. Namely, consider the associated Serre exact sequence

$$H_1(E(q,G))_G \to H_1(B(q,G)) \to H_1(BG) \to 0.$$

If G is finite and odd order, then the map $H_1(E(q,G))_G \to H_1(B(q,G))$ is not a surjection. However the action of G on $H_1(E(q,G))$ is still not understood well, and an independent proof of the previous fact is not known. One problem is listed below concerning this action.

In addition, recall that Milnor showed that if X is a path-connected CW-complex, then there is a topological group $G(X)$ such that $BG(X)$ has the homotopy type of X [22]. Thus the spaces $B(q, G(X))$ give a new filtration of any connected CW-complex X

$$B(2, G(X)) \to B(3, G(X)) \to \cdots \to BG(X) \simeq X.$$

Properties of this filtration as well as other related filtrations were introduced in [3]. One immediate corollary follows next, see also [3, Theorem 6.3] for a proof.

Corollary 5 *Let $G(X)$ denote a topological group whose classifying space is homotopy equivalent to X, a path-connected finite CW-complex. Then the looped maps $\Omega(B(q, G(X))) \to \Omega(BG(X))$ all have a section up to homotopy.*

The spaces $B(2, G)$ and their generalizations have several further properties described here. The cohomology of $B(2, G)$ with the order of the Weyl group inverted was described in [3]. Examples were given by transitively commutative groups, and Suzuki groups.

Adem and Gomez [2, Definition 2.1] give the next definition.

Definition 5 Assume that X is a CW-complex. A principal G-bundle $q : E \to X$ is *transitionally commutative* provided there is an open cover $\{U_i \mid i \in I\}$ of X such that the bundle $q : E \to X$ restricted to U_i is trivial, and the transition functions

$$\rho(i,j) : U_i \cap U_j \to G$$

commute when they are defined simultaneously.

Theorem 10 (Adem and Gomez [2, Theorem 2.2]) *Suppose that G is a Lie group. Let $f : X \to BG$ denote the classifying map of a principal G-bundle $q : E \to X$ over the finite CW-complex X . Then up to homotopy, f factors through $B(2, G)$ if and only if there is an open cover of X on which the bundle is trivial over each open set and such that on intersections the transition functions commute when they are simultaneously defined, i.e. q is transitionally commutative.*

A beautiful theorem of Adem et al. [4] follows next.

Theorem 11 (Adem et al. [4, Theorem 1.1]) *The spaces $B(q, G)$ for*

$$G = O, SO, U, SU, Sp$$

provide a filtration by E_∞-ring spaces of the classical infinite loop spaces BSU, BU, BSO, BO and BSp, respectively.

Recall the natural maps

$$\Sigma_n \to O(n)$$

induce an infinite loop map

$$\Omega^\infty S^\infty \to \mathbb{Z} \times BO$$

closely tied to the J-homomorphism. Since this map may be chosen to be a homomorphism, there is an induced map

$$B(q, \Omega^\infty S^\infty) \to B(q, \mathbb{Z} \times BO).$$

It is reasonable to ask about the behavior of this map.

The purpose of the rest of this section is to give an extension of the construction $B(2, G)$ where G is replaced by a small category C to obtain a space denoted

$$B(2, C).$$

The methods are based on the article by Weiss [30]. This extension is also a direct consequence of [2].

The classifying space of a topological group G is the geometric realization of the simplicial space gotten from all ordered n-tuples of composable morphisms. The space $B(2, G)$ is analogous where all elements in an ordered n-tuple are required to pairwise commute. Thus given any topological category C, there is an analogous classifying space obtained from ordered n-tuples of composable elements which themselves pairwise commute. Furthermore, these morphisms are endomorphisms of the same object. The associated classifying space is denoted $B(2, C)$.

Weiss considers the contravariant functors from C to the category of sets together with natural transformations between these functors. The space BC classifies representable sheaves over a topological space with stalks which are commutative diagram of composable endomorphisms. In the process of proving this, Weiss refines a C-set as contravariant functors from C to sets with the morphisms given by natural transformations. The C-sets which are of the form $b \to mor_C(b, c)$ are said to be representable.

There is a sheaf theoretic analogue of the definition stated next which Adem and Gomez gave in [2, Definition 2.1].

Definition 6 Assume that X is a CW-complex, and C is a small category. A C-sheaf over $q : \mathcal{E} \to X$ is *transitionally commutative* provided there is an open cover of X $\{U_i \mid i \in I\}$ such that the sheaf $q : \mathcal{E} \to X$ restricted to U_i is trivial, and the transition functions

$$\rho(i,j) : U_i \cap U_j \to q^{-1}(s \in U_i \cap U_j)$$

commute when they are defined simultaneously.

Theorem 12 *If C is a small category, the classifying space B(2, C) classifies representable sheaves of C-sets over a topological space which are transitionally commutative.*

The proof of this is essentially that given in [2, 30].

10 Right-Angled Artin Groups

The purpose of this section is to demonstrate how earlier methods imply stable decompositions for $Hom(\pi, G)$, where π is a finitely generated right-angled Artin group.

Recall that a finitely generated right-angled Artin group is given by the fundamental group of the polyhedral product $Z(K; (S^1, *))$, where K is a finite simplicial complex [5, 13, 28]. In particular, a presentation is given by generators $x_1, \ldots, x_m$ with relations $[x_i, x_j] = 1$ if and only if vertices i and j share an edge in K.

Write $\pi(K)$ for the fundamental group of $Z(K; (S^1, *))$. The fundamental groups of $Z(K; (S^1, *))$ are all right-angled Artin groups. The group $\pi(K)$ is given by a graph product of groups [13, 28]. In addition, if K is a flag complex, then $Z(K; (S^1, *))$ is a $K(\pi, 1)$ [13, 28].

Definition 7 Let σ denote a *maximal full face* with $m(\sigma)$ vertices. Namely, (i) σ is a face which is abstractly a $(j-1)$-simplex $\Delta[j-1]$ with $j = m(\sigma)$ vertices as a simplicial complex, and (ii) if σ is contained in any face τ in K, then $\sigma = \tau$. Let $\pi(\sigma)$ denote the fundamental group of $Z(\sigma; (S^1, *)) = (S^1)^j$. Thus $\pi(\sigma)$ is a free abelian group of rank $m(\sigma)$ the number of vertices of σ. Let $\mathcal{M}(K)$ denote the set of maximal full faces of K.

The next Lemma stated without proof is an observation, where $*_\alpha G_\alpha$ denotes the free product of groups G_α.

Lemma 1 *Let $\pi(K)$ denote the fundamental group of the polyhedral product $Z(K; (S^1, *))$. Then there are surjections*

$$*_{\sigma \in \mathcal{M}(K)} \pi(\sigma) \to \pi(K),$$

and

$$\prod_{\sigma \in \mathcal{M}(K)} \pi(\sigma) \to \mathbb{Z}^m.$$

The next Lemma gives general properties about $Hom(\pi, G)$.

Lemma 2 *Let G be a Lie group, and A, B, C be discrete groups. Then*

1. *The natural forgetful map*

$$Hom(A * B, G) \to Hom(A, G) \times Hom(B, G)$$

is a homeomorphism.
2. *If $f : A \to C$ is a surjection, then the induced map*

$$Hom(C, G) \to Hom(A, G)$$

is an embedding.
3. *The group $Aut(A) \times Aut(G)$ acts naturally on $Hom(A, G)$.*
4. *If G is simply-connected and compact, then $Hom(\mathbb{Z}^n, G)$ is a closed subset of G^n.*
5. *The suspension $\Sigma(Hom(A \times B, G))$ is homotopy equivalent to*

$$\Sigma(Hom(A, G) \vee (Hom(B, G)) \vee W(A, B, G)),$$

where $W(A, B, G)$ depends on A, B, G.

One immediate consequence follows.

Corollary 6 *Let G be a Lie group, and $A_i = \mathbb{Z}^{n_i}$, $1 \leq i \leq m$. Then the space $Hom(*_{1\leq i\leq m}A_i, G)$ is homeomorphic to $\prod_{1\leq i\leq m} Hom(\mathbb{Z}^{n_i}, G)$.*

Corollary 7 *Let G be a Lie group. Let K be a finite simplicial complex, with $\pi(K)$ the fundamental group of $Z(K; (S^1, *))$, and $\mathcal{M}(K)$ the set of maximal faces of K. Then*

$$Hom(\pi(K), G) \subset \prod_{\sigma \in \mathcal{M}(K)} Hom(\pi(\sigma), G)$$

where $\pi(\sigma)$ is isomorphic to $\mathbb{Z}^{m(\sigma)}$.

Furthermore, there is a homotopy equivalence

$$\Sigma(Hom(\pi(K), G)) \to \Sigma(X) \vee \bigvee_{\sigma \in \mathcal{M}(K)} \Sigma(Hom(\mathbb{Z}^{m(\sigma)}, G))$$

for a space $X = Hom(\pi(\sigma), G)/ \bigvee_{\sigma\in\mathcal{M}(K)}(Hom(\mathbb{Z}^{m(\sigma)}, G))$.

Example 1 Let K be a flag complex with 6 vertices $\{1, 2, 3, 4, 5, 6\}$ and with maximal faces $\sigma = \{1, 2, 3\}$, $\tau = \{4, 5, 6\}$, and $\rho = \{3, 4\}$. Then $Hom(\pi(K), G)$ has the following three spaces as a retract: $Hom(\pi(\sigma), G)$, $Hom(\pi(\tau), G)$, and $Hom(\pi(\rho), G)$

It follows by a direct computation that there is a homotopy equivalence

$$\Sigma(Hom(\pi(K),G)) \to \Sigma X \vee \Sigma(Hom(\pi(\sigma),G)) \vee \Sigma(Hom(\pi(\tau),G)) \vee \Sigma(Hom(\pi(\rho),G))$$

for some space X. Furthermore,

$$Hom(\pi(K),G) = \Big(Hom(\pi(\sigma),G)) \times Hom(\pi(\tau),G))\Big) \cap \Big(G^2 \times Hom(\pi(\rho),G) \times G^2\Big).$$

Definition 8 Let $\mathcal{M}(K)$ denote the set of maximal full faces of K which has m vertices. Let σ denote a maximal full face in $\mathcal{M}(K)$. Define

$$G(\sigma) = \{(g_1, g_2, \cdots, g_m) \in G^m | \ [g_i, g_j] = 1 \text{ for all } i, j \in \sigma\}.$$

Observe that

$$Hom(\pi(K),G) \subset G(\sigma)$$

for each $\sigma \in \mathcal{M}(K)$. Thus $Hom(\pi(K),G) \subset \Big(\bigcap_{\sigma \in \mathcal{M}(K)} G(\sigma)\Big)$. Also, observe that $\bigcap_{\sigma \in \mathcal{M}(K)} G(\sigma) \subset Hom(\pi(K),G)$ restated as a Theorem.

Theorem 13 *If G is a compact Lie group, then*

$$Hom(\pi(K),G) = \bigcap_{\sigma \in \mathcal{M}(K)} G(\sigma),$$

where, $G(\sigma)$ is homeomorphic to $Hom(\mathbb{Z}^t, G) \times G^{m-t}$ for some t by a homeomorphism which permutes coordinates. Furthermore, $Hom(\mathbb{Z}^t, G)$ is a retract of $Hom(\pi(K), G)$ for each $\sigma \in \mathcal{M}(K)$, where σ has t vertices.

Corollary 8 *If G is a compact Lie group, then*

$$G^m - Hom(\pi(K),G) = \bigcup_{\sigma \in \mathcal{M}(K)} (G^m - G(\sigma))$$

gives a Mayer-Vietoris spectral sequence abutting to the homology of $G^m - Hom(\pi(K), G)$.

11 Problems

This section is a list of certain problems concerning the spaces constructed here.

1. Let G be a Lie group and $C(G)$ be the space of closed subgroups of G as defined in [11], and developed in Bridson et al. [10].

   Definition 9 The *Chabauty topology* on $C(G)$ is defined by giving a basis of neighborhoods for a closed subgroup $H \in C(G)$ by the subsets

   $$\mathcal{V}_{K,U}(H) = \{D \in C(G) : D \cap K \subset HU \text{ and } H \cap K \subset DU\},$$

   where K is compact in G and U is an open neighborhood of the identity $1 \in G$. A subspace of $C(G)$ which is of interest is the *space of abelian closed subgroups* $\mathcal{A}(G)$.

   Describe the analogue of the Chabauty topology for closed, and maximal abelian subgroups. Does $Comm(G)$ admit a map onto this space? If there is such a map, is it a quasi-fibration?
2. Recall that an infinite loop space has the homotopy type of a topological group. In view of the results of [4], it is natural to ask whether the spaces $B(q, G)$ are infinite loop spaces in more generality.

   Do infinite loop maps induce infinite loops on the level of $B(q, G)$?

   Replace G by a small category C. Does $B(q, G)$ admit a natural extension to $B(q, C)$? If so, what are the properties?

   Is the homotopy type of $B(q, G)$ independent of the multiplication in G where G is an infinite loop space ? Note that BG is not an invariant of the homotopy type of G in general.
3. Any reasonable space X is the classifying space of a group $G(X)$. What does $B(q, G(X))$ look like? This construction thus gives a new filtration of X. It is natural to ask whether this filtration impacts the space X in a useful way. For example, is this filtration a homotopy invariant?
4. Fix a discrete group G, and ask which representations of $Aut(\pi)$ occur in the natural action on $Hom(\pi, G)$. If G is finite, and π is finitely generated, this process gives permutation representations of $Aut(\pi)$.

   An example is given next from unpublished work of Shiu-Chun Wong, Jie Wu and the first author given by families of two-stage nilpotent finite groups Γ which are quotients of $Aut(F_n)$, but where the quotient map does not factor through $GL(n, \mathbb{Z})$ for $n > 1$.

   For instance, the quotient of the free group of rank 2 by the third stage of its descending central series, denoted F_2/Γ^3, is a quotient of $Aut(F_2)$ by the usual action. Requiring the generators of F_2/Γ^3, say x, y, to have order p as well as $[x, y]$ to have order p, gives a group G of order p^3. Checking Magnus' well-known presentation for $Aut(F_2)$, there is an induced epimorphism $Aut(F_2) \to G$ where G is the group of order p^3 defined earlier, for $p > 3$. Thus G is a quotient of $Aut(F_2)$ by a check of the classical presentation, but not a quotient of $GL(2, \mathbb{Z})$.

Which finite groups are quotients of $Aut(F_2)$ obtained from the natural action on the finite set $Hom(F_2, G)$ for G finite? It is classical that $PGL(2, \mathbb{F}_q)$ is such a quotient where $\mathbb{F}_q$ denotes a finite field.

Does anything "new" occur when considering the action of $Aut(F_2/\Gamma^q)$ on $Hom(F_2/\Gamma^q, G)$?

Let $\hat{F}_n$ denote the pro-finite completion of the free group on n letters. Let G denote a finite group. Thus $Aut(\hat{F}_n)$ acts naturally on $Hom(\hat{F}_n, G)$. Are all finite groups given by quotients of this action? Which finite groups occur as quotients of $Aut(\hat{F}_2)$ acting on the set $Hom(\hat{F}_2, G)$ in this setting? Observe that there are finite groups which are quotients of $Aut(F_2)$, but not quotients of $GL(2, \mathbb{Z})$.

There are natural representations of $Aut(\hat{F}_n)$ on the homology of the space $Hom(\hat{F}_n, G)$. What are these representations?

5. Give information about the monodromy representations for

$$B(q, G) \to BG.$$

Show that the map $H_1(E(q, G))_G \to H_1B(q, G)$ is not onto if G is of odd order.

Acknowledgements The authors would like to thank the *Centro di Ricerca Matematica Ennio De Giorgi* at the Scuola Normale Superiore in Pisa, Italy, and the *Istituto Nazionale di Alta Matematica* for their support. The author "Frederick R. Cohen" was partially supported by the Institute of Mathematics and its Applications and "Mentor Stafa" was partially supported by DARPA grant number N66001-11-1-4132.

References

1. A. Adem, F. Cohen, Commuting elements and spaces of homomorphisms. Math. Ann. **338**(3), 587–626 (2007)
2. A. Adem, J. Gómez, A classifying space for commutativity in Lie groups. **15**(1), 493–535 (2015)
3. A. Adem, F. Cohen, E. Torres-Giese, Commuting elements, simplicial spaces, and filtrations of classifying spaces. Math. Proc. Camb. Philos. Soc. **152**(1), 91–114 (2011)
4. A. Adem, J.M. Gomez, J.A. Lind, U. Tillmann, Infinite loop spaces and nilpotent K-theory. Preprint (2015). arXiv:1503.02526
5. A. Bahri, M. Bendersky, F.R. Cohen, S. Gitler, The polyhedral product functor: a method of computation for moment-angle complexes, arrangements and related spaces. Adv. Math. **225**(3), 1634–1668
6. T. Baird, Cohomology of the space of commuting n-tuples in a compact Lie group. Algebra Geom. Topol. **7**, 737–754 (2007)
7. T. Baird, L. Jeffrey, P. Selick, The space of commuting n-tuples in $SU(2)$. Ill. J. Math. **55**(3), 805–813 (2011/2013)
8. M. Bergeron, The topology of nilpotent representations in reductive groups and their maximal compact subgroups. Geom. Topol. **19**, 1383–1407 (2015)

9. A Borel, R. Friedman, J. Morgan, *Almost Commuting Elements in Compact Lie Groups*. Memoirs of the American Mathematical Society, vol. 747 (American Mathematical Society, Providence, 2002)
10. M. Bridson, P. de la Harpe, V. Kleptsyn, The Chabauty space of closed subgroups of the three-dimensional Heisenberg group. Pac. J. Math. **240**(1), 1–48 (2009)
11. C. Chabauty, Limite d'ensembles et gómtrié des nombres. Bull. Soc. Math. France **78**, 143–151 (1950)
12. F.R. Cohen, M. Stafa, On spaces of commuting elements in Lie groups. Math. Proc. Camb. Philos. Soc. arXiv:1402.6309 (to appear)
13. M. Davis, B. Okun, Cohomology computations for Artin groups, Bestvina–Brady groups, and graph products. Groups Geom. Dyn. **6**(3), 485–531 (2012)
14. T. Friedmann, R.P. Stanley, Counting conjugacy classes of elements of finite order in lie groups. Eur. J. Comb. **36**, 86–96 (2014)
15. W.M. Goldman, The symplectic nature of fundamental groups of surfaces. Adv. Math. **54**(2), 200–225 (1984)
16. W. Goldman, Topological components of spaces of representations. Invent. Math. **93**(3), 557–607 (1988)
17. J. Gómez, A. Pettet, J. Souto, On the fundamental group of $\mathrm{Hom}(\mathbb{Z}^k, G)$. Math. Z. **271**(1–2), 33–44 (2012)
18. I. James, Reduced product spaces. Ann. Math. **62**(1), 170–197 (1955)
19. V. Kac, A. Smilga, Vacuum structure in supersymmetric Yang-Mills theories with any gauge group in. *The Many Faces of the Superworld* (World Scientific, Singapore, 2000), pp. 185–234
20. S. Lawton, D. Ramras, Covering spaces of character varieties. New York J. Math. **21**, 383–416 (2015)
21. R.J. Milgram, The bar construction and abelian h-spaces. Ill. J. Math. **11**(2), 242–250 (1967)
22. J. Milnor, Construction of universal bundles, II. Ann. Math. **63**(2), 430–436 (1956)
23. C. Okay, Homotopy colimits of classifying spaces of abelian subgroups of a finite group. Algebr. Geom. Topol. **14**(4), 2223–2257 (2014)
24. A. Pettet, J. Souto, Commuting tuples in reductive groups and their maximal compact subgroups. Geom. Topol. **17**(5), 2513–2593 (2013)
25. G.H. Rojo, On the space of commuting orthogonal matrices. J. Group Theory **17**(2), 291–316 (2014)
26. D. Sjerve, E. Torres-Giese, Fundamental Groups of Commuting Elements in Lie Groups. Bull. Lond. Math. Soc. **40**(1), 65–76 (2008)
27. M. Stafa, On Polyhedral Products and Spaces of Commuting Elements in Lie Groups. Ph.D Thesis, University of Rochester, 2013
28. M. Stafa, On monodromy representations in Denham-Suciu fibrations. J. Pure Appl. Algebra **219**(8), 3372–3390 (2015)
29. N. Steenrod, A convenient category of topological spaces. Mich. J. Math. **14**(2), 133–152 (1967)
30. M. Weiss, What does the classifying space of a category classify? Homol. Homotopy Appl. **7**(1), 185–195 (2005)
31. E. Witten, Constraints on supersymmetry breaking. Nucl. Phys. B **202**, 253–316 (1982)
32. E. Witten, Toroidal compactification without vector structure. J. High Energy Phys. **2**, 006 (1998)

CPSIA information can be obtained
at www.ICGtesting.com
Printed in the USA
BVOW10*1329020916
460988BV00002B/4/P

9 783319 315799